ASTRONOMY AND ASTROPHYSICS LIBRARY

Series Editors: I. Appenzeller, Heidelberg, Germany
G. Börner, Garching, Germany
M. Harwit, Washington, DC, USA
R. Kippenhahn, Göttingen, Germany
J. Lequeux, Paris, France
P. A. Strittmatter, Tucson, AZ, USA
V. Trimble, College Park, MD, and Irvine, CA, USA

Springer
Berlin
Heidelberg
New York
Barcelona
Hong Kong
London
Milan
Paris
Tokyo

Physics and Astronomy ONLINE LIBRARY

http://www.springer.de/phys/

F. Combes   P. Boissé
A. Mazure   A. Blanchard

Galaxies and Cosmology

Translated by Mark Seymour
Second Edition
With 222 Figures

 Springer

Françoise Combes
DEMIRM – Observatoire de Paris
61 Av. de l'Observatoire
75014 Paris, France

Alain Mazure
LAM – Lab. d'Astrophysique de Marseille
Traverse du Siphon, BP 8
13376 Marseille Cedex 12, France

Patrick Boissé
Ecole Normale Supérieure
24 Rue Lhomond
75231 Paris Cedex 5, France

Alain Blanchard
OMP – Observatoire Midi-Pyrénées
14 Av. Edouard Belin
31400 Toulouse, France

Translator:
Mark Seymour
5 Skene Close
Headington
Oxford OX3 7XQ, United Kingdom

Cover picture: The galaxy pair NGC 6872/IC 4970, in the Pavo group of galaxies (photo ESO/VLT), superposed to the fluctuations of the cosmic background radiation (detected by BOOMERANG, de Bernardis et al.: Nature 404, 955 (2000)).

Title of the original French edition: *Galaxies et cosmologie*
© InterEditions, Paris et Editions du CNRS, Meudon 1991

Library of Congress Cataloging-in-Publication Data. Galaxies et cosmologie. English. Galaxies and cosmology/ F. Combes ...[et al.] ; translated by M. Seymour. – 2nd ed. p. cm. – (Astronomy and astrophysics library) Includes bibliographical references and index. ISBN 3540419276 (hardcover : alk. paper) 1. Galaxies. 2. Cosmology. 3. Astrophysics. I. Combes, F. II. Title. III. Series. QB857.G3913 2001 523.1'1–dc21 2001039747

ISSN 0941-7834
ISBN 3-540-41927-6 2nd Edition Springer-Verlag Berlin Heidelberg New York
ISBN 3-540-58933-3 Springer-Verlag Berlin Heidelberg New York

This work is subject to copyright. All rights are reserved, whether the whole or part of the material is concerned, specifically the rights of translation, reprinting, reuse of illustrations, recitation, broadcasting, reproduction on microfilm or in any other way, and storage in data banks. Duplication of this publication or parts thereof is permitted only under the provisions of the German Copyright Law of September 9, 1965, in its current version, and permission for use must always be obtained from Springer-Verlag. Violations are liable for prosecution under the German Copyright Law.

Springer-Verlag Berlin Heidelberg New York
a member of BertelsmannSpringer Science+Business Media GmbH

http://www.springer.de

© Springer-Verlag Berlin Heidelberg 1995, 2002
Printed in Germany

The use of general descriptive names, registered names, trademarks, etc. in this publication does not imply, even in the absence of a specific statement, that such names are exempt from the relevant protective laws and regulations and therefore free for general use.

Data conversion: Frank Herweg, Leutershausen
Cover design: *design & production* GmbH, Heidelberg

Printed on acid-free paper   SPIN: 10834841   55/3141/ba - 5 4 3 2 1 0

Preface

Extragalactic astrophysics and observational cosmology are still in their youth and this is why they are evolving so rapidly!

We had already noted in the previous edition that considerable progress was made between 1991 (1st French edition) and 1995 (1st English edition). The same remains true in 2001! For example:

- About ten very large optical telescopes (8 ∼10m in diameter) are (or will soon be) operating in the world, increasing very significantly our ability to detect and study galaxies at cosmological distances. The Hubble sequence has already been found to evolve considerably with time.
- Catalogues of galaxies have increased by orders of magnitude in size as a result of the large CCD formats now available; this allows significant statistical analyses and confirms the fundamental hypothesis of large-scale global homogeneity.
- There has been a large increase in the look-back time through the Universe: observations of quasars and (primordial?) galaxies at $z = 5 \sim 6$ are feasible now, and hundreds of 'normal' galaxies have been studied at $z = 2$, while such objects were accessible only at $z < 1$ some years ago. This reveals that galaxies were forming stars at a much higher rate in the past, and we are at the point of tackling the history of star formation in the universe.
- Orders of magnitude increases in the computing power and the number of particles in more and more sophisticated numerical simulations have demonstrated the paramount importance of gas and star formation processes in galaxy formation.
- Orders of magnitude increases in sensitivity and resolution already obtained in the description of the CMB fluctuations constrain the cosmological scenario (e.g. inflation versus cosmic strings) as well as the values of cosmological parameters ...

But there is still an unsolved key problem: what is dark matter? Some candidates have been ruled out (such as brown dwarfs), while new tracing methods (e.g. weak lensing) have emerged.

If some doubts remain, it is becoming ever clearer that extragalactic astrophysics and cosmology are physical sciences where our knowledge is increasing very rapidly. In this second English edition, we have not undertaken a thorough systematic revision, but have included some of the major highlights of these recent years.

Paris, July 2001
Françoise Combes, Patrick Boissé,
Alain Mazure, and Alain Blanchard

From the Preface to the First English Edition

Much progress has been made in our understanding of 'galaxies and cosmology' since the publication of the French edition of this book in 1991. Four years later it is interesting to look back on the main advances.

One of the most fundamental issues, the nature of dark matter, has been tackled in many ways: we now know, from numerous observations of atomic gas (H I) in dwarf and giant galaxies, that the relative importance of dark and visible matter varies along the Hubble sequence. Dwarfs have relatively more dark matter than giant galaxies. The deflection of light rays by gravitational forces has been widely used to investigate the total mass and in particular the distribution of dark matter. Gravitational lensing produces luminous arcs and arclets in clusters of galaxies that mark out the potential: a very concentrated distribution of dark matter has been revealed in clusters. Microlensing by stars or compact objects has been observed and simulated in order to probe the nature of dark matter.

The *Hubble Space Telescope* has provided an increased spatial resolution that has allowed the discovery of the long-expected massive black holes at the centres of elliptical galaxies: a candidate in M87 has now received strong support. As the telescope operates outside the atmosphere, it can observe in the ultraviolet and has extended the study of Lyman α absorption lines down to low redshifts. A wealth of new Lyman α lines has been discovered, establishing that the gas responsible for the Lyman α forest has not yet disappeared and allowing us to investigate its relation to galaxies.

The *ROSAT* X-ray satellite with its enhanced sensitivity has uncovered the structure of galaxy clusters, since the hot X-ray gas is a probe of the cluster potential. Contrary to previous belief it now appears that the ratio of dark to luminous matter does not increase at large distances from the cluster centre.

Finally the *COBE* satellite has for the first time detected fluctuations in the cosmic microwave background, which are vestiges of the density fluctuations that gave birth to galaxies.

We have attempted to incorporate these new results into our book as far as possible, or at least to the extent that we estimate they will have a significant impact on extragalactic research. In spite of this, however, the book is not intended to be fully comprehensive and complete; its aim is to serve as an introduction at the graduate level to extragalactic astrophysics and cosmology.

We are very grateful to Mark Seymour for his translation.

Paris, April 1995
Françoise Combes, Patrick Boissé,
Alain Mazure, and Alain Blanchard

From the Preface to the French Edition

The subject of galaxies and cosmology is evolving very rapidly. So, in this book, we have put the emphasis principally on the methods of investigation, the mechanisms involved, and the theories capable of providing unified descriptions; at the same time we have tried our best to review current knowledge and to answer as far as possible the questions outlined in the General Introduction. The book covers nearby galaxies, their dynamics, and their interactions; quasars and distant radio sources; and the formation of galaxies and large structures, which then leads naturally on to cosmology and our present view of the early universe.

A list of constants, notation, and units used appears at the end of the book.

It is with pleasure that we thank Suzy Collin, Alain Omont, and Jacques Rolland for their comments.

Paris, August 1991 *Françoise Combes, Patrick Boissé,*
Alain Mazure, and Alain Blanchard

Translator's Note

Cosmology, the study of the physical universe on the grandest of scales, is not only fascinating and inspiring in its own right but is also, I believe, now becoming still more important and relevant as physics comes full circle, and particle physics and cosmology literally try to join forces. It was therefore a particular pleasure to be able to work on this new book and make it accessible to an English-speaking audience. I thank Professor Wolf Beiglböck at Springer-Verlag for offering me the opportunity to do so and for his generosity and understanding throughout. Besides my family, I should also like to thank Keith Orton and especially Laurent Sauvage for their help, and of course Professor Françoise Combes and her coauthors for their friendly advice and comments.

St Andrews, May 1995 *Mark Seymour*

Contents

General Introduction 1

1. **The Classification and Morphology of Galaxies** 5
   1.1 The Classification of Galaxies 5
       1.1.1 Hubble's Classification Scheme 5
       1.1.2 De Vaucouleurs's Revised Classification Scheme (1959) 7
       1.1.3 Other Elements of the Classification 9
   1.2 Luminosity Distributions 11
       1.2.1 The Luminosity Profile of Ellipticals 11
       1.2.2 The Luminosity Profile of Spirals 14
             The Central Bulge 14
             The Disc 14
             The Vertical Structure 16
   1.3 Stellar Populations, the Distribution of Colour,
       and Evolutionary Models 17
       1.3.1 Stellar Populations 17
       1.3.2 The Distribution of Colour 20
       1.3.3 Evolutionary Models 22
   1.4 The Statistical Properties of Galaxies 23
       1.4.1 The Luminosity Function 23
       1.4.2 Morphological Segregation 27
   1.5 The Evolution of Galaxy Morphology 27
   References 30

2. **The Galactic Interstellar Medium** 33
   2.1 Ionized Gas 33
       2.1.1 Ionized Regions of the Disc 33
       2.1.2 Emission Lines in the Nucleus 37
       2.1.3 Radio Emission 38
             The Nature of the Emission 38
             General Characteristics of the Emission 40
   2.2 Atomic H I Gas 43
       2.2.1 Excitation 43
       2.2.2 Global Properties. Correlations 45

              2.2.3   The Distribution of Atomic Hydrogen   46
                      The Radial Distribution   46
                      Interferometric Observations   49
                      The Spiral Structure   51
              2.2.4   H I Holes and the Connection with Star Formation . . .   51
              2.2.5   Distortion, or Warp, of the Plane   54
       2.3 The Molecular Phase   54
              2.3.1   Excitation and the Conversion Ratio   54
              2.3.2   Cold Dust Emission   58
              2.3.3   Global Properties. Correlations   59
              2.3.4   The Radial Distribution   59
              2.3.5   The Spiral Structure   62
              2.3.6   The Connection with Activity in the Nucleus   65
              2.3.7   Other Molecules. Abundances   67
       References   68

3. **The Kinematics and Masses of Galaxies**   69
       3.1 Optical Determinations   69
              3.1.1   Methods   69
              3.1.2   The General Properties of Rotation Curves   70
       3.2 Radio Determinations   73
              3.2.1   Spiral Arms   75
              3.2.2   Bars and Oval Distortions   79
              3.2.3   Distortion, or Warp, of the Plane   80
              3.2.4   Asymmetries   81
       3.3 Determination of the Mass Distribution   81
              3.3.1   Methods of Analysis   81
                      Disc Galaxies   81
                      Elliptical Galaxies   83
              3.3.2   The Mass–Luminosity Ratio   84
              3.3.3   The Tully–Fischer Law and Its Interpretations   87
       Exercises   88
       References   89

4. **Elliptical Galaxies**   91
       4.1 Spectroscopic Observations   91
              4.1.1   General Remarks   91
              4.1.2   Rotation Velocities   93
       4.2 The Three-Dimensional Shape   95
              4.2.1   Ellipticity Profiles   96
              4.2.2   Other Tests of the Three-Dimensional Shape   97
       4.3 Models of Elliptical Galaxies   101
              4.3.1   The Vlasov Equation and Jeans's Theorem   101
              4.3.2   Models of Spherical Galaxies   102
                      $f(E)$ Systems   103

		$f(E, J)$ Systems 105

        4.3.3 Models of Axisymmetric Galaxies 106
  4.4 The Fundamental Plane 107
Exercises 109
References 110

5. **The Spiral Structure of Galaxies** 113
  5.1 Stellar Dynamics. Stability. Orbits 113
        5.1.1 Stability 114
               Jeans Instabilities 114
               Stability due to Rotation 115
               The Stability Criterion 115
        5.1.2 Stellar Orbits 116
               Epicycles 116
               Lindblad Resonances 119
               Surfaces of Section 119
  5.2 The Density-Wave Theory 121
        5.2.1 The Winding Problem 122
               Stationary Density Waves or Stochastic Spirals? 124
               Kinematic Waves 125
        5.2.2 The Wave Dispersion Relation 127
               Obtaining the Dispersion Relation 129
               Wave Propagation 132
               Swing Amplification 134
        5.2.3 Shock Waves Induced in the Gas 137
               The Continuous Interstellar Medium 138
               Interstellar Clouds and the Warm Medium 140
               Damping of the Waves 141
  5.3 Spiral-Wave Generation Mechanisms 142
        5.3.1 Angular-Momentum Transfer 142
        5.3.2 The Excitation of Spiral Waves by a Companion 144
Exercises 146
References 148

6. **Barred Galaxies** 149
  6.1 Observations 149
  6.2 The Theory of Bar Formation 151
        6.2.1 Orbits in a Barred Galaxy 152
        6.2.2 The N-Body Problem 158
               Numerical Methods 158
               Tests of Stability 160
               The Ostriker–Peebles Criterion 164
        6.2.3 Equilibrium Perpendicular to the Plane 166
  6.3 The Response of the Gas to a Barred Stellar Potential 168
        6.3.1 Theory 168

            6.3.2 Gas-Cloud Simulations 171
                  Torques Exerted by the Bar on the Gas 172
                  Bars Within Bars 176
                  Comparison with Observations 176
      Exercises 179
      References 180

7. **Interactions Between Galaxies** 181
   7.1 Galactic Tides 181
       7.1.1 The Principles of Tidal Action 183
       7.1.2 Numerical Simulations 187
             The Three-Body Problem 187
             Taking Account of Dissipation 187
       7.1.3 The Formation of Filaments and Ring Galaxies 190
             Filaments 190
             'Coring' Galaxies 192
   7.2 Vertical Oscillations and Warps 196
       7.2.1 Differential Oscillations 196
       7.2.2 Vertical Waves and the Dispersion Relation 200
       7.2.3 Tidal Interaction and Warps 202
   7.3 Dynamical Friction 203
       7.3.1 Estimating the Frictional Force 203
             Gravitational Deviation 203
             Calculating the Braking Force 207
             The Limitations of Chandrasekhar's Formula 208
       7.3.2 The Criteria for Merger 209
             The Merger of Two Elliptical Galaxies 209
             The Merger of Two Spiral Galaxies 209
   7.4 Shells Around Elliptical Galaxies 211
       7.4.1 The Shell Formation Mechanism 213
       7.4.2 Sampling the Gravitational Potential
             of an Elliptical Galaxy 216
       7.4.3 The Three-Dimensional Shape of Elliptical Galaxies .. 217
   7.5 The Formation of Ellipticals. Conclusions 219
   7.6 Cosmological Implications 221
   Exercises 221
   References 222

8. **Extragalactic Radio Sources** 225
   8.1 Physical Processes 226
       8.1.1 Radiation Emitted by an Ensemble
             of Relativistic Electrons 226
       8.1.2 The Internal Energy of a Gas 227
       8.1.3 Energy Losses 227
       8.1.4 Polarization of Radiation. Faraday Rotation 228

	8.2	The Various Types of Radio Source	
		and Associated Optical Objects 229	
		8.2.1 Compact Sources. Extended Sources 229	
		8.2.2 The Optical Identification of Radio Sources 229	
	8.3	Extended Sources 231	
		8.3.1 Observed Morphologies 231	
		Extended Edge-Brightened Double Sources 231	

8.2 The Various Types of Radio Source and Associated Optical Objects 229
    8.2.1 Compact Sources. Extended Sources 229
    8.2.2 The Optical Identification of Radio Sources 229

8.3 Extended Sources 231
    8.3.1 Observed Morphologies 231
           Extended Edge-Brightened Double Sources 231
           Extended Edge-Darkened Double Sources 232
           Wide Double Sources 232
           Sources with Two Tails 232
           Sources with a Single Narrow Tail 232
    8.3.2 Morphological Classification 232
    8.3.3 The Intrinsic Size of Radio Sources 234
    8.3.4 The Spectrum and Polarization of Extended Lobes ... 234
           Diffuse Emission 235
           Hot Spots 236

8.4 Radio Jets 236
    8.4.1 Symmetry, Shape, and Size 237
    8.4.2 The Spectrum and Polarization 239
    8.4.3 Lateral and Longitudinal Variation
           of the Radio Emission............................ 239
    8.4.4 Optical and X-ray Emission
           Associated with Radio Jets 241

8.5 Compact Sources 242
    8.5.1 The Radio Spectrum.............................. 243
    8.5.2 Variability....................................... 244
    8.5.3 Morphology, Changes of Structure,
           and Superluminal Velocities 246

8.6 Radio-Source Modelling 249
8.7 Radio-Source Counts. Evolution 253
    8.7.1 Expected Counts in the Absence of Evolution 253
    8.7.2 Observed Counts and Their Consequences 254
    8.7.3 The Size–Redshift Relation. Size Evolution 255

Exercises 256
References 257

9. Quasars and Other Active Nuclei 259
9.1 Emission from the Nucleus 260
    9.1.1 Continuous Emission.............................. 260
           The Radio Domain 260
           The Infrared Domain 261
           The Optical and Ultraviolet Domains 262
           The X-ray Domain 263
    9.1.2 Emission Lines 263
           Permitted Lines 263

		Forbidden Lines 266

        9.1.3 Variability and Polarization 266
  9.2 Systematic Quasar Searches 268
        9.2.1 The Radio Domain 268
        9.2.2 The Visible Domain 268
        9.2.3 Selection Based on X-ray and Infrared Emission 270
  9.3 The Spatial Distribution of Quasars 270
        9.3.1 The Distribution Projected on the Sky 271
        9.3.2 The Luminosity Function of Quasars 271
        9.3.3 Quasar Evolution 272
  9.4 Gravitational Lenses 274
        9.4.1 Characteristics of the Phenomenon 274
        9.4.2 An Example:
               The Gravitationally Lensed Quasar 0957+561 277
  9.5 The Quasar Environment. The Nature of the Redshifts 279
        9.5.1 The Host Galaxies of Quasars 279
        9.5.2 The Nature of the Redshifts 280
  9.6 Other Classes of Active Nuclei 282
        9.6.1 Seyfert Galaxies 282
        9.6.2 BL Lac Objects 283
        9.6.3 Radio Galaxies 283
        9.6.4 Modelling Active Galactic Nuclei 283
  Exercises 287
  References 287

10. Quasar Absorption-Line Systems 289
  10.1 General Remarks 290
        10.1.1 The Information Contained in Line Profiles 290
        10.1.2 The Identification of Absorption-Line Systems 292
        10.1.3 The Empirical Classification
              of Absorption-Line Systems 293
  10.2 Narrow-Metal-Line Systems 294
        10.2.1 The Redshift Distribution of the C IV
              and Mg II Systems and Its Implications 294
        10.2.2 Physical Properties of the Absorbing Gas 296
              The Velocity Dispersion 296
              The Ionization State 297
              The Size of the Haloes 298
        10.2.3 Damped Lyman α Systems 299
              Detection: Lyman α and 21 cm Absorption 299
              The Search for Associated Molecules: H_2 and CO 301
              Dust 302
        10.2.4 Direct Searches for Absorbing Galaxies at Low z 303
        10.2.5 Narrow-Line Systems at $z_a \approx z_e$ 304
  10.3 Broad Absorption Line Systems 306

       10.3.1 The Characteristics of QSOs
              with Broad Absorption Lines 306
       10.3.2 Modelling the Broad-Line Systems................... 307
  10.4 Lyman α Systems... 308
       10.4.1 The Redshift Distribution 308
       10.4.2 Column Densities and the Velocity Dispersion........ 311
       10.4.3 The Heavy-Element Abundance 311
       10.4.4 Other Properties. Possible Models 312
  Exercises 313
  References 314

11. **The Universe on a Large Scale**............................. 315
    11.1 Structure and Homogeneity 315
    11.2 The Distance of the Galaxies: Probing the Universe 320
         11.2.1 Parallaxes and Trigonometric Methods 320
         11.2.2 Cepheids and Standard Candles 322
         11.2.3 The Tully–Fischer Relation 324
         11.2.4 The Sunyaev–Zel'dovich Effect 327
         11.2.5 The Surface Brightness Fluctuation Method 328
    11.3 The Third Dimension 332
    11.4 Statistical Methods: Correlation Functions and Percolation .. 338
         11.4.1 Correlation Functions 338
         11.4.2 Percolation 346
    11.5 Estimating the Mass of Groups and Clusters of Galaxies..... 348
                The Baryon Catastrophe 352
    11.6 Large-Scale Motions. The Virgo Infall 353
         11.6.1 The Great Attractor 353
         11.6.2 Motion with Respect
                to the Cosmic Microwave Background Radiation 355
    Exercises 358
    References 358

12. **The Formation of Galaxies and Large Structures
    in the Universe** 361
    12.1 The Jeans Mass and the Growth of Perturbations 362
         12.1.1 The One-Component Model 362
         12.1.2 The Two-Component Model 365
         12.1.3 Dark Matter 366
    12.2 The Origin, Spectrum, and Nature of the Fluctuations 367
         12.2.1 The Origin of the Fluctuations 367
         12.2.2 Isothermal and Adiabatic Fluctuations 368
    12.3 The Linear Evolution of Perturbations 369
    12.4 Nonlinear Evolution 375
         12.4.1 The Pancake Model 375
         12.4.2 The Hierarchical Scenario: The Spherical Model 375
         12.4.3 Numerical Simulations 377

XVIII Contents

    12.4.4 Semi-analytical Approaches 379
 12.5 The Models in the Light of Observations 382
 12.6 The Quest for Primordial Galaxies 386
 12.7 Conclusions 390
 Exercises 391
 References 391

13. Cosmology 393
 13.1 The Geometrical Description of the Universe 395
    13.1.1 The Redshift 397
    13.1.2 The Concept of Distance 398
    13.1.3 The Concept of Horizon 400
    13.1.4 The Evolution
       of the Cosmic Microwave Background Radiation 400
 13.2 Friedmann–Lemaître Models 402
    13.2.1 The Matter-Dominated Universe 403
    13.2.2 Models with a Zero Cosmological Constant 405
       Case 1: $\Omega_0 = 1$ 405
       Case 2: $\Omega_0 < 1$ 405
       Case 3: $\Omega_0 > 1$ 406
    13.2.3 Models with a Nonzero Cosmological Constant 407
    13.2.4 The Radiation-Dominated Universe 410
 13.3 The Hot Phase of the Universe 412
    13.3.1 The Thermal History of a Particle 413
    13.3.2 A Description of the Initial State of Equilibrium .. 413
    13.3.3 The Chemical Decoupling of a Particle 415
    13.3.4 Primordial Nucleosynthesis 416
 13.4 The Very Early Universe 421
    13.4.1 Problems of the Classical Big-Bang Theory
       and Grand Unified Theories 421
    13.4.2 The Theory of Inflation 425
       The Horizon Problem 425
       The Flatness Problem 426
       Primordial Fluctuations 426
       The Problem of Magnetic Monopoles 427
       A Particular Phase Transition: Inflation 427
    13.4.3 Nonbaryonic Dark Matter 430
    13.4.4 Recent Observational Confrontations 432
       Nucleosynthesis 432
       Inflation and Cosmic Background Anisotropies 433
 Exercises 434
 References 435

List of Constants, Notation, and Units Used 437

Index 439

General Introduction

Look at the night sky while you are sheltered from any interfering light, and on the dark background, strewn with points of light, you will see a fairly broad whitish streak: the Milky Way. It is a huge conglomeration of stars, dust, and gas, to which our Sun belongs: the Galaxy. A few of the many light points are the planets of the Solar System (although only Venus, Mars, Jupiter, and Saturn are clearly visible with the naked eye), but the vast majority of them are stars belonging to our Galaxy, the Milky Way. Even on a very clear night, careful observation will enable only several thousand to be discerned. However, our Galaxy contains around 100 billion.

The Milky Way is a very ordinary spiral galaxy, similar to thousands of others compiled in catalogues. To clarify matters, several orders of magnitude follow. The unit of mass used by astronomers is the mass of the Sun, an average star. A solar mass ($1\,M_\odot$) equals 2×10^{30} kg and the mass of our Galaxy is then $10^{11}\,M_\odot$, that is, 2×10^{41} kg. The Milky Way is a flattened disc approximately 30 kpc in diameter: the unit here is the parsec (pc), that is, 3.26 light-years. The thickness of the plane of stars is of the order of 1 kpc. The average star density is thus around 0.1 star per pc^3.

Now galaxies have been recognized as separate entities equivalent to our Milky Way only since the beginning of the century. In 1924 Hubble was able to resolve the Andromeda 'nebula' into stars and therefore to prove that it was another galaxy and not a nebulosity of gas like that in Orion, which shines from the radiation of ionized hydrogen.

Interstellar gas today represents only a small fraction of the total mass of the galaxies, of the order of 10% in spiral galaxies. Nevertheless it is from gas that stars were formed and are being formed now; hence at their formation galaxies are just balls of gas. This gas is essentially hydrogen in atomic or molecular form, or sometimes ionized, mixed with helium (25% by mass) and other trace elements.

The forces of gravitation are responsible for the cohesion and morphology of galaxies. Why do the galaxies not collapse in on themselves? In spiral galaxies all the stars and interstellar matter rotate about the axis of the disc and the corresponding centrifugal forces balance the gravitational forces. Perpendicular to the galactic plane the velocity dispersion of the stars is equivalent to a pressure whose gradient is able to balance the gravitational attraction. In elliptical galaxies, which do not rotate, this 'pressure' alone is able to compensate for selfgravity.

Galaxies are grouped together in structures on various scales: small groups to very rich clusters and superclusters. The Milky Way thus belongs to the Local Group, which contains the Magellanic clouds and the Andromeda galaxy, the spiral galaxy closest to us. On average the distance between two galaxies is about 3 Mpc. Occasionally, during an encounter, galaxies interact by tidal effects: they are then noticeably deformed because of the gradient of gravitational forces across their length and breadth, which are not negligible compared with intergalactic distances.

The whole of the electromagnetic spectrum is used to acquire knowledge about galaxies. The visible region is of course favoured, because it is here that the sensitivity and spectral resolution are greatest and the stars radiate a large part of their energy. However, the near infrared is equally important since it reveals stars that remain hidden in the visible by interstellar dust. The thermal radiation of the dust grains heated by young stars and prototype stars, like the general interstellar radiation, lies in the far infrared. Owing to the transparency of the interstellar medium at these wavelengths, this frequency domain brings us vital information about the formation of stars in galaxies. In turn radio astronomy enables us to determine the properties and distribution of interstellar gas, be it atomic, molecular, or ionized. In all wavelength regions spectra give us information about the kinematics of matter. In the visible, absorption lines in particular, formed in the atmospheres of stars, enable us to piece together the kinematics of the stars within the galaxy, while emission lines give the speed of the ionized gas (emission lines of a hot gas ionized by young stars).

The dynamics of spiral galaxies, which are actually thin, rotating discs, appears to be well understood; nevertheless away from their centres their rotation velocity does not decrease, whereas the light of the stars does. Is this a proof of the existence of dark matter? According to kinematic measurements, close to 80% of the mass of the universe would in this way escape observation.

Elliptical galaxies at first seem to be simple objects, flattened to a greater or lesser extent according to their degree of rotation. Yet since the beginning of the 1980s astronomers have come to the conclusion noted above that these galaxies do not rotate! How do these objects form? What do they look like in three dimensions: flattened or elongated? More and more precise observations are making evident secondary components in these apparently very simple objects: discs, boxes, and dust lanes. Are elliptical galaxies the result of mergers between two spiral galaxies?

The spiral structure of galaxies has for a long time posed a serious theoretical problem for dynamicists. The spiral arms cannot be a permanent accumulation of material since they would quickly wind up owing to differential rotation. Only density waves are able to explain such a coherent structure and such little winding-up over the whole galaxy. But how are these waves generated and amplified? Are the tidal interaction between galaxies and the formation of barred waves by gravitational instability alone able to explain the observed structures?

The tidal interaction, which today seems a decisive factor in the evolution of galaxies, without doubt played an even more fundamental role at the beginning of the universe. Is it, through triggering an outburst of star formation or the appearance of active nuclei, the origin of most of the ultraluminous galaxies?

According to some researchers the merger of two spiral galaxies leads to the formation of quasars. In the course of these violent events a large quantity of gas is driven towards the centre of the system, feeding the central supermassive object (a black hole?).

Quasars are the most distant objects that we are aware of in the universe. As such they form valuable probes that reveal all the matter interposed in the foreground. Thus new entities, apparently without equivalents in our local universe, are revealed, such as the clouds responsible for the 'Lyman α forest'. If massive objects – galaxies or clusters of galaxies – are found very close to the line of sight of a quasar, the light emitted by them is deviated: we see gravitational mirages. This phenomenon of deflection and amplification makes it possible to observe very distant galaxies, which then appear in the form of circular arcs in clusters of galaxies.

How were the galaxies formed in the primordial universe? The 2.7 K cosmic background radiation, a remnant of the epoch where matter was in equilibrium with radiation, constitutes a very strong constraint for theories of galaxy formation. Indeed it really seems difficult to reconcile its remarkable degree of isotropy with the highly structured distribution of galaxies that we see in the local universe. Recently the *COBE* satellite has confirmed that the spectrum of the cosmic background radiation is, to a high degree of precision, that of a black body: this, moreover, is an argument in favour of the big-bang theory. The early universe is a unique laboratory: indeed it is only there that the unification of the forces of physics can have taken place. The scenarios describing the primordial universe are still very uncertain; recent theories make appeal to an inflationary phase that would provide an explanation for our universe being so uniform and homogeneous.

In this book we are going to be concerned first of all with the details of galaxies, which are, in a way, the building blocks of the universe, reviewing their nature, structure, and kinematics. Next a leap into space leads us on to a consideration of those distant galaxies manifesting 'peculiar activity' (radio galaxies and quasars) before we describe the large-scale structure of galaxies: clusters and superclusters. The organization of the universe and its evolution and the formation of galaxies lead us then quite naturally on to cosmology, the first instant of the universe, and possible models that describe the formation of the universe from the initial explosion (the 'big bang') where all the forces of physics were originally unified.

1. The Classification and Morphology of Galaxies

The discovery of galaxies as such goes back to 1924. As a result of observations he made at the 2.5 m telescope at Mount Wilson, Edwin Hubble demonstrated definitively that certain nebulae do not form part of our Galaxy but are 'island universes', independent conglomerations of stars, gas, and dust. The historical confusion between the nebulae of ionized gas in the Milky Way and the external galaxies derives from the use of general catalogues, such as Messier's (1794), which contains 39 galaxies out of 109 nebular objects or clusters of Galactic stars, and especially the New General Catalogue (NGC)(Dreyer 1890), which contains 7840 objects, of which 3200 are galaxies, and the Index Catalogue (IC)(1895–1910), which contains 5836 objects, of which 2400 are galaxies. The Harvard catalogue (Shapley and Ames 1932) contains only the brightest galaxies, with apparent magnitudes $m < 13$ (1249 objects in total). The apparent magnitude is equal to $-2.5\log$ (luminosity). The eye can only perceive stars up to a magnitude $m = 6$. Up to an apparent magnitude $m = 17.5$ there are 500 000 galaxies, and up to $m = 23$, 10^9. The first step in getting to know the galaxies better is to describe the various types and classifications. We shall see throughout the book how important the classifications are, the different types of galaxy corresponding to different formation mechanisms and different environments.

1.1 The Classification of Galaxies

1.1.1 Hubble's Classification Scheme

The principal morphological types are represented by Hubble's celebrated 'tuning-fork' diagram (Fig. 1.1); the galaxies are separated into three principal classes: ellipticals (E), spirals (S), and irregulars (Irr). Spirals are made up of two families, so-called normal spirals and barred spirals (SB), and are further divided into types Sa, Sb, and Sc, which correspond to a gradual evolution rather than to distinct classes. From the left to the right of Hubble's tuning-fork diagram the disc, which does not exist in ellipticals, becomes more important, as does the proportion of gas and young stars.

Elliptical galaxies are seen projected onto the sky as more or less flattened ellipses. The ratio of the axes (a and b) varies from only 1 to 3; the

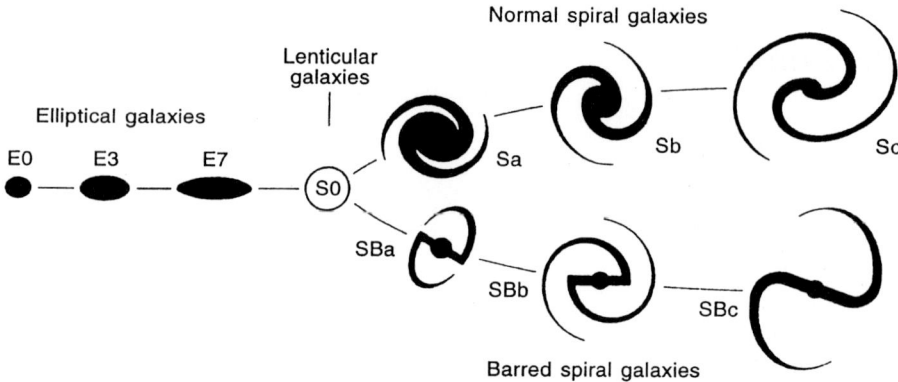

Fig. 1.1. The principal morphological types are represented by Hubble's classic 'tuning-fork' diagram. From the left to the right the disc (nonexistent in the ellipticals) becomes more important, as does the proportion of gas and young stars. The lenticulars (S0), which have a disc but very little gas, lie at the transition from the spheroidal systems to the spiral galaxies proper. On the one side of the two branches of the fork are the normal spirals, on the other, the barred spirals

ellipticity ϵ, defined by $\epsilon = (a - b)/a$ (a being the major axis), thus varies from 0 to 0.7. Ellipticals are classified by an index equal to 10ϵ (from E0 to E7). The effects of projection hinder us greatly in our attempts to determine the intrinsic shape of these galaxies. For a very long time the hypothesis of an axisymmetric ellipsoid was accepted without a problem, the flattening of the galaxies being attributed to the rotation. We shall see later on that this hypothesis is now questioned. Just as an E0 galaxy might be a very flattened galaxy but seen face-on, so an E7 galaxy might only be seen edge-on: indeed no galaxies more flattened than E7 exist. The most massive galaxies known are the elliptical galaxies, the majority of which are more luminous than the spirals. There nevertheless exists a class of dwarf ellipticals, which in general are companions of larger galaxies. Elliptical galaxies have no particular substructure. Their luminosity decreases very regularly from centre to edge. They have no, or very little, gas but a population of relatively old stars, and globular clusters, which are just very dense and generally old stellar clusters.

Spiral galaxies (S) take their name, of course, from the spectacular spiral arms that wind around in a very thin disc; these structures are very flattened, as shown by certain specimens seen edge-on: the ratio of their axes can reach 100! The spiral arms meet towards the central bulge, the brightest component, whose shape is ellipsoidal and much less flattened than the disc; it is this bulge that gives rise to the swollen shape of the centre of a spiral galaxy seen in profile. In barred spiral galaxies (SB) the nucleus is crossed by a bar of stars at the ends of which the spiral arms begin. These are in general less tightly wound and are known as open spirals.

The Hubble sequence, from Sa ('early') towards Sc ('late'), is formed according to several criteria:

1. the relative importance of the central bulge: the ratio of the size of the bulge to the size of the disc decreases steadily from Sa to Sc;
2. the resolution and predominance of the spiral arms;
3. the presence of gas and dust, ionized regions, and regions of young stars, which grows steadily towards Sc;
4. the spiral arms being more and more open from Sa to Sc;
5. the total luminosity decreasing from Sa to Sc.

Following Hubble's original classification, the sequence after Sc was extended to Sd and Sm by de Vaucouleurs in 1959 and united with the class of irregulars, Irr: these are not well-defined structures; they are amorphous (with no nucleus, no disc, no spiral arms, and so on) but contain much gas and are often the location of major outbursts of star formation. The Sm or SBm galaxies that form the transition are Magellanic-type objects (the prototypes being the Small and Large Magellanic Clouds).

The transition from ellipticals to spirals is confirmed by the S0 galaxies, which are known as lenticulars: these are galaxies with a very large central bulge and also a flattened disc of stars; this disc distinguishes them from the ellipticals but does not contain spiral arms, nor, in general, gas or dust.

The merit of Hubble's classification is that it ignores or leaves in the background the many superficial details that make galaxies complex (much more difficult to classify than stars, for example) but retains the main features: in this way every galaxy is included in the classification, even when it is perturbed by an interaction with companions (the adjective 'peculiar' (Pec) is then added to its type, for example Sa Pec). Around two-thirds of galaxies are spirals; the remainder are divided into 10% ellipticals and 20% lenticulars. The irregulars constitute only a small percentage. These populations are, however, the average, since we shall see that the relative percentages of S, E, and S0 galaxies vary according to the environment, whether rich groups or clusters of galaxies, or isolated galaxies, known as field galaxies.

1.1.2 De Vaucouleurs's Revised Classification Scheme (1959)

De Vaucouleurs introduced in 1959 some intermediate types and finer classifications, taking account of secondary characteristics that in fact reveal much of importance. His revisions lie at the heart of the modern classification commonly used today and visualized in the form of a 'volume' of classification (Fig. 1.2):

1. Between the strongly barred galaxies, SB, and the 'normal' spirals, SA, can be distinguished weakly barred or distorted oval spirals, SAB. By and large the classes SA, SAB, and SB each represent a third of all spiral galaxies.

8   1. The Classification and Morphology of Galaxies

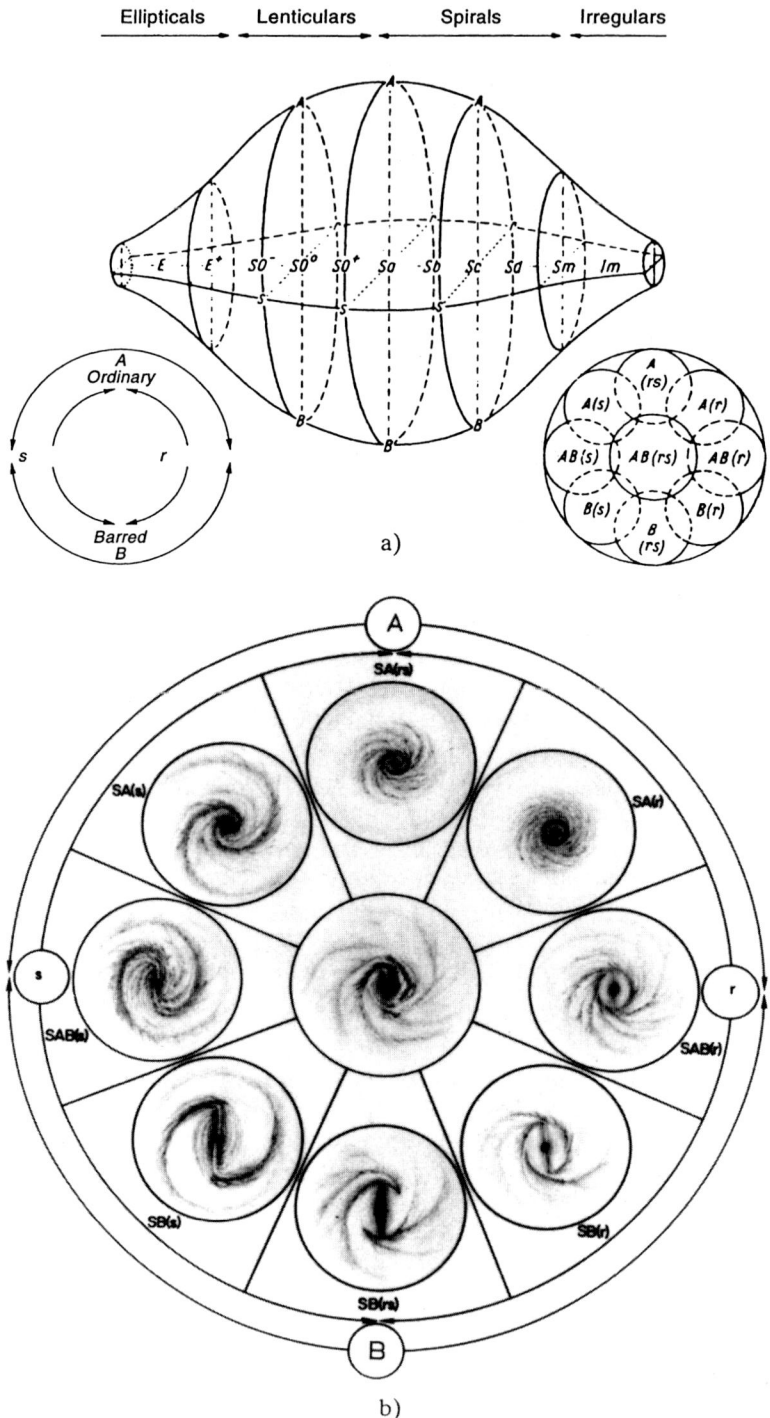

2. Certain galaxies have an internal ring surrounding the nucleus of the galaxy and where the spiral arms begin; they are type r, denoted S(r). When the spiral arms begin directly at the nucleus, they are known as type s, written Ṡ(s).
3. The spiral galaxies sometimes have an external ring, that is to say, a ring surrounding the whole disc together with its spiral structure. In many cases this is just the spiral arms joined together on the outside to form a circle (the pseudoring), and hardly any internal spiral structure can be distinguished. The presence of such an external ring is expressed by the letter R preceding the type (for example RSB(r)a). The rings are present predominantly in early-type spiral galaxies (Sa, Sb).

1.1.3 Other Elements of the Classification

Finally with regard to the structure of the spiral arms themselves we can distinguish two very different classes: galaxies with filamentary spiral structure, where the apparent regularity is in fact constituted by a series of spiral-arm fragments sufficiently short that none continues to form a spiral on a large scale (Fig. 1.3); and 'grand-design' galaxies with a massive and ordered spiral structure, most often forming two continuous spiral arms that can be followed from centre to edge; barred spirals are in general of this type.

Beyond simply distinguishing between all the possible shapes, we shall see that this classification may in fact be interpreted in terms of the different formation processes, as a result of theories of stellar dynamics; each component is a necessary consequence of the evolution of the galaxy and also the cause of subsequent evolution: the spiral arms, for example, generated by the gravitational potential of the bar, transfer angular momentum from the centre to the edges of the galaxy, as we shall see later on, and are the origin of the ringlike structures.

Note that the more luminous and thus massive a galaxy is (for a given mass–luminosity ratio), the more well-defined structures it has. This is the principle behind van den Bergh's classification, which arranges the galaxies according to their intrinsic luminosity. As this decreases, first the rings become invisible, then the spiral arms, and finally the bar and disc. In the least luminous class only dwarf ellipticals and irregulars rich in gas are to be found.

Evolution may also be a consequence of interactions with the environment, and we single out the class of so-called anaemic galaxies between the lenticulars (S0) and the ordinary spirals (SA) (the whole series exists: Aa–Ab–Ac). Their gas content is very low, and, like the S0s, they are much more

◀ **Fig. 1.2.** De Vaucouleurs's volume of classification (1959), introducing the intermediate case between nonbarred spirals, SA, and strongly barred spirals, SB. The galaxies possess either two distinct spiral arms or an internal ring (r) or external ring (R), where the spiral structure is less dominant

10   1. The Classification and Morphology of Galaxies

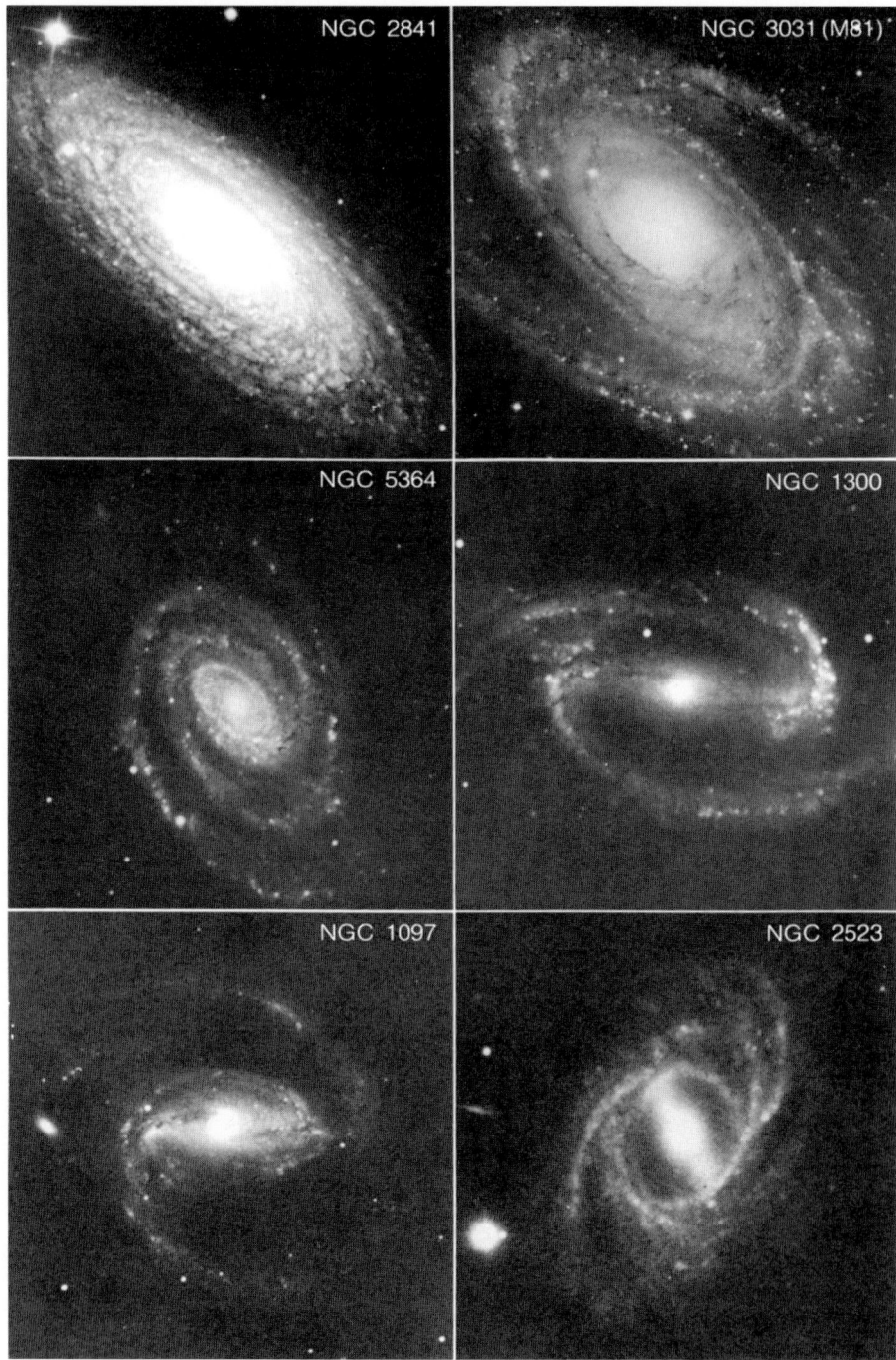

commonly found in clusters of galaxies than in the field. In clusters we further identify galaxies referred to as 'cD', supergiants, which are elliptical with a very extended envelope. Whereas the anaemic galaxies lose matter in interactions with the rest of the cluster, the cD galaxies at the centre of clusters accrete gas and stars. To understand the nature of the dynamic equilibrium of galaxies better, and to be in a position to propose formation processes, our first step will be to determine the distribution of the stars and then the luminosity distribution, and finally to identify several components, such as a central bulge, a disc, a dense core, or a diffuse envelope.

1.2 Luminosity Distributions

Surface photometry and the examination of luminosity distributions are a first step towards identifying the various components of a galaxy and determining their masses. In fact because the mass–luminosity ratio, M/L, varies from the centre to the edge of galaxies, and to a lesser extent from one galaxy to another, kinematic observations are required to determine the true mass distribution of galaxies.

1.2.1 The Luminosity Profile of Ellipticals

For elliptical galaxies, which have an apparently very simple structure, the luminosity distribution is the crucial observational parameter, enabling us to determine the structure and dynamics of these galaxies. Its principal characteristic is a pronounced maximum at the centre and a rapid and uniform decrease with distance from the centre, following a quasi power law. Two analytic functions are at present used to represent this distribution:

- *Hubble's law (1930):*
$$I/I_0 = [(r/a) + 1]^{-2},$$
I being the intensity emitted per unit area at a distance r from the centre. The two parameters of the model are the intensity per unit area at the centre, I_0, and the radial extension (or core radius), a.
- *De Vaucouleurs's $r^{1/4}$ law (1948):*
$$\log(I/I_e) = -3.33\left[(r/r_e)^{1/4} - 1\right].$$

◀ **Fig. 1.3.** Some examples of spiral galaxies: NGC 2841 is the prototypical filamentary galaxy, formed from fragments of incoherent spiral arms. M81 and NGC 5364 are two 'grand-design' galaxies, where the two spiral arms are continuous and coherent. NGC 5364 has arms that wrap around more than 360°, which never occurs in a barred spiral. (Photographs from Sandage 1961)

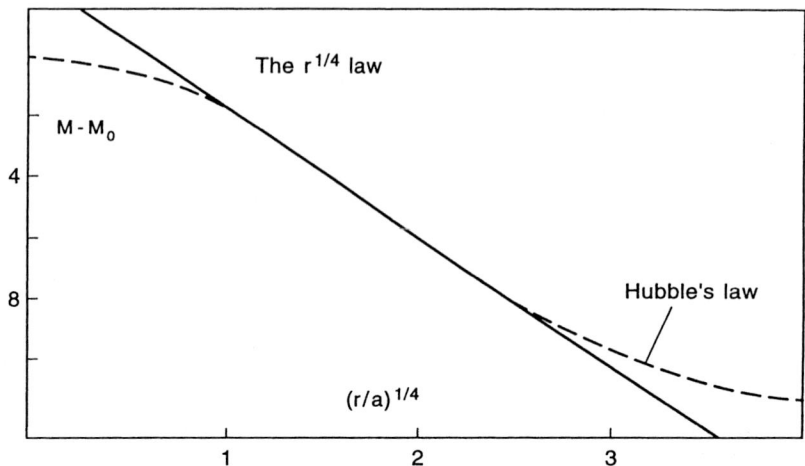

Fig. 1.4. A comparison of two luminosity distributions often used to represent elliptical galaxies: Hubble's law, $I/I_0 = [(r/a) + 1]^{-2}$, and de Vaucouleurs's $r^{1/4}$ law, $\log(I/I_e) = -3.33[(r/r_e)^{1/4} - 1]$

This is also a model with two parameters: r_e is the radius containing half of the total luminosity and I_e is the intensity at a distance r_e from the centre.

These two representations are in fact equivalent, but the second is more widely used in practice because it is a power law for the observed magnitude (this being equal to $-2.5 \log I$). Each becomes invalid both close to and far from the centre. Hubble's law, for instance, represents the profiles of elliptical galaxies between $r = a$ and $r = 10a$ very well (Fig. 1.4). It is easy to find the correspondences between the parameters of the two laws:

$$a = 0.093 r_e,$$
$$-2.5 \log I_0 = -2.5 \log I_e - 5.27 \text{ mag arcsec}^{-2}$$

(see Kormendy 1982).

The success of these luminosity functions derives from the fact that the luminosity profiles of elliptical galaxies are given in practice by power laws with respect to the radius; this is also the case, for example, for the asymptotic form of Hubble's law, $I = kr^{-2}$, when $r \gg a$. But a power law has only one constant of proportionality k. What then is the physical meaning of the two parameters obtained by using this model? In fact the profiles are not power laws exactly, and the second parameter only serves to limit the deviations from this law. As these are very small, the modelling is very tricky. If the agreement is not rigorous, the parameters obtained have no physical meaning; they are only very strongly coupled to one another. For the same reason the search for a model with three parameters in the same radius interval would prove illusive. This property of elliptical-galaxy profiles approaching a power law for intermediate radii may be interpreted physically, notably by King's model (1966), as we shall see in Chap. 4, and constitutes an invaluable clue in the search for theories of formation.

The distribution of colour is almost constant from the centre to the edge for elliptical galaxies, with, however, a slight reddening of several hundredths of a magnitude towards the centre.

Another important characteristic of elliptical galaxies revealed by photometry is the shape of the contours traced for given luminosities (isophotes). These isophotes are in general relatively uniform ellipses, whose ellipticity can be determined; the position angle in the sky is defined by the orientation of the major axis of the ellipse (the position angle is defined with respect to north and is positive towards the east). For around half of all elliptical galaxies the major axes of the isophotes are not aligned: there is a uniform and continuous rotation as a function of the distance to the centre, particularly for the outer isophotes. The apparent ellipticity also varies slowly with distance from the centre. The position angle of the isophotes varies especially for those galaxies which are only slightly flattened in projection, and its total rotation may reach up to 60°. Typically the mean variations in ellipticity and position angle recorded on the scale of a characteristic radius r_e are respectively 0.04 and 5° (Leach 1981). The phenomenon has recently been carefully studied, because it might represent a proof of the triaxial nature of ellipticals (Chap. 4). When the three axes of an ellipsoid are different and the ellipticity varies from the centre to the edge, the effects of projection make the apparent major axes rotate with respect to the real major axes by an angle that is larger the rounder the contour is. Figure 1.5 shows, in an extreme way, in two dimensions (one of the axes of the ellipsoid is therefore zero), how projection effects alone may make isophotes rotate that are in

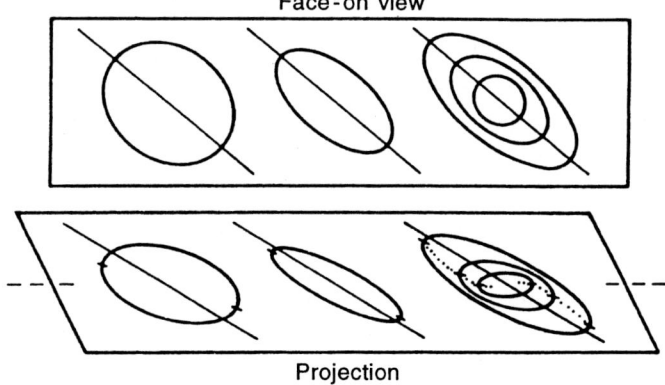

Fig. 1.5. Projection effects make the apparent major axes rotate with respect to the real major axes by an angle that is larger the rounder the contour is. The observed rotation of the isophotes does not therefore automatically imply that the galaxy is deformed. (From Kormendy 1982)

reality aligned. Nevertheless a large number of these distortions may be real and attributable to tidal effects between galaxies. More than a third of apparently deformed ellipticals have nearby companions, and the proximity of the companion is correlated with the amplitude of rotation of the isophotes. But there remain several cases of isolated galaxies that are without doubt triaxial. The statistical study of the amplitude of rotation as a function of the apparent ellipticity shows that, on average, the triaxiality is weak (the ratio of the two axes being between 0.8 and 1).

1.2.2 The Luminosity Profile of Spirals

Spiral galaxies comprise a central bulge or spheroidal component and a very flattened disc that extends to a large distance from the centre.

The Central Bulge. The bulges of spiral galaxies closely resemble elliptical galaxies. They are also similar to them in terms of morphology, luminosity profile, and stellar content (colour). However, certain features are rather of something intermediate between ellipticals and discs:

- the rotation is greater for the bulges than for elliptical galaxies; their gravitational equilibrium is not the same;
- the luminosity profiles are well represented by an $r^{1/4}$ law along the major axis but not along the minor axis;
- the bulges are not as dense and luminous as elliptical galaxies; in comparison with ellipticals of the same total luminosity they have a much smaller surface luminosity and a much greater characteristic radius;
- the bulges are on average more flattened in projection than ellipticals.

The influence of the disc (and its rotation) may be seen in the bulge luminosity profiles. According to the bulge-to-disc ratio, the bulge profile in magnitude obeys an $r^{1/n}$ law, with n around 1 (exponential) to 6 from late- to early- type galaxies (Andredakis et al 1995).

The Disc. Freeman's study (1970) of the photometry of 36 spiral galaxies (including S0s) concluded with an exponential representation of the luminosity profile:

$$I(r) = I_0 e^{-r/r_0},$$

where I_0 is the luminosity extrapolated to the centre and r_0 the characteristic radius. The central surface brightness of discs appears to saturate at a constant value, but there is no minimum (e.g. de Jong 1996).

However, modelling is more difficult here than for elliptical galaxies, on account of the superposition of many components. The distribution of light in the disc depends on the model chosen for the central bulge, and the analysis is only certain when the bulge does not dominate the luminosity. This is the case for late-type spirals, but then another factor contributes significantly to the luminosity: the spiral arms (Fig. 1.6). What is more there are very often

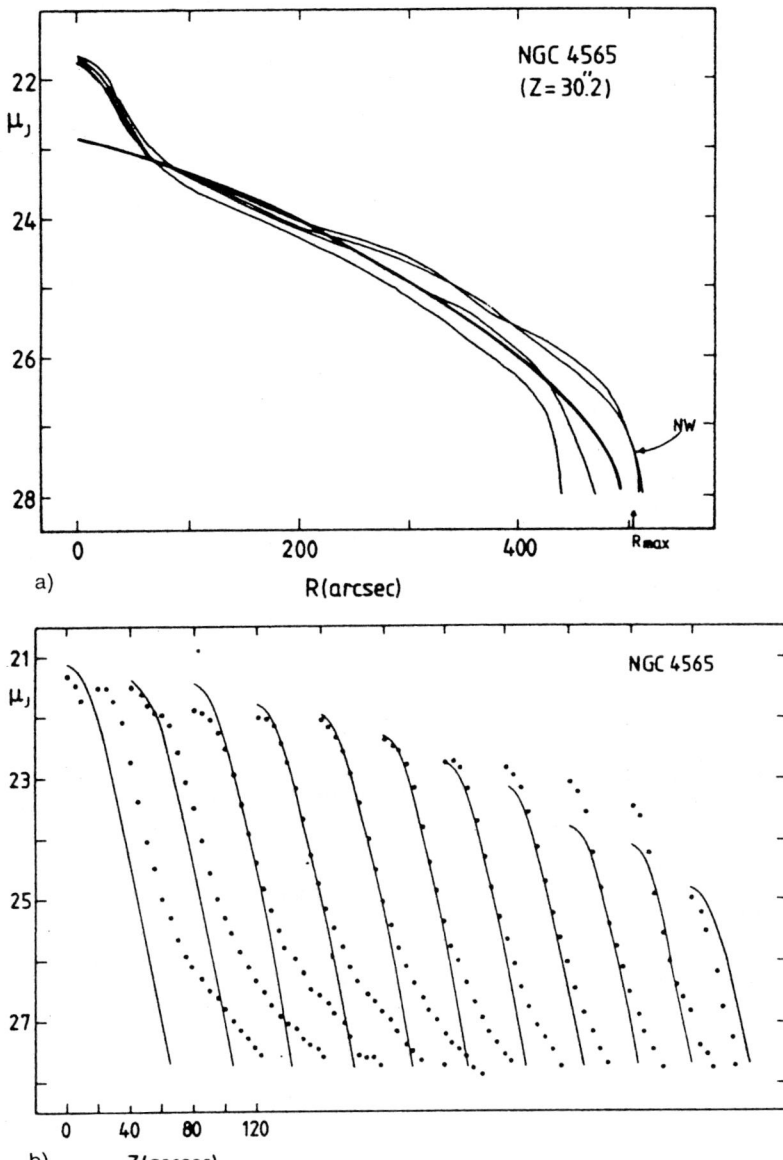

Fig. 1.6. (a) Radial luminosity profiles at a fixed altitude z below the Galactic plane for the galaxy NGC 4565 seen edge-on. The four faint curves correspond to the distributions observed in the four quadrants of the image, and the bold curve is the exponential model of the disc cut off at $R_{\max}$. (b) The dots represent the luminosity distributions observed as a function of z at several radii. The curves represent the model of the disc (a $\mathrm{sech}^2(z/z_0)$ law). Towards the centre the bulge dominates and the model does not represent the data. Towards the exterior the differences with the model are interpreted as being due to the presence of a thick disc. (From van der Kruit and Searle 1981)

many secondary components (bars, lenses, rings, and so on), or even simply regions of star formation, which prevent an unambiguous analysis; there is also coupling between the various components that creates intermediate components. As a consequence there are many exceptions to the exponential disc law.

The Vertical Structure. The vertical structure of discs has been well studied in spiral galaxies viewed edge-on, where the effects of projection are minimized (the inclination being between 80 and 90°; by convention a galaxy seen face-on has an inclination $i = 0°$). The luminosity profiles in the direction Oz perpendicular to the disc are exponential for intermediate radii, avoiding the central bulge, with a tendency to be rounder in the neighbourhood of the plane, approaching a Gaussian. This distribution corresponds to the vertical structure of a disc of selfgravitating stars in equilibrium owing to a velocity dispersion along z but independent of z. Consider the problem in one dimension of an infinite disc of stars of density ρ subject only to gravitation: the equivalent pressure of this gas of stars, of vertical velocity dispersion σ_z, is $P = \rho \sigma_z^2$. The gravitational potential Φ is given by the Poisson equation: $\Delta \Phi = 4\pi G \rho$, where G is the gravitational constant; hydrostatic equilibrium ($\boldsymbol{\nabla} P = -\rho \boldsymbol{\nabla} \Phi$) is thus given by

$$\frac{\mathrm{d}}{\mathrm{d}z}\left(\frac{1}{\rho}\frac{\mathrm{d}\rho}{\mathrm{d}z}\right) = -\rho\, 4\pi \frac{G}{\sigma_z^2}.$$

This equation can be easily integrated since $\sigma_z =$ constant for an isothermal model, the solution then being of the form

$$\rho = \rho_0 \operatorname{sech}^2(z/H) = \frac{\rho_0}{\cosh^2(z/H)},$$

where ρ_0 is the density at $z = 0$ and where the characteristic height H is given by

$$H = \left(\frac{\sigma_z^2}{2\pi G \rho_0}\right)^{1/2},$$

or

$$H = \frac{\sigma_z^2}{\pi G \mu(r)}.$$

In this last equation $\mu(r)$ represents the surface density ($\approx 2H\rho_0$).

This density distribution agrees exactly with observations of discs viewed edge-on; it has two types of asymptotic behaviour:

$$\operatorname{sech}^2(z/H) \approx \begin{cases} 1 - (z^2/4H^2) & \text{for} \quad z \ll H \\ 4\mathrm{e}^{-2z/H} & \text{for} \quad z \gg H \end{cases}.$$

Moreover an interesting observational result is that in spirals, and also in S0s, the characteristic stellar height H is independent of the radius. As H

depends directly on the density, which decreases exponentially with z, this result implies that the velocity dispersion σ_z decreases in the same way with the radius r (van der Kruit and Searle 1981).

In certain spiral galaxies with a large central bulge (in particular, S0s), after the luminosity profile of the exponential disc has been modelled and subtracted there remains a flattened distribution of stars between the disc and the central bulge: this component has been christened the 'thick disc' (Burstein 1979). It might be due to the perturbation exerted by the central bulge on the flattened potential of the disc (Fig. 1.7).

1.3 Stellar Populations, the Distribution of Colour, and Evolutionary Models

In our quest for clues to determine the dynamic structure and formation processes of galaxies, the age of the constituent stars is a fundamental parameter. One has to be able to explain, for example, how spiral galaxies are still actively forming stars today, unlike elliptical galaxies.

1.3.1 Stellar Populations

The idea of classifying stars into populations was introduced by Baade in 1940. He distinguished just two categories of stars: population I, of which the most luminous stars are blue, very hot, massive, and therefore young; and population II, of which the most luminous stars are cool, red, giant, and very old. In a colour–magnitude diagram (the Hertzsprung–Russell diagram, Fig. 1.8) population II stars would occupy the same region as the globular clusters of the Milky Way, for example, which are dense clusters of old stars, whereas population I correspond to the clusters of young stars well known in the Galactic plane. In our own Galaxy, population I stars are found exclusively in the disc, the majority in the spiral arms with very few in the regions in between. Population II stars are concentrated in the central bulge and globular clusters.

Baade interpreted Hubble's classification sequence E–Sa–Sb–Sc–Irr as the continuous variation in relative importance of the two populations (at one extreme, the ellipticals, composed uniquely of population II stars, and at the other, the irregulars composed of population I stars). This of course corresponds to the increasing fraction of gas along the Hubble sequence.

Ever since, the classification has been refined: the stars have been regrouped in 5 classes according to their temperature, luminosity, and metallicity, and there is a continuity from one class to another. Table 1.1 brings together the various characteristics of the 5 populations of our Galaxy (the extreme halo population II is sometimes called population III). The same populations are of course found in all external galaxies.

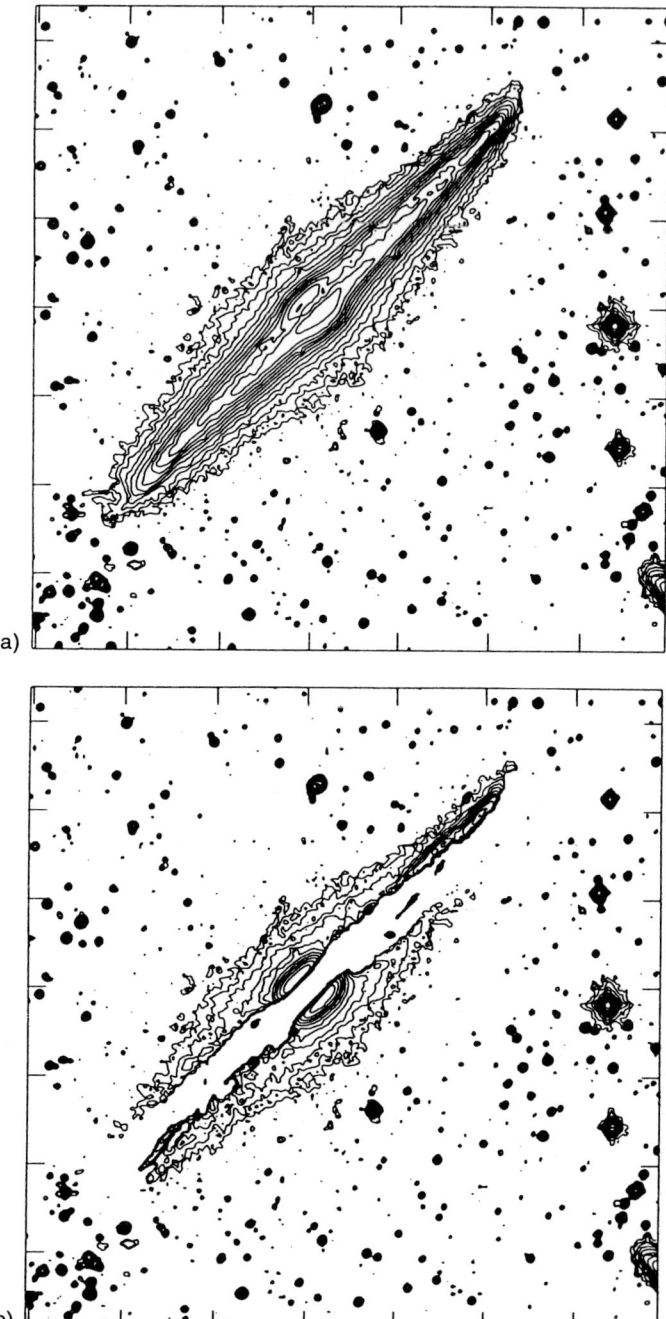

Fig. 1.7. Isophotes of the galaxy NGC 4565. The image in (**b**) represents the central bulge (and the thick disc) remaining after the subtraction of the best model of the disc found from the profiles (**a**). (From van der Kruit and Searle 1981)

1.3 Stellar Populations, the Distribution of Colour, and Evolutionary Models 19

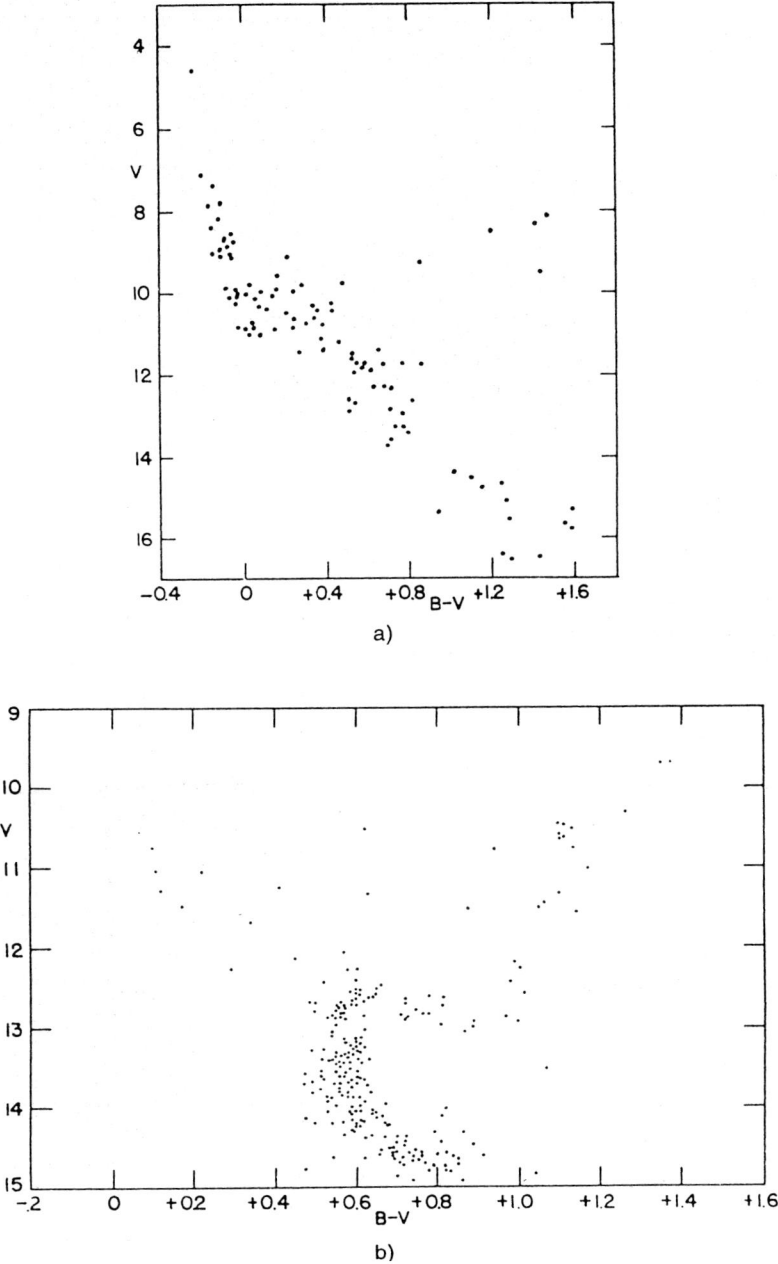

Fig. 1.8. Hertzsprung–Russell diagrams of two extreme stellar clusters in our Galaxy: (**a**) the very young cluster NGC 2264 (from Walker 1956); and (**b**) the very old cluster M67 (from Johnson and Sandage 1955)

Table 1.1. Stellar Populations

Population	Category of Star (Typical)	Velocity Dispersion ΔV [km s^{-1}]	Shape of System	Abundance of Light Elements with Respect to Hydrogen
Halo population II	Globular clusters, red giants	130	Spherical	0.003
Intermediate population II	Stars with high velocities	50	Intermediate	0.01
'Disc' population	Stars with weak lines	30	Intermediate	0.02
Intermediate population I	Stars with strong lines	20	Intermediate	0.03
Extreme population I	Blue supergiants	10	Flat	0.04

As one can see in Table 1.1, the stellar populations suggest that the disc of a spiral galaxy is flattened more and more as the formation of stars proceeds. Imagine a protogalaxy in the form of a ball of gas. The first stars that form out of the primordial gas have a very low metallicity: the metals or elements heavier than carbon are only present in the gas following the explosions of stars, where these elements are synthesized by nuclear reactions. The older a star, the less metallic it is. On the other hand once the stars are formed they no longer interact with the gas; no longer undergoing collisions, the system of stars formed will retain its initial geometry, whereas the gas continues to cool down and flatten by gravitational collapse. The first stars formed therefore retain a spherical geometry known as the 'halo'. The next stars formed have a more flattened spheroidal symmetry. In the course of the flattening and gravitational collapse of the gas the degree of rotation increases. In the halo the rotation is weak but by contrast the velocity dispersion of the stars is high.

1.3.2 The Distribution of Colour

Every star has a specific colour according to its spectral type, determined by the nature of the lines appearing in its spectrum, its mass, and its age; this multicolour photometry of a galaxy can give us information about the stellar populations, star formation, and the evolution of the galaxy. Many questions are raised. Does star formation vary from one region to another in the galaxy as a function of the morphology, of the dynamics, or of the presence of gas? We can only approach these questions by constructing models: by combining the different stellar populations and calculating the corresponding colours.

The colours can be defined in the standard UBV photometric system (U = magnitude in the ultraviolet, B = magnitude in the blue, and

1.3 Stellar Populations, the Distribution of Colour, and Evolutionary Models

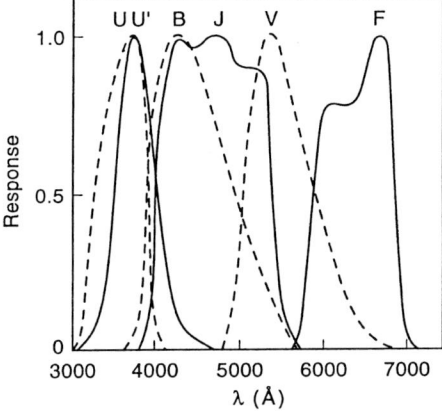

Fig. 1.9. Examples of the definition of a photometric system. The response curves of U, B, and V filters (the standard system) are indicated by dashed lines; in comparison the response curves of the system $U'JF$ are shown as continuous lines

V = visible magnitude) or some other equivalent system (an example is shown in Fig. 1.9). They correspond to differences in magnitude: $U - B$ and $B - V$; these quantities are plotted in a UBV diagram, which combines the data, as in Fig. 1.10. There are great variations in colour from one galaxy to another for the same position in the Hubble sequence. Nevertheless it is by colours that we can discern the effects of evolution: irregular galaxies are the least evolved systems, containing a lot of gas and appearing blue, while ellipticals are today composed almost entirely of old red stars.

In elliptical galaxies the nucleus is in general a redder colour than the outer regions, which appear bluer. How can this colour gradient be explained? As the luminosity profile is very concentrated (an $r^{1/4}$ law), if star formation still takes place, owing to the residual gas continually thrown out by the old stars, the models predict bluer cores, the new stars being formed at the centre. We can in fact consider the ellipticals to be like systems of a single population of old stars, the colour gradient being due solely to the observed metallicity gradient: for a given global age a stellar population appears bluer when the metal abundance is lower.

In spiral galaxies the colour indices do not seem to vary significantly from the centre to the edge, at least if just the disc is taken into account (actually the apparent colours change globally from red to blue, owing solely to the presence of the central bulge, which is made up of old-population stars). In fact the colours show variations across the specific morphological components in the spiral arms; for example onto the medium-age disc population there is superposed a blue and young population. Likewise the presence of a 'lens', an oval component at the centre of galaxies, is revealed by bluer colours.

If the colours of spiral galaxies are relatively uniform over the whole disc, this implies a practically constant star formation rate in the region, inde-

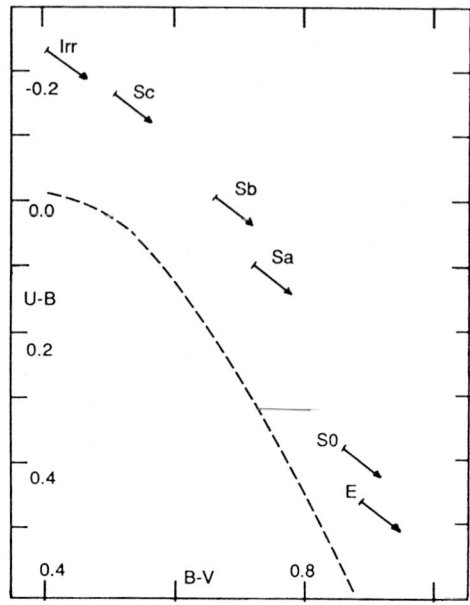

Fig. 1.10. A colour–colour $(U - B, B - V)$ diagram representing the position of normal galaxies (*points*) and the main sequence of stars (*dashed line*). The arrows indicate the effect obtained by correcting for galactic reddening, $E(B - V) = 0.06$

pendent, in particular, of the gas density. This surprising result has been interpreted as a progressive accumulation of gas in the disc, an almost continuous inflow from the exterior (perhaps provided by a halo encircling the galaxy or by another galaxy by tidal interaction). The present distribution of gas is only instantaneous, thus providing no information on its past distribution, and some data suggest that the gaseous component moves radially over the course of time (the metal abundance, asymmetries of stellar distribution leading to gravitational torques, and so on).

1.3.3 Evolutionary Models

To understand the integrated colours of galaxies and deduce implications for the history of star formation and galactic evolution, it is necessary to try to reproduce the actual luminosity of galaxies from the contributions of several successive generations of stars, whose evolution can be followed on theoretical curves (the evolutionary tracks in the Hertzsprung–Russell (HR) diagram). The various basic necessary ingredients for these models are:

- the rate of formation of new stars;
- the mass spectrum of newly formed stars (an initial mass function: $dN_*(M)/dt$; that is, the number of stars formed of mass M);
- the evolutionary track in the HR diagram;

- the enrichment of gas in metals by massive stars at the end of their evolution, notably supernovae (more precisely the rate of enrichment normalized to the rate of consumption of gas);
- the nonstellar contributions to the luminosity (the emission lines of the gas, dust, nonthermal sources, and so on).

It has thus been possible to interpret observations of the colours of elliptical, spiral, and irregular galaxies with a star formation rate decreasing exponentially as a function of time. The models suggest that all galaxies are the same age, around 10^{10} years old, and have a mean initial mass function corresponding qualitatively to the Salpeter function: $dN_*(M)/dt = CM^{-\alpha}$, with $\alpha \approx 2.5$. More sophisticated models make use of a different power law for several mass intervals, but the most important thing is to choose the upper and lower bounds of the initial mass function: particularly important is the choice of the upper bound, since stars that are very massive and very luminous evolve very quickly; typically the bounds are around 0.25 and 35 $M_\odot$. The observed sequence of galaxies in a colour–colour diagram is reproduced only by making the rate of decrease as a function of time of the star formation rate vary. If it varies as $e^{-\beta t}$, a value of β tending towards infinity (a single outburst of star formation 10^{10} years ago, that is, a single generation of stars) represents the colour of elliptical galaxies; a value of β tending towards 0 (an almost constant star formation rate over time) represents the colour of late-type spiral galaxies. Furthermore certain irregular blue galaxies will from time to time undergo random bursts of star formation, and these 'starbursts' will form a greater proportion of massive stars.

On this simple schema for 'normal' galaxies are going to be superposed more violent phenomena, when galaxies interact. The colour–colour diagram has the advantage of showing how the dynamics of galaxies influences their star formation rate; evolution also occurs significantly as a result of sudden events. It even becomes difficult ultimately to distinguish in the integrated colours of a galaxy what is due to slow evolution from what is due to a series of starbursts averaged over time. Figure 1.11 shows, however, that galaxies that have recently interacted, which is manifested by the vestiges of tidal interaction (tails and bridges between the galaxies, deformations, and so on), occupy a particular position in the colour–colour diagram. Starbursts lasting between 2×10^7 and 5×10^8 years reproduce the phenomenon well.

1.4 The Statistical Properties of Galaxies

1.4.1 The Luminosity Function

The luminosity function characterizes the number distribution of galaxies as a function of their intrinsic luminosity L, for a volume-limited sample. The remarkable fact is that this function is relatively constant, both for

24   1. The Classification and Morphology of Galaxies

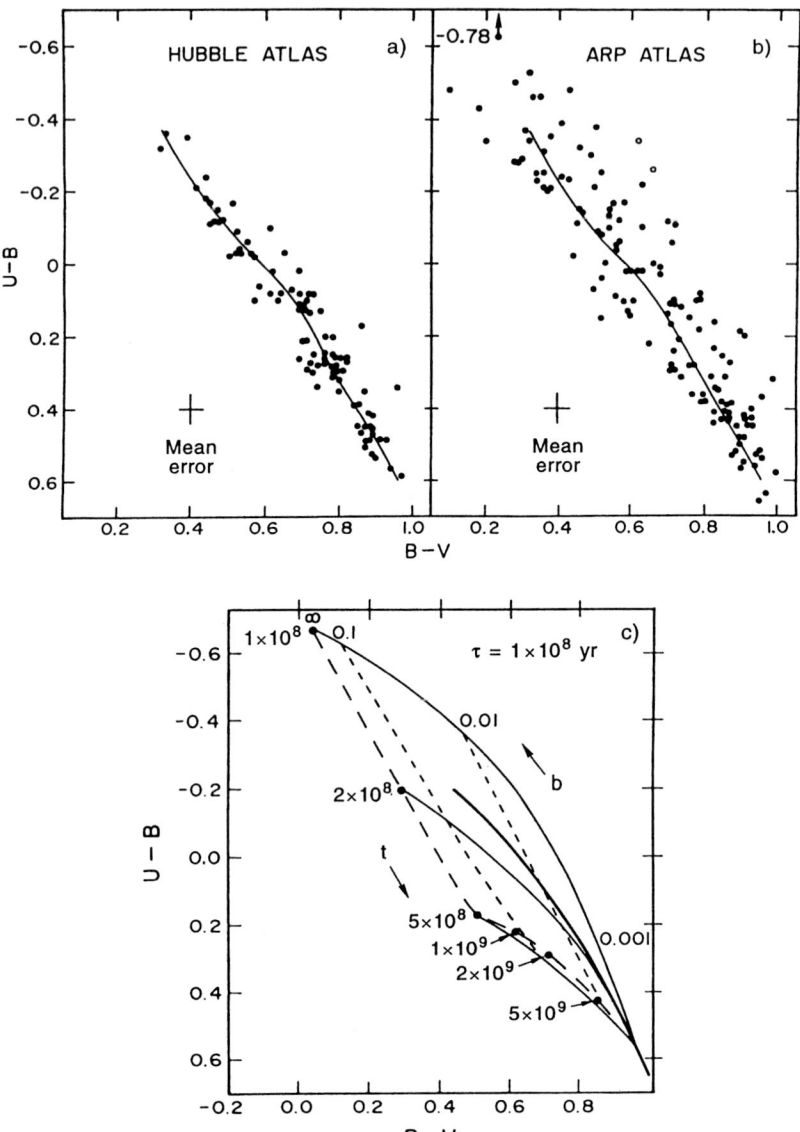

Fig. 1.11. *UBV* diagrams for two samples of galaxies: (**a**) normal galaxies, and (**b**) 'interacting' galaxies (all chosen at high Galactic latitude, $b > 20°$). The curve in each is a visual model of normal galaxies for comparison. (**c**) Theoretical *UBV* diagrams. The bold curve corresponds to normal galaxies, for which the star formation rate decreases uniformly, from 5×10^9 to 2×10^{10}. The faint curves correspond to galaxies that have undergone a burst of star formation for 10^8 years, with a variable intensity b and an age indicated on each curve. The dashed curves represent the evolution over time t of several starbursts (according to their intensity b). (From Larson and Tinsley 1978)

bright field galaxies and for distant clusters, which allows us to define a
'universal' function. This function is very useful, equally for determining the
correlation function between galaxies, the distance of clusters, the frequency
of absorption lines in the spectrum of quasars, or corrections of evolution.
This universal function, the so-called Schechter function, may be written as

$$\frac{\mathrm{d}N(L)}{\mathrm{d}L} = C\left(\frac{L}{L^*}\right)^\alpha \mathrm{e}^{-L/L^*},$$

where C is a constant, L^* is a characteristic luminosity corresponding to the
absolute magnitude (in the blue, B) $M^* = -20.6$ (this function is represented
in Fig. 1.12), and α is a constant around -1. The luminosity function appears to be slightly different for bright galaxies, for which α is higher (around
-0.5), than for the faint end of the galaxy distribution, where α decreases, to
converge towards -1.8. The slope of the Schechter profile is also a function
of the environment, since mergers have taken place more thoroughly in clusters; the slope is then flatter. There is also an observed trend for the slope
to steepen towards the outskirts of clusters, confirming that the latter are
relatively more populated by faint galaxies. The number of galaxies brighter
than L^* falls dramatically, and the number of galaxies less bright forms a
plateau. For a given cluster it is possible to determine the characteristic apparent magnitude m^* with a precision of 0.25 mag and thus to be able to find
its distance knowing the absolute magnitude M^*.

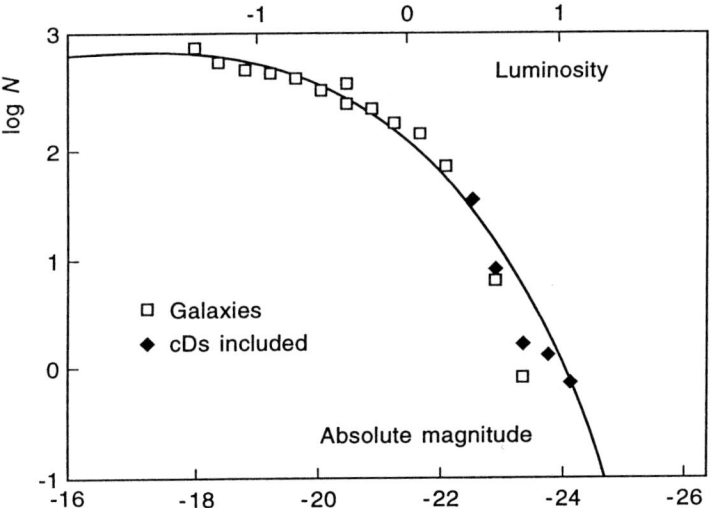

Fig. 1.12. The luminosity distribution of galaxies in clusters: the open squares
represent observations of several clusters. The continuous curve is the best model
(Schechter's law). The solid diamonds show the effect obtained by including cD
galaxies

Although to the first order the luminosity function is universal, it is possible to the second order to distinguish the effects of the environment and maybe those of evolution. Clusters rich in spiral galaxies are distinguished from the others by their flatter luminosity function. Clusters rich in lenticulars and ellipticals have a very steep distribution towards high luminosities; this phenomenon is in general accompanied by the presence of a cD galaxy, a superluminous giant elliptical galaxy situated at the cluster's core. One current interpretation of this phenomenon is the merger of the largest galaxies of the cluster into one, the cD; the presence of a superluminous galaxy that is very distant according to its spectrum together with a deficiency of neighbouring galaxies would explain the sharp fall in the number of galaxies beyond L^* in rich clusters.

Let us consider the so-called cannibal cD galaxies further. This type of galaxy has been known of since the 1960s, when bright elliptical galaxies of a particular type found in clusters and surrounded by extended stellar envelopes were christened D galaxies. The supergiant cD galaxies ('cluster D') are characterized above all by their very extended luminous envelope, of the order of 100 kpc – 1 Mpc; they are located at the centre of rich clusters, but what matters is the local density of galaxies: sometimes they are found in clusters globally poor in the total number of galaxies but with a very high density (several galaxies per Mpc^3), which suggests that the local environment alone is fundamental. Their three-dimensional form is very flattened, and sometimes their central surface luminosity is very weak, unlike in normal ellipticals. Furthermore cDs very often have multiple nuclei, which agrees with the idea of the fusion of galaxies. It should nevertheless not be forgotten that, in the very dense regions where these galaxies were discovered, the effects of projection simulate systems with several nuclei along the same line of sight; the radial velocity of the galaxies, however, allows this difficulty to be resolved. The phenomenon of cannibalism is due to the tidal interaction between galaxies and the braking by dynamical friction that results (Chap. 7). The more a galaxy has an extended envelope, the more the dynamical friction is effective, and the more it pulls towards it the companions of the cluster. The process then amplifies by itself, mainly at the centre of the cluster, where a galaxy might preserve its envelope from destruction by the tidal effects of the cluster, the reason being that the tidal effects grow in proportion to the square of the distance to the centre. Moreover all the debris resulting from the tidal interactions of the galaxies falls to the centre of the potential well of the cluster and will sustain the envelope of the cD. Likewise, perhaps, the very hot intercluster gas, which is observed in the X-ray domain, cools in the neighbourhood of the cD to form the filaments observed around some of these galaxies.

1.4.2 Morphological Segregation

Galaxies are not distributed uniformly in the sky but are mostly in small groups or clusters. About 69% of known galaxies up to 25 Mpc away belong to groups, 20% are associated with other galaxies, and 10% are in loose 'clouds'. There remain only 1% of galaxies that are isolated. Of course, these figures are only indicative of and dependent upon the criteria adopted to define the groups: here two galaxies are said to belong to the same group if their separation R gives to the density of the group L/R^3 a value greater than the threshold, approximately $2.5 \times 10^9 \, L_\odot \, \mathrm{Mpc}^{-3}$ (L is the greater luminosity of the two). With this definition the identified groups are in general made up of entities that are gravitationally bound; the mean velocity dispersion of these groups is $100 \, \mathrm{km \, s^{-1}}$.

The proportion of galaxies belonging to rich clusters where the density of galaxies exceeds 1 galaxy per Mpc^3 is, however, small, of the order of 5%. It is in these regions that the environment has marked effects on the morphology and evolution of the galaxies. In particular beyond this critical density the proportion of elliptical galaxies grows dramatically and exceeds that of spirals. Figure 1.13 shows the proportion of the various types of galaxy observed as a function of the mean density and the number of galaxies covered in each interval of mean density. It is striking to note the disappearance in clusters of gas-rich galaxies, mainly to the advantage of lenticulars, and then to ellipticals, at very high local densities. One of the main interpretations is that the gas of galaxies is swept up in clusters by the hot intergalactic wind; interferometric observations of the H I component of some galaxies in the Virgo cluster have confirmed this hypothesis, revealing a wild system of gas swept in the opposite direction to the cluster centre. Tidal effects are superposed, however, on the intergalactic wind to transform the spiral into lenticular galaxies, thickening the discs and enhancing the central bulges. This effect is produced on a slower time scale than the sweeping up. At very high densities the fusions of galaxies intervene, the merger of two disc systems giving birth to a spheroidal system, which becomes an elliptical galaxy after relaxation. Other interpretations, however, have been proposed: they postulate an intrinsic morphological segregation into clusters: regions of high density in the universe will only be able to form a particular type of galaxy. This is the 'nature' theory as opposed to the 'nurture' theory.

1.5 The Evolution of Galaxy Morphology

Galaxies are expected to evolve and change morphology on a time scale much shorter than the present age of the universe: they have formed most of their stars, and they have undergone tidal interactions and mergers even more frequently in the past. Realistic galaxy simulations including gas and stars confirm that galaxies should evolve quite rapidly and change morphology on

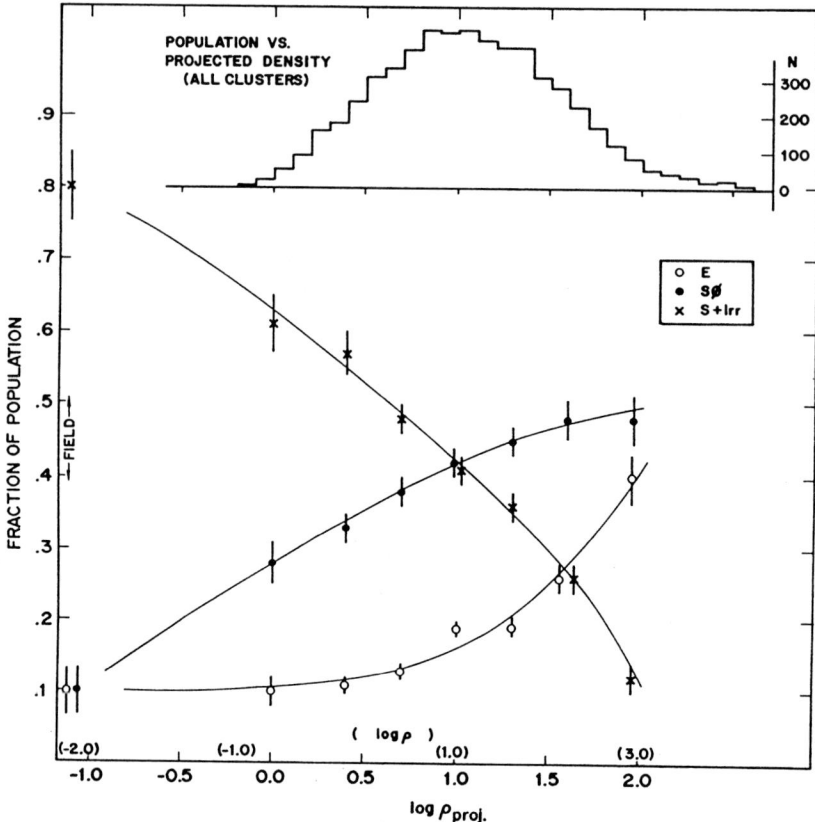

Fig. 1.13. The fraction of elliptical (E), lenticular (S0), and spiral and irregular galaxies (S+I) as a function of the projected density of galaxies along the line of sight (revealingly transformed into volume density on the abscissa scale). The histogram at the top shows the number of galaxies observed together with the corresponding density. (From Dressler 1980)

a time scale much shorter than the Hubble time: strong asymmetries in their mass distribution (of which spiral structure is a frequent example) drive the concentration of matter and the building of bulges. It is therefore possible to learn about galaxy formation and evolution through observations of galaxy morphology as a function of redshift.

In recent years, it has become possible to detect galaxies at increasingly high redshift and investigate their properties. In particular, the Hubble Space Telescope (HST), with its high resolution of 0.1 arcsecond, and its deep fields (HDF, north and south, observed during 10 continuous days each, in four passbands) has enabled a breakthrough to be made in this domain, by allowing morphological classification of remote galaxies. Although more schematic than for nearby galaxies, this classification can capture the essential properties, such as the concentration of light and its asymmetry (Abraham et al. 1996) or the preponderance of discs/spheroids, through their luminosity profiles ($r^{1/4}$ or exponential, see Marleau & Simard 1998). The redshifts are known either directly, through spectroscopy with large telescopes or photometry. The latter method is quicker, being based on templates of spectral energy distribution of galaxies and derived colors, but its reliability depends on redshift.

The most striking result of these morphological evolution studies is the much larger percentage of irregular, peculiar, and interacting galaxies at high redshift. This confirms previous observed trends that the galaxy interaction rate increases as fast as $(1+z)^m$, with m of the order of 4 (e.g. Carlberg 1991). These trends are expected in hierarchical cosmological scenarios, where galaxies form through merging. Also, the percentage of spheroid-dominated and elliptical galaxies seems to be lower at $z > 2$, and the size of galaxies smaller, which is consistent with the picture of galaxies forming out of the merging of subclumps. However, the results should be taken with caution, given the various biases and incompleteness problems encountered in these pioneering studies.

At redshifts greater than 2.5, distant galaxies are very efficiently identified by the presence in their spectrum of a Lyman continuum break, due to cumulative absorption by the many H I clouds along the line of sight, already known from QSO absorption lines (Chap. 10). With only three-colour photometry, it is then possible to detect a considerable number of galaxies between $z = 2.4$ and 3.4, through their drop in flux in the bluer color (while normal galaxies have a rather flat UV spectrum). This U-dropout technique, used first by Steidel et al. (1996) is now generalized at much higher redshifts, identifying B-, or V-, and even R-dropout galaxies. Hundreds of galaxies have been catalogued at $z \sim 3$, and statistical studies are possible. In particular, galaxies at $z \sim 3$ appear very clustered on the sky, at least as clustered as nearby galaxies, constraining large-scale structure formation scenarios (Chap. 12).

These techniques open the possibility of investigating the star formation history of the universe (Madau et al. 1996). Through the luminosity density and colours as a function of z, and on the basis of standard galaxy evolutionary models and initial stellar mass function (IMF), the metal ejection rate and the star formation rate can be estimated as a function of redshift (see Fig. 1.14). At lower redshifts ($z < 1.3$), the CFRS (Canada-France redshift survey) has revealed the strong evolution of galaxy populations (Lilly et al. 1995): this evolution is selective in colour, in the sense that the luminosity function of red galaxies does not change with z, while that of blue galaxies evolves considerably, brightening and steepening. The general trend of the recent deep redshift surveys is that early-type galaxies remain practically unevolved from $z \sim 1$ to the present, while late-type gas-rich galaxies strongly evolve in luminosity and/or number during the same period. The peak around $z = 1.5$ of the star formation rate seems to correspond to a

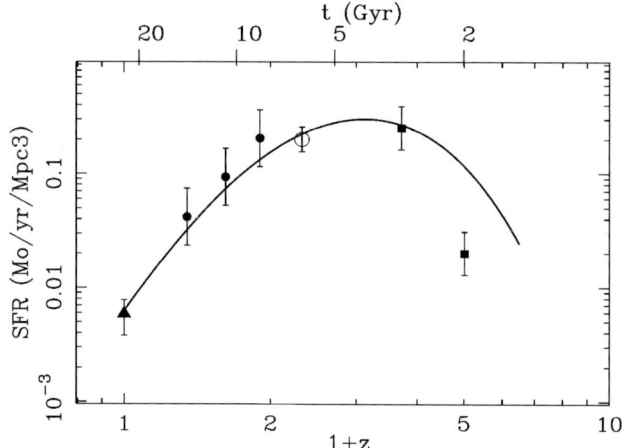

Fig. 1.14. Cosmic history of star formation. The various data points, coming from different surveys, give the universal metal ejection rate, or the star formation rate SFR (*left-scale*), as a function of redshift z. The first point (*triangle*) is from Gallego et al. (1995), the three following points (*full circles*) from Flores et al. (1999), the empty circle from Yan et al. (1999), and the two high-redshift points (*filled squares*) from Pettini et al. (1998). The history at high redshift is still very uncertain, since a significant part of the star formation could be hidden by dust

brightening of discs, which are still being formed at higher redshifts. Yet, many interpretations are still possible, since the apparent non evolution of a given morphological population can also be explained in part by an equal number of systems joining this class or disappearing from it (by merging, for example).

References

Abraham, R., et al. (1996) *Mon. Not. Roy. Astron. Soc.* **279**, L47.
Andredakis, Y.L., Peletier, R.F., Balcells, M., (1995) *Mon. Not. Roy. Astron. Soc.* **275**, 874.
Burstein, D. (1979) *Astrophys. J.* **234**, 829.
Carlberg, R., (1991) *Astrophys. J.* **375**, 429.
de Jong, R., (1996)*Astron. Astrophys.* **313**, 45.
De Vaucouleurs, G. (1959) *Handb. Phys.* **53**, 275.
Dressler, A. (1980) *Astrophys. J.* **236**, 351.
Flores, H.et al. (1999) *Astrophys. J.* **517**, 148.
Freeman, K. C. (1970) *Astrophys. J.* **160**, 811.
Gallego, J., Zamorano, J., Aragon-Salamanca, A., Rego, M., (1995) *Astrophys. J.* **455**, L1.
Hubble, E. P. (1930) *Astrophys. J.* **71**, 231.
Johnson, H. L., and Sandage, A. R. (1955) *Astrophys. J.* **121**, 616.
King, I. R. (1966) *Astron. J.* **71**, 64.

Kormendy, J. (1982) in *Morphology and Dynamics of Galaxies* (Saas-Fee Advanced Course 12, Geneva Observatory, Geneva), p. 113.
Larson, R. B., and Tinsley, B. M. (1978) *Astrophys. J.* **219**, 46.
Leach, R. (1981) *Astrophys. J.* **241**, 485.
Lilly, S., et al. (1995) *Astrophys. J.* **455**, 108.
Madau, P., et al. (1996) *Mon. Not. Roy. Astron. Soc.* **283**, 1388.
Marleau, F., Simard, L. (1998) *Astrophys. J.* **507**, 585.
Pettini M., Kellogg M., Steidel C.C. et al. (1998) *Astrophys. J.* **508**, 539.
Sandage, A. R. (1961) *The Hubble Atlas of Galaxies* (Carnegie Institution of Washington, Washington DC).
Steidel, C., Giavalisco, M., Dickinson, M., Adelberger, K., (1996) *Astron. J.* **112**, 352.
Van der Kruit, P. C., and Searle, L. (1981) *Astron. Astrophys.* **95**, 105.
Walker, M. (1956) *Astrophys. J.* **23**, 1.
Yan L., McCarthy P.J., Freudling W. et al. (1999) *Astrophys. J.* **519**, L47.

2. The Galactic Interstellar Medium

Galaxies are formed mainly of stars, but these stars are immersed in a relatively diffuse and cold gaseous medium. Its density is on average of the order of 1 particle per cm^3, 10 cm^{-3} in clouds of atomic hydrogen, and 1000 cm^{-3} in molecular clouds. Its temperature goes from 5 K in these latter regions up to 10^4 K in ionized regions heated by young stars. Hydrogen forms the bulk of the interstellar gas, helium around 25%. Other elements are present in trace amounts. The interstellar gas is enriched by heavy elements ejected by stars (in supernova explosions and stellar winds).

The distribution of gas in galaxies is obtained from a group of complementary data in various wavelength domains: in the optical domain emission lines are the signatures of ionized regions or H II regions (notably the Hα line); in the centimetre radio domain thermal continuum radiation also indicates the presence of H II regions, and the 21 cm line of atomic hydrogen (H I) allows the mapping of galaxies on a large scale and at very great distances from the centre, well beyond the optical radiation of the galaxy; in the millimetre radio domain the 2.6 mm line of the carbon monoxide molecule, CO, is a tracer of the H$_2$ molecule and dense molecular clouds, which are much more concentrated towards the centre than are clouds of atomic gas.

2.1 Ionized Gas

While absorption lines, which dominate the spectrum of a galaxy, provide information about the stellar content (the lines come from the atmospheres of the stars), it is emission lines that enable the distribution of ionized gas to be studied.

2.1.1 Ionized Regions of the Disc

In a spiral galaxy's disc the H II regions are ionized by the radiation from young and massive stars, associations of OB stars[1]. In the vast majority of cases they are concentrated along the optical spiral arms. Their radial distribution has a maximum at about a quarter of the distance from the

---

[1] These are the most massive stars in the OBAFGKM classification.

centre of the outermost H II region. This is a large concentration compared with the distribution of atomic hydrogen, H I, whose maximum density almost always lies outside the optical spiral structure.

The observed emission-line flux allows us to obtain approximately the mass of ionized hydrogen. In the H II regions of our Galaxy the observed emission lines are, in addition to the Balmer lines (Hα, Hβ, Hγ), the 'forbidden' lines[2] N$_1$ and N$_2$ (so called because they were originally attributed to 'nebulium') of O^{2+}: [O III] at 4959 and 5006.8 Å, the lines corresponding to the isoelectronic ion N II (6548.1 and 6583.6 Å), and the O II lines at 3726.1 and 3728.8 Å. A typical spectrum is given in Fig. 2.1. Corrections for extinction due to dust are made following a comparison of the observed and theoretically predicted Balmer decrement. The emissivity of the Balmer lines varies proportionally with the product of the volume density of electrons and protons $n_e n_p \approx n_e^2$ (if one neglects the density of other ions) and a factor dependent on the temperature. By defining the emission measure $E_m = n_e^2 l$ (in pc cm^{-6}), that is, the integral along the line of sight (l) of the square of the electron density, and assuming a given temperature (usually of the order of 10^4 K), one can deduce the observed flux in the Balmer lines and the value of the root mean square of the electron density n_e (rms). The mass of ionized hydrogen in a given volume V is obtained from

$$M(\text{rms}) = M_H \left[1 + 4\frac{N(\text{He})}{N(\text{H})}\right] \int f\, n_e(\text{rms})\, dV,$$

where $N(\text{He})/N(\text{H})$ is the ratio of the densities of helium and hydrogen, m_H is the mass of atomic hydrogen, f is the ratio of the number of protons to electrons, and V is the volume. The calculation takes account of the presence of helium, about 25% by mass.

The ratio of intensities of different lines other than hydrogen serves to determine n_e and also the temperature more directly. The ratio of the O II lines at 3729 and 3726 Å, produced by levels separated in energy by just 0.002 eV (≈ 20 K), depends only on the electron density. We therefore obtain a linear mean of the electron density, which we call $n_e(3727)$, and a mass defined in the same way as above: $M(3727)$.

The masses determined by these two methods would only be equal if the electron density were perfectly homogeneous. In fact the gas is distributed in small dense clouds, embedded in a more diffuse medium. If we consider as a first approximation that the density is n_{ec} in the clouds and zero outside, we can define a filling factor η such that

$$n_e(\text{rms}) = \sqrt{\eta}\, n_{ec} = \frac{n_e}{\sqrt{\eta}};$$

---

[2] The forbidden lines are by convention written between square brackets. These lines do not occur in the laboratory, since the levels have too long a lifetime and deexcite by collision. However, at low interstellar densities these lines are possible.

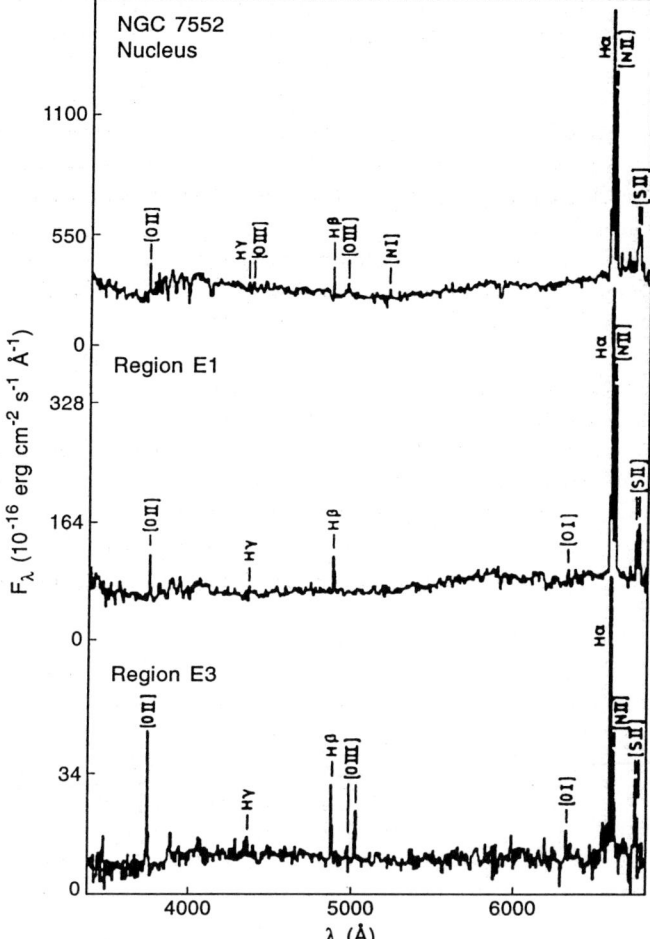

Fig. 2.1. The optical spectrum obtained along the gaps in the barred spiral galaxy NGC 7552 towards the nucleus (at the top) and in two other regions of the disc (E1 and E3). The relative intensity of the emission lines gives information about the excitation ([O III]/Hβ) or extinction (Hα/Hβ). (From Durret and Bergeron 1988)

n_{ec} is known from the ratio of the O II lines, $n_{ec} \approx n_e(3727)$. In Orion the ratio $n_{ec}/n_e(\mathrm{rms})$ is equal to 6, which allows us to obtain a filling factor of $\frac{1}{36} = 0.03$. In external galaxies this factor is still greater, up to 0.12 to 0.21, for example, in H II regions of the Large Magellanic Cloud.

The temperature can be obtained from ratios of pairs of emission lines, i.e. those emitted by a single ion from two levels with very different excitation energies. For example, the ratio of O III lines and N II lines are good indicators. Independent information on temperature can be obtained through radio-continuum observations. An excitation parameter that is very useful is the ratio of the intensity of the lines of [O III] to those of Hβ, [O III]/Hβ. But in the most studied H II regions the various methods give different temperatures, which may reflect the spatial variations of the density, the degree of ionization, and so on.

Once the temperature is known it is possible to obtain the chemical abundances from the ratio of the emission lines. But to determine the total abundance of a given element one must take account of all the other states, ionized or not, of that element. For most of the time the corresponding correction factor is much greater than one, which leads to great uncertainties.

Also, the intensities of the most prominent oxygen lines are indirectly correlated to metallicities owing to the cooling mechanism. Paradoxically, for a certain range in metallicity, the [OIII] line is the stronger as the abundance is lower. Indeed, when O is less abundant, cooling through the optical forbidden lines is less efficient, and the temperature is larger, considerably increasing the [OIII] intensity. Empirical abundance indicators are then frequently used, built from the more readily observable lines. The ratio R_{23} = log (([OII] λ 3727+ [OIII] $\lambda\lambda$ 4959, 5007)/Hβ) has been calibrated through many studies (cf Fig. 2.2). In the ambiguous region of this diagram (the turn-over), the

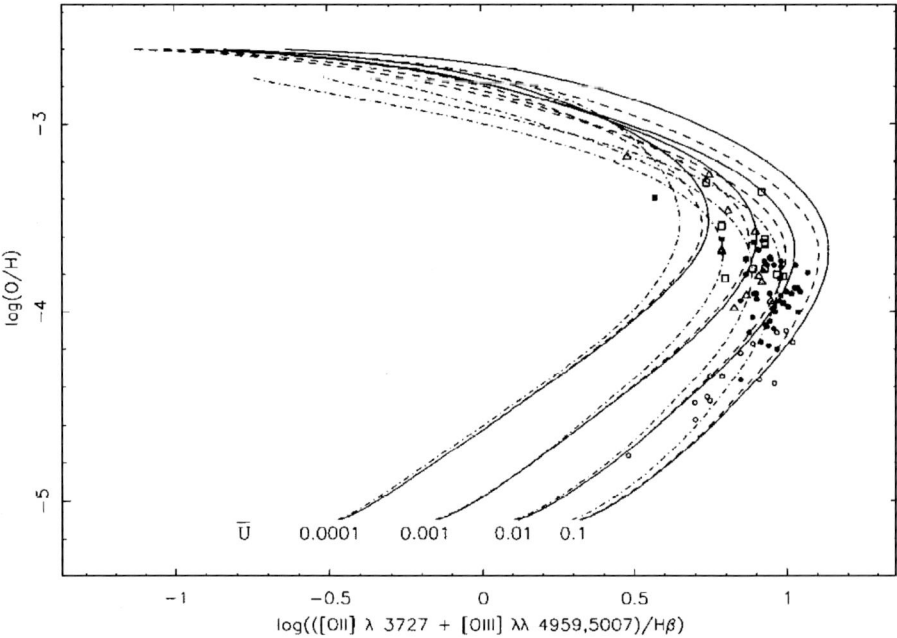

Fig. 2.2. The ratio R_{23} = log (([OII] λ 3727+ [OIII] $\lambda\lambda$ 4959, 5007) /Hβ) is an indicator of metallicity (O/H). Curves are the models for different averaged ionization parameter U =ratio of ionizing photon density to gas density, and the symbols are some H II regions data for comparison. Note the double-valued nature of the relation (from McGaugh, 1991)

diagnostic can be completed through the N II/Hα ratio, which increases with O/H (albeit with some scatter).

The extragalactic H II regions are called giant H II regions, in comparison with the H II regions in the solar neighbourhood. This is only due to bias in observational selection; if we take into account the distance of these galaxies, it is impossible to detect the smaller H II regions. The masses of ionized gas deduced from the observed flux are from 10^5 to 10^7 $M_\odot$. The abundance ratios (N II/Hα, for example) are in general very similar to those of H II regions in the Milky Way, which immediately implies that the abundances and degrees of excitation are similar. Very often the corrections for extinction (determined by the degree of reddening of received radiation) are greater than what one might get along the line of sight that crosses our own Galaxy, which indicates that a large proportion of the reddening is due to dust within extragalactic H II regions.

In irregular galaxies, and in particular in dwarf galaxies, which are true isolated extragalactic H II regions, the heavy elements are very deficient, which is explained by the absence in the past of star formation, which would have enriched the interstellar medium with elements. The abundance ratio O/H might reach $\frac{1}{40}$ to $\frac{1}{100}$ of its value in the solar neighbourhood.

Radial abundance gradients from the centre to the edge of a galaxy are found in almost all large spiral galaxies. The abundances in the outer HII regions are about 10 times lower than in the centre. Chemical evolution models explain these gradients through several factors. First, the gaseous density profile also strongly decreases with radius, and the star formation rate (SFR) is a non linear power law of the gas density ρ_g (the Schmidt law is writen as SFR $\propto \rho_g^n$ with $n > 1$). Second they invoke possible radial gas flows. And third there may be a variation with radius of the yield and of the IMF (initial mass function). It has been observed that in barred galaxies, where the radial gas flows are strong (see Chap. 6), the abundance gradients are flatter, owing to dynamical mixing.

2.1.2 Emission Lines in the Nucleus

Emission lines are also detected in the direction of the nuclei of spiral galaxies. In many cases, it is possible to retrieve the typical spectrum of H II regions, where the gas is photoionized by hot stars. In particular, most starburst galaxies, forming one or two orders of magnitude more stars than normal, have a peak of concentrated star formation in their nucleus. But in addition, a small fraction of spiral galaxies have a nuclear emission-line spectrum with a wider range of ionization than any H II region. They are no longer due to conventional H II regions excited by OB stars. The ratio N II/Hα there is much greater than 1, for example, which is not the case in general for the H II regions of the disc. Usually, these emission-line spectra are very broad, much broader than for starburst nuclei. These nuclei are called active galactic nuclei (AGN); they are Seyfert galaxies and share their properties with the more luminous radio galaxies and quasars (see Chaps. 8–9). Most of the ionized gas in AGNs is photoionized, but the source of ionization is no longer hot stars but probably an accretion disc around a black hole. The ionizing spectrum extends to much higher energies than for hot stars. Some of the gas can be collisionally excited in shocks.

Line ratios can be used as good diagnostics to distinguish between H II-region galaxies and active galactic nuclei (see Osterbrock 1989). In particular the ratio [OIII]/Hβ is an indicator of the mean level of ionization and temperature, while the [OI]/Hα and [SII]/Hα ratios serve as indicators of the extent of high-energy photoionization. When [OIII]/Hβ is plotted versus one of these ratios, or [NII]/Hα, the HII-region nuclei and AGNs are clearly separated.

2.1.3 Radio Emission

The Nature of the Emission. The ionized interstellar medium is also revealed in radio astronomy by continuous radio emission and recombination lines. In addition to the new information that this brings, this emission has the advantage of ridding us of the problems of extinction by dust, which affects only the visible and ultraviolet domains. Extragalactic continuous radio emission may be due to two mechanisms:

- *Thermal emission, or bremsstrahlung,* of the ionized gas. The electrons of the interstellar plasma experience the electrostatic attraction of ions: they travel on hyperbolic trajectories in which they are accelerated and radiate proportionally to the square of this acceleration. In the course of this pseudotransition, they go from a 'free' state of a virtual atom to another 'free' state, and the name 'free–free' is given to this emission.
One can show that if the electrons have a Maxwellian velocity distribution, the emissivity j_ν of the free–free transition is proportional to $T_e^{-1/2} n_e^2 g(T_e, \nu)$, where $g(T_e, \nu)$ (the Gaunt factor) is a slowly varying function of frequency ν and electron temperature T_e (n_e is the electron density); the coefficient of absorption κ_ν thus varies as $T_e^{-3/2} \nu^{-2} n_e^2 g$, following Kirchhoff's law $j_\nu = \kappa_\nu B_\nu$. One can deduce that at the lowest frequency the radiation becomes opaque: the brightness temperature T_b is therefore close to the temperature $T = T_e$ of the interstellar medium, the radiative-transfer equation being $T_b = T_{b0} e^{-\tau} + T(1 - e^{-\tau})$. In this formula T_{b0} is the temperature of the radiation source and τ is the optical depth of the medium.
- *Nonthermal synchrotron emission* is superposed on the thermal radiation of H II regions. This is radiation emitted by relativistic electrons spiralling in a magnetic field $\boldsymbol{B}$. In its helical trajectory around the field $\boldsymbol{B}$ the electron is accelerated and it radiates. The emission is mainly attributed to plasma ejected by supernovae; this enables us to obtain the energy distribution of the relativistic electrons and the intensity of the magnetic field.

Measurements of the polarization enable us to determine the dominant orientation of the field $\boldsymbol{B}$, notably with respect to the spiral arms.

The nature of the emission, thermal or nonthermal, is determined by comparison of the radiation at various frequencies. The radiation spectrum is a signature of the emission. The thermal radiation of H II regions is characterized by its flux density in the radio region with a form illustrated on the M82 spectrum in Fig. 2.3:

$$S = 2kT_\mathrm{b}\nu^2 \frac{\Omega}{c^2}$$

(where k is the Boltzmann constant, c the speed of light, and Ω the solid angle), with T_b, the brightness temperature, varying as ν^{-2} for a small optical thickness $\tau \ll 1$ (if $\tau \gg 1$, T_b is almost constant). On the other hand, synchrotron radiation depends on the energy distribution of the electrons, which is in general a power law. The frequency spectrum of global synchrotron

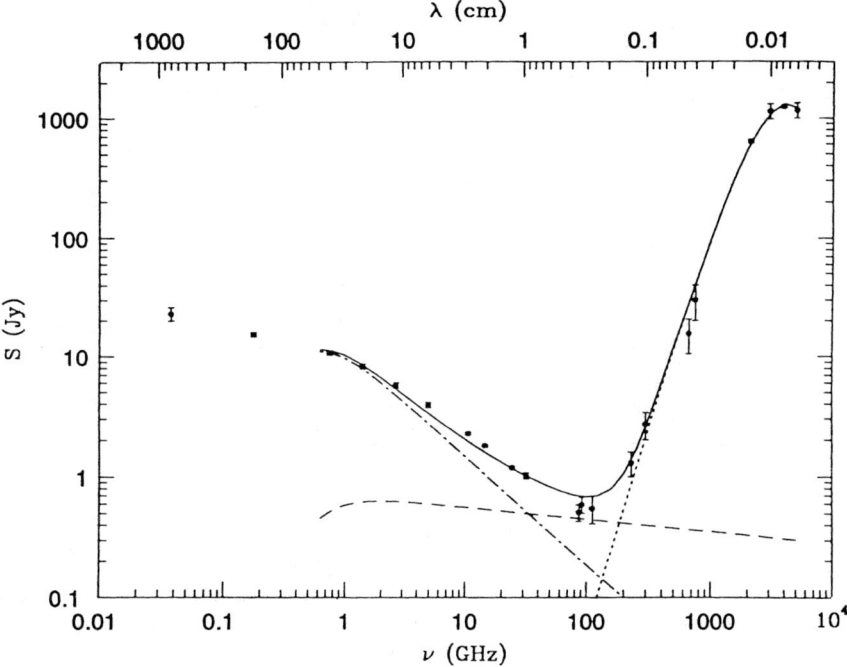

Fig. 2.3. The radio and far-infrared spectrum of the galaxy Messier 82. Observation points are marked by symbols. The free-free emission is modeled by the dashed line: the H II regions in this bright starburst galaxy start to become opaque at $\nu \sim 1$ GHz. The synchrotron emission is modeled by the dot-dashed line, and the dotted line is the emission from dust heated by star light. All three components are combined to yield the full-line fit to the data (from Condon 1992)

radiation is thus also a power law, with a spectral index $\alpha > 0.5$ (0.7 on average: $S_\nu = \nu^{-\alpha}$), always much steeper than that for thermal emission.

General Characteristics of the Emission. To be able to resolve the structure of the radio emission in an external galaxy and in particular to distinguish the emission from the disc from that from the nucleus, it is necessary that the half-power beam width of the radio telescope is less than 15% of the Holmberg radius of the galaxy (the radius where the surface brightness falls to 26.5 mag arcsec^{-2}). In general this requires interferometric observations. When the emission of the disc is detectable, its spatial extension coincides with the optical disc and its radial distribution is broadly the same: exponential with the same characteristic radial scale. On the other hand the mean brightness of the disc in the radio spectrum does not seem to be correlated with the total mass or total luminosity of the galaxy. This distribution suggests that the emission has its origin in a young population (population I) coinciding with that of the H II regions. Nevertheless the mean spectral index of the radio emission of galactic discs indicates a high proportion of nonthermal emission ($\alpha = -0.7$ to -1.0) between 408 and 1420 MHz. Moreover there is a certain correlation between the radio power emitted by the nucleus and the surface brightness of the disc, which suggests a reinforcement of the electron density (or of the magnetic field?) by the activity and supernovae in the nucleus.

In certain galaxies the size of interferometer is sufficient to resolve the spiral structure, and the radio emission in general follows the spiral arms. One of the most beautiful examples is given by the galaxy M51, whose radio isophotes at 6.2 cm are superimposed on the optical photograph in Fig. 2.4. The crests of radio emission trace the spiral arms perfectly and, more precisely, coincide with the dust lanes situated at the concave edge of the arms. A considerably increased rate of supernovae or the confinement of electrons in the arms is insufficient to explain the large emission in the arms. We have to consider the compression of matter in the spiral arms by density waves (Chap. 5). The magnetic field, which is 'frozen' in the interstellar medium, has its lines of force pulled about in the compression, and the field's intensity is considerably increased in the arms. It can be seen in Fig. 2.5 that the magnetic field lines are oriented parallel to the spiral arms.

Note, however, that the rate of compression of the magnetic field lines is not as large as predicted by the theory of a compression in one dimension: the reason is that, owing to the Parker instability, the field lines are stretched out and deformed in a direction perpendicular to the plane of the galaxy. This phenomenon also provides an explanation of the thickening of the radio disc, which is greater than that of the optical disc.

In spiral galaxies less regular than M51, if the spiral structure is clearly visible in the radio region, it is nonetheless perturbed by emission maxima corresponding to giant H II regions in which 'core-envelope' structures or

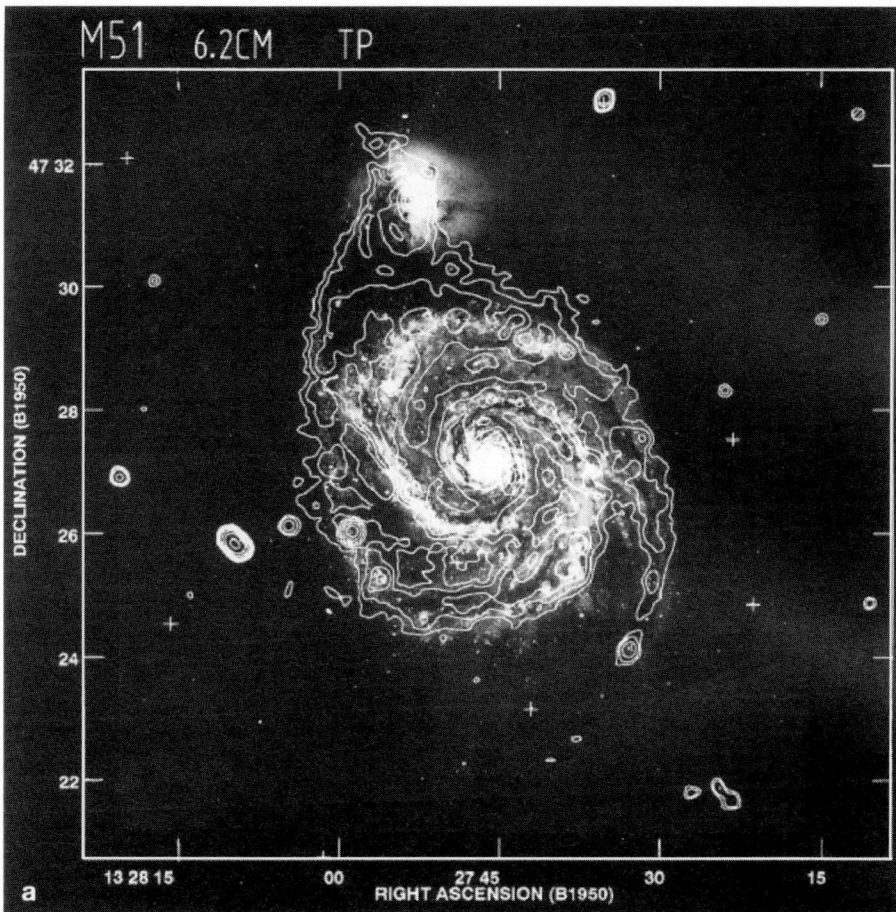

Fig. 2.4. Isophotes of radio-continuum emission at 6.2 cm for the galaxy M51 and its companion NGC 5195 (from Horellou et al. 1990). The map was obtained with the VLA interferometer, with a resolution of $12''$. As well as a very powerful nucleus, M51 displays two very marked spiral arms, with crests of emission displaced towards the internal edge of the arms, in perfect correlation with the dust lanes. The pointlike source on the eastern arm is a supernova remnant; the source still more intense and farther east is a background source not correlated with the galaxy

condensations ('clumps') can be resolved. This is notably the case for M101 and M33.

Some galaxies have a halo of radio-continuum emission, whose emissivity per unit volume is very small (about 10 times less than in the disc), but whose total luminosity is nonneglible compared with that of the disc. This emission is entirely nonthermal, its spectrum being much steeper than that of the disc, and it is not detectable at high frequencies. In several galaxies

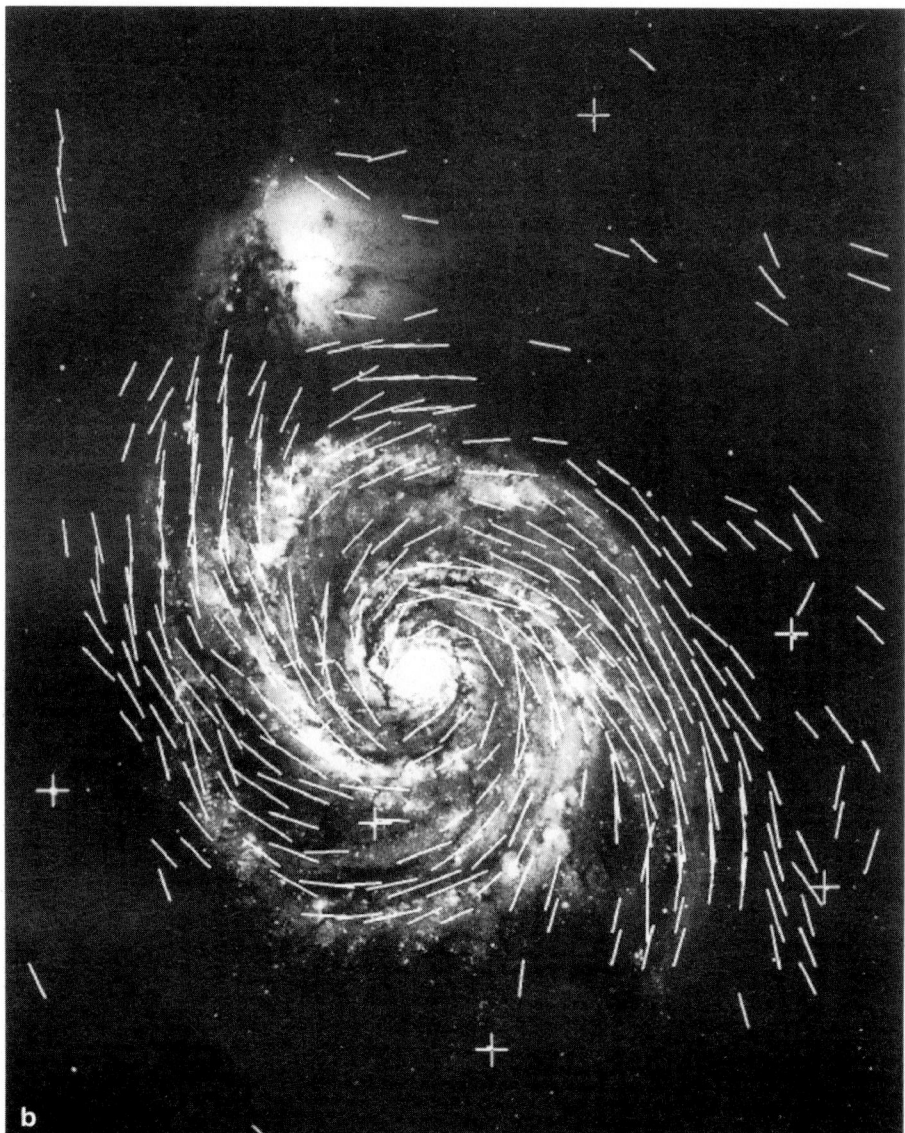

Fig. 2.5. The direction of magnetic field vectors, as determined from polarized radio-continuum emission at 2.8 cm in M51 (from Neininger 1992). The length of the vectors is proportional to the polarized intensity. The magnetic field on a large scale is oriented parallel to the spiral arms

seen edge-on, notably NGC 891, a flattened halo or 'thick disc' highlights this: in fact the radio emission is decomposed into a thin disc (0.5 kpc thick) coinciding with the H I disc, which might be thermal in origin, and a disc 4 kpc thick of nonthermal origin.

2.2 Atomic H I Gas

Neutral atomic gas on a large scale is almost exclusively known owing to the radio line at 21.1 cm, the wavelength of hydrogen (a frequency of 1420.4 MHz). This line corresponds to the transition between the two hyperfine levels of the ground state of the hydrogen atom H ($n = 1$, $l = 0$, $J = S = \frac{1}{2}$, $I = \frac{1}{2}$). The hyperfine structure is due to the interaction between the spins of the electron (S) and the proton (I). The transition $F = 1 \to 0$ ($\boldsymbol{F} = \boldsymbol{I} + \boldsymbol{J}$) is 'forbidden' – it is a magnetic dipole transition – which explains the very small probability of spontaneous transition $A_{10} = 2.9 \times 10^{-15}\,\text{s}^{-1}$. An atom deexcites by radiation every 10 million years, but this low rate of radiation is compensated by the considerable number of H I atoms along the line of sight, typically $N_{\text{H I}} \approx 10^{21}\,\text{cm}^{-2}$. The existence of the line was predicted in 1944 by van der Hulst and its first detection dates from 1951 by Ewen and Purcell. After 1952, radio astronomers began making the first maps of the Milky Way in neutral hydrogen, which revealed the spiral structure of our Galaxy.

2.2.1 Excitation

In view of the very small probabilities for radiative transitions, collisions dominate the excitation, and in the majority of cases we can consider that local thermodynamic equilibrium (LTE) is achieved. The difference in energy between the levels $F = 1$ and $F = 0$ corresponds to just 0.07 K, and at temperatures normal for interstellar atomic clouds of around 100 K the levels are populated in the statistical ratio

$$\frac{N_1}{N_0} = \frac{g_1}{g_0} e^{-h\nu/kT} \quad (g = 2F + 1),$$

giving

$$N_1 \approx 3 N_0 \quad \text{and} \quad N_{\text{H I}} = N_1 + N_0 \approx 4 N_0.$$

The transfer of radiation involved with the 21 cm line is very simplified, since in the majority of cases the optical thickness is small. Recall the expression for the absorption coefficient in the Rayleigh–Jeans approximation ($h\nu/kT \ll 1$):

$$k_\nu = \frac{c^2}{2\nu^2} A_{10} \frac{g_1}{g_0} n_0(\nu) \frac{h\nu}{kT_{\text{ex}}},$$

where T_{ex} is the excitation temperature and $n_0(\nu) = n_0 g(\nu)$ represents the fraction of the population of level 0 that effectively participate in radiative transfer at the frequency ν. $g(\nu)$ represents the form of the normalized line ($\int g(\nu)\,d\nu = 1$). As the frequencies received are directly converted into speeds ($\Delta v/v = \nu/c$), we can introduce the form of the normalized line $g(\nu)$ as a function of the radial velocity. The optical thickness τ at the velocity v ($\tau = \int k_\nu \, ds$) is thus written as

$$\tau_v = 5.49 \times 10^{-19} \frac{N_{H\,I}}{T_{ex}} g(v),$$

where $N_{H\,I}$ is the column density of atomic hydrogen in cm^{-2} and T_{ex} is the excitation temperature in K (v is in km s^{-1}).

Integrating over all velocities, we can then obtain the expression for the column density:

$$N_{H\,I} = 1.823 \times 10^{18} \int T_A \frac{\tau_v}{1 - e^{-\tau_v}} \, dv \quad [\text{cm}^{-2}],$$

where T_A is the antenna temperature in K, v is in km s^{-1}, and we have used the Rayleigh–Jeans transfer equation:

$$T_A = (T_{ex} - T_{bg})\left(1 - e^{-\tau}\right),$$

where T_{bg} is the temperature of a black body at 2.7 K, negligible here compared with $T_{ex} \approx 100$ K.

In the majority of cases the 21 cm line is optically thin ($\tau \ll 1$), and the expression simplifies to $N_{H\,I} \propto \int T_A \, dv$. The column density is thus directly proportional to the integrated intensity of the line. This small optical thickness is due in large part to the fact that the clouds all emit at different frequencies. However, there exist a number of cases of absorption, whether the selfabsorption of interstellar clouds or the absorption of hydrogen in the spectrum of sources of radio-continuum radiation. Selfabsorption in general comes from very dense clouds that are most often associated with molecular clouds. These lines have been intensively studied in order to determine the ratio between the H I and H II components. On a large scale for an external galaxy the 21 cm line of atomic hydrogen becomes optically thick only in certain cases where projection effects add greatly to the column density: galaxies seen edge-on, for instance.

The observation of H I absorption lines in the spectrum of a radio-continuum source (of temperature T_s), whether belonging to our Galaxy or not, enables us to obtain an estimate of the excitation temperature of the atomic clouds. The optically thin emission lines give only the column density but not the temperature T_{ex} ($T_A = T_{ex}\tau$). Another equation is required to obtain the two unknowns T_{ex} and τ. The principle is to observe successively in the direction of the source, where $T_A(\text{ON}) = (T_{\text{ex}} - T_s)(1 - e^{-\tau})$, and away from the source, where $T_A(\text{OFF}) = T_{\text{ex}}(1 - e^{-\tau})$ (we assume that T_s, the radiation temperature of the source, is obtained by comparison with neighbouring frequencies). The above two equations giving $T_A(\text{ON})$ and $T_A(\text{OFF})$ allow us immediately to deduce τ and T_{ex} once the T_A are measured.

In the Milky Way, and more generally in 'normal' spiral galaxies, observations at 21 cm highlight two components of the interstellar medium:

- A cold medium composed of relatively dense atomic clouds (seen in emission or absorption): the temperature here is of the order of $T \approx 100$ K, and the density is of the order of $n_H \approx 10 \text{ cm}^{-3}$ on average. The clouds are several parsecs in diameter (≈ 10 pc). The column density (integrated along the line of sight across a cloud) ranges from 10^{19} to 10^{21} cm^{-2} ($\tau \approx 0.2$). These clouds are concentrated in the plane of the Galaxy with a typical thickness of 350 pc and a mean density throughout the volume corresponding to 0.3 cm^{-3} in this plane. The corresponding line profiles are several km s^{-1} in size, which implies turbulent motion inside the clouds (the thermal velocity at a temperature $T = 100$ K is $\sqrt{2kT/m} \approx 1.5$ km s^{-1}).
- A hotter and more tenuous medium between the clouds, observed solely in emission: $T \approx 10^4$ K and $n_H \approx 0.15 \text{ cm}^{-3}$ on average. The corresponding profiles are larger (several tens of km s^{-1}). Its concentration in the plane is less pronounced (with a typical thickness of 600 pc).

We shall see in what follows that these two media coexist with a denser and colder phase, molecular clouds, and a less dense and hotter phase, the coronal medium (10^6 K).

2.2.2 Global Properties. Correlations

There exists a strong correlation between the mass of atomic hydrogen in a galaxy and its morphological type, the late types (Sc and Irr) being the richest in H I with respect to their total mass. Elliptical galaxies are extremely poor in neutral gas. This absence of gas suggests the existence of galactic winds powerful enough to rid the galaxy of the residual gas ejected by stars. Otherwise this might be cooled and formed into a more abundant interstellar medium. With the increased sensitivity of radio telescopes a growing number of elliptical galaxies have been detected in H I. The richest in general have gas provided by accretion from a spiral companion. An example is given by NGC 5128, which has an important band of dust and a dense interstellar medium containing many molecules.

It is very interesting to compare the mass of hydrogen $M(\text{H I})$ of a galaxy with its luminosity, since the ratio M/L (total mass to luminosity) is at least close to a constant for all spiral and irregular galaxies ($M/L \approx 4$ in solar

46   2. The Galactic Interstellar Medium

units $M_\odot/L_\odot$ in the interior of the visible galaxy), but also because the ratio M/L is independent of the distance D to the galaxies ($M(\mathrm{H\,I})$ and L are both proportional to D^2), and because D is in general poorly known. The ratio $M(\mathrm{H\,I})/L$ grows along the Hubble sequence and reaches a maximum for irregular blue galaxies (especially blue compact ones). $M(\mathrm{H\,I})/L$ is also correlated with the luminosity, suggesting a greater star formation rate in galaxies rich in gas. The proportion of H I to the total mass of a galaxy varies from 1 to 2% in early-type spirals (Sa) up to 30% in irregulars (Irr) and is 10% on average in spiral galaxies. This relation is one of the most marked and suggests that the types at the end of the Hubble sequence represent the least-evolved systems. The variation of H I properties as a function of morphological type is illustrated in Fig. 2.6.

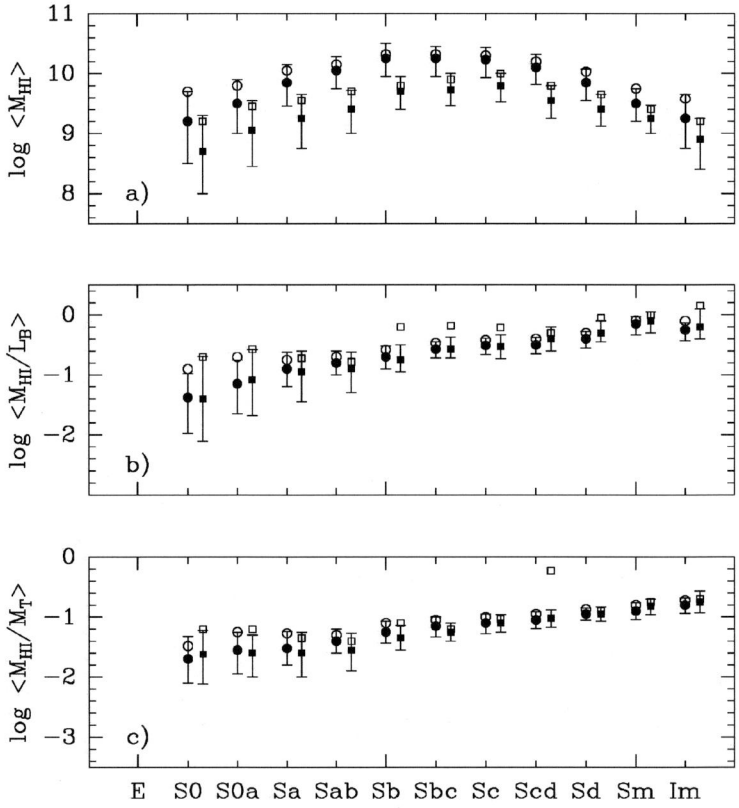

Fig. 2.6a–c. Global properties of the H I component of galaxies as a function of their morphological type. Circles and squares represent two samples of galaxies. The filled symbols are median, open ones are mean values. Bars are drawn between the 25th and 75th percentiles values. (**a**) Total H I mass (in $M_\odot$); (**b**) H I mass to blue luminosity ratio; (**c**) H I mass fraction (from Roberts & Haynes 1994)

The spatial extension of the H I component with respect to the optical disc is likewise an increasing function of morphological type. In the late-type spirals and irregulars the diameter of H I (determined, for example, by the point where the surface density of H I decreases to below 1 $M_\odot$ pc^{-2}) exceeds the Holmberg diameter by a factor of 2 or more.

Close to half of all galaxies show a strong asymmetry on a large scale, the centre of H I emission being clearly displaced with respect to the optical centre of the galaxy. Interferometric observations confirm that relatively symmetric galaxies are the exception. This is evidence that the galactic interstellar medium is not in equilibrium, because the differential rotation would become homogeneous in some 10^8 years. This might be due to the galaxy's own evolution (stochastic bursts of star formation), to perturbations caused by tidal effects with companion galaxies, or to continuous gas accretion.

2.2.3 The Distribution of Atomic Hydrogen

The Radial Distribution. Although atomic hydrogen is part of the young-population (population I) components of a galaxy, its radial distribution does not at all resemble that of the other population I components, that is, the molecular clouds, the young stars, the ionized gas, the supernovae, and so on, which have much more concentrated distributions. Figure 2.7 compares all these distributions in the case of the Milky Way. Atomic hydrogen extends very far from the centre, beyond the optical limit of the galaxy. It is nevertheless the reservoir of gas from which molecular clouds and then stars are formed (and therefore H II regions, supernovae, and so on), but the mechanism that triggers its condensation to a colder and denser molecular component is complex. This mechanism might be accompanied by a flow of neutral gas towards the centre of the galaxy, the H I reservoir remaining in the outer regions.

In external spiral galaxies the deficiency of H I gas towards the centre is an almost general phenomenon. In our nearby neighbour, the Sb galaxy M31 (Andromeda), the distribution of H I is well represented by a ring of mean radius 10 kpc. In galaxies farther away the problem of spatial resolution limits the number of galaxies where the structure of the atomic component is known. With a large single-dish antenna (Nancay, Effelsberg, or Arecibo, with beams at 21 cm of $4' \times 24'$, $8' \times 8'$, and $3' \times 8'$ respectively) when the diameter of the galaxy corresponds to just 2 or 3 observational beams, only a model can give us information on the large-scale structure. It has been determined by this method that about 30% of galaxies can be represented by annular distributions, such as that of M31. But this method can of course only detect rings of large radius, and we can imagine that a much larger fraction of spiral galaxies are deficient in atomic gas at the centre. On the basis of fewer interferometric observations (with a spatial resolution of about $20''$), 70% of spiral galaxies have a relative minimum of H I at the centre (Fig. 2.8).

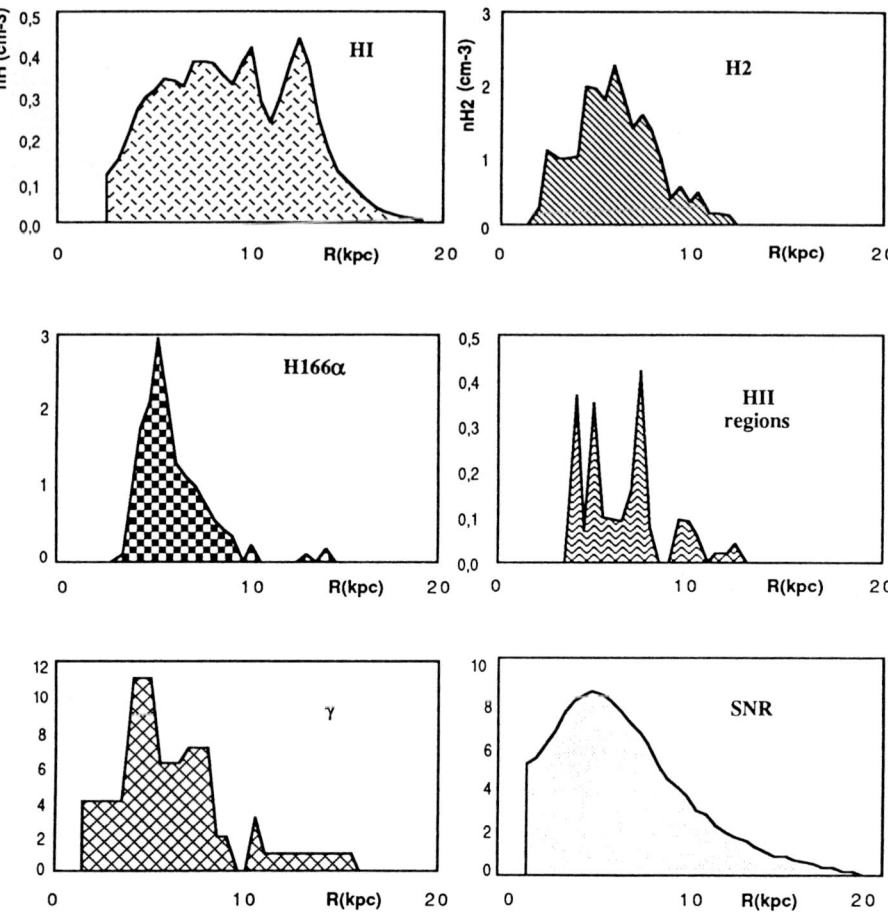

Fig. 2.7. A comparison of radial distributions of various components in the Milky Way. Note that only atomic hydrogen (H I) extends significantly beyond 10 kpc. The molecular clouds (observed in CO), ionized gas (H II regions, the H166α recombination line), and supernova remnants are all associated with star-formation regions. γ rays trace the nucleus, that is, matter in all its forms: they are due to the interaction of cosmic rays with the matter. Their falling-off beyond 9 kpc leads to the relative weakness of $n(\text{H I})$ with respect to $n(\text{H}_2)$ (compare the respective scales)

In most cases this deficiency in H I gas at the centre is compensated by a condensation of molecular gas, with the result that the interstellar gas is just as concentrated; but there are notorious exceptions (M31 and M81, for example) where the centre of the galaxy is totally devoid of gas. It has likewise been suggested that as soon as the surface density of H I gas exceeds a certain threshold, from 5 to 10 $M_\odot\,\text{pc}^{-2}$, the rest of the interstellar gas ends up in the form of molecular clouds.

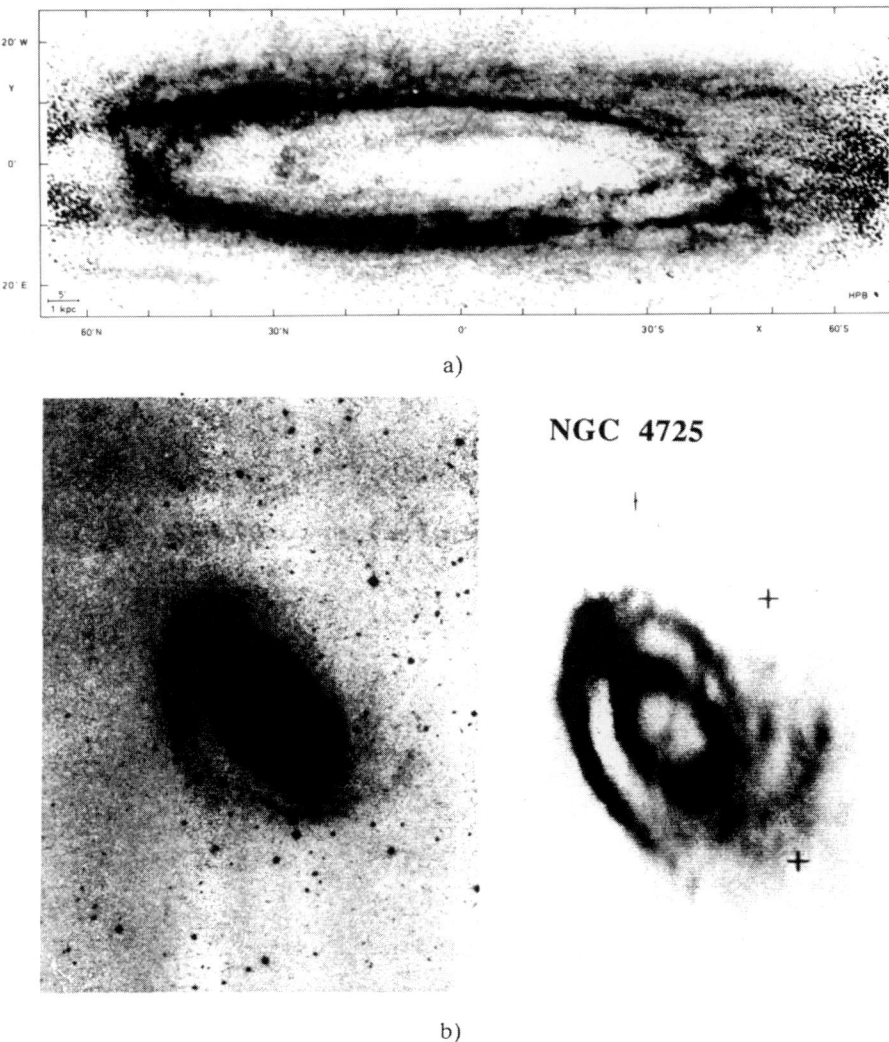

Fig. 2.8. (a) A map of atomic hydrogen in the Andromeda galaxy (M31), obtained using the Westerbork interferometer with a resolution of $24'' \times 36''$ by Brinks (1984). Note the marked annular distribution inside which spiral arms can be distinguished. (b) A comparison between the optical photograph (in J) and the H I map of the galaxy NGC 4725. (From Wevers 1984)

Interferometric Observations. Atomic gas is not homogeneously distributed over the whole galactic disc; this is visible in the Milky Way, where the concentration in the spiral arms, or even pieces of spiral arms, is very clear. In external galaxies where the angle of view is more favourable for studying the spiral structure than in our Galaxy, the problem is to have sufficient spatial resolution, and this is only possible with interferometric observations. To resolve the spiral arms a beam size of less than one minute of arc is needed, which is the order of magnitude of the angular separation between arms in nearby galaxies. While the techniques of aperture synthesis were initially developed at Cambridge, UK (using the 'Half-Mile' telescope), many galaxies were mapped in H I at Westerbork, the Netherlands (12 antennae with a maximum separation of 3 km). Today the VLA (Very Large Array; 7 antennae with a maximum separation of 36 km) near Socorro, New Mexico, offers the greatest sensitivity for the best spatial resolution. These aperture-synthesis instruments use techniques that we shall give only the principal features of here.

Observations are made in several series where the telescopes are fixed in a given configuration. In one configuration several separations between telescopes are used (if there are more than two telescopes) and the source is held steady over a certain number of hours. Owing to the rotation of the Earth, the apparent separation, from the point of view of the source, of each pair of telescopes thus describes a portion of an ellipse, and spatial-frequency space may in this way be sampled sufficiently rapidly in a particular configuration. Luminosity maps in the plane of the sky are obtained by Fourier transforms of the data. The spatial resolution of the observations, the synthesized beam, is obtained from the largest baseline D obtained (the solid angle subtended by the beam is $(\lambda/D)^2/\cos\delta$, δ being the declination of the source).

The field of view of the observations corresponds to the primary beam, the beam of one antenna of the network, which is approximately λ/d (if d is the diameter of each antenna). If the observations are made with the spacing between antennae equal to a or multiples of a, the network will have secondary beams at λ/a in right ascension and $(\lambda/a)/\cos\delta$ in declination. This is why the interval has to be sufficiently small to reject the secondary beams far from the centre of emission. Moreover the antennae cannot come closer together than less than their approximate diameter, and interferometric observations always fail for short spacings; this leads to the absence of low spatial frequencies in the synthesized maps: the interferometer is thus only very sensitive for point sources or for highly contrasted sources and not for slowly varying background emission. It is thus necessary to complete the map by observations made with a single-dish antenna whose diameter is equal to at least twice d, the diameter of the individual antennae of the interferometer. This procedure requires delicate calibrations. What is more, the sensitivity of the interferometer to an extended source decreases very rapidly when the resolution increases. For a given integration time the signal-to-noise ratio in the antenna temperature grows as D^2 (D is the maximum baseline). It is in general the sensitivity that limits the maximum resolution of observations. One can thus seek to lower this resolution to be able to detect extended H I components and better determine the size of the galaxy, for example.

2.2 Atomic H I Gas

The Spiral Structure. Interferometric maps of nearby spiral galaxies reveal a very pronounced spiral structure in the H I component. The spiral arms constitute a gravitational potential well, which traps the atomic gas, which is relatively cold compared with the old stellar population, with a strong velocity dispersion. The arm–interarm contrast in H I may reach 8–10. A beautiful example of H I structure is that of the galaxy M81, observed at Westerbork by Rots and Shane (1975): the central regions are deficient in H I, as in the majority of galaxies; then, between 3 and 10 kpc from the centre, the gas forms into two very regular spiral arms, from which two external spiral arms extend from 10 kpc outwards (Fig. 2.9). Inside a radius of 13 kpc the eastern half of the galaxy contains 50% more H I gas than the western half, which clearly illustrates the asymmetries in the gaseous component of galaxies, even in this particular galaxy, which is nevertheless very regular overall. Beyond the optical radius the extensions of H I gas form bridges, linking M81 with its companions NGC 3077 and M82.

2.2.4 H I Holes and the Connection with Star Formation

Good spatial resolution allows us to study in addition to the spiral structure the local details of the H I distribution, such as the 'holes' in M31, our neighbour. These holes represent a notable absence of H I (10^3 to $10^7 M_\odot$) over the range 100 pc – 1 kpc; they appear roughly circular in projection on the map of Fig. 2.10 and are thus probably spherical in form. These holes are clearly visible owing to the velocity resolution of $8.2\,\mathrm{km\,s^{-1}}$ of the observations. Some of them are large enough to enable edges very rich in H I to be distinguished, like shells. This type of H I shell (or supershell) is also observed in our own Galaxy and in the Large Magellanic Cloud. Figure 2.10 shows the distribution of 141 H I holes detected by Brinks (1984) in M31; these holes are all situated in the ring of mean radius 10 kpc where the spiral arms can be distinguished. There is no molecular emission detected from H I shells, although not all H I holes have been observed at millimetre wavelengths (since observations are time consuming). The large-scale distribution of holes is correlated with that of H II regions, which suggests that the holes are due to intense star formation. Moreover H I gas bordering the holes has been observed with an expansion velocity of the order of 10 to $20\,\mathrm{km\,s^{-1}}$, which gives the approximate age of the holes as 10 to 20×10^6 years, comparable to the age of OB associations in M31. If the formation of holes is really due to supernovae, their formation rate in M31 would be 10^{-3} per year, an order of magnitude less than that for the Milky Way (one supernova every 30 to 100 years).

The large-scale star formation rate is assumed to be dependent on the volume density of the gas ρ_g, and a power law is very often used, of the form

$$\frac{\mathrm{d}\rho}{\mathrm{d}t} = \text{constant} \times \rho_\mathrm{g}^n,$$

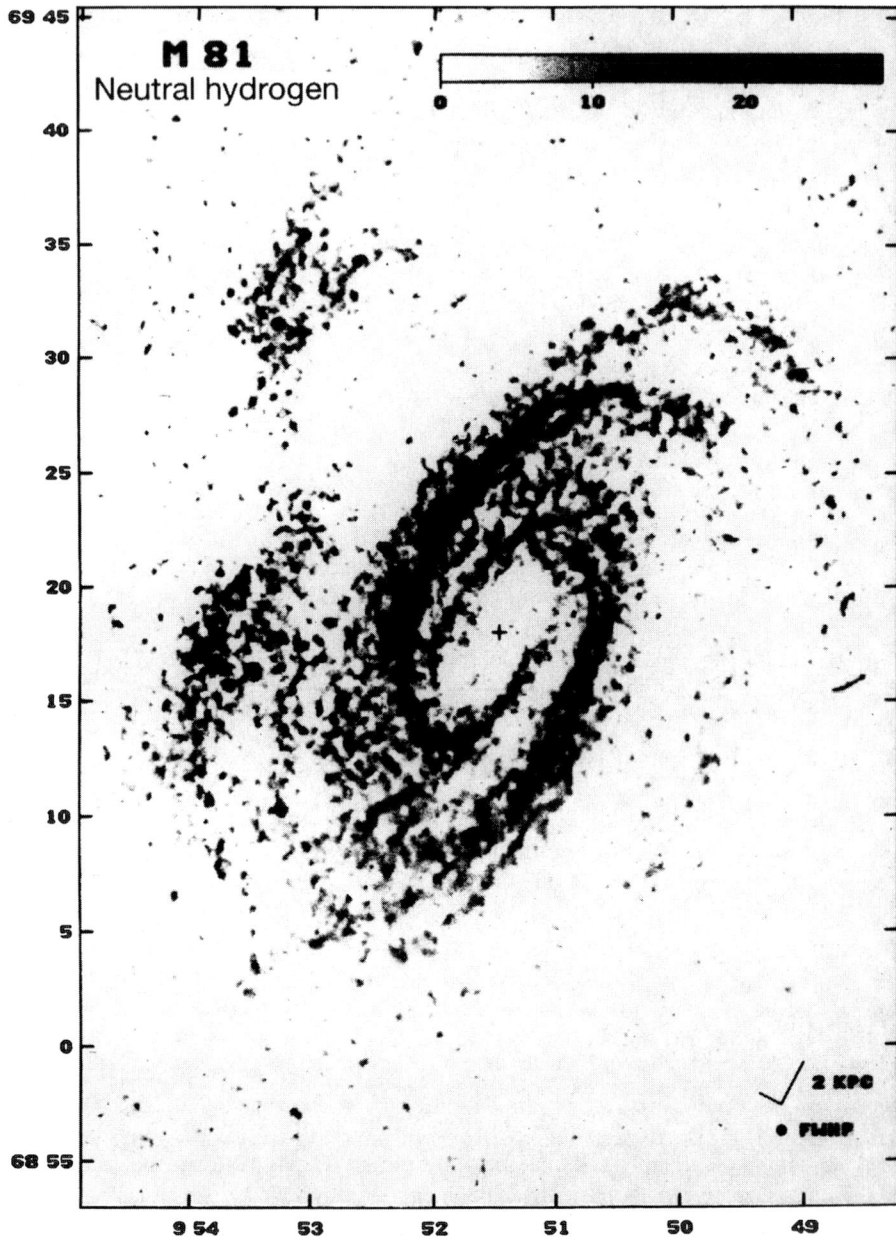

Fig. 2.9. An H I map of M81 obtained at Westerbork with the spatial resolution indicated by the beam at the lower right. (From Rots and Shane 1975)

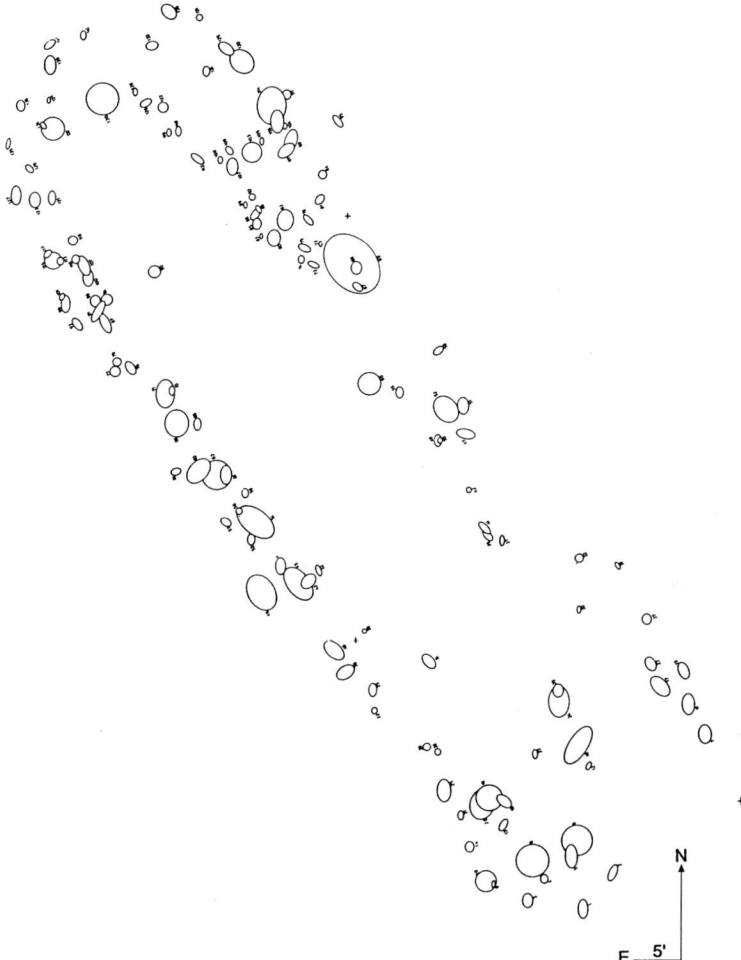

Fig. 2.10. The distribution of H I holes in M31 (from Brinks 1984). These holes are found in the spiral arms and are due to the formation of young stars

with $n \approx 2$ (Schmidt's law). The distribution of H I hydrogen does not seem a priori to support such a law, its surface density showing no correlation with the luminosity distribution of a galaxy. Nevertheless what is important for the star formation rate is the volume density, and although H I hydrogen may be very abundant in the external parts of a galaxy, it is true that its volume density there is very low: the thickness of the H I plane grows linearly as a function of the radius. This is related to the decrease of the mass surface density $\mu(r)$, since the height H varies as μ^{-1} for an isothermal plane.

2.2.5 Distortion, or Warp, of the Plane

Not only does the thickness of the plane in the majority of spiral galaxies increase as a function of the radius, but also the mean plane becomes more and more distorted as one moves away from the centre. This feature has of course been observed in the Milky Way, where away from the solar neighbourhood the plane lifts up at one edge like a pancake and goes down at the other, the deformation being dependent on $\cos\theta$ with respect to the azimuth θ. Many galaxies seen edge-on have a very marked distribution like an integral sign $\int$, the H I gas rising above the central plane, beginning just outside the optical disc. Deformation of the optical disc itself occurs but is relatively rare; it is above all visible owing to dust lanes in very inclined galaxies. For moderately inclined galaxies it is only by kinematic perturbations that a warp can be detected (and one then speaks of a kinematic warp).

In the Andromeda galaxy the warp is immediately visible in optical photographs and manifests itself by two velocity components markedly separated in the atomic and even molecular gas (CO emission), as can be seen in Fig. 2.11. The separation between the two components is difficult in some regions. The gas lies above the plane defined in the central region up to a height

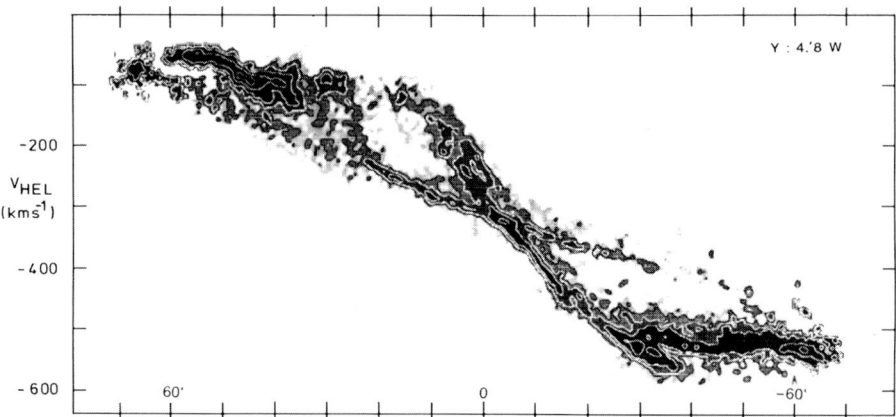

Fig. 2.11. Position–velocity diagrams in a direction parallel to the major axis of M31 (X indicates the position on this major axis; see Fig. 2.8a) for $Y = 4.8'$ W. Two components are clearly visible, forming a structure reminiscent of a figure 8: (1) The normal component, represented by the initially steepest curve. The velocity curve is characteristic of that of spiral galaxies: it climbs linearly (indicating rigid rotation) and then splits towards a flat rotation curve. (2) The component corresponding to the warp, or distorted part, of the plane of M31. The velocity is linear from one end to the other, like that of a rotating ring seen in projection. This indicates that this component exists only at a large distance from the centre. The velocity is less because the plane of the ring is less inclined with respect to the plane of the sky. (From Brinks 1984)

of 4 kpc. This excursion out of the plane might be useful for determining the existence and the shape of an invisible mass component.

Such deformation of the plane is so frequently observed in galaxies that it poses theoretical problems. This is not a static deformation but a dynamic one: the plane of the galaxy ripples like a flag in the wind. The gravitational force of attraction due to the stars of the optical disc makes the gas oscillate about the plane, and the oscillatory motion precesses about the centre of the galaxy. Nevertheless the precession is differential and the warp dies down over time more or less rapidly as a function of the flattening of the mass distribution. As tidal interaction is one of the mechanisms for inducing such warping, we shall return to this problem in Chap. 7.

2.3 The Molecular Phase

2.3.1 Excitation and the Conversion Ratio

The molecular component in external galaxies is observed by means of the rotation lines of the CO molecule, the most abundant after H_2 ($CO/H_2 \approx 6 \times 10^{-5}$). Molecular hydrogen cannot be observed in the radio domain, the reason being that since the H_2 molecule is symmetric, it does not possess an electric dipole. The CO molecules are excited by collisions with H_2 molecules and therefore constitute a tracer of the molecular component. Moreover since the dipole moment of CO is relatively weak ($\mu = 0.1$ debye), the spontaneous deexcitation rate (Einstein's coefficient A_{21} being proportional to μ^2) is small, and the molecules remain excited in regions of average density, from about 300 molecules cm^{-3}. More generally for a molecule to be excited beyond the black-body temperature 2.7 K (the excitation level for the CO(1–0) line corresponds to 5 K) collisions must be effective with respect to radiative transitions; that is, the density n_{H_2} of the medium should be sufficient for the ratio C_{21}/A_{21}, which varies as $n_{H_2}\sqrt{T}/\nu^3\mu^2$ (C_{21} is the collision rate), to become significant. In Table 2.1 the critical density n_{Hc} is given for which the ratio $C_{21}/A_{21} = 1$ (meaning that the collisional and radiative excitations are equal). The corresponding lines are first detected when $n_{H_2}/n_{Hc} \approx 10^{-2}$.

Table 2.1. Dipole Moments and Critical Densities for some Interstellar Molecules

Molecule	CO	NH3	CS	HCN
μ [debye]	0.1	1.5	2.0	3.0
n_{Hc} [cm^{-3}]	4.0×10^4	1.1×10^5	1.1×10^6	1.6×10^7

The CO molecules, and especially the $J = 1$–0 line at 2.6 mm wavelength, then allow complete mapping of the molecular component, while HCN or CS enables the existence of dense cores to be determined.

As the rotational transitions of the CO molecule are not in general optically thin, how can the total mass of molecular hydrogen be deduced? Resolving the problem of radiative transfer in galaxies in detail is out of the question in view of the number of free parameters (density distribution, temperature distribution, and the abundance of clouds). Several methods, however, seem to indicate a constant ratio of proportionality between the mass of H_2 and the integrated intensity of CO emission (T_A^* is the antenna temperature):

$$I_{CO} = \int T_A^* \, dV \quad [\text{K km s}^{-1}].$$

- *First method:* Observation of the isotopic molecule ^{13}CO, which is 90 times less abundant than ^{12}CO. A typical molecular cloud in our Galaxy has a mean optical thickness of the order of 7 to 10 in the ^{12}CO ($J = 1$–0) line. The ^{13}CO ($J = 1$–0) line is then very probably optically thin on a large scale in galaxies ($\tau_{13} = \tau_{12}/90 \ll 1$) and could serve to trace the molecular mass. Now the mean ratio observed between ^{12}CO and ^{13}CO emission is equal to about 10, to within a factor of 2, in the majority of spiral galaxies. By successive calibrations, knowing the abundance ratio $^{13}CO/H_2$ in the solar neighbourhood, as a result of direct observation of molecular clouds (through the ultraviolet lines of H_2 and CO), one obtains the conversion ratio of the integrated intensity of CO to the mass of H_2 for the external galaxies.
- *Second method:* Modelling of CO emission received in the beam of the radio telescope by a large number of 'standard' clouds. The line at 2.6 mm wavelength of CO ($J = 1$–0) is, for each standard molecular cloud, optically thick, but the surface filling factor of these clouds in the beam is, at a given frequency, less than 1: the clouds do not overlap, account being taken of their different velocities and their velocity dispersions. In a sentence, the global molecular component that constitutes the clouds is assumed to be optically thin, and the ratio for the column density of hydrogen N_{H_2} to the emission I_{CO}, N_{H_2}/I_{CO}, is constant. This constant agrees with that of the first method to within a factor of 2.
- *A constant ratio N_{H_2}/I_{CO} on a large scale in galaxies can be theoretically justified with only two hypotheses:* (1) the integrated emission I_{CO} can be regarded as the sum of the emission of independent clouds (the surface filling factor $\ll 1$); and (2) the emitting clouds are in virial equilibrium. The first hypothesis is justified by the low antenna temperatures observed in external galaxies (of the order of 0.1 K with a resolution of 1 minute of arc, compared with several K for a molecular cloud in our own Galaxy), which clearly shows that the clouds at a given velocity undergo a strong spatial dilution in the observational beam. The second hypothesis has been

confirmed by statistical studies of the number of molecular clouds in our Galaxy, revealing correlations between the mass M and size R of the cloud ($M \approx R^2$) and velocity dispersion σ and size R ($\sigma \approx R^{1/2}$).

The method of calculation based on virial equilibrium has the advantage of not relying on the abundance ratios ^{13}CO/H$_2$ and ^{12}CO/H$_2$. These are poorly known and moreover may vary from one galaxy to another or even inside the same galaxy. The conversion factor N_{H_2}/I_{CO} resulting from this model depends, however, on the radiation temperature T_R and the density n of the clouds. Consider the area A of the beam in the plane of the sky at the distance D of the galaxy ($A = (\pi/4)(\theta D)^2$ if θ is the diffraction beam) in which are observable N spherical clouds from a projected area $a = \pi d^2/4$ of diameter d, velocity dispersion ΔV, and radiation temperature T_R. The emission I_{CO} is equal to the sum $\sum_1^N T_A \Delta V$, and each cloud contributes to the antenna temperature T_A with a dilution factor a/A;

$$I_{CO} = A^{-1} N \left(\frac{\pi d^2}{4}\right) T_R \Delta V.$$

The mean surface density is

$$N_{H_2} = A^{-1} \frac{N \pi d^3 n}{6},$$

where n is the volume density of each cloud. The ratio of these two quantities gives

$$\frac{N_{H_2}}{I_{CO}} = \frac{2}{3} \frac{dn}{T_R \Delta V}.$$

The virial relation gives $\Delta V = (2\alpha GM/d)^{1/2}$ with a numerical coefficient α of the order of 1, where G is the gravitational constant and M the cloud mass. ΔV then varies as $n^{1/2} d$ and the conversion ratio N_{H_2}/I_{CO} as $n^{1/2}/T_R$ to within a numerical factor. This factor is such that the conversion ratio is 2.8×10^{20} cm^{-2} (K km s^{-1})$^{-1}$ for $T_R = 10$ K and $n = 200$ H$_2$ cm^{-3}. It is also possible to develop a more sophisticated model with a variable density n and a size distribution for the clouds.

In this model we expect a very weak dependence of the conversion ratio on the metallicity: since the molecular clouds are optically thick, their contribution to the emission depends only on their excitation temperature and their size; this does not vary much if the abundance of CO is reduced. Nevertheless if the conversion ratio appears to be relatively constant for all spirals, several indicators (notably the abundance of dust, infrared emission, star formation, and weak CO emission) tend to show that the ratio does not apply to irregulars that are very deficient in metal. The CO molecule is especially affected by a deficiency of heavy elements, C/O varying as O/H and leading to CO/H$_2 \approx$ (O/H)^2. Now the ratio $M_{H_2}/M_{H\,I}$ has been observed in the deficient galaxies to follow an (O/H)^2 law, which suggests that the standard ratio N_{H_2}/I_{CO} very greatly underestimates the mass of H_2 in these galaxies (in particular in the Magellanic clouds).

On the other hand at the centre of certain starburst galaxies there might exist an important hot and optically thin molecular component for which the standard conversion could overestimate the mass of H_2. This may be the case for M82, which has a molecular ring 400 pc in diameter at its centre; the ratio of 2 between the CO emissions CO(2–1) and CO(1–0) implies the presence of hot and optically thin gas, which seems to be confirmed by the very greatly increased ratio of emission $^{12}CO/^{13}CO$. However, starburst galaxies in general also have denser gas (as seen in HCN and CS), and this could compensate in the conversion ratio, which varies as $n^{1/2}/T_R$.

2.3.2 Cold Dust Emission

Another tracer of cold molecular clouds is the emission of cold dust, in the millimetre wavelength domain. At $\lambda = 1$ mm, the spectrum of the dust at $T_d = 20$ K is in the Rayleigh-Jeans domain, where the Planck function reduces to

$$B(\nu, T) \approx \frac{2kT\nu^2}{c^2}$$

and therefore the flux is quasi linear in temperature. This cold dust is associated with the cold interstellar medium, which is mostly molecular. Since this emission is in general optically thin, the estimation of the molecular column density should be more directly derived from observations. The only bias is the dust-to-gas ratio, which is likely to be directly proportional to the metallicity. Therefore, the radial distribution of the dust emission should drop much more rapidly than the gas distribution in the outskirts of galaxies. The comparison between the CO-line tracer and the dust emission tracer is therefore quite informative.

Recently galaxies have been mapped with a bolometer at millimetre wavelengths with increased sensitivity (IRAM -30 m in Spain, SCUBA at JCMT -15 m in Hawaii). In many galaxies rich in CO, the dust emission map is perfectly superposable on the CO map. This is surprising, since the dust varies quasi linearly with metallicity, and the CO emission was not believed to be as sensitive (this is possible, however, if molecular clouds do not have sharp boundaries; with decreasing metallicity, the size of CO clouds at optical depth unity decreases considerably). In some galaxies, however, the dust emission is even more extended than the CO emission (see Fig. 2.15), almost as extended as the HI emission. This tends to show that the CO line is a poor tracer of the cold interstellar gas. This gas could be distributed mainly in more diffuse clouds, which are not dense enough for the lines to be collisionally excited.

2.3.3 Global Properties. Correlations

Millimetric receivers are still far from being sensitive enough to be able to detect the CO emission of all the catalogued spiral galaxies. In several galaxies observed, one out of two or three have been detected in this wavelength domain. In this case there still exist many observational selection effects in the detected galaxies. Observations, for example, always begin with the nucleus of the galaxy, favouring the detection of galaxies having central molecular concentrations. There nevertheless are galaxies where only the disc has molecular clouds (M31, M81) and they are thus definitely undersampled. Likewise the galaxies seen face-on whose velocity profiles are narrower and therefore simpler to detect on an irregular baseline are favoured.

There does not seem to be a marked correlation between the global H I and H_2 contents of galaxies. On the other hand, CO emission is clearly correlated with the intensity of nonthermal radio-continuum emission and with far-infrared emission. The presence of an important molecular component implies a high OB star formation rate; these objects have short lifetimes and become supernovae, which constitute the source of relativistic particles producing synchrotron emission. These correlations between CO, far-infrared and radio-continuum emissions are illustrated in Figs. 2.12 and 2.13.

Likewise the dense molecular clouds are always associated with the dust in galaxies. The far-infrared emission $L_{\rm FIR}$ (from observations made using the *IRAS* satellite) of external galaxies, which is due to dust heated by young stars newly formed in molecular clouds, shows a certain correlation with the integrated CO emission. But the ratio $L_{\rm FIR}/M_{H_2}$ varies by a factor of almost 100 from one galaxy to another and depends strongly on the temperature T of the dust. For a given mass of dust the infrared luminosity varies roughly as T^5. The temperature of the dust, determined by the ratio of emissions in the 60 and 100 μm wavelength bands, is increased all the more since the density of forming stars is large; the ratio $L_{\rm FIR}/M_{H_2}$ is used as an indicator of the efficiency of star formation: in 'normal' spiral galaxies this ratio varies from 1 to 3; in starburst galaxies the ratio is 20–30, although in galaxies that are ultraluminous in the infrared, discovered by *IRAS*, it is 100–200. These galaxies emit more than 95% of their total luminosity in the infrared. Figure 2.13 shows that the infrared emission is perfectly correlated with the radio-continuum emission.

2.3.4 The Radial Distribution

Galaxies that are rich in molecules and have a strong central CO emission have a quasiexponential radial distribution $I_{\rm CO}$ as a function of radius, similar to the distribution of blue light. This characteristic has for a long time been interpreted as indicating a mean star formation rate proportional to the density of the molecular gas. Nevertheless this exponential distribution might often be an artefact of the lack of spatial resolution of the observations.

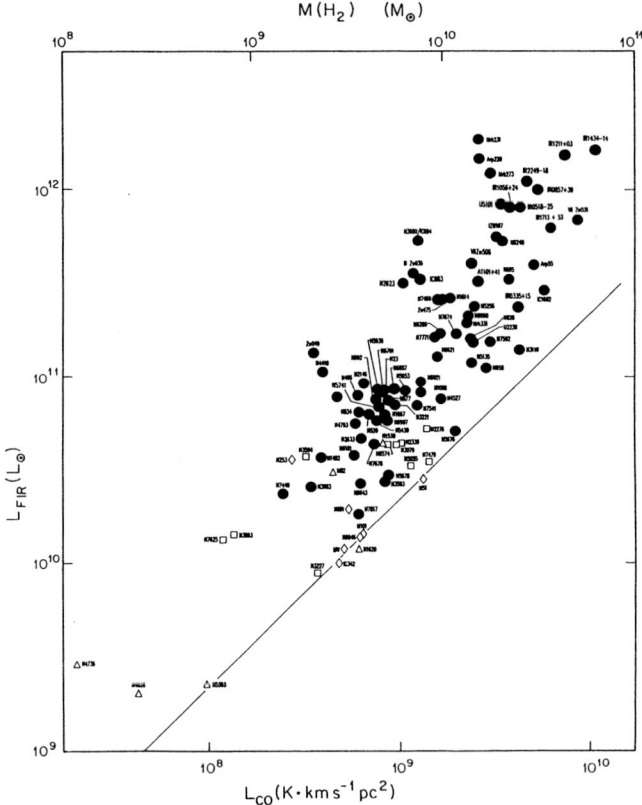

Fig. 2.12. The correlation between luminosity in the far infrared (*IRAS*) and CO luminosity (tracing molecular hydrogen, H_2). The solid line represents where points would be for which the ratio $L_{\rm FIR}/M_{H_2}$ is equal to that of the Milky Way ('normal' star formation rates). The points at the top right, well above this curve, represent starburst galaxies

Indeed it might not be possible to resolve an annular distribution with an intense central source, as in the Milky Way, in a distant galaxy with a resolving power of the order of minutes of arc. The large millimetric instruments of today benefit from a resolution of the order of 10 seconds of arc, which is thanks to a few single-dish antennae (the 30 m telescope of the Franco-German Institute, IRAM in Spain, and the 45 m telescope of Nobeyama Observatory in Japan) and interferometers (the five 10 m antennae of Caltech in Owens Valley, the five Berkeley–Illinois–Maryland 6 m antennae, the ten 10 m antennae of Nobeyama, and the four 15 m antennae of IRAM). High-resolution maps are beginning to reveal structures in CO emission: bars and rings at the centre; spiral arms in the disc. Take the example of the galaxy NGC 6946, a galaxy where the CO emission is very concentrated. The emission from the

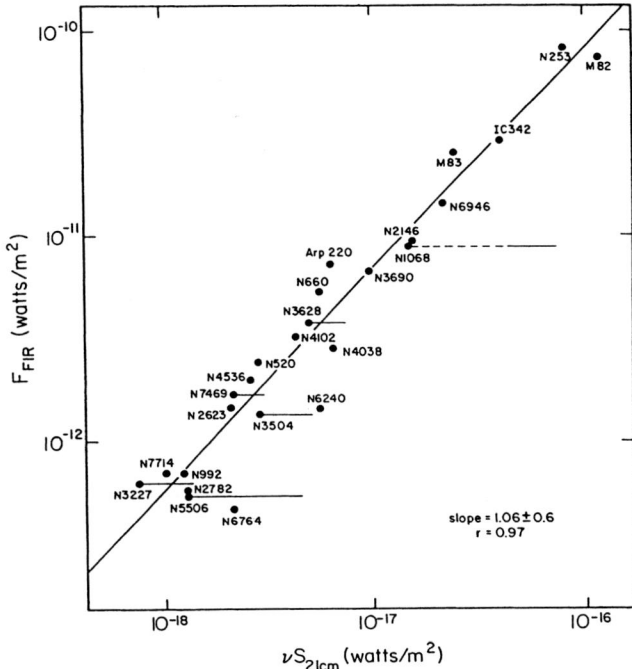

Fig. 2.13. The correlation between the far-infrared flux (*IRAS*) and the radio-continuum flux at 1.4 GHz for several bright galaxies. (From Sanders and Mirabel 1985)

nucleus does not have entirely the same appearance as that of the disc, as is clearly shown in Fig. 2.14. Although the antenna temperature is comparable, and sometimes even less, the velocity (200 km s^{-1}) is considerable in the nucleus. Moreover the nucleus–disc transition is rather abrupt. The central profiles are much larger than we would like for a simple rotation curve with a very moderate inclination of 30°; this is due to noncircular motion corresponding to the molecular bar.

Molecular bars are often revealed in the centres of galaxies, even in galaxies not classified as barred according to their optical morphology. This is due to the cold character of the molecular component, because the collisions between molecular clouds are very dissipative; it thus remains trapped in gravitational potential wells. It is an excellent tracer of density waves, which even enables us to discover weak oval distortions in the stellar potential of galaxies.

The starburst galaxy M82 (Fig. 2.16) has at its centre a ring seen edge-on and detected also in H I and OH (18 cm) absorption lines in the spectrum of the continuum source E of the nucleus. The very strong ^{12}CO(2–1) lines and very weak ^{13}CO lines suggest the existence of a diffuse molecular component

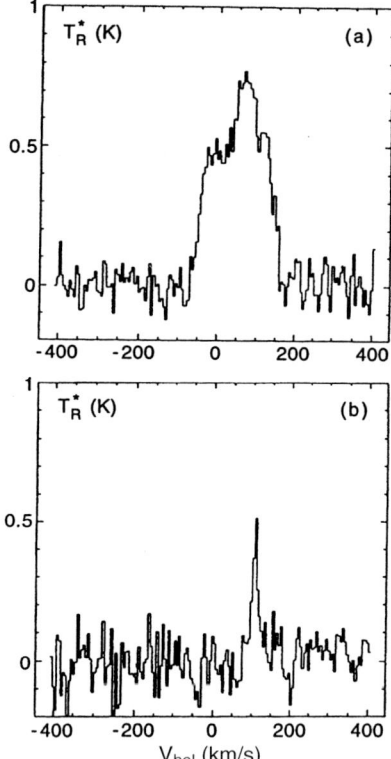

Fig. 2.14. A comparison between the CO spectra of the nucleus (**a**) and the disc (**b**) of a spiral galaxy (here NGC 6946; Weliachew et al. 1988). The antenna temperature is in general the same; however, the velocity width is clearly greater in the nucleus

caused by the destruction and heating of dense clouds by ultraviolet radiation from newly formed stars. Dense cores, however, are not absent, as attested for by the HCN, HCO^+, and CS maps.

Other molecular rings have been observed in many galaxies; they can be small, closed, and often inside a barred structure (NGC 1097); or the beginning of spiral structure with the ring only just closing (M51, NGC 1068). The smallest rings (M82) have been interpreted as matter swept up by the outburst of a series of supernovae corresponding to the central starburst. All the others seem to correspond well with the accumulation of matter associated with the inner Lindblad resonance (Chap. 5).

2.3.5 The Spiral Structure

Very high spatial resolution is required to resolve the spiral arms, as well as great sensitivity, because they are clearly separated in the disc only a long way from the centre where CO emission is weaker. An arm–interarm contrast

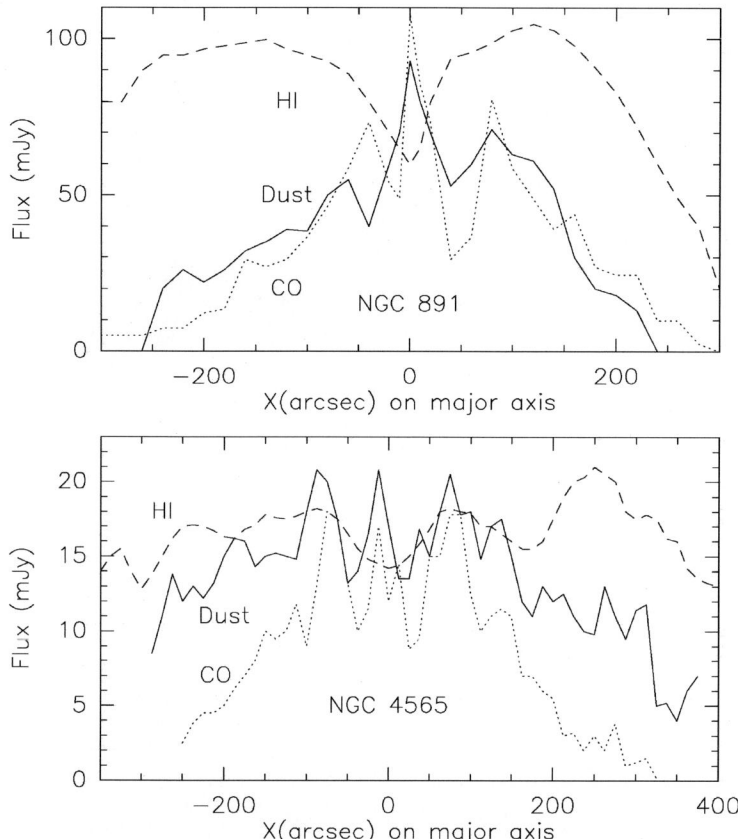

Fig. 2.15. Radial profiles of CO(1-0) emission and cold dust emission at 1.3 mm in the edge-on galaxies NGC 891 and NGC 4565, compared with the H I profile. In NGC 891, the dust and CO profiles are almost identical, while the dust emission is more extended than the CO in NGC 4565, and follows more the H I profile (from Guélin et al. 1993, Neininger et al. 1996)

greater than that observed in H I (> 10) is obtained in CO in Andromeda, and giant molecular complexes within the arm can be distinguished, surrounded by an envelope of H I, the ensemble being gravitationally bound. In barred galaxies like NGC 1530 (see Fig. 2.17), the molecular clouds are confined to the spiral dust lanes leading the bar, and often to the nuclear rings, when they exist. In M51, interferometric observations combining eight fields of view, shown in Fig. 2.18, trace one of the northern spiral arms and show a spatial shift between CO and H I condensations.

One important observational fact remains unexplained: why do spiral galaxies have a bimodal distribution from the point of view of their molecular gas content? There are galaxies that are rich in CO emission and that

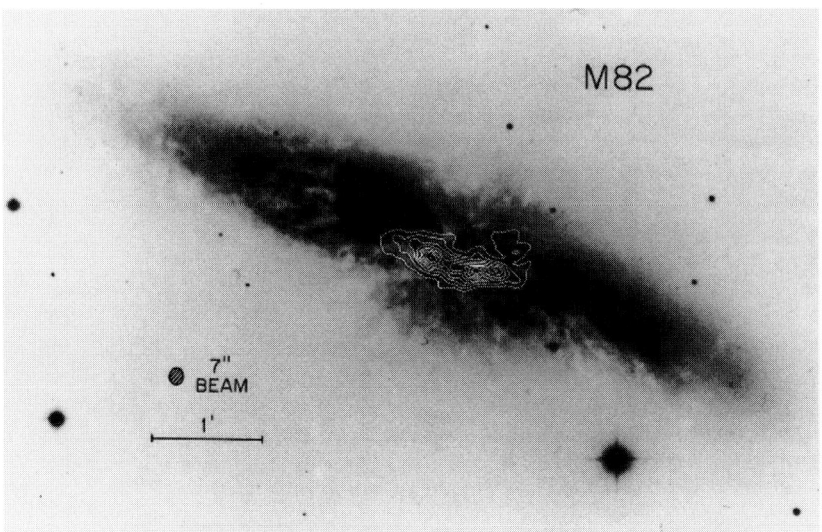

Fig. 2.16. Isophotes of CO emission at the centre of the galaxy M82, obtained with the Caltech interferometer by Lo et al. (1987), in a field of 1′ with a resolution of 7″. The molecular clouds are confined to a ring, here viewed edge-on

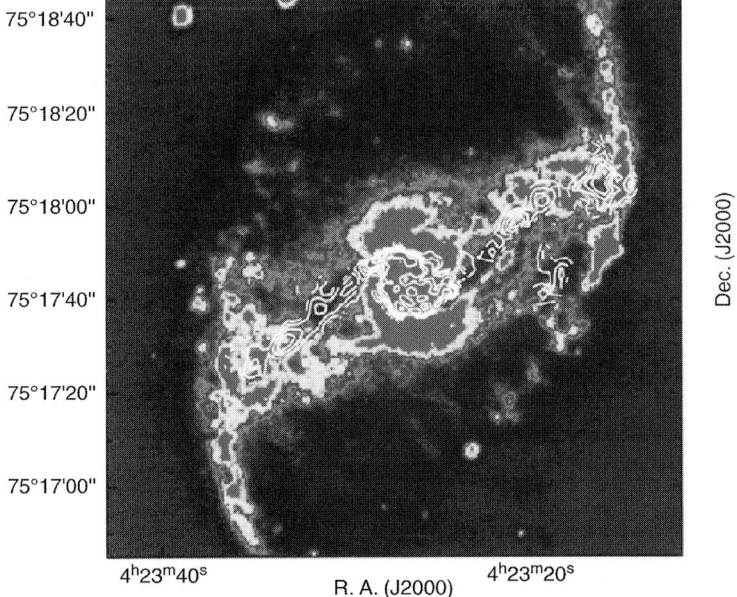

Fig. 2.17. CO(1-0) contours of the barred galaxy NGC 1530, tracing the molecular component, superposed on an optical image of the galaxy (colour). The gas is confined to the dust lanes leading the stellar bar and in the nuclear ring (From Reynaud & Downes 1997)

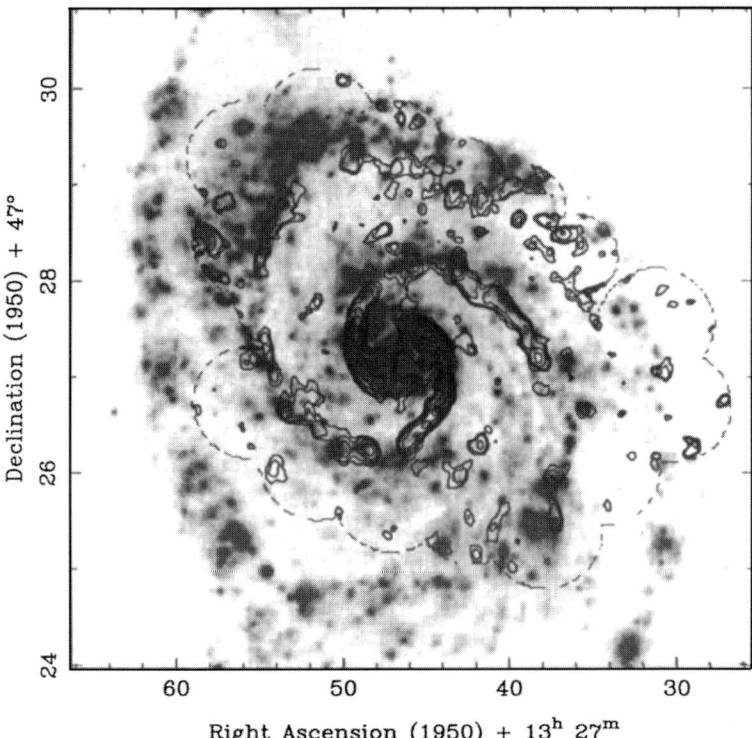

Fig. 2.18. A CO map of M51 obtained with the Caltech interferometer by Rand and Kulkarni (1990) superimposed on an optical image. The molecular clouds trace the spiral arms perfectly. The interferometric data reveal small displacements perpendicular to the arm between the various components of the interstellar medium (H I atomic gas, ionized gas, molecular gas)

have molecular clouds up to, or close to, the centre; on the other hand, galaxies are also found that are weak in CO, where it can only be found in the disc, close to the dust lanes. In certain galaxies of the second type (M33, M101), only the H I–H II complexes have molecular clouds associated with them.

Elliptical galaxies are difficult to detect in CO emission, except for some rare exceptions (NGC 5128, for example, which is rich in gas owing to the accretion of a spiral galaxy.)

2.3.6 The Connection with Activity in the Nucleus

Many galaxies with strong CO emission at the centre also exhibit nuclear activity at the same time as considerable star-formation activity. Recently galaxies that are ultraluminous in the infrared, detected by the *IRAS* satellite (up to more than 95% of their total luminosity is in the far infrared; $L_{\rm FIR} > 10^{12} L_\odot$ sometimes), have helped to determine the cause of this phenomenon. In practice all of these very active galaxies result from the merger of galaxies; they have the vestiges of interaction between two galaxies: deformation of matter; tails and bridges brought about by tidal interaction; and a very irregular and complex structure, sometimes with two nuclei. The more moderate starbursts (such as those in M82) are also due in general to the interaction between galaxies, but without merger having already taken place. That the interaction between galaxies is the origin of star formation is a fact already established by the presence of H II regions and the blue colour of interacting galaxies; a brilliant confirmation comes from *IRAS* data. Figure 2.19 shows that those galaxies bright in the infrared are galaxies that interact tidally.

How can the two phenomena, intense star formation and nuclear activity, be linked? For there to be such a considerable outburst of star formation at the centre of a galaxy (the rate may reach up to 100 times that observed for a normal galaxy), a necessary condition is that almost all the interstellar gas, the main material for forming these stars, must be transported to the centre. This has to take place in a very short time: in a starburst all the interstellar medium of a galaxy ($4 \times 10^9 \, M_\odot$) would be exhausted in about 10^8 years. Is this influx of material towards the centre not also necessary to account for active nuclei? Nuclear activity is assumed to be due to the accretion

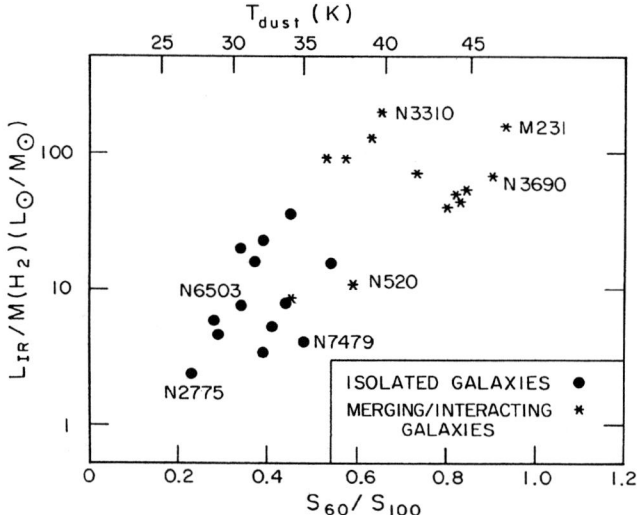

Fig. 2.19. A comparison of luminosities in the far infrared and in CO for normal and interacting galaxies (from Young et al. 1986). The ratio $L_{\rm FIR}/M_{\rm H_2}$ is higher for interacting galaxies, and their dust temperature (obtained from the ratio of the flux at 60 and 100 μm) is also larger

of material by a compact massive object (a black hole) situated within the nucleus. Its luminosity is caused by the gravitational energy of the accreting matter, while the nature and spectra of the emission depend on the density and optical thickness of the matter surrounding the compact object.

In some of these ultraluminous galaxies the OH radical is detected in such intense and extended OH maser emission (18 cm) – up to $L_{OH} = 10^3 \, L_\odot$ – that the phenomenon is termed 'megamaser'. Unlike masers that are very localized about a star evolving in our Galaxy, this emission corresponds to a very extended region of star formation; the powerful infrared flux produced plays an important role in the pumping of the maser.

Galactic mergers could have been much more frequent in the past, especially at the epoch of formation of the first galaxy clusters (at a redshift of around $z = 2$). Recently a very powerful infrared object has been identified at $z = 2.28$: its far-infrared luminosity reaches $10^{14} \, L_\odot$, the luminosity of the most powerful quasars. It appears to be powered almost entirely by a giant starburst and not by a central supermassive black hole, as in quasars. The amount of molecular gas it contains, derived from the detection of the CO lines, is between 10^{11} and $10^{12} \, M_\odot$, that is, 100 or 1000 times that of our own Galaxy.

2.3.7 Other Molecules. Abundances

The observation of molecules other than CO may lead to information about not only the abundance of the elements but also the excitation of lines and the nature of extragalactic molecular clouds. HCO^+ and HCN, for example, have been discovered in many galaxies; the results suggest types of cloud and density distributions comparable to those of the Milky Way. HCO^+ has also been detected in the Large Magellanic Cloud, with a ratio HCO^+/CO comparable to that of galactic clouds, indicating a normal abundance of ionized cosmic rays.

The absorption of OH in the spectrum of extragalactic continuous sources is a means of exploring the molecular gas close to the nucleus. The advantage of these observations is that the feasibility of detection is independent of the distance of the galaxy (which is not the case for CO emission); it depends only on the optical thickness and the intensity of the continuous source. The observation of OH absorption in a very excited level in Arp 220, for example, enables us to deduce the presence of a starburst. The detection of H_2 emission at $2 \, \mu m$ testifies to the existence of shock waves (the temperature of the medium is approximately 2000 K) accompanying the star-formation regions.

In summary, observations of the molecular phase have enabled us to identify galaxies where stars are actively being formed as galaxies rich in gas, where this is concentrated in very precisely defined regions. It remains for us to understand by what mechanisms such concentrations of gas are suddenly released.

References

Brinks, E. (1984) Ph.D. Thesis, University of Leiden.
Condon, J.J., (1992), *Annu. Rev. Astron. Astrophys.* **30**, 575.
Durret, F., and Bergeron, J. (1988) *Astron. Astrophys. Suppl.* **75**, 273.
Guélin, M., et al. (1993), *Astron. Astrophys.* **279**, L37.
Horellou, C., Beck, R., and Klein, U. (1990) in *Galactic and Intergalactic Magnetic Fields*, edited by R. Beck (IAU Symposium 140, Kluwer, Dordrecht), p. 211.
Lo, K. Y., Cheung, K. W., Masson, C. R., Phillips, T. G., Scott, S. L., and Woody, D. P. (1987) *Astrophys. J.* **312**, 574.
Mathewson, D. S., van der Kruit, P. C., and Brouw, W. N. (1972) *Astron. Astrophys.* **17**, 468.
McGaugh, S.S., (1991), *Astrophys. J.* **380**, 140.
Neininger, N., et al.(1996), *Astron. Astrophys.* **310**, 725.
Neininger, N. (1992) *Astron. Astrophys.* **263**, 30.
Osterbrock, D.E., (1989), *Astrophysics of Gaseous Nebulae and Active Galactic Nuclei* Mill Valley, Univ. Science Books.
Rand, J., and Kulkarni, S. (1990) *Astrophys. J.* **349**, L43.
Reynaud, D., Downes, D. (1997), *Astron. Astrophys.* **319**, 737.
Roberts, M.S., Haynes, M.P., (1994), *Annu. Rev. Astron. Astrophys.* **32**, 115.
Rots, A. H., and Shane, W. W. (1975) *Astron. Astrophys.* **45**, 25.
Sanders, D. B., and Mirabel, I. F. (1985) *Astrophys. J.* **298**, L31.
Weliachew, L., Casoli, F., and Combes, F. (1988) *Astron. Astrophys.* **199**, 29.
Wevers, B. M. H. R. (1984) Ph.D. Thesis, University of Groningen.
Young, J. S., Kenney, J. D., Tacconi, L., Claussen, M. J., Huang, Y. L., Tacconi-Garman, L., Xie, S., and Schloerb, F. P. (1986) *Astrophys. J.* **311**, L17.

3. The Kinematics and Masses of Galaxies

In the 1920s, several years before the spiral nebulae were even identified as entirely separate galaxies, it was discovered that they rotated. This discovery came about as a result of the inclination of absorption lines in the spectra of the central regions of, in particular, the galaxies M81 and M104. Up to the 1970s, rotation curves, or in other words the law describing how the rotation velocity V varies as a function of the distance to the galactic centre, were obtained solely at optical frequencies from the absorption lines of stars in the central regions and from the emission lines of H II regions in the outer regions. Subsequently interferometric radio observations at 21 cm (the H I line) with the Westerbork telescopes (in the 1970s) and the VLA (in the 1980s) enabled a large number of rotation curves to be quickly determined, at the same time with greater spectral resolution and an unequalled radial extension, the gaseous H I component generally exceeding the optical limit in the outer regions of galaxies. This confirmed in a striking way that the rotation curves remain flat at a large distance from the centre: does an enormous quantity of invisible mass then exist beyond the visible limits of a spiral galaxy?

3.1 Optical Determinations

3.1.1 Methods

Absorption lines, which give information about the kinematics of the stellar component, are difficult and time-consuming to obtain and never extend over the whole disc. Nevertheless they are invaluable, in particular for comparing the rotation velocities of stars and gas. In many galaxies it has been possible to make the comparison from the centre up to a radius of about 20 kpc and to show that the velocities of the stars and gas do not differ by more than the measurement errors (about $30 \, \text{km s}^{-1}$) in the discs of spirals. This result is fundamental and allows us more generally to determine the dynamics of galaxies from the rotation curves observed for gas. In the central bulge, however, the stars have a rotation velocity equal to only half the velocity of the gas: the gas is a cold component that traces the mass distribution in the manner of test particles, whereas the large velocity dispersion of the stars' velocity in the bulge significantly reduces their stellar group rotation velocity.

The comparison, however, is tricky, owing to important noncircular velocities also present in the stellar and gaseous components (various perturbations, absorptions, bars, and so on); ionized gas rarely follows a circular orbit at the galactic centre and its rotation velocity is often less than that expected from the visible mass.

Rotation curves from the emission lines of ionized gas (essentially Hα and [N II]) can be determined by the method of long-slit spectroscopy: in particular for distant galaxies it is easy to arrange many H II regions along the same slit, usually along the major axis of the galaxy. To make the velocity field complete (including noncircular velocities, orientation parameters, and so on) it is necessary to observe many slits spread out across the disc, and the integration time required may sometimes be long. The Fabry–Perot method of interferometry, on the other hand, enables one to obtain large fields at a single time and is being developed further. The reduction of the data, however, is more time-consuming and difficult than for slit spectroscopy.

Determination of the velocity fields of a galaxy from the lines of ionized gas is very useful and complements the H I radio determination. In the central regions a very high spatial resolution is required, and moreover atomic hydrogen is very often absent from the central regions. In the disc regions the H I component is more uniformly distributed than the H II regions, but it is essential for theories of star formation to consider all systematic differences between the two components (for example those due to the spiral structure).

3.1.2 The General Properties of Rotation Curves

Figures 3.1, 3.2 and 3.3 bring together a large number of rotation curves obtained for spiral galaxies of different Hubble types, Sa, Sb, and Sc. These curves were obtained from the emission lines of ionized gas, by taking the mean of the values obtained for either side of the major axis, and 'deprojected' (that is, account was taken of the inclination of the galaxy to the plane of the sky). The most striking characteristic is that these curves climb rapidly (in less than 5 kpc) towards a maximum and then reach a plateau that is maintained up to the end of the measurements, that is, a radius R_{25} where the surface brightness becomes less than $25\,\mathrm{mag\,arcsec}^{-2}$, or a significant fraction of R_{25}. Of course, the plateaux reveal fluctuations, some wavelike, some slightly sloping, more often positively than negatively. Note that there is a large variation in size and luminosity between the galaxies, which can lead to great disparities in their evolution, the outer parts of the largest galaxies having made only 10 rotations in the Hubble time, the smallest, however, having made 300.

The properties of the rotation curve vary uniformly with the luminosity: as the luminosity increases, the central velocity gradient increases, as does the amplitude of the maximum velocity. There also exists a marked difference between the types: at a given luminosity the rotation velocities are greatest for Sa, then Sb, and finally Sc galaxies. This is probably due to the central

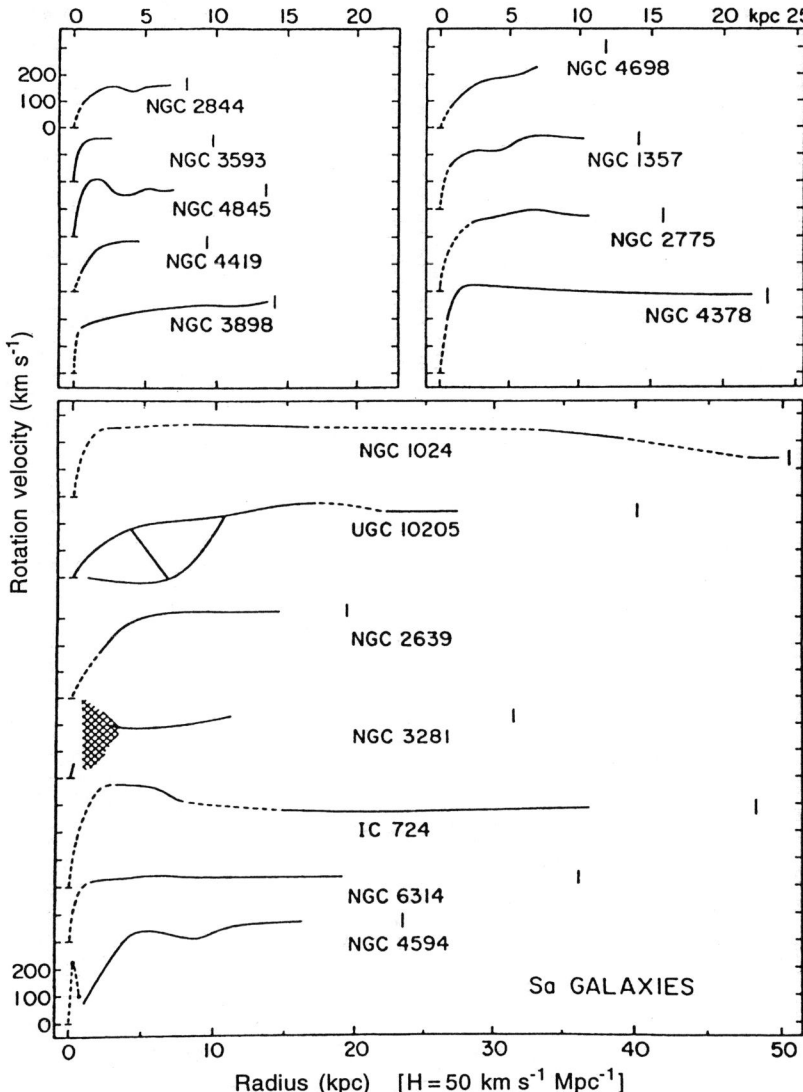

Fig. 3.1. Rotation curves obtained at optical frequencies from the emission lines of ionized gas for spirals of type Sa. A striking characteristic of all these curves is their flatness at a large distance from the centre. (From Rubin et al. 1980, 1982, 1985)

bulge–disc ratio increasing from Sc to Sa: matter is more concentrated in the earlier types, leading to strong rotation for the same total mass. The normalized form (at characteristic radius and velocity) of the rotation curves nevertheless (almost) no longer depends on the type or luminosity, as shown in Fig. 3.4.

72   3. The Kinematics and Masses of Galaxies

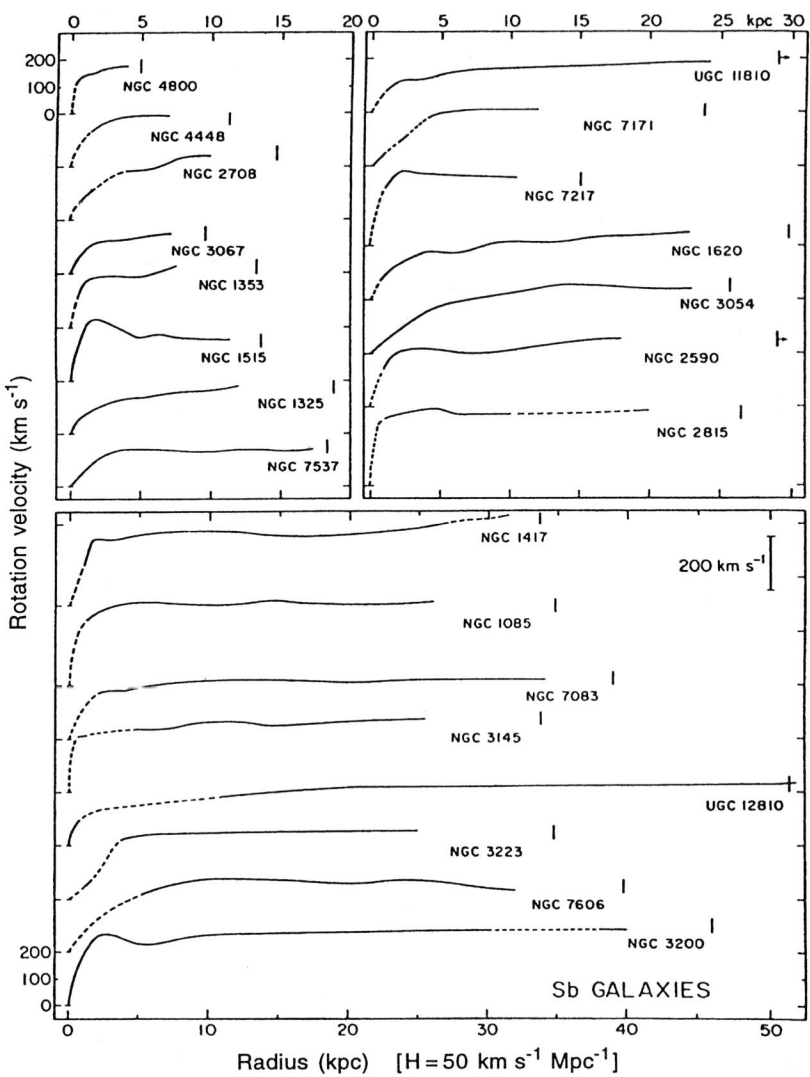

Fig. 3.2. As in Fig. 3.1, but for Sb galaxies

A correlation exists between the maximum rotation velocity $V_{\max}$ of a galaxy and its luminosity L, which can be written as roughly $L = \text{constant} \times V_{\max}^4$, known as the *Tully–Fischer* relation and illustrated in Fig. 3.5. This law allows a good determination of the distances for faraway galaxies. Galaxies of different morphological types seem to satisfy the same law but with slightly different constants.

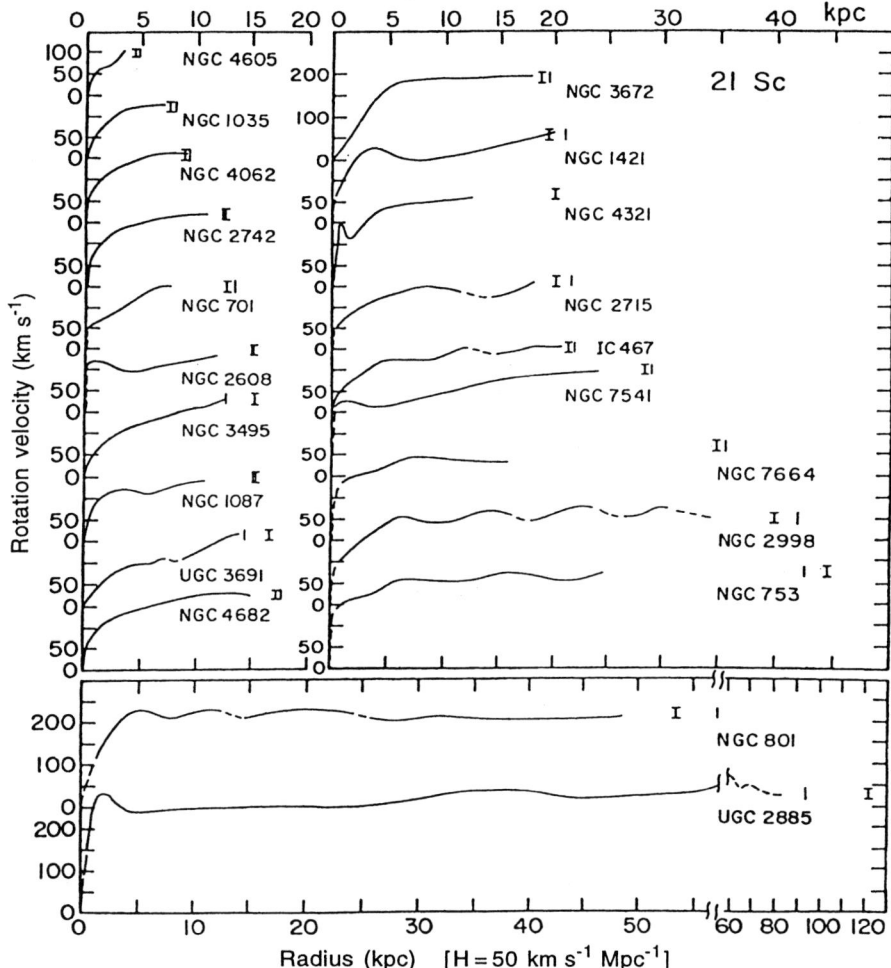

Fig. 3.3. As in Fig. 3.1, but for Sc galaxies

3.2 Radio Determinations

The determination of velocity fields from the H I component in general suffers from poor spatial resolution, the velocity at a point being the mean of many velocities weighted by the signal at that velocity, so producing spurious effects when high density contrasts are present (spiral arms, for example). The spectral resolution, however, leads to great precision, of the order of several km s^{-1}. The velocity fields are particularly simple to read in the form of isovelocities, which resemble the famous 'spider' diagram that can be seen in Fig. 3.6 for the case of a field of pure rotation velocities.

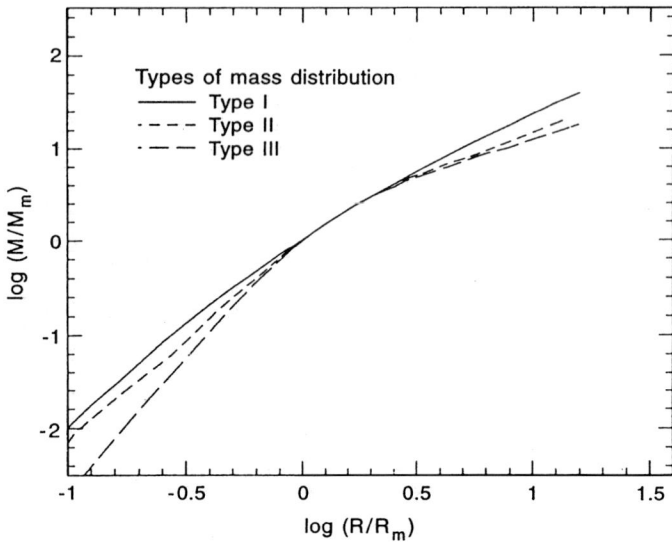

Fig. 3.4. Normalized mass distributions (the radius R_m containing about $M_m = 10^{10} M_\odot$ but adjusted so as to minimize the distance from the universal curve). Sa, Sb, and Sc spiral galaxies divide into three distribution types, increasingly curved from I to III. Sa and Sb galaxies are found among the three types more or less equally, whereas Sc galaxies are mainly of type I. (From Burstein and Rubin 1986)

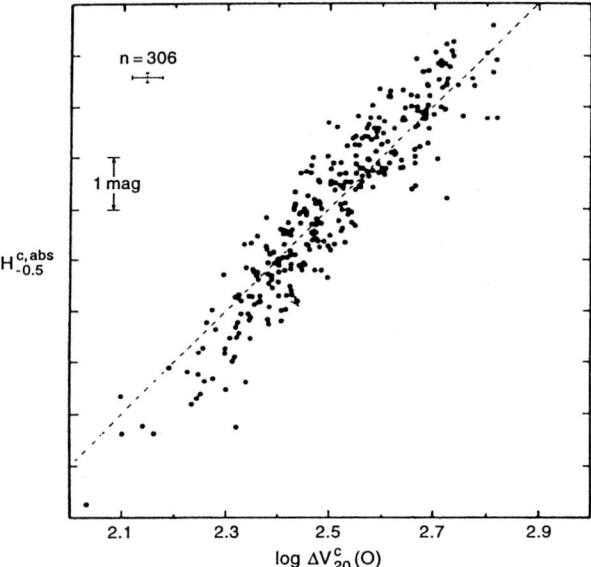

Fig. 3.5. Infrared magnitudes (H) as a function of the velocity width of the H I profile for 306 nearby galaxies. The dashed curve is a power law of power 10, corresponding in luminosity to a power 4: the galaxies satisfy the Tully–Fischer law ($L \propto V^4$) remarkably well. (From Aaronson et al. 1982)

Fig. 3.6. The classic so-called spider diagram representing the isovelocities of a uniformly rotating galaxy with no noncircular perturbation

Consider a disc inclined to the plane of the sky at an angle β and for which the line of nodes (the apparent major axis of the galaxy) has a position angle Φ_0; the systemic velocity of the centre of the galaxy is V_{sys}. The observed radial velocity will be $V_{\text{rad}} = V_{\text{sys}} + V_{\text{rot}} \sin\beta \cos\theta$ ($+V_r \sin\beta \sin\theta$ if there are noncircular velocities V_r in the plane of the galaxy).

If these velocities are purely circular, the minor axis has to be joined with the isovelocity $V = V_{\text{sys}}$, the axis of symmetry of the spider diagram. When the diagram is relatively regular, it is more precise to derive the position parameters of the galaxy (Φ_0, β, V_{sys}, and the centre of gravity) from the velocity field than from photometric observations.

In fact if the isovelocity diagrams all had the general appearance of a differentially rotating disc, one might only be struck by the large number of asymmetries and kinematic perturbations, several examples of which are given in Figs. 3.7 and 3.8. There are either local irregularities of velocity (which can also lead to irregularities in the density) or perturbations recognizable on a large scale, such as manifestations of a particular structure or 'density waves': spiral arms, stellar bars, or the warp of the galactic plane.

3.2.1 Spiral Arms

H I gas is a component with weak velocity dispersion that should be a good tracer of gravitational potential wells. It can serve as a test for models of density waves proposed to take account of the spiral structure of galaxies (Chap. 5). On crossing a spiral arm, systematic velocity perturbations are to be expected corresponding to a slowing-down of the gas between the arms. The galaxy M81 provides one of many beautiful examples of undulating isovelocities due to the presence of spiral arms. In Fig. 3.9 isovelocities observed and modelled from a spiral gravitational potential are superimposed on the H I density map. Modelling has the advantage of showing that the nonlinear theory of density waves for the gas, including shock waves crossing the

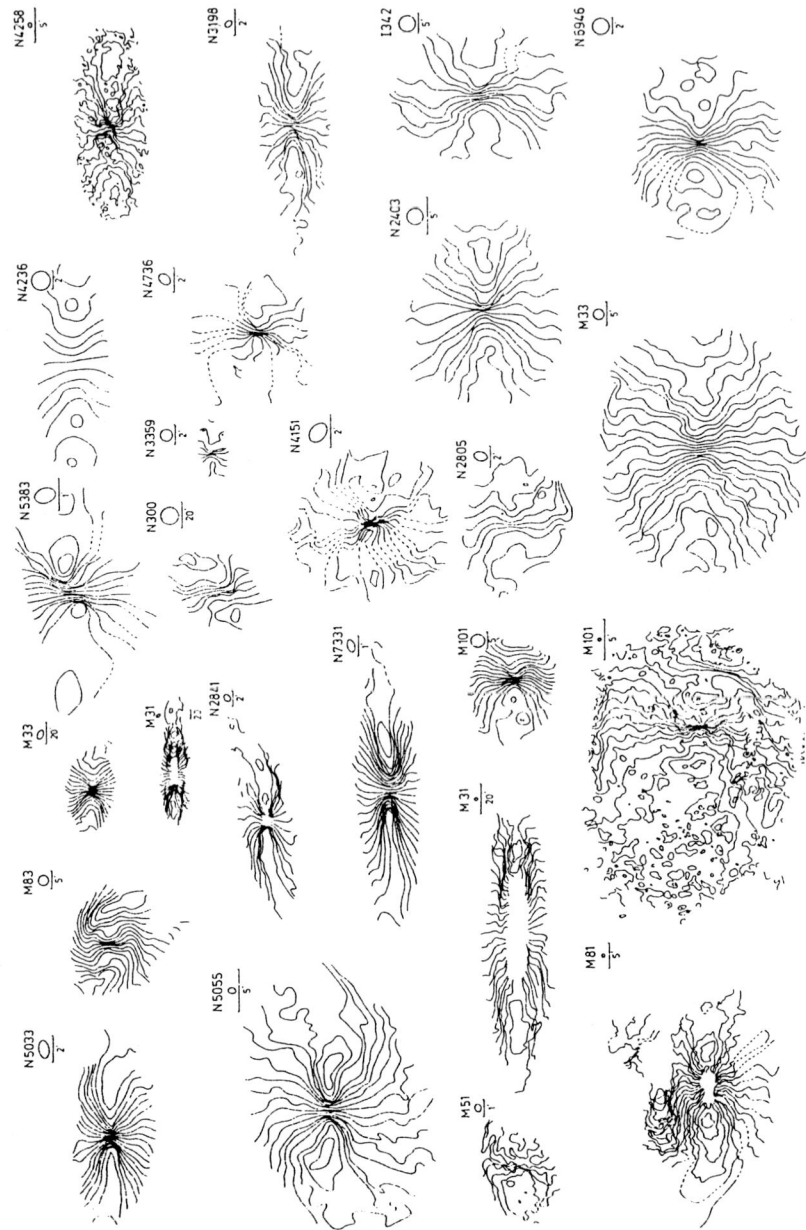

Fig. 3.7. H I velocity fields for 22 spirals, observed with different telescopes and different resolutions (indicated by ellipses at the upper right of each image). The galaxies have been reorientated so that all the major axes are horizontal. Note the many deformations with respect to the classic spider diagram. The central deformations (S-shaped) are due to the presence of a bar; the outer deformations (kinematic rotation of the major axis) are due to warping of the plane. (From Bosma 1981)

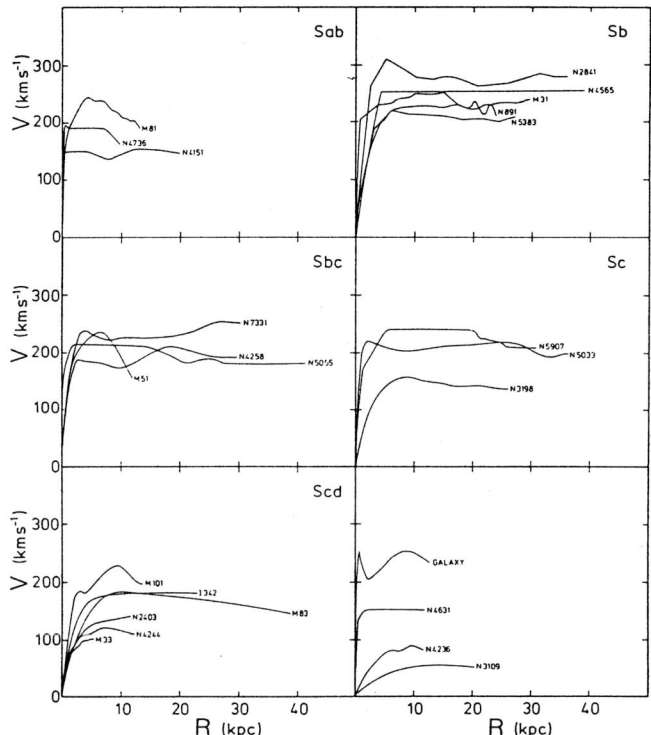

Fig. 3.8. H I rotation curves for several types of spiral galaxy, confirming the flatness of the optical rotation curves at far greater distances still from the centre. (From Bosma 1981)

arms, is necessary to take account of observations. Jumps in velocity from 30 to 50 km s^{-1} can be produced in the shocks. Likewise the effects of spatial resolution can be quantified: the perturbations due to density waves are not observable as soon as the size of the antenna beam is greater than half the separation between the arms. On the other hand the rotation curve is modified by the density waves, even once the wavelets are smoothed.

The problems of spatial resolution, the inclination of the galaxy to the plane of the sky, and sometimes the weakness of the shock waves themselves mean that kinematic perturbations are rarely as significant as in M81 and all in all trace the spiral structure less well than the H I distribution itself. Observations of the molecular gas (the CO molecule), a better tracer of the density waves than H I gas (because it is colder), will enable the kinematic perturbations to be studied in greater detail, as soon as large-scale interferometric observations become available.

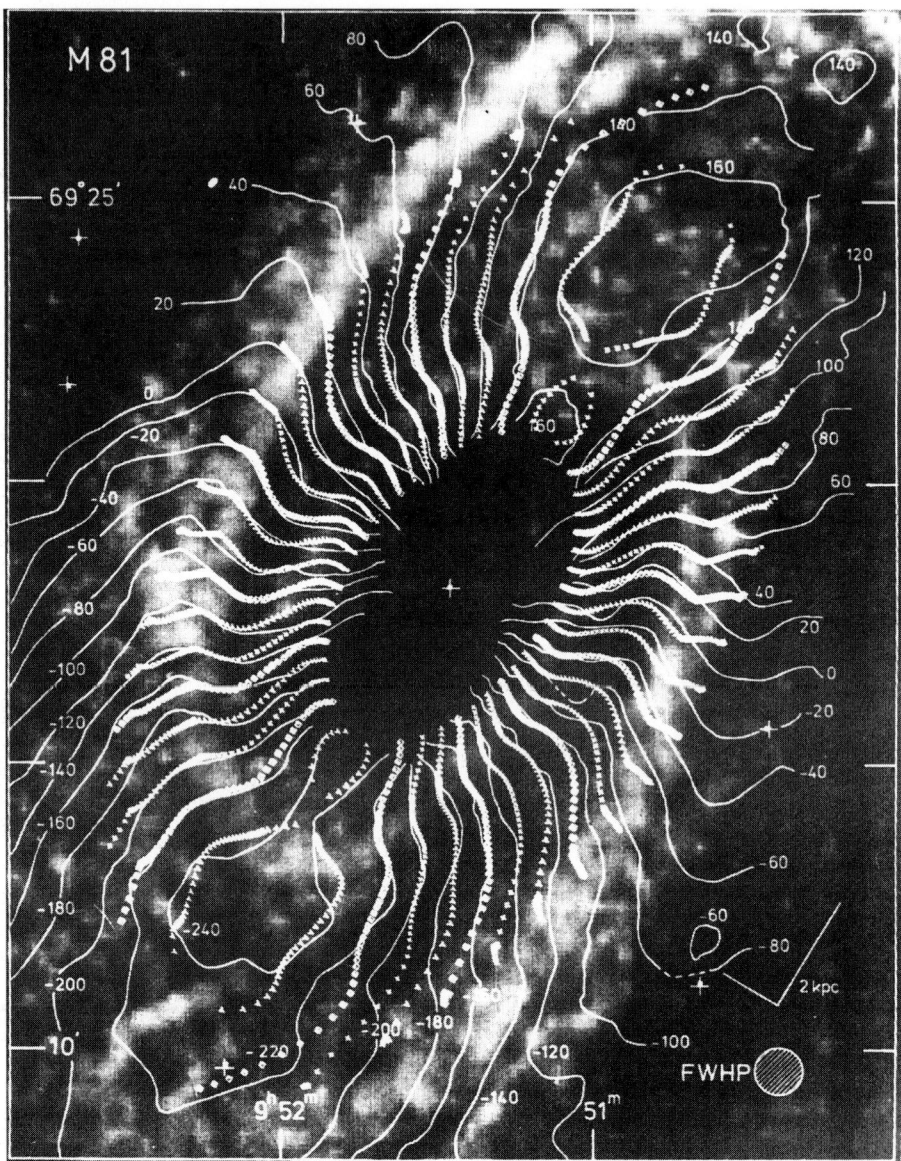

Fig. 3.9. Isovelocities of M81 (*thin lines*) observed in H I with the Westerbork interferometer, superimposed on an H I density radio map. The curves made of symbols represent the model constructed from the theory of density waves. (From Visser 1980)

3.2.2 Bars and Oval Distortions

When a bar is clearly observable in the stellar component (SB galaxies), the gas dynamics show very distinctive characteristics, corresponding to elliptical and noncircular orbits. The gas follows the potential of the bar in its rotating reference frame, and the orbits are elongated in accordance with the bar (Chap. 6). It is then easy to show that the isovelocities are deformed symmetrically into an S shape, provided, however, that the bar is not situated on the major axis or minor axis: the isovelocity V_{sys} is no longer parallel to the minor axis; as the velocities are no longer circular, the gas along the minor axis has a radial velocity towards the observer different from V_{sys}. The inclination of this central isovelocity along the minor axis is one of the main features used to identify a bar. One example is given in Fig. 3.10. The distortions concern only the central regions of galaxies. When the galaxy is not classified as SB, that is, if the bar is not sufficiently strong to be visible, the kinematics of H I gas can all the same reveal a weak bar, which is also known as an oval distortion of the potential.

The bars or oval distortions are often accompanied by the presence of internal or external rings, especially for early-type galaxies. The internal rings are the site of pronounced noncircular velocities, probably due to the elliptical orbits of the gas. The rings are not circular in the plane of the galactic disc.

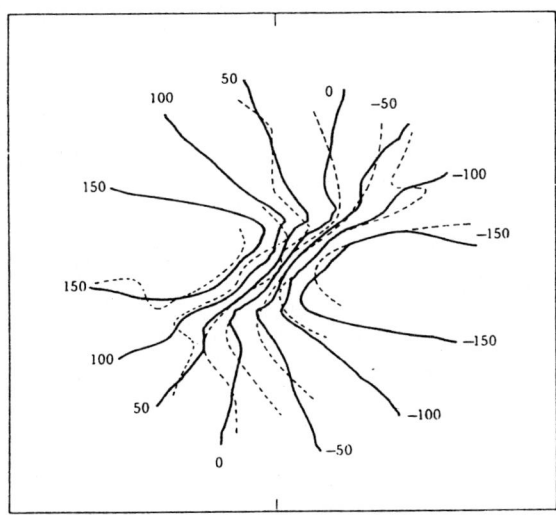

Fig. 3.10. In a barred galaxy the gas rotates about the centre in an elliptical orbit elongated along the bar. Isovelocity curves thus reveal a characteristic S-shaped perturbation: the kinematic minor axis is no longer perpendicular to the kinematic major axis. The bold curves are the isovelocities of NGC 5383 observed in optical emission lines by Peterson et al. (1978). The dashed curves represent the hydrodynamic model of Sanders and Tubbs (1980)

In the outer regions there are often 'pseudorings' formed by two spiral arms that tend to close on themselves: the perturbations associated with the arms are then dominant.

3.2.3 Distortion, or Warp, of the Plane

When the plane of a galaxy is deformed, its inclination β and its position angle Φ vary with radius, which perturbs the velocity field and resembles noncircular velocities, even if the rotation remains uniform in the galactic disc. The kinematic perturbation can be modelled by dividing the galactic disc into concentric rings of a thickness equivalent to the spatial resolutions sought. Each ring has its own orientation in the sky, determined by β_i and Φ_i. These parameters can be found by comparing the radial velocity obtained in the projected ring with the velocity field observed.

One such model, shown in Fig. 3.11, gives isovelocities whose major axis progressively changes orientation with radius. This rotation of the major axis affects only the outer regions of galaxies, since the warp concerns only the gas beyond the optical disc. It is this characteristic that allows us to distinguish between this type of perturbation and the oval distortions at the centre; also, in the case of a warp, the kinematic major and minor axes are always perpendicular. The two perturbations can of course coexist, two-thirds of spiral galaxies being barred and the warp phenomenon being very common.

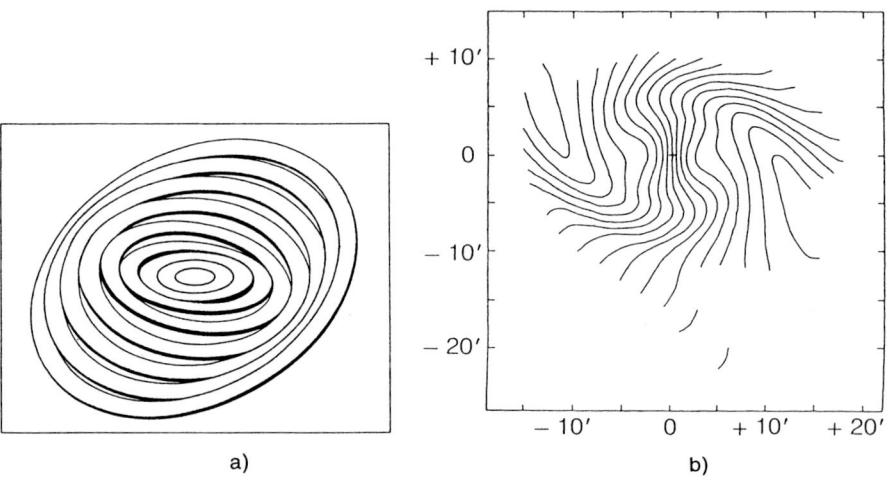

Fig. 3.11. (a) A perspective view of a model for the planar distortion, or warp, of the galaxy M83. The view is at 90° to the actual line of sight. (b) Isovelocities constructed from the model (compare them with observations – Fig. 3.7). (From Rogstadt et al. 1974)

3.2.4 Asymmetries

The perturbations we have just mentioned in general affect a galaxy symmetrically with respect to the centre. But in many cases very marked asymmetries are present in the velocity field and in the density map at the same time. These asymmetries can be attributed to tidal interactions with companions in the majority of cases.

3.3 Determination of the Mass Distribution

Obtaining a mean rotation curve for galaxies enables us to determine the distribution of mass and, eventually, by extrapolation, the total mass of the galaxy. The big problem, however, is that it does not seem that the rotation velocity decreases significantly at large distances: if the mass of the galaxy beyond a certain limiting radius is constant, as for all finite systems, the rotation velocity would decrease in Keplerian fashion, according to an $r^{1/2}$ law, which is not observed: the mass of the galaxy seems always to grow, beyond the Holmberg radius of the galaxy, without levelling off. For a flat rotation curve $V(r) =$ constant, $M(r)$ increases as r, and the density ρ of a supposedly spherical invisible component varies as r^{-2}.

3.3.1 Methods of Analysis

Disc Galaxies. Assume that the gas in the galactic disc is in dynamic equilibrium and that the forces of pressure are negligible, as is the velocity dispersion of the gas. This is justified since the dispersion, being of the order of $10\,\text{km}\,\text{s}^{-1}$, is just a small fraction of the rotation velocity ($\approx 200\,\text{km}\,\text{s}^{-1}$). The rotation velocity $V(r)$ thus allows us to deduce the gravitational potential $\Phi(r)$ (assumed to have an axial symmetry) and, from the Poisson equation, the mass density at every point on the disc. The reduced equations may be written as

$$V^2(r) = r\,\partial\Phi/\partial r; \quad \Delta\phi(\boldsymbol{r}) = 4\pi G \rho(\boldsymbol{r}).$$

Right up to the 1970s the flattened-spheroid method was commonly used. In this method the mass distribution in the galaxy is modelled by a superposition of many flattened shells of density $\rho(a)$, where a is the semimajor axis of the shell. The circular velocity $V(r)$ is thus written as

$$V^2(r) = r\frac{\partial\phi}{\partial r} = 4\pi G \int_0^r \frac{k\rho(a)a^2\,\mathrm{d}a}{(r^2 - k^2a^2)^{1/2}},$$

where the flattening of the shell is equal to $(1 - k^2)^{1/2}$, k being the ratio of the two axes. The advantage of this model is that the velocity $V(r)$ depends only on $\rho(a < r)$, because the potential Φ inside a shell is constant. The above equation for $V^2(r)$ can be inverted for infinitely flattened shells, using

$$k\rho(a) = \frac{1}{2\pi^2 G a^2} \frac{d}{da} \int_0^a \frac{V^2(r) r \, dr}{(a^2 - r^2)^{1/2}} \quad (k \to 0).$$

This formula enables us to deduce the mass distribution at each point; but more simply the rotation curve can be parametrized, giving access to known solutions. The Brandt curve is often used:

$$V(r) = V_{\max} \left(\frac{r}{r_{\max}}\right) \bigg/ \left[\frac{1}{3} + \frac{2}{3}\left(\frac{r}{r_{\max}}\right)^n\right]^{3/2n},$$

where $V_{\max}$ is the maximum of the rotation curve, reached at a radius $r_{\max}$, and n is a form parameter, determining at what radius the velocity begins to fall in a Keplerian way. Of course, this parametrization assumes a priori that the system is finite, even if the rotation velocity only falls off beyond the optical disc. With the hypothesis that this parametrization can be extrapolated beyond the measured curve, the total mass of the galaxy is $M_{\text{tot}} = (3/2)^{3/n} V_{\max}^2 r_{\max}/G$. The surface density of the disc $\mu(r)$ in the spheroidal model can be obtained from

$$\mu(r) = 2 \int_r^\infty \frac{k\rho(a) a \, da}{(a^2 - r^2)^{1/2}},$$

where $k\rho(a) \to 0$ when $k \to 0$. Because the model makes use of the (unknown) mass distribution perpendicular to the plane of the galaxy, Toomre's model of the disc is now often preferred, since it allows one to deduce directly the physical quantity $\mu(r)$ of the rotation curve $V(r)$. Toomre's solution of the Poisson equation for an infinitely thin disc is based on Fourier–Bessel integrals. A solution can be written down, provided that $V(0) = V(\infty) = 0$:

$$\mu(r) = \frac{1}{2\pi G} \int_{u=0}^\infty \frac{dV^2(u)}{du} A(u, r) \, du,$$

where

$$A(u, r) = \begin{cases} \dfrac{2}{\pi r} K\left(\dfrac{u}{r}\right) & \text{for} \quad u < r, \\ \dfrac{2}{\pi u} K\left(\dfrac{r}{u}\right) & \text{for} \quad u > r. \end{cases}$$

K is the complete first-order elliptical integral. Toomre gave a simple formulation for a series of infinitely thin selfgravitating discs, the simplest of which is of the form $\mu(r) = \mu_0(1 + r^2/a^2)^{-3/2}$, with the potential $\Phi(r) = \phi_0(1 + r^2/a^2)^{1/2}$ and the rotation velocity

$$V^2(r) = \frac{V_{\max}^2}{2} 3^{3/2} \frac{r^2}{a^2} \left(1 + \frac{r^2}{a^2}\right)^{-3/2},$$

where $V_{\max}^2 = 4 \times 3^{-3/2} \pi G a \mu_0$.

3.3 Determination of the Mass Distribution

The other models are parametrized by an integer n, and are expressed by

$$\mu(r) = \mu_0 \left(1 + \frac{r^2}{a^2}\right)^{-[n+(1/2)]}.$$

Toomre's method can be used to find the surface density and consequently the mass included within the radius r for any rotation curves.

On the other hand the invisible mass in the outer parts of galaxies, the presence of which is suggested by the flat rotation curves, might not be distributed in a disc: the flattening of this component of mass is unknown and the total mass $M(R)$ deduced from the rotation curve would become larger the less flattened this component is. To quantify the domain of variation in the extreme cases of a sphere and a disc, we have for the total mass

$$M_{\text{tot}} = V_{\text{rot}}^2 R/G \quad \text{(sphere)},$$

$$M_{\text{tot}} = (2/\pi) V_{\text{rot}}^2 R/G \quad \text{(disc)},$$

for a disc cut off at $r = R$ and whose rotation curve is $V = \text{constant} = V_{\text{rot}}$ inside the disc ($r < R$). The necessary mass M_{tot} for the hypothesis of a disc is less by a factor of $2/\pi = 0.64$, but the rotation curve on the outside ($r > R$) then decreases more quickly than for the sphere.

Elliptical Galaxies. For spheroidal systems that have only a little rotation, what are the methods of calculating the total mass? Chapter 4 gives the details of the dynamics of these systems, whose stability with respect to gravitational collapse only depends on the velocity dispersion of the stars. One of the methods of analysis is therefore based on the virial theorem:

$$\frac{1}{2}\frac{d^2 I}{dt^2} = 2T + W,$$

where I is the moment of inertia of the system, T its kinetic energy, and W the gravitational potential energy. For a galaxy in equilibrium the moment of inertia does not evolve as a function of time, and we can write

$$M \langle V^2 \rangle + W = 0,$$

where $\langle V^2 \rangle$ is the mean square of the velocities, weighted by the mass of the stars, and M is the total mass. In practice it is relatively difficult to estimate all these parameters: in particular, $\langle V^2 \rangle$ requires knowledge of the velocity dispersion as a function of the radius in the galaxy. In general this dispersion decreases with r, but it is only well estimated in the neighbourhood of the centre (where the absorption lines are intense enough). To estimate W a mass distribution law is required: the most common hypothesis is to assume that the luminosity traces the mass and to choose the de Vaucouleurs $r^{1/4}$ profile, which describes the luminosity distribution quite well (Chap. 1). The potential energy is then written as

$$W = -\frac{1}{3}\frac{GM^2}{r_e},$$

where r_e is the radius containing half the mass (or luminosity). Besides the differences from the de Vaucouleurs profile observed in the outer regions (notably for galaxies in a rich environment and, in particular, cD galaxies), this estimate can be inexact if the mass in the outer regions no longer follows the luminosity, as in the spiral galaxies.

This is why other methods have been developed, based on theoretical models of elliptical galaxies (in particular, King's model; see Chap. 4). Since the observed luminosity profiles are calibrated well by these models, we can then deduce the relations between the ratio M/L and the physical quantities observed: the velocity dispersion at the centre, σ_{v0}; the surface brightness at the centre, I_0; and the core radius (where the surface brightness is $I_0/2$) or effective radius r_e.

On the other hand it is sometimes possible to make use of test particles for probing the potential and thus the mass distribution: around lenticulars, or even ellipticals with dust lanes, a rotating disc of gas can allow a more exact study of the mass distribution; likewise the presence of shells, which are the stellar vestiges of the infall of a small companion onto the central galaxy. When X-ray emission is observed, owing to a cloud of gas assumed to be in hydrostatic equilibrium in the potential well of the elliptical galaxy, the total mass up to a very large radius can be estimated.

3.3.2 The Mass–Luminosity Ratio

For spiral galaxies the surface brightness decreases exponentially with the radius, and the surface density deduced from the velocity field is proportional to $\mu(r) \approx r^{-1}$; the ratio M/L will then increase with the radius. Great caution, however, is in order, since the total mass will possibly be enormous if the rotation curve is assumed by extrapolation to be flat even beyond the measurements. The estimation of M/L also depends on the form of the invisible component adopted (spheroidal or disc-shaped), possibly being 35% less in the case of the disc (see Sect. 3.3.1). It has not yet been possible to determine with certainty the three-dimensional shape of the invisible mass: it could be distributed in a spherical halo, or even flattened into a disc. Arguments have been proposed in favour of a spherical shape:

- the frequent presence of warps in the external regions of a galaxy, for which a highly flattened distribution accelerates the differential precession and therefore the destruction of the warps; but other explanations exist for the persistence of warps – see Chap. 7;

- the high rotational velocity of the gas orbiting in polar rings, that is, in orbits perpendicular to the plane of some peculiar spiral galaxies; however, polar-ring galaxies never yield simultaneously estimates of rotational velocities in the plane and perpendicular to it, since the existence of two perpendicular gaseous rings of the same radius is not a stable situation; also the mass of the polar ring itself locally perturbs the gravitational potential.

On the other hand the flaring of the gaseous plane in the outer parts of galaxies, that is, the fact that the thickness of the plane increases roughly linearly with radius, supports a flattened distribution for the dark matter.

This uncertainty in the shape of the invisible component explains why mass–luminosity ratios are always estimated using the spherical hypothesis, for the sake of simplicity.

How does the total mass–luminosity ratio vary with morphological type? M/L evolution is shown in Fig. 3.12 as a function of colour, which is also an evolution of type, the bluer colours being found in late-type galaxies (on the left of the figure). The total ratio M/L is roughly constant, while the expected ratio $M_{\rm vis}/L$ from stars varies along the sequence. In Fig. 3.12 the

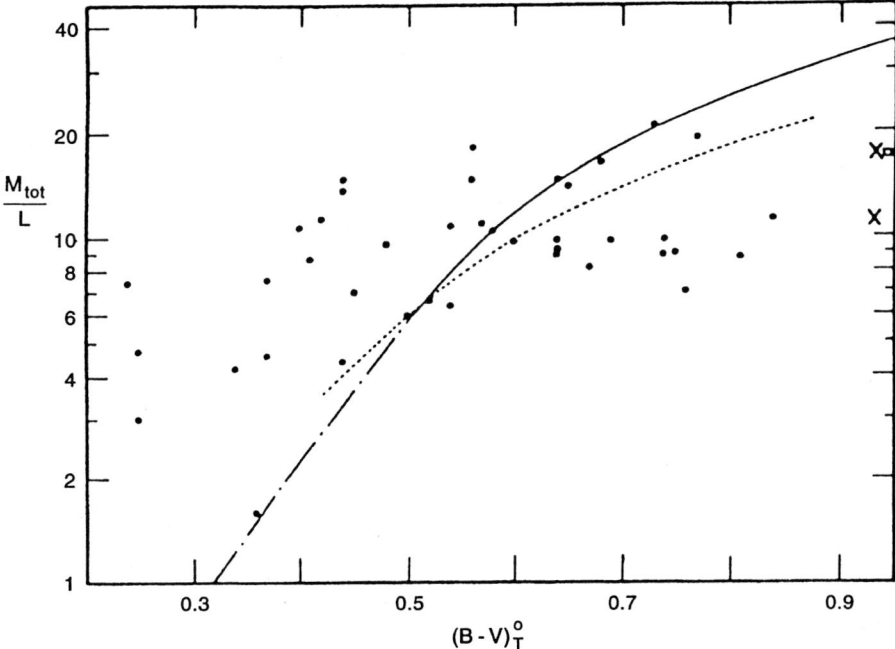

Fig. 3.12. The mass–luminosity ratio versus colour. Data are for spiral galaxies (*dots*), two S0 galaxies (*crosses*), and the core of a giant elliptical (*square*). The two theoretical lines are $M_{\rm vis}/L$, taken from stellar-population models. (From Tinsley 1981)

curve reproduces the expected values of M_{vis}/L computed by Tinsley (1981) from stellar-population studies. The mean colours for the three types Sa, Sb, and Sc are $B - V = 0.75$, 0.64, and 0.52 respectively, and from these colours the theoretical values of M_{vis}/L_B are 3.1 (Sa), 2.0 (Sb), and 1.0 (Sc). Only for very blue galaxies, that is, for late-type galaxies rich in gas and star formation, does the total ratio M/L clearly exceed the visible mass–luminosity ratio. For very early-type galaxies no dark matter seems required inside the Holmberg radius. This result has been recently confirmed by the observation of falling rotation curves in early-type galaxies. These galaxies being gas deficient, the determination of their rotation curve requires great sensitivity, and more observations are still needed to improve the statistics.

Since the total mass (or blue luminosity) increases from late-type to early-type galaxies (see Chap. 1), the ratio $M_{\text{vis}}/M_{\text{tot}}$ is also an increasing function of total mass, as shown in Fig. 3.13. Dwarf galaxies have proportionally more dark matter than massive galaxies.

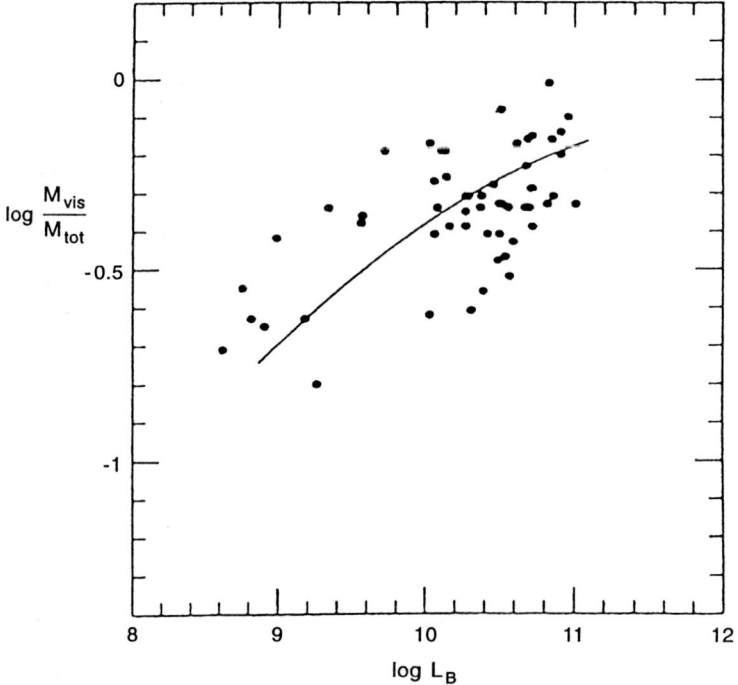

Fig. 3.13. The fractional visible mass $M_{\text{vis}}/M_{\text{tot}}$ at R_{opt} as a function of luminosity. (From Persic et al. 1993)

3.3.3 The Tully–Fischer Law and Its Interpretations

The relation between the total luminosity of a galaxy and its rotation (or dispersion) velocity, $L \propto V^4$, is much better satisfied when the H luminosity (the infrared luminosity in the H band, L_H) is taken into account, since then the visible mass–luminosity ratio does not depend on morphological type: M_{vis}/L_H becomes approximately 2 for all types. This can be understood when one realizes that L_B measures mainly the contribution of the young population of stars (population I), whereas L_H is a much more universal tool, tracing the old population common to all types of galaxy. In Fig. 3.5 it can be seen that the Tully–Fischer law is quite well satisfied in the H band, and ΔV can therefore serve as a distance indicator.

What could be the physical significance of this relation, which is satisfied just as well by rotating discs as by spheroidal systems? It can be shown that the virial theorem taken together with a universal law for the mass distribution implies the relation $M \propto V^4$, which reduces to $L \propto V^4$ if the ratio M/L is constant for all galaxies, which seems to be the case for L_H. The virial theorem is written as $MV^2 \propto M^2/R$. In using $M \propto V^4$ to eliminate the velocity, we get $M \propto R^2$. For elliptical galaxies the mass distribution is well represented by King's model, with the density of the form $\rho(r) = \rho_0 f(r/r_0)$ ($f(x)$ being a universal function). The total mass is therefore $M = \rho_0 r_0^3 \int f(x) 4\pi x^2 \, dx$. Now the central density ρ_0 varies roughly as $1/r_0$, the central brightness being greater for compact galaxies with a small equivalent radius r_0. It is easy to deduce that the mass of galaxies M is proportional to R^2. For spiral galaxies the universal form of the mass distribution law leads to a surface density $\mu(r) = \mu_0 \exp(-r/r_0)$ and a total mass $M = \mu_0 r_0^2 \int 2\pi x f(x) \, dx$, giving the relation $M \propto R^2$ if the surface density is a universal constant; this is what is observed: the surface brightness at the centre of spiral galaxies is almost always equal to 21.65 B-mag arcsec^{-2}. To what physical phenomenon are these quasiuniversal laws due? It should be noted that similar relations link the masses, radii, and velocity dispersions of molecular clouds in the Galaxy: a hierarchical ensemble of selfgravitating clouds satisfying these relations. These properties could also correspond to galactic systems formed by collisions with subsystems. In all cases these relations are fundamental for theories of galaxy formation.

Other interpretations have been advanced, but up to now none has been entirely convincing and the subject remains very much open.

The formation of the two mass components, one visible, the other invisible, has to be extremely closely linked to end up with flat curves: how do we explain that an extended mass component can take over from the visible component without a single break or perturbation in the velocity field? It is this 'conspiracy' that has given rise to theoretical models that explain the flat curves by a modification of Newton's law of gravitation: at large distances the gravitational attraction of a given mass m does not decrease as quickly as $1/r^2$. The acceleration (and thus the rotation velocity) of gas for a given stellar mass will be greater. These models encounter many difficulties, however.

Another possibility is that the invisible matter is exactly of the same composition as the visible one: it could be in the form of cold and dense molecular gas, in thermal equilibrium with the cosmic background radiation (see Chap. 13). It would be progressively transformed into stars and enter the visible disc.

Exercises

3.1 We want to estimate the error made in the determination of the total mass of a galaxy owing to the hypothesis of a spherical distribution of mass. This error is a maximum when the mass is entirely confined to an infinitely thin disc. We assume for simplification that the rotation curve is flat up to a radius R that delimits the galactic disc ($V_{\rm rot} = V$ = constant for r, R) and that there is empty space beyond R.

(a) Calculate the total mass of the galaxy in the case of a spherical distribution; what is the law for the density $\rho(r)$?

(b) Use the spheroidal model to show that in the case of a flat distribution the total mass is less by a factor of $2/\pi$; what is the surface density $\mu(r)$ then?

3.2 The use of the virial theorem, together with the hypothesis of a universal mass distribution in spiral galaxies and, importantly, of a central surface density μ_0 = constant, allows us to give a physical interpretation of the Tully–Fischer relation $L_H \propto V_m^4$ (or $M \propto V_m^4$, since $M/L_H \approx$ constant).

(a) By hypothesis the density in spiral galaxies can be written

$$\rho(r) = \rho_0 f\,(r/r_0) = \rho_0 f(x).$$

Show that the surface density can therefore be written as

$$\mu(r) = \mu_0 g\,(r/r_0) = \mu_0 g(x)$$

and likewise the rotation curve as

$$V(r) = V_{\rm max} h(x),$$

where $f(x)$, $g(x)$, and $h(x)$ are universal functions.

(b) Express the total kinetic energy T as a function of the parameters ρ_0, r_0, and $V_{\rm max}$ and a numerical factor dependent only on the universal functions.

Likewise express the total potential energy.

(c) Write down the virial theorem. Furthermore integrate the surface density $\mu(r)$ to obtain the total mass M. Show then that

$$M = k\mu_0^{-1} V_{\rm max}^4,$$

where k is a numerical factor.

3.3 *Galactic-Disc Models.* The disc of a spiral galaxy is considered infinitely thin, and a model of the surface density $\mu(r)$ is proposed of the form

$$\mu(r) = M \frac{a}{2\pi} \left(r^2 + a^2\right)^{-3/2},$$

where M is the total mass of the disc and a is a scale typical for the size of disc. The density is written $\rho(r,z) = \mu(r)\delta(z)$.

We want to calculate the potential $\Phi(r,z)$ and the rotation velocity at each point of the disc $v(r)$.

Write down the Poisson equation in cylindrical coordinates and find Φ in the form

$$\Phi(r,z) \approx \left[r^2 + (a+|z|)^2\right]^n.$$

What is the value of the power n for which the equation is satisfied? Hence give the rotation curve of the selfgravitating disc.

References

Aaronson, M., et al. (1982) *Astrophys. J. Suppl.* **50**, 241.
Bosma, A. (1981) *Astron. J.* **86**, 1791.
Burstein, D., and Rubin, V. C. (1986) *Astrophys. J.* **297**, 423.
Faber, S. M., and Gallagher, J. S. (1979) *Annu. Rev. Astron. Astrophys.* **17**, 135.
Persic, M., Salucci, P., and Ashman, K. M. (1993) *Astron. Astrophys.* **279**, 343.
Peterson, C. J., Rubin, V. C., Ford, W. K., and Thonnard, N. (1978) *Astrophys. J.* **219**, 31.
Rogstadt, D. H., Lockhardt, I. A., and Wright, M. C. H. (1974) *Astrophys. J.* **193**, 309.
Rubin, V. C., Ford, W. K., and Thonnard, N. (1980) *Astrophys. J.* **238**, 471.
Rubin, V. C., Ford, W. K., Thonnard, N., and Burstein, D. (1982) *Astrophys. J.* **261**, 439.
Rubin, V. C., Burstein, D., Ford, W. K., and Thonnard, N. (1985) *Astrophys. J.* **289**, 81.
Sanders, R. H., and Tubbs, A. D. (1980) *Astrophys. J.* **235**, 803.
Tinsley, B. (1981) *Mon. Not. Roy. Astron. Soc.* **194**, 63.
Visser, H. C. D. (1980) *Astron. Astrophys.* **88**, 159.

4. Elliptical Galaxies

Elliptical galaxies are the simplest galaxies with the most regular structure of all the Hubble sequence. For a long time theoreticians were misled by this deceptive simplicity and until the end of the 1970s believed that they had succeeded in giving a complete mathematical description of the structure of these galaxies. Ellipticals were then thought to be axially symmetric isothermal ensembles, increasingly flattened the more rapidly they rotate around their axis of symmetry.

Observations did not contradict these models until 1977, when sufficient spectroscopic data (stellar absorption lines) established for certain that elliptical galaxies do not rotate globally, or so little that this cannot cause their flattened shape. These systems, which are apparently very regular, are in fact not isothermal, and perhaps do not even have any axial symmetry, and their velocity dispersion is anisotropic. But then what can their three-dimensional shape be? Do these stellar systems have an origin of their own or are they just the residue of the merger between two (or more) spiral galaxies?

The central bulges of spiral galaxies are spheroidal systems and have always been compared to elliptical galaxies. Do they really have the same dynamics?

4.1 Spectroscopic Observations

4.1.1 General Remarks

These observations rely solely on the measurement of stellar absorption lines (the H and K lines of Ca II, the double D lines of Na I, Mg I lines, and so on), which explains why it is so difficult to obtain rotation curves. The absorption lines are relatively weak, so one can measure only the central profile easily, the position of the lines giving the systemic velocity of the galaxy, and the profile width giving the velocity dispersion at the centre. Away from the centre the luminosity decreases very quickly, and therefore measurements are always made within a radius of 10 kpc. Moreover the width of the observed lines in the spectrum of a galaxy results from the convolution of the stellar velocity dispersion throughout the galaxy and from the intrinsic width of the

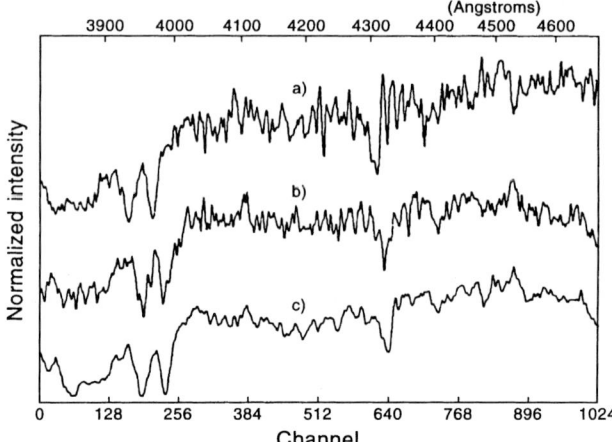

Fig. 4.1. Examples of optical absorption spectra, which trace the stellar component, showing clearly the difficulty of determining the kinematics of elliptical galaxies. (**a**) The spectrum of the reference star HR5709 (spectral type K0 III). (**b**) The spectrum of the elliptical galaxy M87 (at the centre of the Virgo cluster), 70″ W of the centre. (**c**) The spectrum of M87, 4″ E of the centre. The width of the lines is about $200\,\mathrm{km\,s^{-1}}$ for (b) and $300\,\mathrm{km\,s^{-1}}$ for (c). (From Sargent et al. 1978)

line emitted by a single star. But this line is already very broad (corresponding to 50–$100\,\mathrm{km\,s^{-1}}$) and makes the measurement of velocity dispersions extremely uncertain (as shown in Fig. 4.1). All this explains the imprecise results obtained, until the introduction towards the end of the 1970s of automatic algorithms that reduce the spectra by means of Fourier transforms. In Fourier-transform space, convolutions become simple multiplications, and deconvolutions divisions; we can also use an algorithm using cross-correlations, another method giving similar results.

Detailed observations of the kinematics of an elliptical galaxy, determining the speed and the dispersion as a function of the position from the centre to the edge, have been made only for some tens of galaxies. Most of the time, one has only the central profile, which gives V_{sys} and the dispersion σ_0 at the centre. There exists a strong correlation between the luminosity L in the galaxy and σ_0: $L \propto \sigma_0^4$, and this is known as the Faber–Jackson relation, the analogue of the Tully–Fischer relation for the spirals. However, this relation does not have exactly the same slope for all luminosities: for example cD galaxies have a velocity dispersion σ_0 less than the predictions of the $\approx L^{1/4}$ curve. The central velocity dispersion of the bulges of the spiral galaxies also follows the Faber–Jackson relation.

The ratio M/L for which one can measure the velocity dispersion (Chap. 3) does not seem to depend on the luminosity L, even though some tendencies to this behaviour have sometimes been announced, the subject remaining much debated. Stellar-population models indicate that M/L should depend weakly on L ($L^{0.13}$), because luminous galaxies are also richer in metals.

More precisely and quantitatively the global properties of elliptical galaxies (luminosity, velocity dispersion, surface brightness, and effective radius) have been shown to form a two-dimensional family: the power-law relations between three of these fundamental quantities define a fundamental plane (Djorgovski and Davis 1987). The equations of this plane can be used for more precise distance indicators than the Faber–Jackson relation.

4.1.2 Rotation Velocities

Figure 4.2 shows the rotation curves and velocity-dispersion curves for some spheroidal galaxies. Even though the measurements are possible only for a radius less than some tens of arcseconds ($r < 10\,\text{kpc}$), they are sufficient to detect the maximum of the curve, at about $r = 2\,\text{kpc}$. Beyond this radius $r_{\max}$ the slope of the curves decreases slowly.

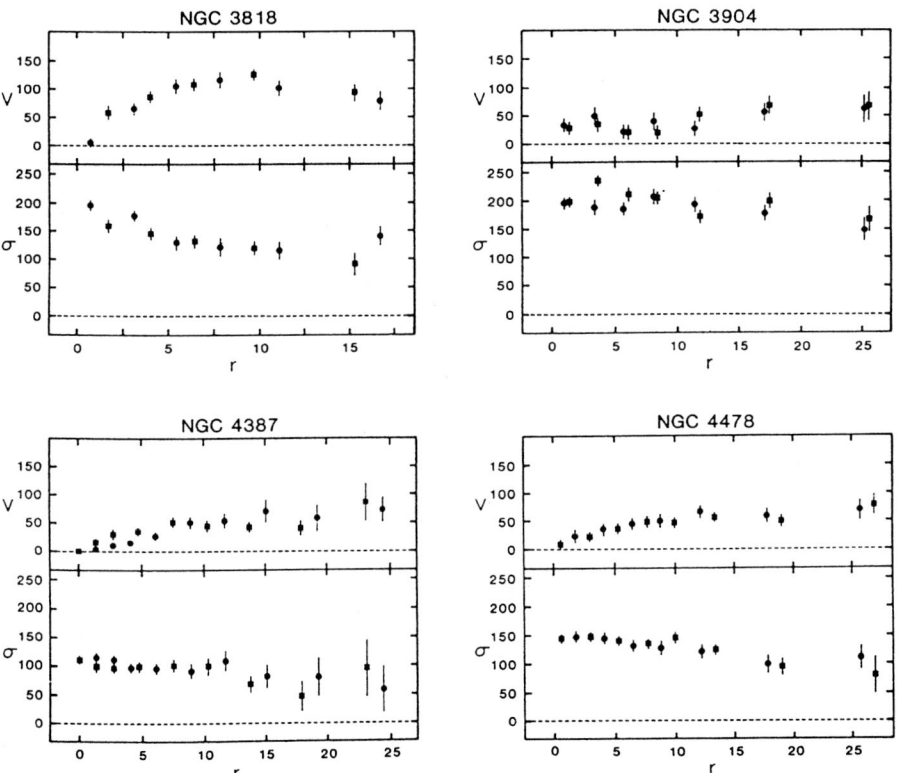

Fig. 4.2. Examples of rotation curves V and velocity-dispersion curves σ for some elliptical galaxies. (From Davies et al. 1983)

For elliptical galaxies knowledge of the true rotation velocity, corrected for projection effects, is impossible. We do not know, as we do for spirals, the intrinsic flatness of the galaxy and we cannot deduce the angle of inclination of the apparent ellipticity ϵ (defined as $\epsilon = 1 - b/a$; $b/a =$ the ratio of the axes at the isophote $25\,\mathrm{mag\,arcsec^{-2}}$). Thus it is also helpful to show the variations of the ratio $V_{\max}/\sigma$ of the maximum rotation velocity observed to the central dispersion velocity σ as a function of ϵ, as we can see in Fig. 4.3. No obvious correlation exists between V/σ and ϵ. This result allows us to eliminate the traditional model (pre-1980) for elliptical galaxies of systems flattened like pancakes (oblate), where the velocity dispersion is isotropic and the rotation is responsible for the flatness (the 'isotropic oblate', or IO, model). The IO curve in Fig. 4.3 corresponds to the increase in the apparent rotation velocity with inclination β; the curve in fact represents systems at $\beta = 90°$ with various intrinsic oblatenesses, but statistically the projected oblate systems should be found approximately together on this curve. Since most of the points are located under the IO curve, it is clear that the rotation is far too slow to be responsible for the flatness.

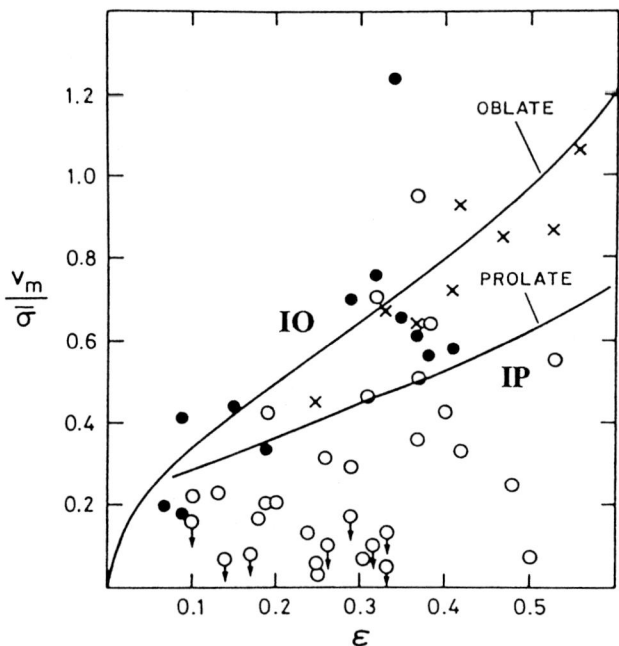

Fig. 4.3. A graph of $V_{\max}/\bar{\sigma}$ ($V_{\max}$ is the maximum rotation velocity and $\bar{\sigma}$ the average dispersion of the velocities within $\frac{1}{2}r_{\mathrm{e}}$), as a function of apparent ellipticity ϵ. The brightest elliptical galaxies are represented by empty circles, those less bright by full circles. The crosses correspond to the central bulges of spirals. The solid curve is the theoretical relation predicted for galaxies flattened by rotation with isotropic velocity dispersion. (From Davies et al. 1983)

Other models have also been proposed: in particular, elongated ('prolate') systems like cigars whose global structure tumbles around one of the minor axes, a little like a bar in a barred spiral galaxy. This model is supported by numerical simulations of the collapse of a spherical system of stars with a certain angular momentum: the result clearly shows a system in the form of a rotating bar, the stars moving inside the bar with systematic streaming motions much greater than that of the bar. Moreover spectroscopic observations reveal rotational motions along the apparent minor axis of elliptical galaxies, suggesting the existence of these systematic proper motions; however, there are relatively few observations of rotation on the minor axis that are unambiguous. If the elliptical galaxies were 'rotating cigars', would a correlation between V/σ and ϵ exist? When the line of sight is parallel to the rotation axis, V/σ is small and the system appears to be very flattened (high ϵ). On the other hand when the line of sight is parallel to the major axis, the system is almost round ($\epsilon = 0$). If we were to consider all the possible directions of the line of sight, we should find the points distributed in the lower-left corner of the V/σ–ϵ diagram (like a triangle), and in particular we should see an accumulation along the axis V/σ, where in fact there is hardly a single point. With the added hypothesis of an isotropic velocity dispersion, the virial theorem allows us as before to identify these systems in the V/σ–ϵ diagram more precisely. This is the curve marked IP ('isotropic prolate') in Fig. 4.3: half the galaxies should be located above this IP curve, provided that they are rotating bars with an isotropic velocity dispersion.

In Fig. 4.3 the central bulges of spirals are represented by crosses. To within the error bars the bulges seem to correspond exactly with the IO model of systems flattened by rotation; the velocity dispersion seems isotropic, in agreement with the dispersion curves. These systems rotate like a rigid superposition of coaxial spheroids, as the model predicts. But there are exceptions where the rotation is cylindrical, as in the discs of spirals; this is particularly the case for bulges with a box shape or peanut shape; these structures will be described in Chap. 6.

4.2 The Three-Dimensional Shape

The last section clearly shows at what point our understanding of the three-dimensional shape of elliptical galaxies is incomplete. Several models have been proposed, and we are now going to review all the theoretical and observational means of approaching the problem. One point is clear: since rotation is not enough to explain the flattening of these galaxies, the velocity dispersion is necessarily anisotropic; stability with respect to gravitational collapse is ensured in all directions by the dispersion, this being stronger in the direction of the major axes. The origin of this anisotropy could be primordial; that is, the anisotropies in the velocities of the protogalactic medium could have survived the collapse or could be vestiges of the formation of these systems

96   4. Elliptical Galaxies

by accretion, according to the collision parameters, for example. Moreover if the dispersion is anisotropic, there is no reason to think that two of the principal components of the velocity ellipsoid should be equal, or in other words that the axial symmetry is respected. The elliptical galaxies could be triaxial, especially those that do not rotate at all.

4.2.1 Ellipticity Profiles

The hypothesis of triaxiality, despite its complexity, is very often evoked because it allows us to explain the rotation of the major axes of the isophotes of elliptical galaxies. How does this phenomenon present itself? The brightness contours of an observed galaxy can be modelled by ellipses, as shown in Fig. 4.4. The ratio of the axes (b/a) gives the apparent ellipticity of the galaxy ($\epsilon = 1 - b/a$) for the corresponding mean distance (the equivalent radius $r = (ab)^{1/2}$). Thus one can represent the variation of the ellipticity

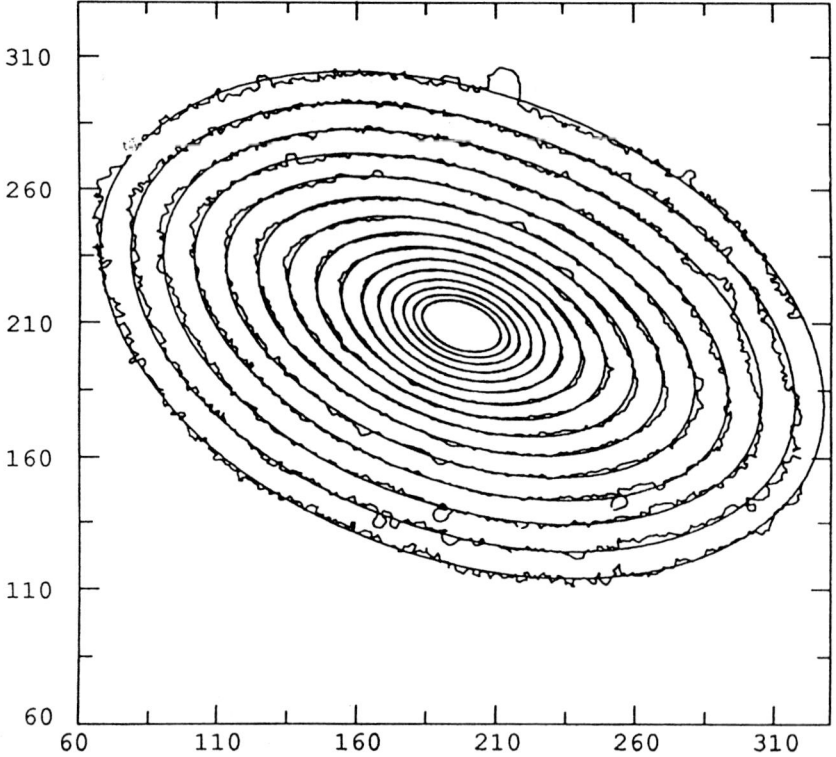

Fig. 4.4. The elliptical galaxy NGC 4697 illustrates the power of modelling the luminosity contours by ellipses. Here the major axis of the successive ellipses does not rotate, but one can discern a systematic brightness excess along the major axis, suggesting the presence of a luminous disc. (From Carter 1987)

from the centre to the edge: Figure 4.5a shows a few ellipticity profiles thus obtained. Note that the ellipticity varies by only a small amount from the centre to the edge. The faint variations are moreover correlated with the richness of the environment: for the majority of isolated galaxies ϵ and the radius decrease monotonically. When there are other galaxies nearby, elliptical galaxies in general have a constant ellipticity, which even becomes a slowly growing function in the centre of rich clusters. With this variation of ϵ with the radius, in the majority of cases the major axes also rotate (see Fig. 4.5b). One must remember, however, that many observational effects can simulate such a rotation: instrumental problems such as the diffusion of light in a CCD camera, or the presence of neighbouring stars, or even the miscorrection of the halo effect due to a neighbouring galaxy.

The triaxial structure of elliptical galaxies could be the origin of this rotation of the isophotes (see the simple example in Chap. 1, Fig. 1.5). The problem is that there are other interpretations of the phenomenon that are just as plausible; the rotation of the isophotes does not allow us to conclude that all elliptical galaxies are triaxial. For instance the rotation of the isophotes is particularly severe for elliptical galaxies under the obvious influence of tides and increases closer to the centre of the clusters. These tidal interactions are strong enough to perturb the brightness profile of the corresponding ellipticals, which is demonstrated in Fig. 4.6; the variation in the ellipticity would then be intrinsic to the deformed galaxy and would not be a pure effect of projection. On the other hand it has recently been shown from multicolour photometric measurements that approximately half of all elliptical galaxies have a dust lane that could perturb the profiles and in particular artificially induce rotation of the isophotes.

4.2.2 Other Tests of the Three-Dimensional Shape

In spite of all this, even if the observed rotation of the isophotes is entirely attributed to the triaxiality of the systems, in most cases this triaxiality does not need to be very great: elliptical galaxies could be considered practically biaxial (with a few exceptions), one of the three axes being significantly different from the other two, thus preserving the axial symmetry. Statistical studies of the apparent-flatness distribution, however, do not allow us to settle the matter and know if most of the galaxies are elongated ('prolate') or flattened ('oblate'). If the chosen systems do not suffer from the problems of absorption, the observed surface brightness should be more important when the line of sight is parallel to a major axis rather than a minor axis; thus there would be a correlation between the apparent ellipticity and the brightness of the surface if most of the galaxies are oblate and an anticorrelation if they are prolate. The same test can be used for the velocity dispersion, this being much greater if the galaxy is seen with $\epsilon = 0$ in the case of a prolate galaxy. The negative result of these tests seems to show that there exists among elliptical galaxies a mixed population of prolate and oblate galaxies.

98   4. Elliptical Galaxies

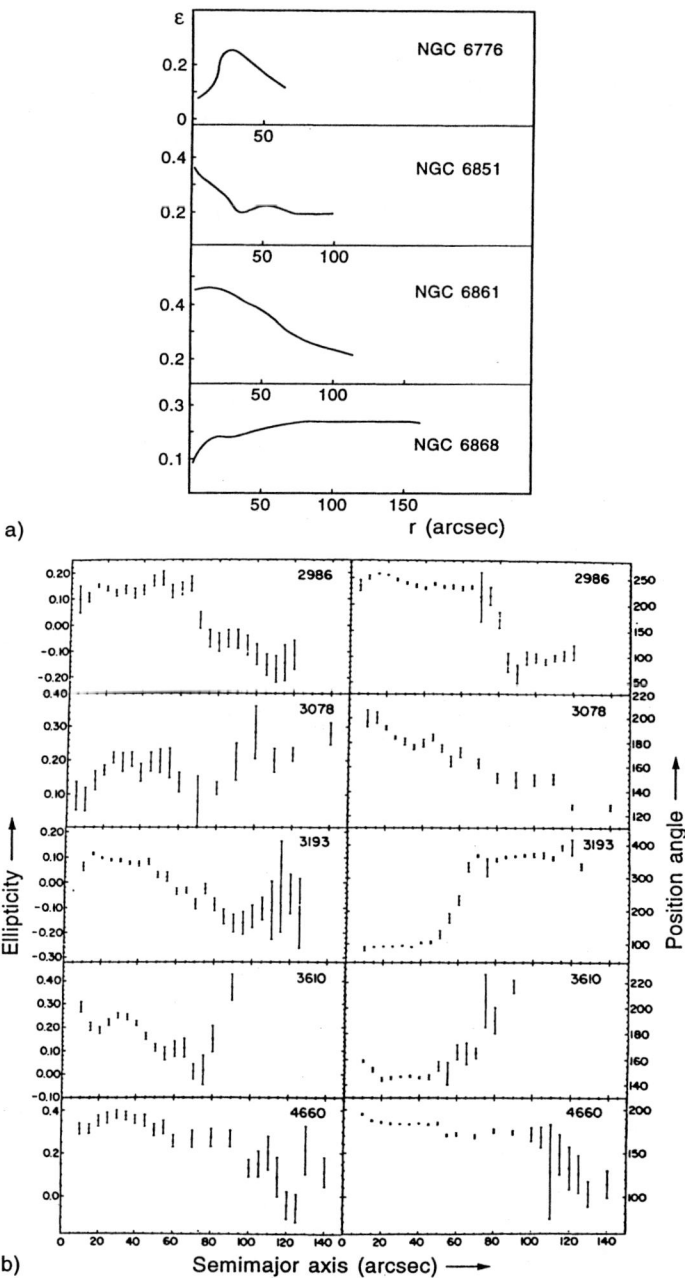

Fig. 4.5. (a) Ellipticity profiles for some elliptical galaxies. (From di Tullio 1979.) (b) Ellipticity profiles and position angles of the major axes of five elliptical galaxies for which the rotation of the isophotes is particularly fast; two of them show a change in sign of the ellipticity near a sudden variation of the direction of the major axis. (From Leach 1981)

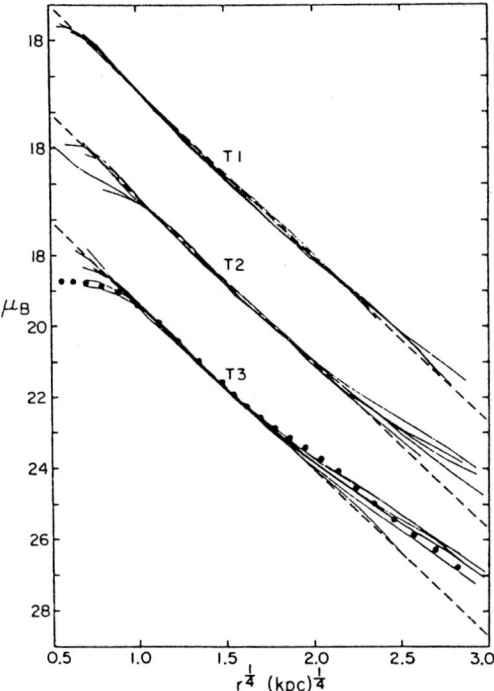

Fig. 4.6. Typical deviations in profiles of elliptical galaxies according to de Vaucouleurs's $r^{1/4}$ law: galaxies can be classified into three groups according to their environment. T1 galaxies are isolated, whereas T3s have close neighbours. (From Kormendy 1977)

A recent test of the three-dimensional form of elliptical galaxies comes from the discovery in 1984 of the frequent presence of *shells* around these galaxies. These shells are very thin and faint stellar structures forming circular arcs around the galaxies: up to 26 shells can fit together, the last being found at 100–200 kpc from the centre (as is the case for NGC 3923, shown in Fig. 4.7). Special spatial filtering techniques are required to detect them: this filtering can be done by analogue methods (the unsharp-masking technique) or digital methods (CCD imaging). With the unsharp-masking technique, for example, a fuzzy negative image is obtained from the photographic plate by defocusing the optical system. Then this 'unsharp mask' is added to the original image, in this way subtracting the low spatial frequencies and bringing out all the fine and contrasted structures. Shells are observed with this technique in at least 40% of elliptical field galaxies, this number being a lower limit in view of the difficulty of detection.

These shells can be very nicely explained by a stellar-oscillation phenomenon during the accretion by the elliptical galaxy of a small companion in a quasiradial orbit (the theory is developed in Chap. 7). Shells formed by

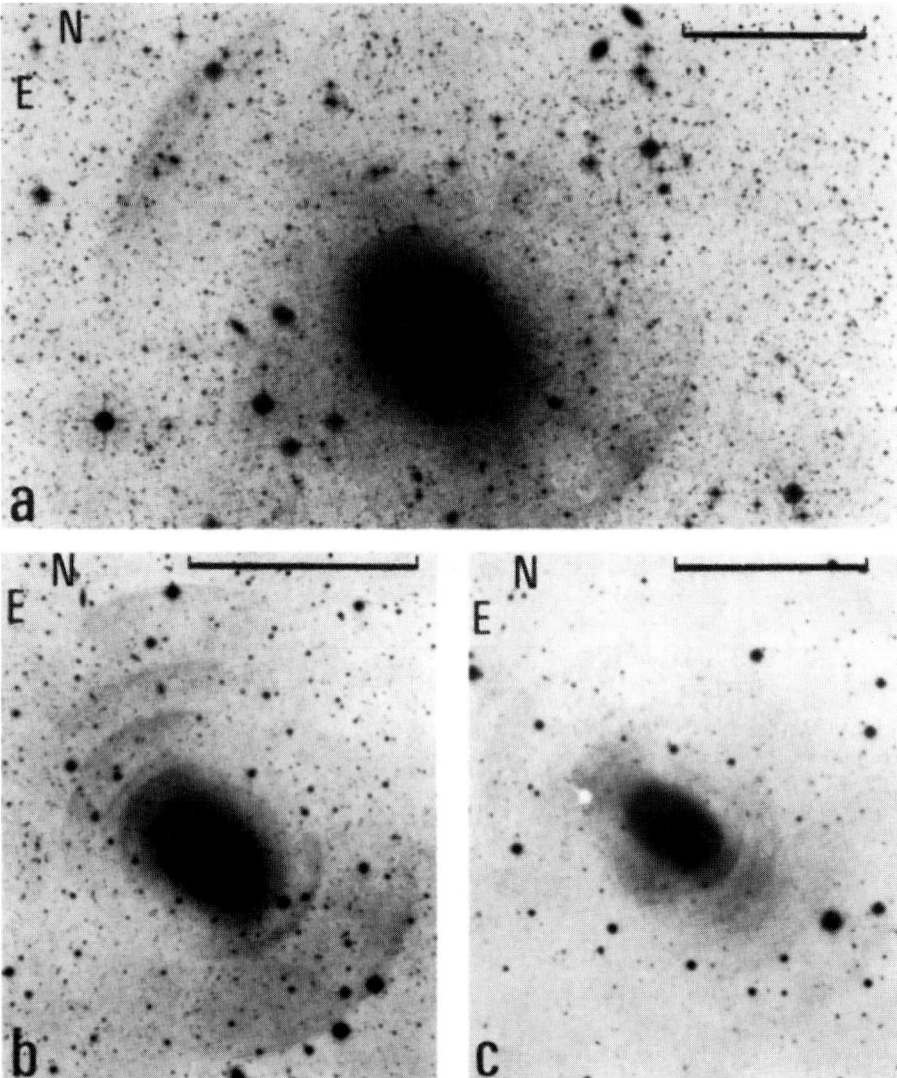

Fig. 4.7a–c. Three pictures taken by Malin and Carter (1983) of the elliptical galaxy NGC 3923 with shells. The scales and the exposure times are different: the bar at the top represents 10′, at the lower left 5′, and at the lower right 2′. About twenty shells can be counted using the 'unsharp-masking' technique

the stars of the companion could thus be used as test particles enabling the gravitational potential of the elliptical galaxy and in particular its three-dimensional shape to be probed. Numerical simulations indeed show that flattened and elongated galaxies can have shell systems with different geometries: for an elongated system the shells are like brackets aligned along the apparent major axis of the galaxy; no alignment is visible for a flattened galaxy. This method seems to indicate that the proportion of flattened galaxies is twice that of elongated galaxies.

Such a distribution is also suggested by the observation of dust lanes. The dust is probably due to gas carried along by a spiral galaxy during a collision. The hypothesis in this method is that the absorbing dust forms a disc of test particles, which settle in equilibrium in the plane of symmetry of the elliptical galaxy. If the dust lane is parallel to the apparent minor axis (major axis), the galaxy is prolate (oblate).

4.3 Models of Elliptical Galaxies

The dynamics of elliptical galaxies is defined essentially by the equations of stellar dynamics. We shall now detail the basic principles that will also be used in subsequent chapters. The application of these principles to spheroidal systems allows models for elliptical galaxies to be constructed.

4.3.1 The Vlasov Equation and Jeans's Theorem

The stars of a galaxy constitute a *collisionless* dynamic ensemble. By 'collision' we obviously do not mean a real collision between two stars, which is almost impossible with the mean separation between stars being of the order of 1–2 pc, approximately 10^8 times the stellar radius. A collision is defined here in a gravitational sense as a significant deviation of the stellar trajectories during an encounter. The time of 'collision' between stars, also known as the *relaxation* time, is defined as the time after which the gravitational encounters between stars (taken two at a time) begin to perturb the trajectories; or, if we want to quantify this perturbation, the time when the accumulation of variations in energy dE in two-body encounters becomes equal to the energy E of a single star. The relaxation time can thus be evaluated as

$$T_{\text{relax}} = V^3/8\pi n G^2 m^2 \log(R/b),$$

where V is the mean relative speed between the stars, n the number of stars per unit volume, m the mean mass of a star, R the total radius of the galaxy, and b the minimum impact parameter during the encounter between two stars (see Exercise 1). With regard to the time t_c taken to cross through the galaxy by a star ($t_c \approx R/V$), the relaxation time varies approximately as $0.2(N/\log N)t_c$, if N is the total number of stars in the system. Well t_c for a galaxy is typically of the order of one hundredth of the age of the universe, that is, 10^8 years, and the relaxation time is about 10^{17} years, well above the age of the universe. Collisions therefore have no importance for the stars in a galaxy.

The stars form a collisionless medium because, paradoxically, their number N is large. During its motion a star is subjected to the mean potential created by distant stars (or 'smoothed') onto which are superposed the fluctuations due to neighbouring stars. The importance of these fluctuations is obviously the less the larger the number N of stars the potential is averaged over. This absence of two-body interactions is fundamental for the dynamics of galaxies and will have to be taken into account, especially in simulations. However, it should be noted that there are some stellar systems for which the relaxation time is comparable to the crossing time: globular clusters, for example, where $N \approx 10^5$; indeed encounters between stars play an important role at the centre of these clusters. In the same way, in a rich cluster of galaxies where the total number of galaxies is $N \approx 10^3$ and $t_c \approx 10^8$ years, encounters between galaxies can play a fundamental role.

The distribution of stars within a galaxy at each instant t can be represented by a continuous, always positive function $f(\mathbf{r}, \mathbf{v}, t)$ in phase space $(\mathbf{r}, \mathbf{v})$ known as the *distribution function*. If the stars move in a mean potential $U(r)$, conservation of mass implies that f satisfies the continuity equation

$$\frac{df}{dt} = \frac{\partial f}{\partial t} + \mathbf{v} \cdot \frac{\partial f}{\partial \mathbf{r}} - \frac{dU}{d\mathbf{r}} \cdot \frac{\partial f}{\partial \mathbf{v}} = 0,$$

where $-dU/d\mathbf{r}$ is the gravitational force exerted by a unit mass at the point $\mathbf{r}$. This is the collisionless Boltzmann equation, also known as the *Vlasov equation*. The potential U of the density distribution $\rho(\mathbf{r})$ of the stars is deduced from the Poisson equation

$$\Delta U(\mathbf{r}) = 4\pi G \rho(\mathbf{r}).$$

A distribution function which satisfies both the Vlasov equation and the Poisson equation is a selfconsistent solution of the problem. Most of the time, we are interested in only the stationary distributions of the system for which $\partial f/\partial t = 0$. The time-independent Vlasov equation shows that f is constant along the phase-space curve followed by the stars: f is an integral of the equations of motion. Jeans's theorem shows that if we can find a complete system of integrals of motion $I_n(\mathbf{r}, \mathbf{v})$, the function f can be written as a function of these integrals $f(I_n)$. But how many integrals can be found for a given potential? There is always at least one, and that is the energy E (the stationary case). For example if the potential is spherical, the angular momentum $\mathbf{J}$ is also a vector integral (that is, three integrals). The maximum number of integrals is five because phase space is six-dimensional: the intersection of the five-dimensional hypersurfaces associated with each of the five integrals must contain the one-dimensional orbit of a star. A fifth integral exists only for specific potentials, such as $U(r) \approx 1/r$ or $1/r^2$.

4.3.2 Models of Spherical Galaxies

The brightness, mass-density, and velocity-dispersion profiles as functions of the radius in an elliptical galaxy are modelled well by systems having a spherical symmetry. The ellipticity of real galaxies perturbs the profiles only weakly. Most of the time it is thus unnecessary to elaborate more sophisticated models.

If the potential has a spherical symmetry, we have seen that there are at least four integrals: E, $\boldsymbol{J}$. The distribution function is therefore $f(E, \boldsymbol{J})$; if the galaxy is symmetric for all of its properties, f does not have to depend on the directions of $\boldsymbol{J}$ but only on its modulus, and we can write $f = f(E, |\boldsymbol{J}|)$. The simplest and most often used models are those whose distribution function depends only on the energy $f(E)$.

$f(E)$ Systems. The energy per unit mass is written as $E = (1/2)V^2 + U$, and the global equation to be solved is

$$\Delta U = 4\pi G \rho = 4\pi G \int f \, d\boldsymbol{V},$$

or in spherical coordinates

$$\frac{1}{r^2} \frac{d}{dr}\left(r^2 \frac{dU}{dr}\right) = 4\pi G \int f\left(\frac{1}{2}V^2 + U\right) d\boldsymbol{V}.$$

If we call the velocities in the three directions V_r, V_θ, and V_ϕ, it can be easily proved that the velocity dispersions are the same:

$$\langle V_r^2 \rangle = \frac{1}{\rho} \int V_r^2 f\left(\frac{1}{2}\sum_i V_i^2 + U\right) dV_r \, dV_\theta \, dV_\phi = \langle V_\theta^2 \rangle = \langle V_\phi^2 \rangle.$$

Hence the velocity dispersion is isotropic. Most of the spherical models used derive from the isothermal sphere:

$$f(E) = \left(2\pi\sigma^2\right)^{-3/2} \rho_0 e^{-E/\sigma^2},$$

where ρ_0 is the central mass density and σ is the velocity dispersion, assumed constant at every point of the galaxy. This distribution looks like the repartition at equilibrium of a gas of isothermal particles at a temperature T: the Maxwell–Boltzmann function $f(E) \approx \exp(-\beta E)$, where $\beta = 1/kT$, the equivalent temperature of the gas of stars being the velocity dispersion. Nevertheless there is a fundamental difference: the energy E in that case represents the energy per unit mass. Indeed we remarked above that collisions between stars do not play any role within galaxies; and collisions alone can make the system relax towards equipartition of the energy and a Maxwellian distribution such that $mV^2/2 \approx kT$. In a galaxy such relaxation, which should slow down the massive stars and lead to them 'settling' towards the centre, will not yet have taken place; on the other hand the period of formation of a galaxy is probably accompanied by a violent relaxation (described by Lynden–Bell in 1967) whose time scale is the crossing time t_c. This corresponds to a complete mixing of particles in phase space. This violent relaxation leads to an equipartition of energy per unit mass and not per particle.

4. Elliptical Galaxies

By integrating the isothermal distribution function over the velocities, we obtain the density at every point:

$$\rho(r) = \rho_0 e^{-U(r)/\sigma^2}.$$

On the other hand the Poisson equation, with this density law, leads, for $r \gg r_c$, to

$$U(r) \approx 2\sigma^2 \log(r/r_c),$$

which defines the core radius $r_c = 3\sigma(4\pi G\rho_0)^{-1/2}$, where the density is divided by two: $\rho(r_c) \approx \rho_0/2$. The asymptotic behaviour of $\rho(r)$ much farther away than the core radius is, if we combine the last two equations, $\rho(r) \propto r^{-2}$.

The isothermal function $f_I(E) \approx \exp(-\beta E)$ extends spatially to infinity and so cannot still accurately represent an elliptical galaxy. Galaxies are not isolated, and those stars that are least bound to the galaxy ($E \approx 0$) will be swept away by tidal interactions with companion galaxies, mainly in the outer regions. An acceptable function is thus that of the isothermal cutoff at the energy E_0:

$$f_K(E) = \begin{cases} 0 & \text{for } E \geq E_0, \\ (2\pi\sigma^2)^{-3/2} \rho_0 \left(e^{(E_0-E)/\sigma^2} - 1\right) & \text{for } E < E_0. \end{cases}$$

This is King's model (1966). The density of King's model can be obtained after integration over the velocities:

$$\frac{\rho(r)}{\rho_0} = e^y \mathrm{erf}\left(y^{1/2}\right) - \left(\frac{4y}{\pi}\right)^{1/2} \left(1 + \frac{2y}{3}\right),$$

where erf is the error function,

$$\mathrm{erf}(x) = \frac{2}{\sqrt{\pi}} \int_0^x e^{-u^2} du,$$

and $y = (E_0 - U)/\sigma^2$. The Poisson equation, linking $U(r)$ and $\rho(r)$, solves the problem by eliminating U. Assuming that the $E_0 - U(r)$ function cancels out at a certain distance from the galaxy, it is easy to see that there exists a value r_t of the radius (the 'tidal radius') beyond which the density is equal to zero; this allows us to define the radius of the galaxy $R = r_t$ and its mass $M = M(r_t)$. The threshold energy E_0 also has a physical meaning: $E_0 = -GM/R$.

The scale factors of King's model are the density ρ_0 and a core radius $r_c = (4\pi G\rho_0/9\sigma^2)^{-1/2}$. Once these factors are fixed there is only one free parameter left, and this measures the importance of the tidal forces experienced by the system, or the importance of the cutoff. This parameter, which could equally be the threshold energy E_0, is in a more physical way the concentration parameter c of King's model,

$$c = \log \frac{r_t}{r_c}.$$

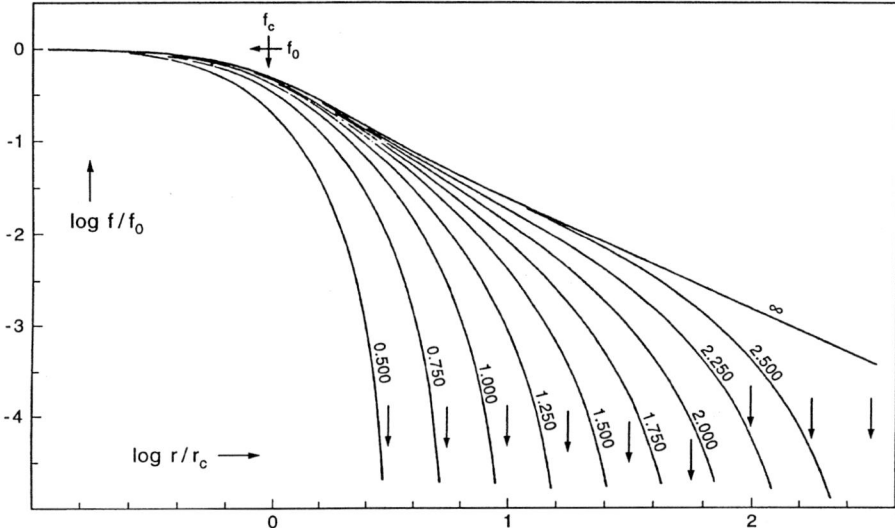

Fig. 4.8. Surface-density profiles for a variety of King's models, normalized to f_0, the central surface density, and r_c, the core radius. The curves are marked by the concentration $c = \log(r_t/r_c)$, where r_t is the tidal radius of the galaxy, indicated by arrows. (From King 1966)

Figure 4.8 shows several density profiles corresponding to King's models with different concentrations: the profile tends towards that of an isothermal sphere when $c \to \infty$ (no cutoff). For elliptical galaxies the concentration parameter $c \approx 2.2$ models the brightness profiles perfectly, as shown in Fig. 4.9. For globular clusters, for example, $c \approx 1$, which shows the greater importance of tidal interactions for these stellar systems.

It should be noted here that in cut-off isothermal systems if the galactic core always behaves isothermally, this is no longer the case for the envelope. For systems with greater energy cutoff than King's models (Wilson's models, for example) the envelope is much more extended. Do the tidal interactions tend to produce more extended envelopes? This is indeed what is suggested by the observations (class T3 in Fig. 4.6).

f (E, J) Systems. These systems, unlike the previous ones, allow us to describe anisotropic velocity dispersions, which is why they are of interest. They are described by the functions $f(E)$ discussed above multiplied by a factor $\exp(-J^2/2r_a^2\sigma^2)$, where r_a is a characteristic radius of anisotropy. The anisotropy parameter $\beta = 1 - \sigma_\theta^2/\sigma_r^2 = [1 + (r_a/r)^2]^{-1}$ in fact never approaches 1 in these models. Eddington's models choose the isothermal function $f(E) = f_I(E)$, and Michie's models choose King's function $f_K(E)$. Some examples are given in Fig. 4.10.

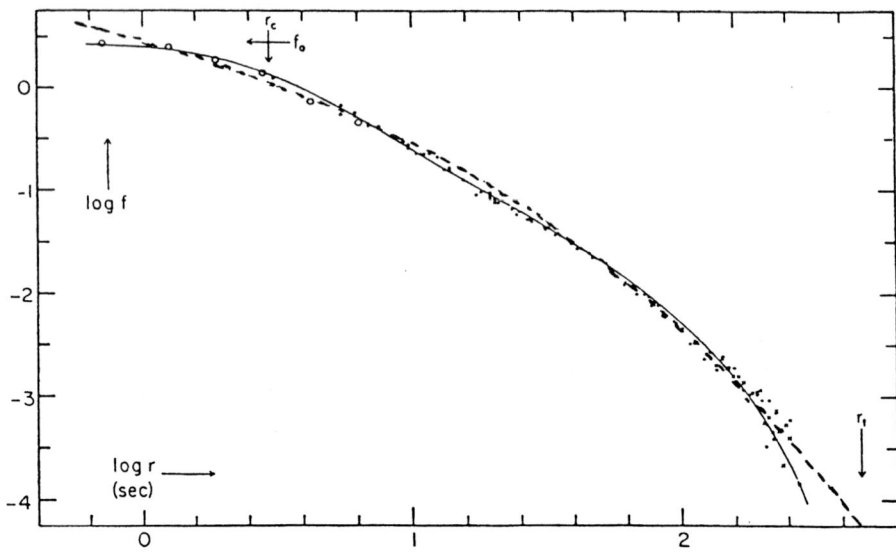

Fig. 4.9. Comparison of the observed brightness profile of the elliptical galaxy NGC 3379 (*dots*) to one of King's models (*solid line*) with a concentration $c = 2.20$. The $r^{1/4}$ profile is indicated by the dotted line. The core radius (r_c) and the tidal radius (r_t) are indicated by arrows. (From King 1966)

4.3.3 Models of Axisymmetric Galaxies

Although elliptical galaxies do not owe their shape to rotation, their rotation velocities are not all zero, and moreover they are probably flattened or elongated. It is also necessary to introduce into these models a privileged axis. Such models have been developed using distribution functions of the form $f(E, J_z)$. However, these models have the feature that they require the velocity dispersions along the radial direction and parallel to the axis of the system to be equal: $\langle V_r^2 \rangle = \langle V_z^2 \rangle$. This is undoubtedly far from reality: in the solar neighbourhood, for example, stars belonging to the galactic halo satisfy the relation
$$\langle V_r^2 \rangle = 4 \langle V_z^2 \rangle.$$
An example of a function representing a rotating galaxy is
$$f(E, J_z) = \begin{cases} f_I(E)\, e^{\Omega J_z/\sigma^2} & \text{for } E < E_0, \\ 0 & \text{for } E > E_0. \end{cases}$$

Towards the centre of these galaxies the rotation is cylindrical and increases up to values Ωr; towards the outer regions the rotation is constant within the spheroids.

More realistic models can be constructed directly from stellar populations having a given type of orbit, or by another completely different technique, which we will speak about again: simulations of the N-body problem.

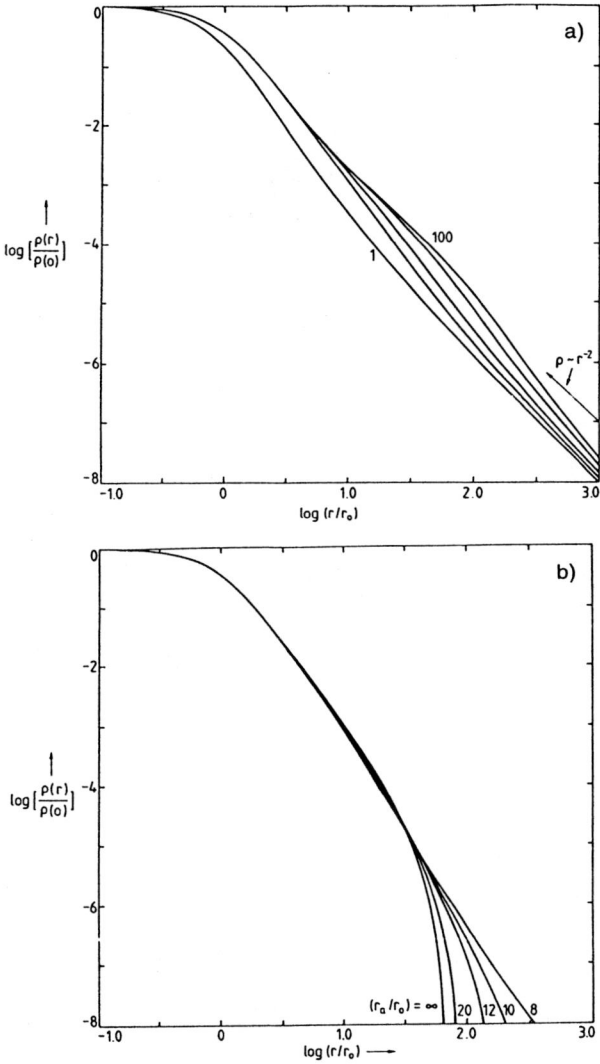

Fig. 4.10. (a) Density profiles for Eddington's models for several values of r_a/r_0. At large distances the density continues to decrease as r^{-2}. (b) Density profiles for Michie's models, for several values of the anisotropic radius r_a. (From Binney 1982)

4.4 The Fundamental Plane

Unlike spiral galaxies, for which the rotational velocity is well determined by one parameter, the luminosity (cf. the Tully–Fisher relation, TF), elliptical galaxies form at least a two-parameter class. The TF-equivalent Faber–Jackson relation, relating luminosity and velocity dispersion ($L \propto \sigma^4$, Faber and Jackson 1976) has residuals that are clearly correlated to galaxy prop-

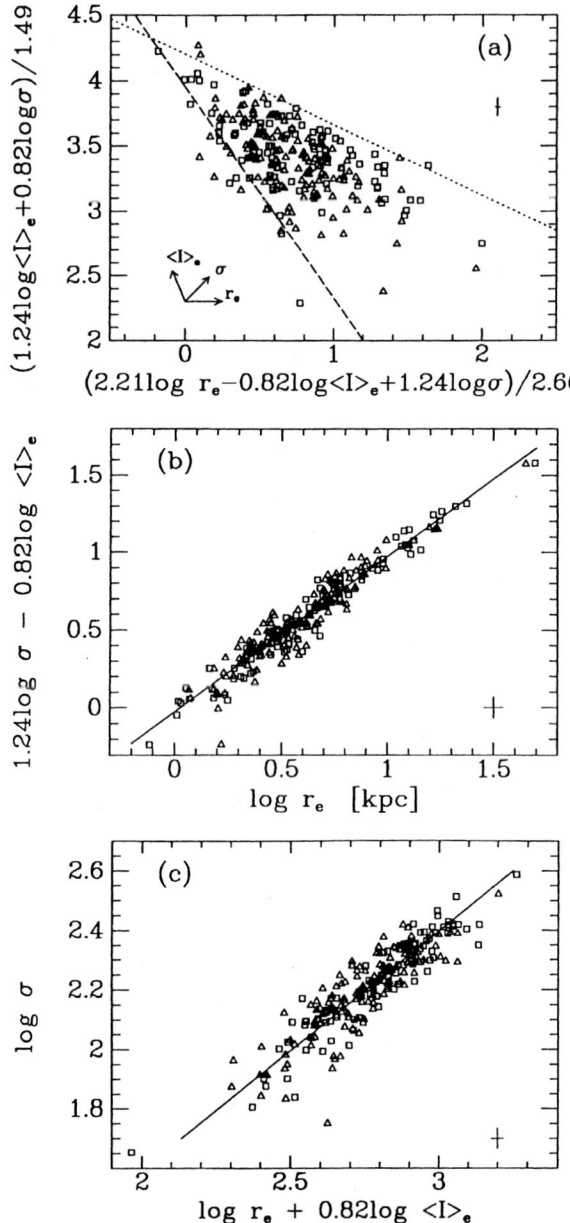

Fig. 4.11. (a) The fundamental plane seen face-on: the relation between mean surface brightness I_e, central velocity dispersion σ, and effective radius r_e (the directions of variation of these three parameters are indicated by the arrows at the bottom left). The dashed line corresponds to the limiting magnitude of the sample (the dotted line is an intrinsic effect); (b) and (c) The fundamental plane seen end-on; elliptical galaxies are represented by boxes, and lenticulars by triangles. (From Jorgensen et al. 1997)

erties. In fact, there exists a linear relation between the logarithm of three main properties of ellipticals that form what is called the fundamental plane (Djorgovski and Davis 1987; Dressler et al. 1987). The three quantities can be the effective radius r_e, the central velocity dispersion σ and the mean surface brightness, or some other related quantities (such as D_n the diameter enclosing an integrated surface brightness, as in Dressler et al. 1987). The relation has very low scatter, resulting in about 17% uncertainty in the radius, and can then serve as a good distance indicator (see Fig. 4.11).

What is the physical origin of the fundamental plane (FP)? The virial theorem also corresponds to such a relation, but one involving the total mass instead of the total light. Assuming the virial theorem is satisfied (since elliptical galaxies are in dynamical equilibrium), the FP relation implies that the mass-to-light ratio M/L depends on the three variables of the FP; the exact orientation of the FP implies that M/L depends essentially on the mass of the galaxy ($M/L \propto L^{0.2}$, if the dynamical structures of the galaxies are the same, for instance; but if the latter vary, a constant M/L can also be fitted).

The main interest of the FP has been as a distance indicator. This assumes that the FP is universal (same slope and same zero point), but this has been disputed; in particular, the FP might be different in clusters (Guzman et al. 1992). The scatter of the FP could be reduced when the metallicity index M_{g_2} is taken into account, to yield more exact distances (Jorgensen et al. 1995). A second interest of the FP is that it provides statistical information on the elliptical galaxies themselves and, for instance, could trace their evolution with redshift. In particular, the M/L ratios of galaxies in clusters at $z \sim 0.4$ were found to be lower than for nearby galaxies (Kelson et al. 1997).

Exercises

4.1 *The Relaxation Time of a Stellar System.* Given a set of stars with the same mass m we would like to evaluate the influence of the gravitational encounters of two of their members.

(**a**) Calculate the energy lost by a star with an incident velocity $\boldsymbol{v}_i$ during a collision with a stationary star whose impact parameter is b. To do this write the differential equation of motion of the effective particle in the centre-of-mass reference frame (using the polar coordinates r and θ). Deduce from this an integral equation that enables the deflection angle α to be obtained (see figure).

(**b**) Calculate α. By returning to the fixed reference frame of the ensemble of stars calculate the kinetic energy gained by the target star, which was initially at rest. Show that the energy δE lost by the first star can be written as

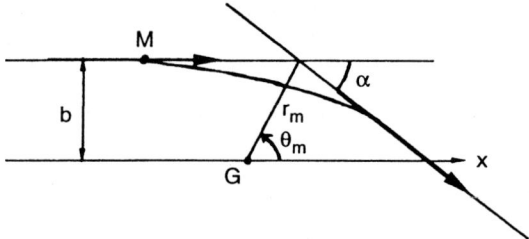

Fig. 4.12. for Exercise 1

$$\delta E = -\frac{v_i^2}{2}\left(1 + \frac{b^2 v^4}{4G^2 m^2}\right)^{-1}$$

(v is the relative velocity between the stars).

This is the formula obtained by Chandrasekhar in 1960.

(c) Calculate the number of encounters for impact parameters lying between b and $b + db$ for a density of stars n. Integrating over the impact parameters lying between $L = b_{\min}$ and $R = b_{\max}$ (the radius of the stellar system), give the mean energy lost by a star for each gravitational encounter (assume that $b^2 v^4/G^2 m^2 \gg 1$ and justify this assumption).

If T_E is the relaxation time after which $\delta E \approx E$, show that this relaxation time is

$$T_E = v^3/8\pi n G^2 m^2 \log(R/L).$$

Calculate T_E numerically for a galaxy (radius $= 30\,\mathrm{kpc}$).

4.2 Solve the Poisson equation exactly for a spherical and isothermal distribution of stars $f(E) \propto \exp(-\beta E)$ and calculate the distribution for the potential $U(r)$. Determine the density $\rho(r)$ at every point.

References

Binney, J. (1982) in *Morphology and Dynamics of Galaxies* (Saas-Fee Advanced Course 12, Geneva Observatory, Geneva), p. 1.
Carter, D. (1987) *Astrophys. J.* **312**, 514.
Davies, R. L., Efstathiou, G., Falls, S. M., Illingworth, G., and Schechter, P. L. (1983) *Astrophys. J.* **266**, 41.
Di Tullio, G. A. (1979) *Astron. Astrophys. Suppl.* **37**, 591.
Djorgovski, G., Davis, M. (1987) *Astrophys. J.* **313**, 59.
Dressler, A. et al., (1987), *Astrophys. J.* **313**, 42.
Faber, S. M., Jackson, R.E., (1976), *Astrophys. J.* **204**, 668.
Guzman, R., Lucey, J. R., Carter, D., Terlevich, R. J. (1992), *Mon. Not. Roy. Ast. Soc.* **257**, 187.
Jorgensen, I., Franx, M., Kjaergaard, P. (1995), *Mon. Not. Roy. Ast. Soc.* **276**, 134.

Kelson, D., et al. (1997) *Astrophys. J.* **478**, L13.
King, I. R. (1966) *Astron. J.* **71**, 64.
Kormendy, J. (1977) *Astrophys. J.* **218**, 33.
Leach, R. (1981) *Astrophys. J.* **248**, 485.
Lynden-Bell, D. (1967) *Mon. Not. Roy. Astron. Soc.* **136**, 101.
Malin, D. F., and Carter, D. (1983) *Astrophys. J.* **274**, 534.
Sargent, W. L. W., Young, P. J., Boksenberg, A., Shortridge, K., Lynds, C. R., and Hartwick, F. D. A. (1978) *Astrophys. J.* **21**, 731.

5. The Spiral Structure of Galaxies

About two-thirds of all galaxies are spiral galaxies, and a large number of them, more than two-thirds, have a regular spiral structure with two arms that can be followed continuously from the centre of the galaxy (the central bulge) to the extremities of the disc. This structure has for a long time posed a serious theoretical problem concerning its origin and persistence in galaxies. The density-wave theory and the amplification mechanism of these waves provide a beautiful solution to the problem in the majority of cases. Before going into the details of this theory (in Sect. 5.2), we must first tackle the problem of the gravitational stability of a galactic disc and define the main characteristics of the orbits of stars in a rotating disc (the theory of epicycles).

5.1 Stellar Dynamics. Stability. Orbits

In this section we shall be concerned with the stability conditions of a galaxy with a disc. First we consider just the stars: the gas represents only a small fraction of the mass ($< 10\%$) and merely plays the role of witness to events. Stars are only subjected to mutual gravitational forces, but there are so many of them in a galaxy that the effects of short-distance interactions are negligible and the motion of each star can be followed as though in a smoothed mean gravitational potential. The dominant forces are in fact all the long-distance attractions, and it is their collective effect that makes the study of selfgravitating stellar systems so difficult.

The paradox we have just alluded to, that is, that forces due to nearby stars are unimportant even though gravitational attraction decreases with the square of the distance, has already been tackled by means of the relaxation-time calculation for two bodies (Sect. 4.3.1). For a galaxy with a disc of thickness h and radius R the relaxation time T_{relax} can be calculated in the same way, and the ratio of this time to the crossing time through the galaxy $t_c = R/v$ is

$$\frac{T_{\text{relax}}}{t_c} \approx \left(\frac{h}{R}\right)\left(\frac{N}{8 \log N}\right),$$

where N is the number of stars in the galaxy ($N \approx 10^{11}$); typically $T_{\text{relax}}/t_c \approx 10^8$, with $t_c \approx 10^8$ years.

As the relaxation time due to collisions between two bodies is considerably greater than the age of the universe, collisions between stars are completely negligible. The ensemble of stars constitutes a collisionless medium, and its equilibrium state in a given potential is governed by the Vlasov equation (Chap. 4). Note that the ratio $T_{\text{relax}}/t_{\text{c}}$ increases as $N/\log N$; that is to say, paradoxically one can increasingly neglect the collisions the greater the number of stars: a small number of stars creates an irregular potential, which deflects the paths of neighbouring objects, whereas a potential created by a large number of stars is smoothed out.

5.1.1 Stability

Once a possible equilibrium state has been found, it is not at all certain that it will be stable, that is, that the system will be maintained in the same state. The simplest example is that of a 'cold' disc representing a spiral galaxy in rotational equilibrium. Each star moves with a circular motion with an angular velocity such that $\Omega^2 = (1/r)(\partial \Phi/\partial r)$. Left to itself this system is violently unstable and forms condensations; a certain degree of velocity dispersion is necessary to avoid the Jeans instability. This kind of perturbation is called 'axisymmetric', as opposed to instabilities in the form of bars or spiral arms, which can still occur in a disc that is 'hot' enough to be stable with respect to the Jeans instability.

Jeans Instabilities

Jeans's method to determine the stability criterion of a system assumes a homogeneous and infinite medium in equilibrium: it consists in weakly perturbing the system's density ρ_0 by a wave of the type $\rho_1(r,t) = \alpha \exp[\imath(kr - \omega t)]$ and linearizing the system equations to find the dispersion relation $\omega(k)$ of these waves. The system is unstable when $\omega^2 < 0$, that is, if there is a solution that increases exponentially with time: $\exp(\sqrt{-\omega^2} t)$. Note that the initial hypothesis – a homogeneous and infinite medium in equilibrium – is only a very rough local approximation (density gradients and the forces due to gravity, pressure, and so on over short distances are neglected).

For a fluid with a density ρ_0 and a pressure $P_0 = \rho_0 v_{\text{s}}^2$ (v_{s} being the velocity of sound) the Jeans criterion guarantees that perturbations with a size $\lambda > v_{\text{s}}/\sqrt{G\rho_0} = \lambda_{\text{J}}$ (the Jeans length) are unstable. This result can be easily derived: the system is unstable over a distance λ if a sound wave does not have enough time to cross through this region in the freefall time (or dynamical time) $t_{\text{ff}} = 1/\sqrt{G\rho}$. Pressure forces are thus negligible compared to gravity. The crossing time for the sound is r/v_{s}, from which $\lambda_{\text{J}} = v_{\text{s}}/\sqrt{G\rho_0}$. For a stellar system Jeans's method can be used by linearizing the Poisson and collisionless Boltzmann equations. The calculation is more complicated but gives the same Jeans length, at least if we regard the random motions of stars as a pressure. This equivalent 'pressure' can be written as $\rho_0 \sigma^2$, where σ is the velocity dispersion. Every element of mass (of size λ) is unstable if the crossing time for this element by a star is longer than the corresponding dynamical time, that is, if $\lambda > \lambda_J = \sigma/\sqrt{G\rho_0}$.

Stability due to Rotation

The rotation of a galactic disc allows stability on large scales, in contrast to the effects of pressure, which stabilize on small scales. The rigorous study of this stability requires the determination of the dispersion relation for the perturbing waves (see below for the dispersion relation for spiral waves with m arms; here $m = 0$ for axisymmetric perturbations). Such a study was made by Toomre (1964), but we give here only a simplified description (in orders of magnitude) sufficient to understand the phenomenon.

Consider a small region of size L on a disc of surface density μ (and thus of mass μL^2), rotating with angular speed Ω, at a distance d from the centre. In the reference frame of this region the forces present are gravitational, centrifugal, and Coriolis. Imagine a perturbation that locally increases the surface density by an amount $\mu \epsilon$. This will have the effect of bringing an element at the periphery of this region ϵL nearer. The increase in the gravitational force that it will experience will be

$$G\mu L^2 \left[(L - \epsilon L)^{-2} - L^{-2}\right] \approx 2\epsilon G\mu$$

per unit mass.

By the conservation of angular momentum per unit mass, $J = d^2 \Omega$, the angular velocity of rotation of the contracted region will also increase proportionally with ϵ, because $\Delta J = 2\Omega d\, \Delta d + d^2\, \Delta\Omega = 0$ and $\Delta\Omega/\Omega = -2\,\Delta d/d = 2\epsilon L/d$. The centrifugal force will therefore also increase: $\Delta(\Omega^2 d) \approx 3\epsilon L\Omega^2$. If this force is sufficient to send the mass element back to its initial position, the system will be stable. Hence we must have that $L > L_{\text{crit}} = 2G\mu/3\Omega^2$. The rotation thus stabilizes on large scales, unlike the pressure forces.

The order of magnitude of L_{crit} is roughly estimated using $\Omega^2 R = V^2/R \approx G\mu R^2/R^2 = G\mu$, and thus $L_{\text{crit}} \approx R$: in other words it is the galactic scales, and not the local ones, that are stabilized.

The Stability Criterion

To guarantee stability on all scales it can be seen that all we need is a certain minimum velocity dispersion to make the critical Jeans length equal to L_{crit}. For a thin disc let us now evaluate the Jeans length again as a function of the surface density; the freefall time is $t_{\text{ff}} = \sqrt{L/G\mu}$, and the crossing time for the system is $t_c = L/\sigma$ (σ is the velocity dispersion). Equating these two time scales gives $\lambda_J = \sigma^2/G\mu$. Stability on all scales is guaranteed as soon as $\lambda_J = L_{\text{crit}}$, that is, when $\sigma = G\mu/\Omega$. Of course, a rigorous calculation is necessary to find the numerical factor, but we have already obtained the dependence as a function of µ and r. The minimum radial velocity dispersion σ_r is in fact $3.36\,G\mu/\kappa$ for an infinitely thin disc, where κ is the epicyclic frequency determined in the next section. In general the ratio of the observed radial dispersion to this critical dispersion is called Q, so that stability is obtained as soon as $Q > 1$. This is the *Toomre criterion*. In the neighbourhood of the Sun the measurement of velocity dispersions of stars gives Q between 1 and 2, which confirms the analysis: stars born within interstellar clouds have, to begin with, a low velocity dispersion ($\sigma_{\text{gas}} \approx 10\,\text{km s}^{-1}$). This weak dispersion probably creates some local instabilities, which have the effect of increasing σ up to a value for which $Q > 1$. Other mechanisms that can increase the velocity dispersion of the stars, for example collisions with molecular clouds, are very slow processes that take over from gravitational instabilities for dispersions greater than the critical one.

5.1.2 Stellar Orbits

The collective behaviour of the stars of a galaxy is in many cases much easier to understand with knowledge of the individual stellar orbits, or at least the main kinds of periodic orbits, which can trap many neighbouring ones. The calculation of orbits in a mean gravitational potential is of course justified by the fact that the stars of a galaxy form a collisionless medium.

We shall now give the main characteristics of the orbits in a disc galaxy, describing the first-order epicyclic theory.

Epicycles

Let us consider a flattened axisymmetric gravitational potential, corresponding to a spiral galaxy, of the form $U(r, z)$, with $U(z)$ symmetric with respect to the plane $z = 0$ (a cylindrical coordinate system (r, θ, z) is chosen). Consider those orbits which are almost circular, undoubtedly the most numerous. We denote by x the radial deviation with respect to a circle of radius R and by y the azimuthal deviation:

$$r = R + x\,,$$
$$\theta = \Omega t + y\,, \qquad (5.1)$$

where Ω is the angular velocity for a circular path:

$$\Omega^2 = \frac{1}{R}\frac{\partial U(R,0)}{\partial r}.$$

In polar coordinates the equations of motion are written as

5.1 Stellar Dynamics. Stability. Orbits

$$\ddot{r} - \dot{\theta}^2 r = -\frac{\partial U}{\partial r},$$

$$r\ddot{\theta} + 2\dot{r}\dot{\theta} = 0,$$

$$\ddot{z} = -\frac{\partial U}{\partial z}. \tag{5.2}$$

Developing the potential $U(r,z)$ in the neighbourhood of the circular trajectory in a Taylor series we get

$$\frac{\partial U(R,0)}{\partial r} = \frac{\partial U}{\partial r} + x\frac{\partial^2 U(R,0)}{\partial r^2} + z\frac{\partial^2 U(R,0)}{\partial r \partial z}.$$

As the last term on the right-hand side is equal to zero, in view of the symmetry of $U(z)$, $\partial U(R,0)/\partial z = 0$. By linearizing (5.2) we find to the first order that

$$\ddot{x} - 2\Omega\dot{y}R - \Omega^2 x = -x\frac{\partial^2 U(R,0)}{\partial r^2},$$

$$R\ddot{y} + 2\dot{x}\Omega = 0,$$

$$\ddot{z} = -z\frac{\partial^2 U(R,0)}{\partial z^2}. \tag{5.3}$$

This approximation is known as the epicyclic approximation. Integrating the second equation of (5.3), we obtain

$$R\dot{y} + 2x\Omega = \text{constant} = a,$$

and inserting this into the first equation, we have

$$\ddot{x} - 2\Omega(a - 2x\Omega) - \Omega^2 x = -x\frac{\partial^2 U(R,0)}{\partial r^2}.$$

This equation can be written in the form

$$\ddot{x} + \kappa^2(x - x_0) = 0$$

and

$$\dot{y} = -\frac{2\Omega x}{R},$$

with

$$\kappa^2 = \frac{\partial^2 U(R,0)}{\partial r^2} + 3\Omega^2 = R\frac{d\Omega^2}{dR} + 4\Omega^2.$$

We can take $a = 0$, whence $x_0 = 0$ (otherwise the small oscillations occur around another point of equilibrium, which in turn alters the initial conditions $r = R$). Thus x and y execute oscillatory motions with the epicyclic frequency κ; likewise z oscillates with the frequency ν_z, such that $\nu_z^2 = \partial U(R,0)/\partial z^2$.

The trajectory of a star projected onto the plane of the galaxy is thus the composite of a circle and an epicycle (a small ellipse in the rotating reference frame), as shown in Fig. 5.1b: in general it has the form of a rosette. The star travels through the epicycle in the reverse direction, the ratio of its axes being $\kappa/2\Omega$. What is the order of magnitude of κ? The rotation curve of a spiral galaxy can be roughly modelled by two domains: one (fixed rotation) close to the centre, where $\Omega = \text{constant} = \Omega_0$ and $V = \Omega_0 r$, the other farther out, where the rotation is differential at $V = \text{constant} = V_0$, with $\Omega = V_0/r$. In the first domain $\kappa = 2\Omega$; then $\kappa = \sqrt{2}\Omega$ (as illustrated in Fig. 5.2). κ/Ω generally lies between 1 and 2 for realistic potentials. The shape of the rosettes can then be determined, the number of lobes per revolution being κ/Ω. When $\kappa = 2\Omega$ exactly, the orbit closes into an ellipse (with two lobes per revolution).

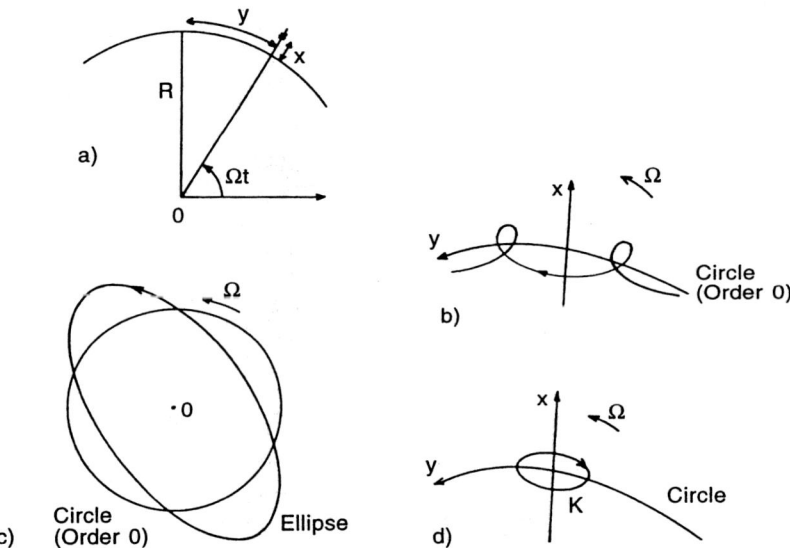

Fig. 5.1. The first-order epicyclic approximation. (**a**) Definitions of the coordinates. (**b**) The epicyclic trajectory in the plane xOy for any star; the epicycle travels in the retrograde sense. (**c**) The special case of an elliptical trajectory when $\kappa = 2\Omega$; here there is a fixed rotation curve ($V = \Omega_0 r$) or a resonant orbit in a reference frame rotating at $\Omega_p = \Omega - \kappa/2$. (**d**) The orbit of a star in a corotating frame, $\Omega_p = \Omega$; the ratio of the axes of the epicycle is $\kappa/2\Omega$

There are three constants of the motion: the angular momentum L_z (conserved because $\partial U/\partial \theta = 0$), and the energies of the two oscillators, in the z direction and epicyclic, giving $E_z = (\dot{z}^2 + \nu_z^2 z^2)/2$ and $E_x = (\dot{x}^2 + \kappa^2 x^2)/2$ (another possibility is the total energy).

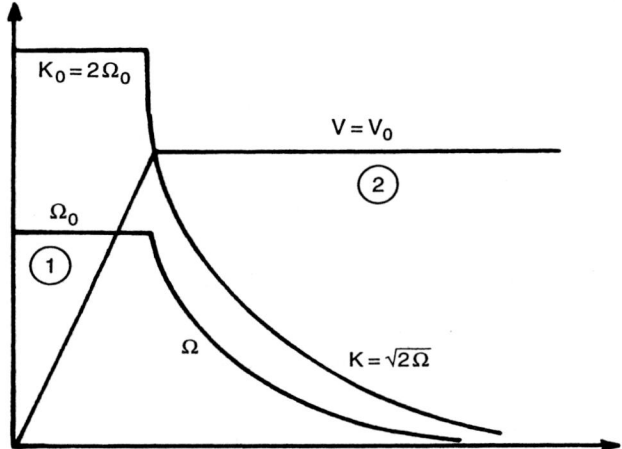

Fig. 5.2. A simplified model of a galactic rotation curve for obtaining the order of magnitude of the epicyclic frequency. Region 1: fixed rotation, $v(r) = \Omega_0 r$. Region 2: flat rotation curve, $v(r) = V_0$. The ratio κ/Ω always lies between 1 and 2

Lindblad Resonances

Whether there is a spiral perturbation with m arms (Sect. 5.2), a bar, or even a companion interacting with the galaxy, there often exists a privileged angular velocity Ω_p that plays a special role for the stars. In the reference frame rotating with the perturbation, the stars themselves rotate with an angular velocity $\Omega' = \Omega - \Omega_p$. Thus regions of the galaxy exist where $\Omega' = \kappa/m$, that is, where epicyclic orbits close up again after m lobes. The corresponding stars align themselves with the perturbation and move along with it; always interacting with it with the same sign, they start to resonate with the perturbation. These regions are known as 'Lindblad resonances' and are illustrated in Fig. 5.3. The most general case is $m = 2$. An obvious resonance is the corotation $\Omega(r_{\rm CR}) = \Omega_p$ ($r_{\rm CR}$ is called the radius of corotation).

In the rotating reference frame the orbits are periodic at the resonances. These periodic orbits play a dominant role in determining the types of stellar population (Contopoulos 1980).

Surfaces of Section

To determine an orbit in phase space we represent 6 coordinates $(x, y, z, \dot{x}, \dot{y}, \dot{z})$ as a function of time. It is necessary to take projections to represent the orbit. A very effective method of representation for visualizing the periodic and quasiperiodic orbits is the method of surfaces of section (due to Poincaré in 1899). This method consists in representing the orbit only at its points of intersection with any plane of phase space, for example $\dot{x} = 0$. To avoid ambiguities due to the sign of x we represent the point only when $x > 0$,

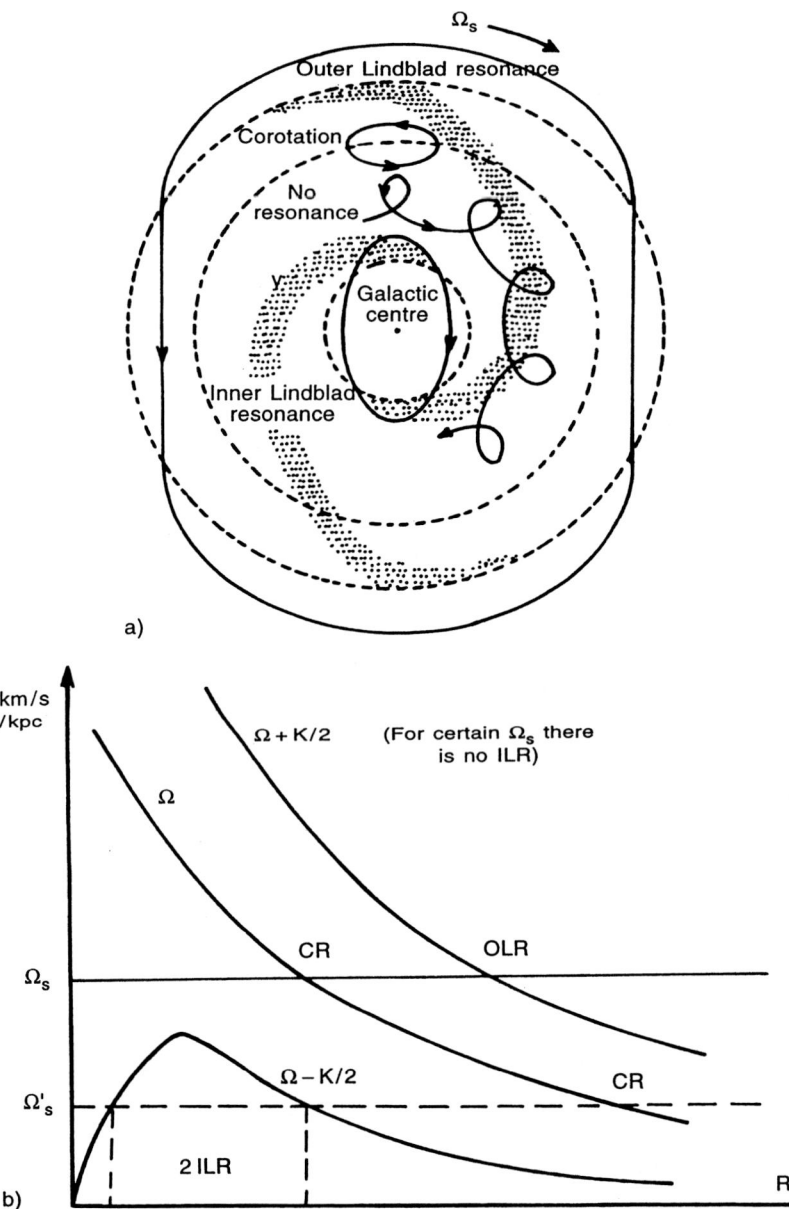

Fig. 5.3. Lindblad resonances. (a) A representation of the resonant orbits in the reference frame rotating with the angular velocity of the spiral perturbation, $\Omega_{\rm p}$: the inner Lindblad resonance, or ILR ($\Omega_{\rm p} = \Omega - \kappa/2$); the corotation ($\Omega_{\rm p} = \Omega$); the outer Lindblad resonance, or OLR ($\Omega_{\rm p} = \Omega + \kappa/2$). (b) Rotation curves $\Omega(r)$ and curves combined with the epicyclic frequency $\Omega_{\rm p} = \Omega \pm \kappa/2$, enabling the location of Lindblad resonances to be determined

for example. Next we project these points in the space $(y, \dot{y})$ or any other choice of two variables. The curves defined by these points are the surfaces of section. They are all included inside an envelope known as the 'zero-velocity curve', because the energy $E > U(r) + \dot{y}^2/2$. The curves have the advantage of immediately giving the integrals of the motion: indeed if the path is subjected to a constraint $I =$ constant, the points of the surface of section form an invariant curve, instead of filling all of the space over time. Periodic orbits appear as points.

5.2 The Density-Wave Theory

Most disc galaxies have a remarkable spiral structure, some being fascinating in virtue of their symmetry and regularity (see the famous galaxy M51 in Canes Venatici or Arp 1, shown in Fig. 5.4 and 5.5 respectively). These structures are not simply magnificent trimmings but an essential element

Fig. 5.4. Regular spiral galaxies with two arms suggest the presence of spiral density waves: this is M51, the well-known galaxy in Canes Venatici, the Hunting Dogs. (From the *Hubble Atlas of Galaxies*, Sandage 1961)

122   5. The Spiral Structure of Galaxies

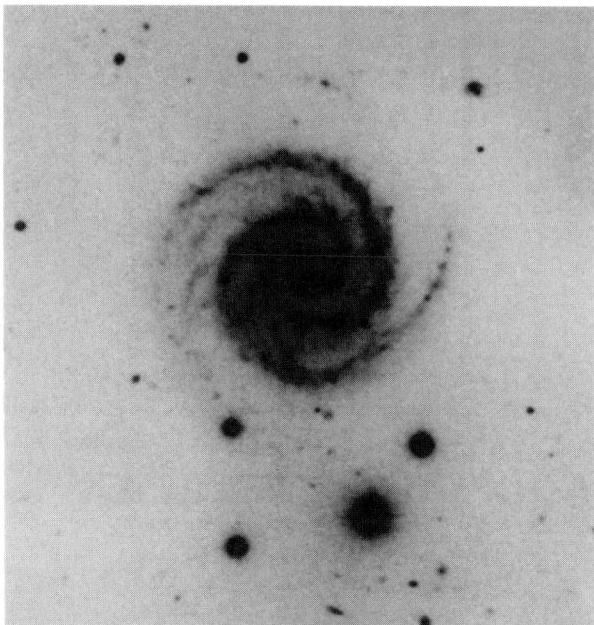

Fig. 5.5. Another very regular spiral: Arp 1 (NGC 2857). (From Arp's catalogue (1966), by permission of the Carnegie Institution of Washington Observatories)

in the formation and evolution of these galaxies: indeed it is in the spiral arms that the majority of stars are formed, and especially the most luminous and massive stars. The spiral is also, as we shall see later on, a means of carrying angular momentum towards the exterior, so forming bars and mass concentrations at the galactic centres.

5.2.1 The Winding Problem

Even the existence of spiral arms in galaxies presents a major theoretical problem. Indeed spiral arms cannot be material arms, according to the intuition one might have from watching milk spiral round in a cup of coffee after having stirred it. If the arms were material, they would wind up over time because of galactic differential rotation: the centre of a galaxy can have made a thousand revolutions since its creation, whereas the outer regions can have made only a few dozen. All spiral structure would quickly become invisible after a few rotations, because of dilution due to winding.

The Swede Lindblad was the first to propose, in 1963, that the spiral arms of a galaxy could be a pattern rotating like a solid body, with the same angular speed Ω_p from centre to edge. At the centre of the galaxy the matter would rotate faster than the spiral, while the outer regions would trail behind. For an intermediate distance there could be a 'corotation resonance' between stars and the spiral. Maintaining this pattern would be due to the selfgravity of the stars.

These ideas are behind the density-wave theory, developed mathematically by Lin and Shu from 1964 to 1970: the spiral arms are regarded as waves that can propagate in a galaxy like waves on the sea. Lin and Shu's hypothesis is to suppose the existence of stationary (or quasistationary) waves that take account of the longevity of the spiral structure: a progressive wave packet could not be maintained for much longer than material arms. Now multicolour observations of galaxies show that spiral arms are visible not only in the young population (with H II regions, interstellar gas, and massive stars), but also in the old population (older than 10^9 years); some examples are given in Fig. 5.6. This lasting quality of the spiral structure for several mean rotations of the galaxy (one rotation is about 2×10^8 years long) lies at the root of the problem, as we shall now make clear.

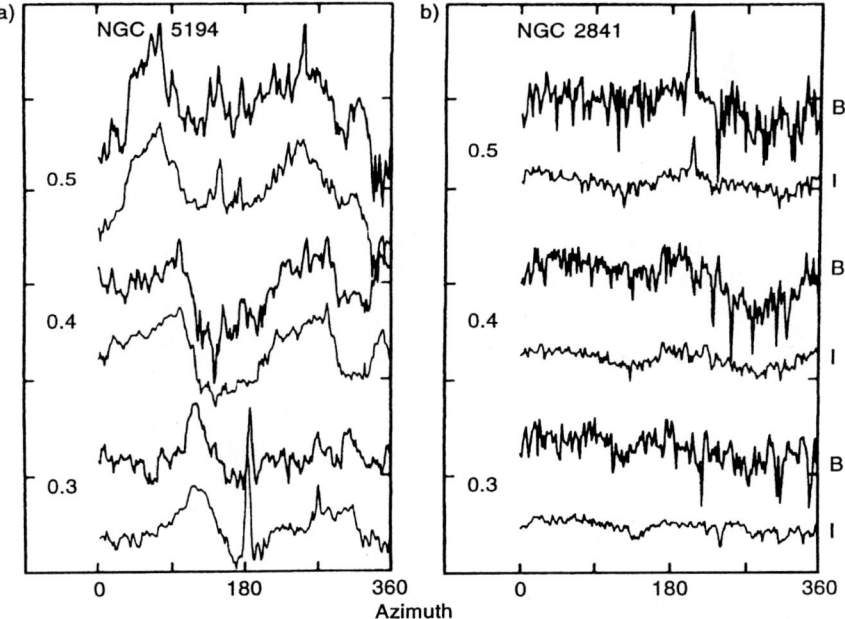

Fig. 5.6. The spiral structure in galaxies having density waves is particularly contrasted in the young population (young stars, ionized regions, and interstellar gas and dust). However, the old population also contributes to the wave: the spiral structure is still clearly visible at red wavelengths. Here azimuthal profiles are shown, corresponding to the distribution of blue (B) and red (I) light, for three galactic radii ($R = 0.3$, 0.4, and $0.5R_{25}$), in the galaxies M51 (**a**) and NGC 2841 (**b**). The two spiral arms (density waves) of M51 are clearly visible in both the blue-light and red-light profiles. On the other hand in the stochastic spiral NGC 2841 there is practically no contrast and the distribution in red light is almost as flat. (From Elmegreen and Elmegreen 1984)

Stationary Density Waves or Stochastic Spirals?

Not all spiral galaxies are as regular as M51 or Arp 1 (Figs. 5.4 and 5.5), which have two spiral arms that are continuous from the centre to the edge. In fact some of them are made up of filaments, of pieces of spiral arms, wound up so as to give the galaxy a spiral appearance: these spirals are known as 'stochastic', and the galaxy NGC 2841 (shown in Fig. 5.7) is a prototype. The persistence of spiral structure in this kind of galaxy is only global; each piece of an arm in fact has a short lifetime, as photometric observations reveal: the old stellar population does not have contrasted structures, and the filaments are only young-population phenomena.

The morphology of these galaxies can be easily explained without resorting to the density-wave theory. Imagine that the formation of stars can propagate like wildfire, over distances from a few parsecs up to several hundreds of parsecs: these pieces of arm are then stretched and wound up by differential rotation to give the observed stochastic spiral structure. Of course, these filaments are only short-lived. This hypothesis of 'contagious' star formation is supported by observations showing young clusters of stars around the shock waves produced by the explosion of supernovae, the expansion of H II regions, and so on. Within a star cluster, chronological sequences of star formation have been identified.

Fig. 5.7. A photograph of the stochastic-galaxy prototype, NGC 2841. The galaxy has only a global spiral appearance, based on multiple pieces of incoherent spiral arms. (From Sandage 1961)

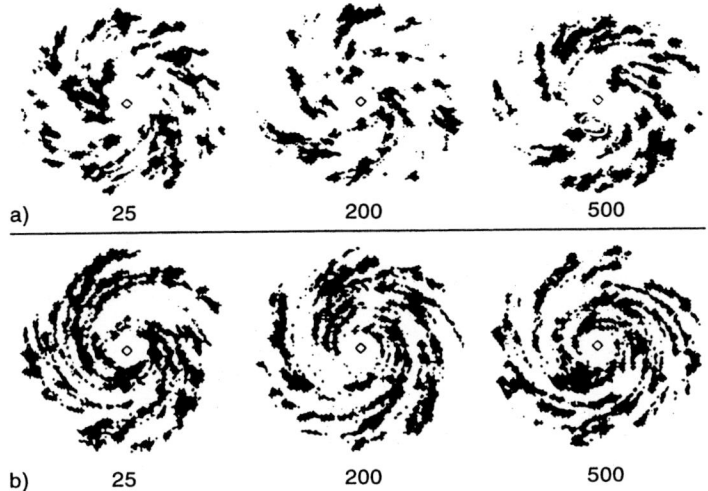

Fig. 5.8. A simulation of the process of 'contagious' star formation and determination of the spiral structure obtained on large scales by differential rotation. In such a numerical simulation the galaxy is divided into cells in which stars form either spontaneously or by being induced by stars that have just formed in neighbouring cells (the Monte Carlo method). Each diagram represents a particular state of the galaxy; the age is indicated for each (units are 15 million years). (**a**) and (**b**) are two simulations corresponding to different rotation curves. (From Gerola and Seiden 1978)

Processes of 'contagious' star formation have been simulated on computers to determine the spiral structures obtained with this mechanism. The result of one of these simulations is shown in Fig. 5.8.

Kinematic Waves

In Sect. 5.1 we studied the orbits of stars in the axisymmetric potential of a spiral galaxy. In the general case the superposition of rotational motion of angular velocity Ω and the epicyclic oscillation of frequency κ leads to a rosette-shaped trajectory that is nonperiodic if there is no rational relation between Ω and κ. For every orbit there exists a reference frame rotating with a frequency Ω_p so that the path closes to form an ellipse, that is, such that the ratio between $\Omega - \Omega_p$, the frequency of rotation in the rotating frame, and κ is $\pm\frac{1}{2}$. In a way, the motion of every star can be described as being like that of a marble sliding quickly down an elliptical tube, itself rotating much more slowly around the centre with an angular velocity $\Omega_p = \Omega - \kappa/2$ (Fig. 5.9). Imagine that this velocity $\Omega - \kappa/2$ is almost constant for most of the stars of the galaxy; in the reference frame rotating with a velocity Ω_p any pattern adopted by the stars at a given time would be quasipermanent. It would not deform but would rotate like a solid body with respect to the fixed

126   5. The Spiral Structure of Galaxies

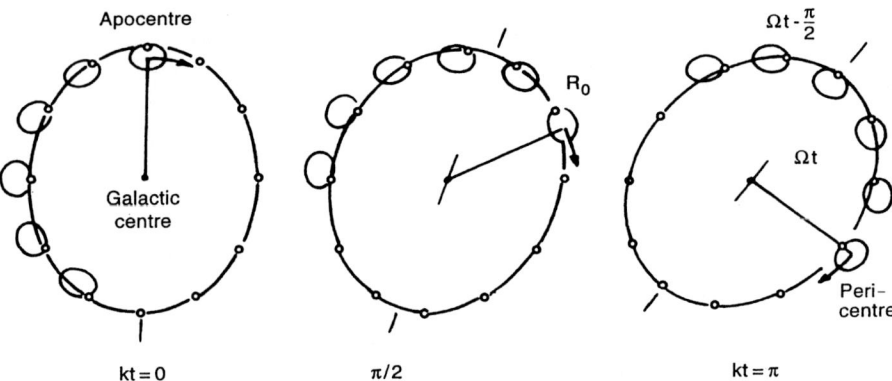

Fig. 5.9. A representation of the rate of precession $\Omega - \kappa/2$ of epicyclic orbits by a ring of identical test particles with angular velocity Ω and epicyclic frequency κ. Each particle (represented by a small circle) makes an epicycle around its guiding centre (represented by a point). The period of time between successive steps is a quarter of the epicyclic period $2\pi/\kappa$. The figure corresponds to the case where $\kappa = \sqrt{2}\Omega$ (a flat rotation curve). The ellipse formed by the ensemble of particles rotates around the centre much more slowly ($\Omega - \kappa/2$) than the particles themselves (Ω). In other words the motion of each particle can be decomposed into a sliding motion along an elliptical tube (with velocity Ω), the tube itself slowly rotating (with a velocity $\Omega - \kappa/2$) around the galactic centre

reference frame. For our own Galaxy the curves Ω and $\Omega - \kappa/2$ determined from observations (Fig. 5.10) reveal that $\Omega - \kappa/2$ remains practically constant over large distances, as Lindblad noted; of course, the centre is excluded, but the spiral arms never wind up to the centre. Notice that no other curve $\Omega - \kappa/m$ varies as slowly as $\Omega - \kappa/2$. This helps to understand the preeminence of two-arm spiral structure in galaxies.

The simplest form adopted by elliptical orbits in a rotating reference frame is a bar (Fig. 5.11a), when all the ellipses are aligned with one another. When the major axis of each orbit is displaced with respect to the previous one by a small angle that varies logarithmically with the radius, it is easy to form a spiral as thick or thin and as loosely or tightly wound up as desired (Fig. 5.11b). These striking examples of kinematic spiral waves were introduced by Kalnajs in 1973. These density waves are said to be 'kinematic' because they only affect massless test particles. It is obvious that, even if $\Omega - \kappa/2$ were a constant for a given axisymmetric potential, the waves of Fig. 5.11, because of their concentration of mass in the arms, would deform the potential, and $\Omega - \kappa/2$ would no longer be a constant. Can the introduction of selfgravity make the rates of precession of the orbits $\Omega - \kappa/2$ coincide, so as to produce quasistationary spiral structures with long lifetimes? This is the problem Lin and Shu tackled in the 1960s.

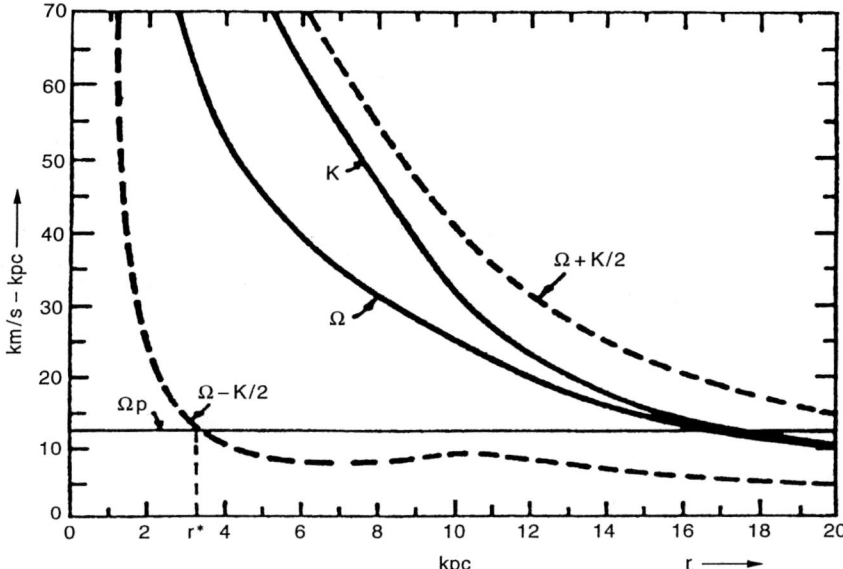

Fig. 5.10. The rotation curve observed for our own Galaxy, the Milky Way, according to Schmidt's model (1965). Lindblad was the first to note that the rate of precession of all the elliptical orbits $\Omega - \kappa/2$ remains almost constant as a function of the radius throughout the galaxy, suggesting that a spiral wave with two arms could precess without any deformation in a galaxy. In addition this explains the predominance of spiral structure with two arms, because the analogous rate of precession $\Omega - \kappa/m$ for orbits with m lobes (which would lead to the formation of m arms) does not have the same constancy throughout the galaxy

5.2.2 The Wave Dispersion Relation

The problem that we are faced with is the coherence of the waves. Suppose that a spiral wave exists in a disc that is considered infinitely thin. The surface density can be written as

$$\sigma = \sigma_0(r) + \sigma_1(r, \theta, t),$$

where $\sigma_0(r)$ is the unperturbed axisymmetric density and σ_1 is the perturbation (r and θ are the polar coordinates in the galactic disc).

This can be expanded in a Fourier series whose terms are

$$\sigma_1 = \sigma_1(r) e^{im[\theta - \theta_0(r)] + i\omega t},$$

which represents a spiral wave with m arms and amplitude $\sigma_1(r)$; the form of the arms is determined by the mean of the function $\theta_0(r)$. Obviously only the real part of this expression has a physical meaning, and we assume in a first approximation that the perturbation has a sinusoidal form. ω/m is the angular velocity of the wave.

128   5. The Spiral Structure of Galaxies

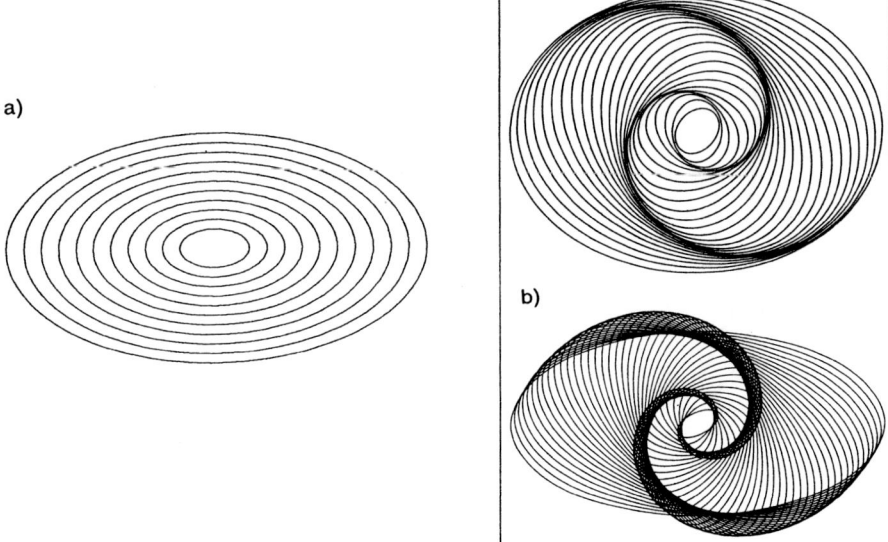

Fig. 5.11. Barred or spiral kinematic waves illustrating the concept of density waves: elliptical orbits, deformed to a greater or lesser extent with respect to a circle, are arranged in a coherent way. (**a**) The major axes are all aligned and the structure formed is a bar. (**b**) The major axes are arranged according to angles θ as a function of their size R: $\theta = -\alpha \log R + \text{constant}$ (so as to form a logarithmic spiral). Orbits become closer at some places to form denser regions with a spiral shape. Each star during its revolution around the centre enters a spiral arm, remains there for a relatively long time, because its path is tangential to the arm, and then leaves, ready to dive into the next arm: this is clearly a density wave (the particles themselves that form the spiral are always changing). These waves are said to be 'kinematic' because they are formed out of test particles whose selfgravity is not taken into account. (From Kalnajs 1973)

The corresponding wave number is $k = \mathrm{d}\theta_0(r)/\mathrm{d}r$. Two cases can be considered.

- $k < 0$: If angles are measured positively in the direction of rotation of the matter (Ω), the particles enter the arm through its concave part. The spiral structure then 'trails'.
- $k > 0$: The particles enter the arm through its convex part. The structure then 'leads' (Fig. 5.12). The winding-up of the spiral is given by the angle i (the pitch angle) between the tangent to the spiral and the circle $r = \text{constant}$:

$$\tan i = \frac{1}{r}\frac{\mathrm{d}r}{\mathrm{d}\theta_0} = \frac{1}{kr}$$

(for a tightly wound spiral, $kr \gg 1$).

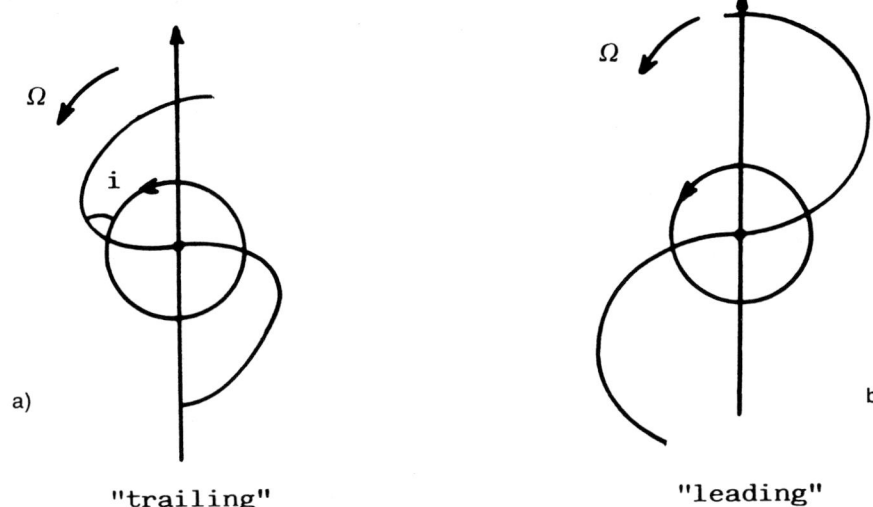

Fig. 5.12. The winding direction of the spiral wave relative to the sense of rotation of the galaxy: when matter crosses through the arms from the concave edge (**a**), the wave is trailing; in (**b**) it is leading. This direction depends on the sign of the wave number $k - \mathrm{d}\theta_0/\mathrm{d}r$, where $\theta_0(r)$ determines the geometrical form of the arms (see the text)

Obtaining the Dispersion Relation

In a first step, the gravitational potential created by the distribution of mass σ_1 can be calculated using the Poisson equation:

$$\Delta \Phi = 4\pi G \rho,$$

where

$$\rho = \sigma(r)\delta(z).$$

How will the stars respond to this potential? The state of the system can be described by its distribution function $f(\boldsymbol{x}, \boldsymbol{v}, t)$, taken from the collisionless Boltzmann equation, in the presence of gravitational forces. In a similar way Lin and Shu used the Liouville equation to determine the probability $f^N(\boldsymbol{x}_n, \boldsymbol{v}_n)$ of finding an ensemble of N stars in the state $(\boldsymbol{x}_n, \boldsymbol{v}_n)$ per unit volume of phase space. f^N is called the distribution function for N particles. With this second step we obtain the density of stars in the disc.

The last step of the calculation consists in identifying the distribution of stars obtained in the second step with the initial distribution, in order to ensure that the hypothesis is coherent. The resulting equation is the dispersion relation for the waves. The calculation is possible analytically only in the approximation of very tightly wound density waves: the wavelength, or radial distance, between two successive windings is small compared with the radius of the galaxy. This approximation of very tightly wound waves is similar to the WKB (Wentzel–Kramers–Brillouin) approximation in quantum mechanics. It allows one to neglect large-scale coupling: the equation is therefore local. Indeed the surface density oscillates quickly when we go far from the point of calculation, which ensures that the distant contributions to the perturbed potential cancel.

In reality a large number of galaxies exist whose spiral structure is not wound up enough to justify this approximation; in particular, late-type spirals are very open. Furthermore in most galaxies the spiral potential represents more than a weak perturbation, and linear theory must be used with caution. Apart from these restrictions the theory of WKB waves remains a fundamental tool for understanding the nature of the spiral waves of galaxies.

The resultant dispersion relation is depicted in Fig. 5.13: the curves represent the absolute value of $\nu = m(\Omega_p - \Omega)/\kappa$, the local frequency of the density waves in units of epicyclic frequency, as a function of the wavelength $\lambda/\lambda_{\text{crit}}$ (in units of the critical wavelength of the Jeans instability:

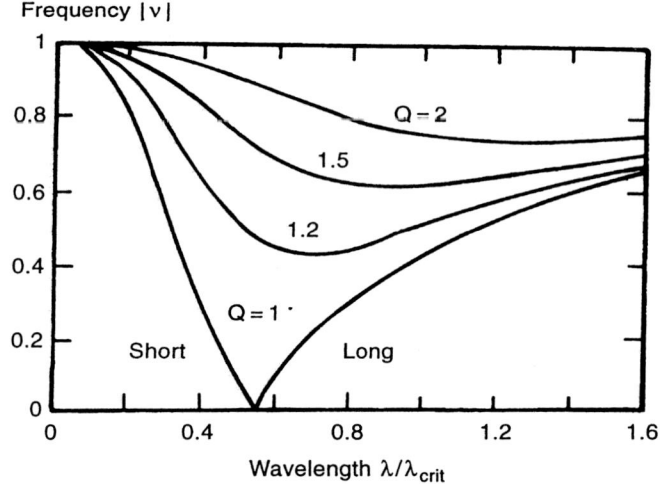

Fig. 5.13. The dispersion relation for WKB waves, obtained by Lin and Shu in 1964. On the y axis is the wave frequency $\nu = (\omega - m\Omega)/\kappa = m(\Omega_p - \Omega)/\kappa$, in units of epicyclic frequency κ. $\Omega_p = \omega/m$ is the angular velocity of the wave. Only the absolute value of ν is given because the diagram can be completed by symmetry. In the same way, the relations are independent of the sign of the wave number $k = 2\pi/\lambda$, and so they are the same for both trailing and leading waves. The wavelength λ is plotted on the x axis in units of the Jeans critical wavelength λ_{crit}. The different curves represent the dispersion relations for several values of the velocity dispersion (Q). For $Q = 1$ the whole of the region between $\nu = -1$ and 1 (between the OLR and the ILR) is allowed, whereas if $Q > 1$, a forbidden region appears around the corotation ($\nu = 0$). For a given frequency, for $Q = 1$, for example, two solutions defining the 'long-wave' and 'short-wave' branches exist (see the text)

$\lambda_{\text{crit}} = 4\pi^2 G\mu(r)/\kappa^2(r)$; see Sect. 5.1). The curves represent only the absolute value of ν, the frequency at which a disc star encounters the wave ($\nu = 0$ corresponds to the corotation), and the curves can be extended by symmetry throughout the corotation. $\nu = \pm 1$ corresponds to the outer and inner Lindblad resonances (OLR and ILR). At these points for a spiral with two arms, for example, $\Omega_{\text{p}} = \Omega \pm \kappa/2$, and in the reference frame of the wave the stars have periodic elliptical orbits. Likewise only the positive wavelengths $\lambda \, (= 2\pi/k)$ have been represented, that is, trailing waves, but the curves can be generalized by symmetry for negative k to leading waves.

The various curves correspond to different values of Q, the Toomre stability parameter. Only curves such that $Q > 1$ are represented, where the velocity dispersions are such that Jeans instabilities are cancelled. The curve $Q = 1$ delimits two regions in the diagram, corresponding to the short and long waves. Normally the WKB approximation taken as the initial hypothesis will restrict the validity of the calculation to the short waves, given that λ_{crit} can be of the order of the radius of the galaxy; however, the long waves are essential to the production of quasistationary spiral waves, and the calculation is always used up into the long branch.

Another fundamental result, shown in the curves in Fig. 5.13, is that there exists for $Q > 1$ a prohibited area around the corotation for which the waves are evanescent (the wave number k is complex). This prohibited area widens out more and more as Q increases, that is, for stars with a greater velocity dispersion.

These results can be easily interpreted if we remember the necessity of adjusting the rates of precession $\Omega - \kappa/2$ of the periodic orbits of stars over large distances to avoid the deformation of the spiral shape formed by these orbits at a given time. When selfgravity becomes negligible, the precession rate is exactly $\Omega - \kappa/2$, and the adjustment occurs only for the Lindblad resonances for $\nu = \pm 1$. This is what is indicated by the evolution of the curves as a function of Q: when Q tends to infinity (the velocity dispersion renders selfgravity negligible), the dispersion relation tends to a horizontal straight line $|\nu| = 1$, and the waves are restricted to the resonance region.

Therefore selfgravity has the effect of decreasing the radial-oscillation frequency of stars from κ to $\nu\kappa$. The new rate of precession of the periodic orbits thus becomes $\Omega - |\nu|\kappa/2$ and can be adjusted. It should be remarked, however, that the adjustment can only be accomplished by varying the wavelength. In the limiting case where $Q = 1$ (selfgravity being predominant) the waves can propagate between the two Lindblad resonances by taking all values on the short branch.

The dispersion relation for the waves allows the shape of the spiral to be calculated. For a given frequency Ω_{p} we obtain $k(r)$ and thus, by integration, $\theta_0(r)$ (since $k = d\theta_0/dr$), the geometrical shape of the corresponding mode. We can already see, from the evolution of the dispersion relation as a function of Q, that for a 'hot' system (one with a high velocity dispersion) the spiral

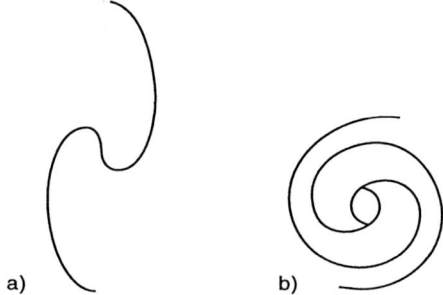

Fig. 5.14. The geometrical shape of waves deduced from the dispersion relation; for a given frequency the wavelength is proportional to Q for the short branch and inversely proportional to Q for the long branch. This is schematized in the figure, where (**a**) corresponds to a star system 'colder' than (**b**) for a long branch

will be more open than for a 'cold' system (Fig. 5.14), for the short branch. The contrary is true for the long branch. Since the long-branch wave has the highest amplitude in the amplification mechanism described in the following paragraphs, it is likely that the observed open spirals indicate the coldest systems.

Given that the dispersion relation of Fig. 5.13 concerns progressive waves, they are going to move along the disc of the galaxy and carry energy and angular momentum. The velocity with which a wave packet moves is known as the group velocity, which in a dispersive medium is written as $v_g = \partial(\nu\kappa)/\partial k = \partial\omega/\partial k$. This velocity can be calculated from the dispersion relation. It has a radial direction. Toomre, in 1969, was the first to calculate this group velocity; he showed that the propagation time of the corotation at the ILR (inner Lindblad resonance) is in fact very short, of the order of a few mean rotations of the galaxy, which creates the problem of maintaining the waves, if they die down at the ILR.

Wave Propagation

In the realistic case of a spiral galaxy Q is clearly greater than 1. Observations of stars in the neighbourhood of the Sun give the result $1.2 < Q < 2.0$, Q generally increasing towards the centre. The waves cannot therefore propagate through the corotation. Nevertheless we can determine the behaviour of a wave packet between the resonances. Figure 5.15 indicates the nature of the waves (trailing or leading, short or long) and the direction of the group velocity as a function of position in the galaxy. It can be seen that in the neighbourhood of the forbidden area around the corotation the group velocity changes sign. This can be interpreted as a reflection of the wave packet. Follow, for example, a very tightly wound, 'leading' wave packet (short waves) as it leaves the ILR. It propagates towards the exterior, progressively opening out (λ increases). When it turns back, just before it reaches the corotation,

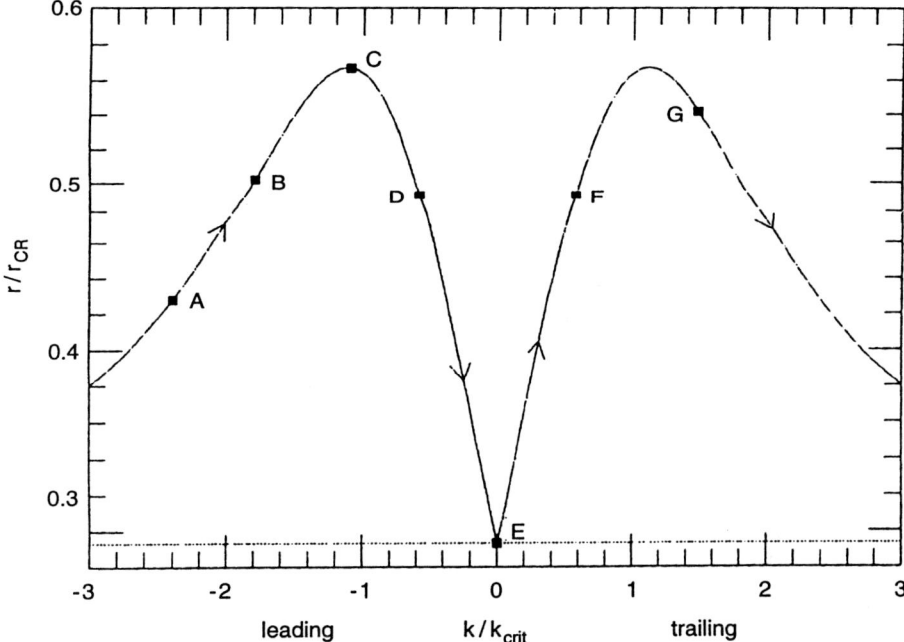

Fig. 5.15. Another form of the dispersion relation, showing the wave number k (on the x axis and in units of the critical wave number $k_{\text{crit}} = 2\pi/\lambda_{\text{crit}}$) as a function of the radius (on the y axis and in units of the corotation radius r_{CR}). Only the region between the ILR and the corotation is represented. The corresponding value of Q is 1.5. The arrows indicate the direction of the group velocity of the corresponding wave packets. From the left to the right we can follow the propagation of a short leading wave: from A to B the wave is propagated to the exterior and reaches the corotation C, opening out more and more. It is reflected, the group velocity changes its sign, and a long leading wave is now propagated towards the interior. At the ILR (at E) leading waves can be reflected into trailing waves (see the text). The long trailing wave (F) is reflected again at the corotation to become a short trailing wave (G) propagating towards the inner region. (From Binney and Tremaine 1987)

it is so open that it topples onto the 'long-wave' branch. What happens after its propagation up to the ILR? There the WKB theory is suspect, because the wave number $k = 0$ and the approximation $kr \gg 1$ is no longer satisfied. However, we can show that if λ_{crit} remains small with respect to the radius R, long leading waves will be reflected at the ILR into long trailing waves. After reflection at the ILR the wave packet trails, and it propagates towards the corotation, tightening up (λ decreases); there, once again, it heads towards the centre along the short branch. Such boundary conditions allow us to imagine the formation of stationary waves.

One of the problems that the relations obtained up until now leave unresolved is the behaviour of tightly wound trailing waves at the ILR. Several authors (such as Mark in 1971 and Lynden-Bell and Kalnajs in 1972) have established that the waves should damp down at the ILR by a phenomenon similar to Landau damping: the interaction between waves and particles at the resonance allows the exchange of energy, and the stars heat up with the weakening of the wave. However, if the function $Q(r)$ grows sufficiently towards the centre (but not so much as to violate WKB approximations), the theory predicts that the wave packet will be reflected before reaching the ILR; more precisely a short trailing wave packet moving in will be reflected as long waves. Such reflections at the ILR and at the corotation lead us to expect a possible amplification of the waves. This amplification is in fact necessary when one considers the absorption in the course of the propagation by shock waves induced in the gaseous component (Sect. 8.2.3) once the perturbation is no longer linear.

To confirm and illustrate the propagation of a wave packet such that the WKB theory is suggested it is important to do a numerical simulation here (Fig. 5.16); analytic theory has sometimes been pushed beyond its limits of validity (especially when used for long waves). The simulation, due to Toomre and Zang (1981), calculates the evolution of a tightly wound leading wave packet from linear perturbation theory (step 1). This allows the propagation of long waves to be studied free of the approximation $kr \gg 1$. Figure 5.16 illustrates the remarkable transformation of the original leading wave packet into a trailing wave packet at the ILR. How can one intuitively interpret this leading–trailing (and trailing–leading) transformation at the centre of the galaxy? Follow a trailing wave at three points A, B, and C in its radial motion towards the centre in the reference frame rotating with the wave (Athanassoula 1984). The later position of these three points can be drawn schematically (Fig. 5.17); as the group velocity is the same for all points of the wave, these points will have covered the same distance, so $AA' = BB' = CC'$. It is easy to see that the direction of winding of the spiral arm is reversed on passing through the centre.

Swing Amplification

The simulation in Fig. 5.16 reveals an important new phenomenon that the WKB theory had not made evident: the remarkable amplification of the amplitude of the original leading wave during its transformation into a trailing wave. When the trailing wave tightens up (step 9), it is more intense than the original leading wave, but the transient amplification as open trailing waves is considerable. This mechanism, known as 'swing' amplification, had already been considered by Goldreich and Lynden-Bell in 1965, but won favour with the work of Toomre in 1981. The principle can be easily understood by referring to Fig. 5.18. The amplification takes place when the open leading wave packet changes into a trailing packet, owing to differential rotation. Thus the motion of the intermediate wave begins to resonate at a frequency κ, and during its epicycle each star interacts strongly with the wave by following it.

Fig. 5.16. The evolution of a tightly wound leading wave packet, calculated using Zang's simulation (Toomre 1981), in a disc of stars with parameters $Q = 1.5$ and $X = 2$ (see the text for the definition of these parameters). The various stages, from 1 to 9, are separated by an interval corresponding to half the period of the spiral. The positions of the resonances (ILR, CR, and OLR) are indicated by circles. This simulation shows the considerable 'swing' amplification of a trailing wave packet from an incident leading packet

Then selfgravity brings the star closer, which amplifies the spiral perturbation that they are forming. As everything happens when the wave changes from leading to trailing, Toomre speaks of conspiracy between the shearing of the differential rotation and the epicyclic oscillation. The amplification energy of the wave is gained to the detriment of the rotational energy.

Several simulations made by varying the parameters, especially the velocity dispersion and the degree of openness of the wave, have allowed the conditions favourable for swing amplification to be determined. Define the parameter X as the ratio between the 'projected' wavelength $\lambda/\sin i$ (in fact it is the projection of the wave number $k \sin i$ that is considered) and the critical wavelength: $X = \lambda/\sin i/\lambda_{\rm crit}$, where i is the angle of inclination of the spiral arms (the pitch angle). Recall that every perturbation larger than $\lambda_{\rm crit} = 4\pi^2 G\mu/\kappa^2$ is stable with respect to Jeans instabilities owing to the

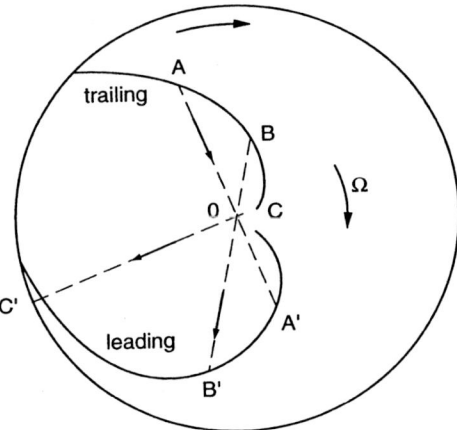

Fig. 5.17. The change in the direction of winding as the wave passes through the centre of the galaxy. On this simplified view the three points of reference A, B, and C of the incident trailing wave after a given propagation time have the positions A', B', and C' so that the paths covered are the same (the group velocity is constant): $AA' = BB' = CC'$. The reference points now trace out a leading wave

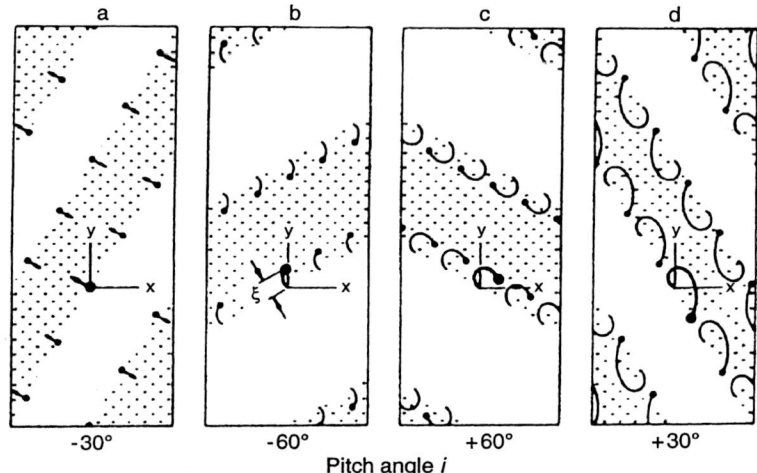

Fig. 5.18. The principle of swing amplification, which derives from the cooperation between differential rotation, epicyclic oscillation, and selfgravity. The diagram represents the motion of stars trapped in a piece of spiral arm (the cross-hatched region). The x direction is radial, towards the exterior of the galaxy; the y direction is tangential, in the direction of the rotation. In (**a**) the wave is leading; it opens up in (**b**). In (**c**) and (**d**) it has become a trailing wave by differential rotation. During this time the selfgravity, which attracts the stars in the arm, and the epicyclic motions conspire to keep the stars in the arm and reinforce the amplitude of the wave. (From Toomre 1981)

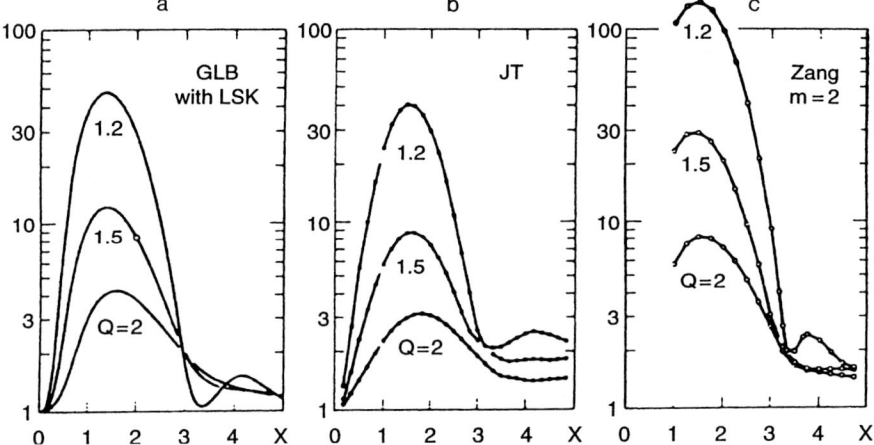

Fig. 5.19. The variation of the swing-amplification rate as a function of the parameters Q and X (defined in the text). The amplification is much less efficient for a 'hot' system (with high Q); on the other hand the amplification rate is a maximum for $X = 2$ and decreases as soon as $X = 3$ (that is to say, when the selfgravity of the disc is cancelled by the addition of invisible mass with a spherical distribution, for example). The three systems of curves compare the results obtained by (**a**) Goldreich and Lynden-Bell (1965), (**b**) Julian and Toomre (1966), and (**c**) Zang's simulation (1981). (From Toomre 1981)

effects of rotation (see Sect. 5.1.1 on rotational stability)(Figure 5.19 shows results only for $X > 1$). The maximum amplification is reached for $X = 2$ and decreases abruptly as soon as $X = 3$. Selfgravity is not strong enough to agglomerate particles over distances much greater than the critical instability size (this is the case when κ increases, and thus the proper motions of dispersion are more important, and also when m is smaller). In the same way it is easy to understand why the mechanism is much more effective in a colder disc where selfgravity is predominant (see the curves in Fig. 5.19, for several values of Q).

5.2.3 Shock Waves Induced in the Gas

The very thin lanes of dust overlying the spiral arms (see M51 and M81), or those still more spectacular streaked across the bar of SB galaxies (such as NGC 5383 and NGC 1097; Fig. 6.4), very quickly suggested the existence of shock waves induced in the gas by the gravity perturbation of density waves. The continuous radio emission is considerably amplified in these dust lanes (Fig. 2.4), which confirms the compression undergone by the interstellar gas and later that of the magnetic field, more or less frozen in the matter.

In 1964 Lin and Shu had already noticed that since the gas was much 'colder' (its velocity dispersion being smaller) than the stars, its response to the density waves would be much stronger (inversely proportional to σ_ν^2). As the arm–interarm density contrast must then greatly exceed that of the stars, the contribution to the gravitational perturbation of the wave itself would be substantial, even though the gas rarely represents more than 10% of the total mass.

The Continuous Interstellar Medium

Until the end of the 1970s the interstellar medium was modelled as a continuous medium with two components: a cold component, consisting of atomic-hydrogen clouds at a temperature $T = 100\,\text{K}$, and a hot, ionized medium with $T = 10^4\,\text{K}$ in the neighbourhood of young stars. As the majority of the mass is in the cold medium, the gas is very sensitive to the gravitational potential well of the spiral arms. More precisely the potential energy of the well corresponds to a kinetic energy such that the relative velocity of the gas becomes supersonic: a very faint shock wave is formed where it enters the arm. The point of entry corresponds to the concave side of the arm if the spiral structure is trailing, and the corotation is located within the limits of the visible disc, as was generally agreed.[1]

The splendour of the spiral structure is due to these shock waves and the compression of gas as it enters the arms. The compression of the gas causes gravitational instabilities inside gas clouds, which were at equilibrium until then, and their collapse into stars or clusters of stars. Obviously stars of all masses can be formed, but it is the massive stars that are essential for the radiation. The luminosity of a star varies roughly in proportion to the fourth power of its mass ($L \propto M^4$), and even if one takes into account the initial mass spectrum of stars ($L \propto M^{-1.5}$), massive stars dominate. Now the lifetime of massive stars is also very short (varying roughly as the ratio M/L, that is, proportional to M^{-3}): for OB associations the lifetime is typically of the order of several 10^7 years, which is the same order as the mean time taken by matter to cross through a spiral arm. Most massive stars whose formation is triggered by the shock wave stop shining when they exit the spiral arm.

The study of the circulation of gas in a galaxy perturbed by a spiral density wave was developed within the context of this model by Roberts in 1969 (Fig. 5.20). The idea that the formation of massive stars is triggered

---

[1] In fact the position of the corotation is not known but is deduced from observations by certain arguments: the position of the dust lanes (on the concave or convex side) can be a sign, although this is very uncertain and is associated with the angle at which we see the galaxy (the dust being visible only from the side nearest the observer). Another argument is the presence of a forbidden region around the corotation. In linear perturbation theory, at least, we should observe a break in the spiral arms in this region, which is not the case, whence the hypothesis that the corotation takes place outside the visible disc. We shall see later on that the position of the corotation is linked to the generating mechanism of the spiral wave (a bar, companion, or some such).

5.2 The Density-Wave Theory

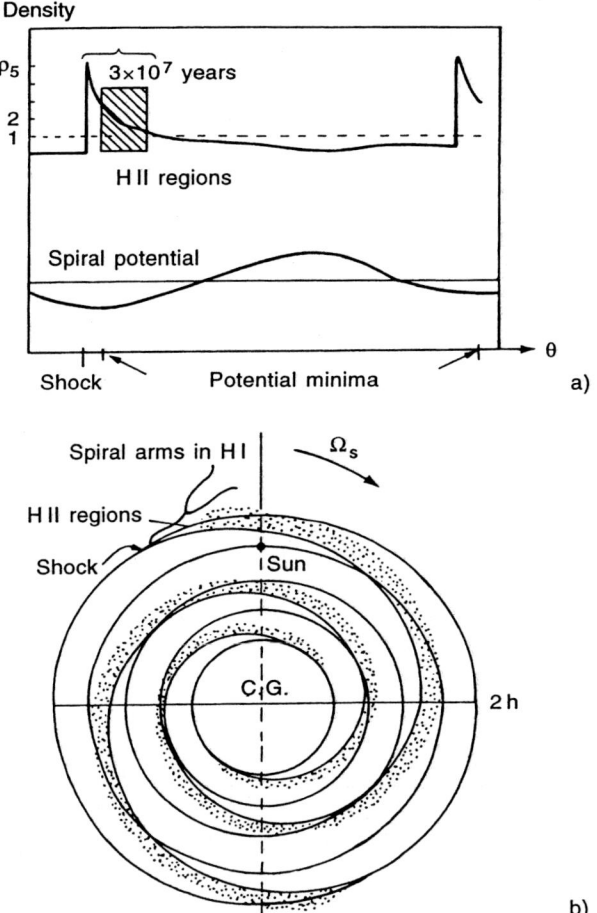

Fig. 5.20. The nonlinear behaviour of interstellar gas, considered as a continuous medium, in a spiral-shaped gravitational potential. Shock waves are formed as the gas enters the spiral arm; the arm–interarm density contrast can reach 5 to 10. (**a**) The compression of gas as it enters the arm sets off the Jeans instability and the formation of new, massive stars. These ionize the surrounding gas, which could explain the presence of many H II regions behind the shock wave (traced by the dust). (**b**) On crossing through the arm the gas experiences large variations in its velocity; systematic deviations are thus represented on the quasicircular trajectories of the gas. (From Roberts 1969)

by the crossing of an arm by gas, even though the mechanisms are not well explained, has proved very rich in terms of the opportunities it offers for observational checks. It enables us to explain the presence of H II regions, 'aligned on spiral arms like pearls on a string' (Baade 1963), the largest arm–interarm contrast in the young component (interstellar gas, ionized regions, and young stars), and the presence of dust lanes. In this way we understand how spiral arms can be very luminous even though the spiral perturbation of the total mass (including stars) is weak (less than 10%).

Interstellar Clouds and the Warm Medium

The first models simulating the response of the gas to a spiral density wave were based on an oversimple view of the interstellar medium, which underwent many revisions in the 1980s. Observations of molecules at millimetre wavelengths (Chap. 2) established that most of the mass of the interstellar medium is contained in dense molecular clouds, whose volume filling factor is very small ($f \approx 3\%$). Moreover X-ray observations and theoretical considerations (McKee and Ostriker 1977) have established that these interstellar clouds are immersed in a tenuous and hot medium, the 'coronal' medium, so-called because of its similarities to the gas of the solar corona (whose temperature is approximately 10^6 K and density 10^{-3}cm^{-3}). This medium would be due to the spatial overlap of bubbles of plasma around the remains of supernovae, and its volume filling factor could reach 80%. Such a gas does not respond at all to the spiral potential, because of its high temperature and its corresponding velocity: that of sound, v_s (v_s^2 is much greater than the energy of the spiral potential well). Only interstellar clouds can be sensitive to it.

Can we regard the ensemble of interstellar clouds as a fluid, whose molecules would be the clouds, with an equivalent 'temperature' measured by the velocity dispersion between clouds? This hypothesis of a fluid would allow the results obtained by Roberts with a continuous interstellar medium to be generalized. However, the ensemble of clouds is only a very imperfect fluid, because the collision time t_{col} is of the same order of magnitude as the crossing time for a spiral arm (t_c can be up to 10^8 years for large molecular clouds). The distribution of the ensemble of clouds does not have enough time to reach equilibrium in the dynamical time (thus we cannot define the equivalent 'temperature'). The clouds must then be regarded as ballistic particles in the spiral gravitational potential, subjected to collisions.

To give a simple and vivid solution to the problem assume that the spiral structure is very tightly wound up ($kr \gg 1$): then the spiral perturbation introduces only radial forces of the form $A\sin(kr + m\Phi)$; as the azimuthal motion of the particles is unchanged, we can write that $\Phi = (\Omega_0 - \Omega_p)t$ in the reference frame rotating at Ω_p. Hence the problem comes down to calculating the motion of N oscillators in one dimension (the radial one) whose proper frequency is the epicyclic frequency κ and which are subjected to a sinusoidal perturbation $A\sin[kr + m(\Omega_0 - \Omega_p)t]$. This simple problem of classical mechanics is illustrated in Fig. 5.21, where one can clearly see the accumulation of particles in the potential well of the progressive wave. In two dimensions this accumulation would correspond to a crowding of the orbits, analogous to that in Fig. 5.11.

In the real problem the collisions between clouds must be taken into account, which favours accumulation if these collisions are inelastic: some larger

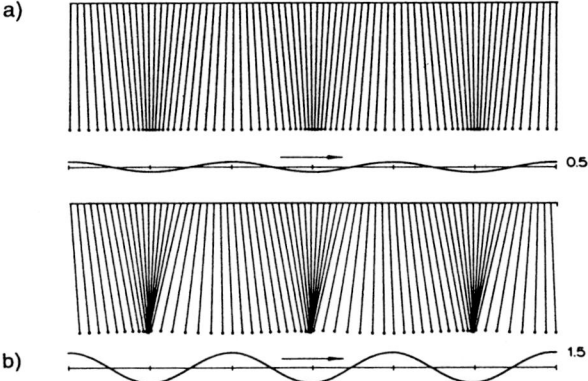

Fig. 5.21. The flow of molecular clouds in a spiral potential, treated by analogy with the simple (one-dimensional) mechanical problem of N oscillators with an epicyclic frequency κ. The sinusoidal perturbation is of the form $A\sin(kx - m\Omega' t)$. The pendulums accumulate in the potential well (**a**). For stronger perturbation amplitudes there can even be collisions between pendulums (**b**). (From Toomre 1977)

and denser clouds will be formed by coalescence. We thus expect the arm–interarm contrast to be more pronounced for the more massive clouds; observations of the giant molecular clouds (GMCs) in nearby spiral galaxies (M31, M51) seem to confirm the predictions. As these GMCs are the birthplaces of new massive OB-type stars, the activation of star formation in the spiral arms is in this way explained. With respect to models of a continuous interstellar medium, all the clouds accumulate in the spiral arms, without shock waves, strictly speaking, but stretch out over a distance of the order of a few free mean paths, or to 1 kpc, the typical total thickness of a spiral arm. In the general case there is no longer a well-defined sequence (as was the case for a continuous medium) between the compressed gas, the star formation, the ionized regions, and the atomic gas (unless the density of molecular clouds is exceptionally high and reduces the mean free path).

Damping of the Waves

Whether by the shock waves created in a continuous medium or more likely by the inelastic collisions of interstellar clouds induced by the crowding of the stream lines in spiral arms, a lot of energy is dissipated at the expense of the wave energy. For example in our Galaxy, given that there is a small proportion of stars with a velocity dispersion low enough to participate actively in the density waves, almost half the wave potential is due to the gaseous medium. The lifetime of the waves can thus be estimated to within one or a few galactic rotations, if no mechanism maintains them indefinitely.

In conclusion the problem of the damping of waves confirms what has already been mentioned about the swing-amplification process: we must give up seeing spiral structures as quasipermanent waves, whose lifetime is of the same order as that of the galaxy, that is, tens of rotations (10^9 to 10^{10} years). After all, the most beautiful examples of spiral density waves are in galaxies that are clearly interacting with a companion (M51 and NGC 5195; M81 and M82) or in barred galaxies. In just a short time from when it formed, a spiral galaxy could amplify several wave packets started by an external perturbation, such as the tidal interaction with a neighbouring galaxy. Between two density waves the galaxy would appear to be of stochastic type. How can the passing of a companion be the origin of density waves in spiral galaxies? This is what we are going to examine now, leaving the problem of barred galaxies for the next chapter.

5.3 Spiral-Wave Generation Mechanisms

Why do spiral density waves trail in the majority of cases? Why is the amplification so efficient for trailing waves and not for leading ones? We have seen in the previous section that the dispersion relation for the waves does not give any answers to these questions, the wave number appearing only as a square: the two kinds of wave thus play equivalent roles. This behaviour must be linked with the fact that if the galaxy is in a stationary state and only stars and gravitational forces are taken into account, the symmetry with respect to time inversion associates a leading solution with any trailing solution. According to this symmetry the morphology of the galaxy is maintained but velocities are reversed, as is the direction of the rotation. This is the basis of the 'antispiral' theorem of Lynden-Bell and Ostriker (1967).

5.3.1 Angular-Momentum Transfer

If there exist mainly trailing waves, this can be explained by the role that waves play in the transfer of angular momentum from the centre to the edge of the galaxy. To decrease its total energy a disc galaxy tends to transfer its angular momentum to the outside. Now trailing waves transfer momentum from the centre to the edge, whereas leading waves do so in the opposite direction.

The natural evolution of an isolated system is to seek states of maximum entropy. These states are those where the kinetic energy associated with the disordered motions is maximal. For a galaxy in a stationary state satisfying the virial theorem the total energy E (which is fixed), the sum of the kinetic energy T and the potential energy W, is such that $W = 2E$ and $T = -E$. T is decomposed as the sum of rotational and random energies, $T = T_{\rm rot} + T_{\rm rand}$, $T_{\rm rot}$ being the only kinetic energy of rotation. As T is fixed, an increase in

T_rand implies a decrease in T_rot. To find the corresponding tendency in the distribution of angular momentum in the galaxy let us write T_rot as a function of the distribution $\mu(j)$, μ being the mass whose angular momentum $j = rv_\theta$ is included between j and $j + \mathrm{d}j$. T_rot is simply defined as $\frac{1}{2}\int j^2 r^{-2} \mu(j)\, \mathrm{d}j$. Therefore a decrease in T_rot implies that the mass must head towards the exterior with its momentum, so that r increases. Of course, there must be some compensation for W to remain constant, but W is strongly balanced by the central masses. There can thus be a transfer of angular momentum to the outer regions with an increase of the outer radius and a small contraction of the inner regions, by means of which $W = $ constant is preserved. More generally the natural evolution of the system will be towards states of minimum energy, the energy gained in this way being used to increase the entropy (disordered motion).

Let us now look for the minimum-energy states for an ensemble of two particles with masses m_1 and m_2, momenta per unit mass j_1 and j_2, and energies per unit mass e_1 and e_2. The energy of the ensemble is $m_1 e_1 + m_2 e_2$; the total momentum must remain constant: $J = m_1 j_1 + m_2 j_2 = $ constant.

Denote by $\epsilon(j)$ the minimum energy per unit mass that a star with a given momentum j can have. Now

$$e = \left(v_r^2 + v_z^2 + j^2/r^2\right)/2 + U(r, z).$$

$U(r, z)$ is minimal when $z = 0$, and the minimum of e corresponds to a circular trajectory with a radius R_j (where $v_r = v_z = 0$). R_j is defined by the relation

$$\frac{\mathrm{d}e}{\mathrm{d}r} = 0 = -\frac{j^2}{r^3} + \frac{\partial U}{\partial r} = 0.$$

The energy sought for, $\epsilon(j)$, is then

$$\epsilon(j) = \frac{1}{2}\frac{j^2}{R_j^2} + U(R_j, 0).$$

Let us return to our ensemble of two particles. The variation of the energy of the ensemble can be written as

$$\mathrm{d}E = m_1\, \mathrm{d}j_1\, \epsilon'(j_1) + m_2\, \mathrm{d}j_2\, \epsilon'(j_2),$$

where

$$\epsilon'(j) = \frac{\mathrm{d}\epsilon}{\mathrm{d}j},$$

and

$$m_1\, \mathrm{d}j_1 + m_2\, \mathrm{d}j_2 = 0.$$

Therefore we must calculate the derivative

$$\epsilon'(j) = \frac{j}{R_j^2} + \frac{\partial R_j}{\partial j}\frac{\partial}{\partial R_j}\left(\frac{j^2}{2R_j^2} + U\right) = \frac{j}{R_j^2} = \Omega(R_j),$$

$$\mathrm{d}E = m_1\, \mathrm{d}j_1\left(\epsilon'(j_1) - \epsilon'(j_2)\right) = m_1\, \mathrm{d}j_1(\Omega_1 - \Omega_2) < 0.$$

For dE to be negative the exchange of momenta must occur to the benefit of the particle with the smallest Ω. Now for a galaxy Ω decreases from the centre to the exterior. It is therefore the outer regions that must receive the angular momentum.

The mean torque exerted by the spiral perturbation was estimated by Lynden-Bell and Kalnajs (1972). The fundamental result is that the sign of this torque is opposite for trailing and leading spirals. The torque exerted on outer regions by inner regions owing to the axial asymmetry is positive in the case of trailing waves. Furthermore if the wave is quasistationary, exchanges of angular momenta between stars and the wave can occur only at the Lindblad or corotation resonances. It is easy to understand that the variations of angular momentum for a nonresonant star will cancel on average, because the star interacts with the wave at phases very different from its orbit. On the other hand the effects will accumulate at the resonances. Detailed calculation by Lynden-Bell and Kalnajs shows that the emission or absorption of j by resonant orbits does not depend on the direction of winding of the spiral: stars give off some angular momentum at the ILR and absorb some at the corotation or the OLR. Thus only trailing waves are able to transfer j from the emitters to the absorbers. The fact that stars lose some angular momentum and energy at the ILR is illustrated in Fig. 5.22, where tangential forces due to the spiral perturbation are illustrated. These forces cancel by symmetry if the path is circular but accumulate for an elliptical resonant orbit. At the OLR the rotation direction is reversed, as is the sense of the exchanges of e and j. The corotation mechanism is different and has certain features in common with Landau damping (in plasmas).

5.3.2 The Excitation of Spiral Waves by a Companion

The passage of a companion creates characteristic tidal forces (Chap. 7). These tidal forces are bisymmetric; they are due to the differential of the force of attraction exerted on the stars by the companion. As the nearer side is attracted more, while the far side is attracted much less, the galaxy is stretched (see Fig. 5.23 in the reference frame where the target galaxy is at rest). The bisymmetry of tidal forces exerted by the Moon and the Sun on the oceans in this way explains the two terrestrial tides every 24 hours. In its Fourier decomposition the tidal potential has a dominant term in $m = 2$. Obviously this estimate is only valid for two companions sufficiently far apart, that is, so that their distance $D \gg R$ (R is the characteristic radius of the galaxies). When the two companions move closer, the term in $m = 1$ can predominate.

Simulations of the encounter of two companions represented by central potentials and test particles (Chap. 7) show in a remarkable way how perturbations can develop in the outer regions. Given that selfgravity is missing from these simulations, waves that develop look like the kinematic waves we described in Sect. 5.2.1: the orbit of a star that was circular becomes elliptical

5.3 Spiral-Wave Generation Mechanisms

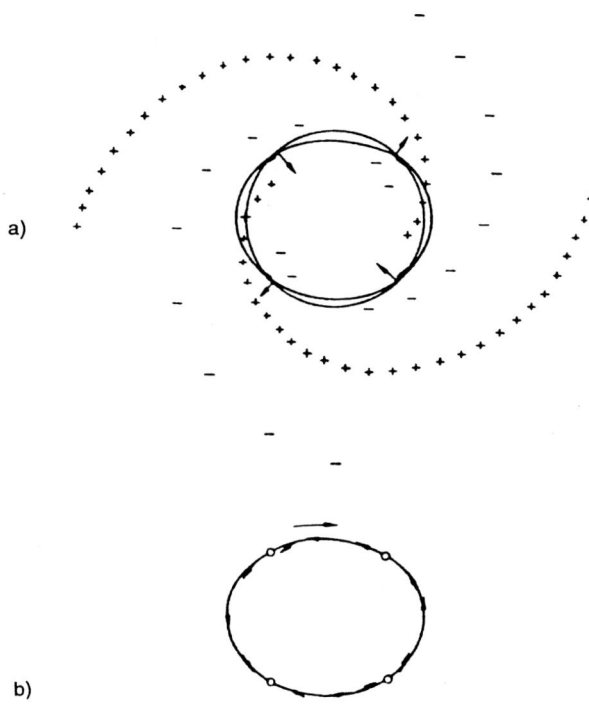

Fig. 5.22. Forces exerted on stars by the spiral perturbation at the inner Lindblad resonance (ILR). (**a**) The excess of radial forces and the corresponding deformation of the circular orbit. (**b**) The excess of tangential forces experienced by the perturbed orbit. The forces are always directed in an opposite sense to the star's motion in its orbit: the star will therefore lose energy and angular momentum. This is all the other way round at the outer resonance. (From Lynden-Bell and Kalnajs 1972)

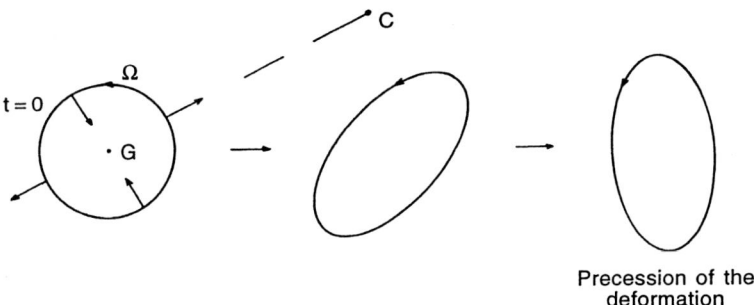

Fig. 5.23. The tidal forces exerted by a companion C at a distance relatively far from the galactic centre G amount to a stretching in the direction GC (Chap. 7). The passage of the companion therefore has the effect of exciting the epicyclic oscillations of the particle, whose orbit becomes elliptical and precesses at the rate $\Omega - \kappa/2$

under the stretching of tidal forces; the ellipse is aligned with the companion at the start. In other words the companion has excited epicyclic oscillations. All the deformed orbits precess at slightly different rates and wind up as a spiral. But how can we explain the presence of a very contrasted internal spiral wave that prolongs the outer spiral arms, for example, as in M51? This problem has for a long time cast doubt on whether the companion NGC 5195 of M51 is able to cause its great spiral wave: tidal forces are zero at the centre of the target galaxy and vary as r^2 with distance from the centre. The wave must therefore be excited within the limits of the galaxy and must propagate towards the centre; but the propagation time is relatively long (a few galactic rotations), and the companion should have had time to disappear. But this reasoning does not take into account the mechanism of swing amplification (Sect. 5.2.2). Even a very weak tidal perturbation at the centre can be amplified and will be much more quickly amplified than in the outer regions, because the causes of the amplification (differential rotation and epicycles) are all faster (see the curves giving Ω and κ, which decrease considerably from the centre to the edge). The cycle of amplification is so fast at the centre that an initially very weak perturbation can become stronger in amplitude than the very strong outer perturbations. All this happens before waves propagate from the centre to the edge (Fig. 5.24; N-body simulations by Zang and Toomre 1981).

Exercises

5.1 Show that circular orbits in a given axisymmetric potential are unstable if the angular momentum per unit mass j decreases towards the exterior. Is this potential possible with a mass distribution that has a spherical symmetry $M(r)$?

5.2 We showed in Sect. 5.1.2 that the epicycles had axial ratios $x/y = \kappa/2\Omega$. In general $\kappa < 2\Omega$, and the epicycle is elongated along the tangential direction. Show that the radial velocity dispersion σ_r and tangential velocity dispersion σ_θ have the inverse ratio:

$$\sigma_r/\sigma_\theta = 2\Omega/\kappa.$$

For this calculate at a radius r_0 the variations of radial and tangential velocities $v_r(r_0)$ and $v_\theta - v_{\text{circ}}(r_0)$ for a group of stars whose orbits have some mean radii r.

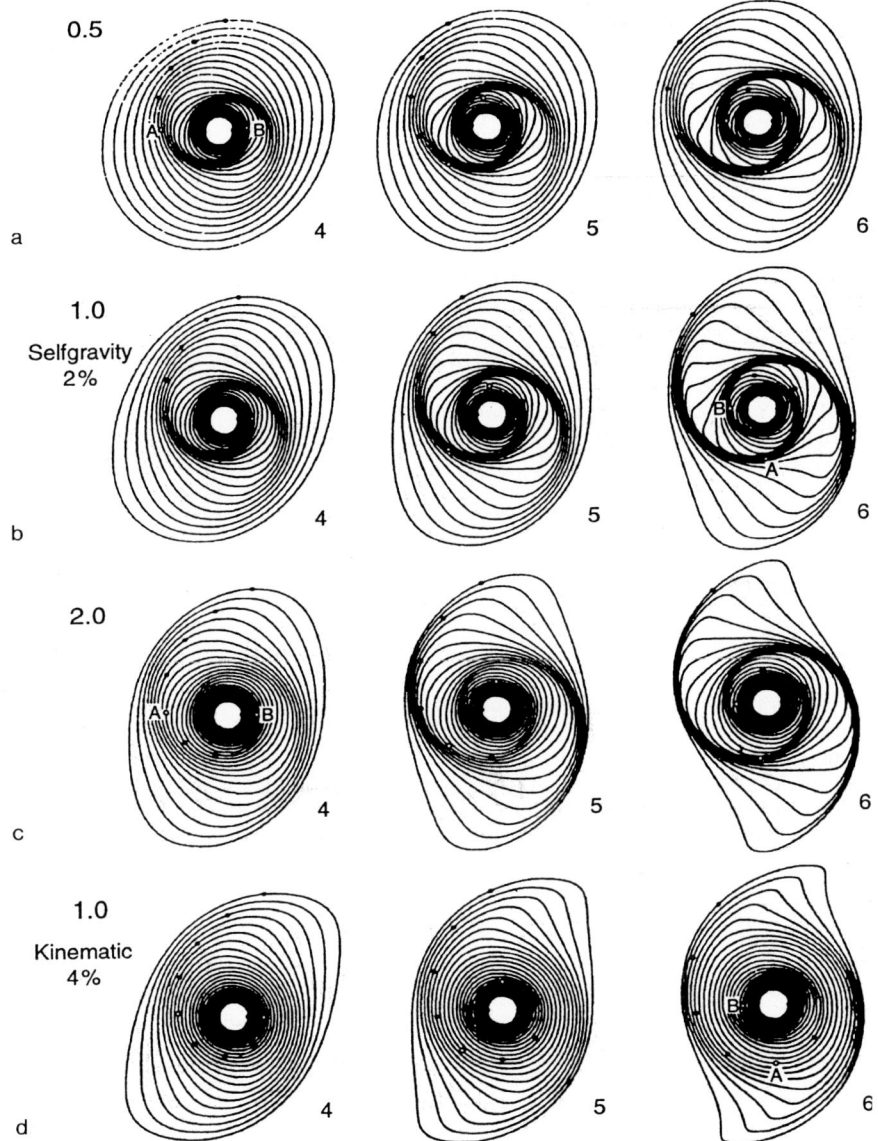

Fig. 5.24. The response of a uniform disc of particles to a tidal impulse of the form $\cos 2\theta$. In (**a**), (**b**), and (**c**) the selfgravity is taken into account and the time of the impulse during which tidal forces act is increased in the ratios 0.5, 1, and 2 respectively; the amplitude of the perturbation is the same (2%) in the three cases. In (**d**) selfgravity is neglected: only kinematic waves develop, and the inner regions of the galaxy are practically unperturbed, even if the amplitude of the tidal forces is doubled (4%). Selfgravity, which enables 'swing' amplification to play its role, is thus essential for a spiral density wave to develop in the inner regions. (From Toomre 1981)

References

Arp, H. (1966) *Atlas of Peculiar Galaxies* (California Institute of Technology, Pasadena CA).
Athanassoula, E. (1984) *Phys. Rep.* **114**, 319.
Baade, W. (1963) in *Evolution of Stars and Galaxies*, edited by C. Payne-Gaposchkin (Harvard University Press, Cambridge MA).
Binney, J., and Tremaine, S. (1987) in *Galactic Dynamics* (Princeton University Press, Princeton NJ).
Contopoulos, G. (1980) *Astron. Astrophys.* **81**, 198.
Elmegreen, D. M., and Elmegreen, B. G. (1984) *Astrophys. J. Suppl.* **54**, 127.
Gerola, H., and Seiden, P. E. (1978) *Astrophys. J.* **223**, 129.
Goldreich, P., and Lynden-Bell, D. (1965) *Mon. Not. Roy. Astron. Soc.* **130**, 125.
Julian, W. H., and Toomre, A. (1966) *Astrophys. J.* **146**, 810.
Kalnajs, A. J. (1973) *Proc. Astron. Soc. Australia* **2**, 174.
Lin, C. C., and Shu, F. H. (1964) *Astrophys. J.* **140**, 646.
Lindblad, B. (1963) *Stockholm. Obs. Ann.* **22**, 5.
Lynden-Bell, D., and Kalnajs, A. J. (1972) *Mon. Not. Roy. Astron. Soc.* **157**, 1.
Lynden-Bell, D., and Ostriker, J. P. (1967) *Mon. Not. Roy. Astron. Soc.* **136**, 293.
McKee, C. F., and Ostriker, J. P. (1977) *Astrophys. J.* **218**, 148.
Mark, J. W. K. (1971) *Proc. Nat. Acad. Sci. USA* **68**, 2095.
Mark, J. W. K. (1976) *Astrophys. J.* **206**, 418.
Roberts, W. W. (1969) *Astrophys. J.* **158**, 123.
Sandage, A. R. (1961) *The Hubble Atlas of Galaxies* (Carnegie Institution of Washington, Washington DC).
Schmidt, M. (1965) in *Galactic Structure*, edited by A. Blaauw and M. Schmidt (University of Chicago Press, Chicago IL), p. 513.
Toomre, A. (1964) *Astrophys. J.* **139**, 1217.
Toomre, A. (1969) *Astrophys. J.* **158**, 899.
Toomre, A. (1977) *Annu. Rev. Astron. Astrophys.* **15**, 437.
Toomre, A. (1981) in *The Structure and Evolution of Normal Galaxies*, edited by S. M. Fall and D. Lynden-Bell (Cambridge University Press, Cambridge).
Zang, T. (1981) quoted in Toomre (1981).

6. Barred Galaxies

The majority of spiral galaxies are barred. How is the bar represented in the dynamics and evolution of these galaxies? The bar and the spiral are without doubt intimately linked, the bar possibly being the origin of the spiral wave (or vice versa). Why, then, are not all spiral galaxies barred?

6.1 Observations

Barred spiral galaxies are by no means the exception, since about two-thirds of spirals are classified as barred. However, only a third are very strongly barred (SB) and another third are classified as 'intermediate' SAB galaxies (having a faint bar). It is possible to detect very small barred perturbations owing to gaseous components: this very 'cold' gas, that is, gas with a very small velocity dispersion, responds more strongly than the stars to any perturbation of the potential. Generally speaking, atomic gas is absent at the centre of spiral galaxies, and it is molecular hydrogen, traced by emission from the CO molecule, that is observed with a barred morphology.

The winding of the arms in a strongly barred galaxy is characteristic. The arms wind round from the ends of the bar and the structure always remains wide open (at about 180°). The presence of a density wave is almost guaranteed in a galaxy once it is barred. According to the work of Bruce and Debra Elmegreen (1983; see Table 6.1) on the fraction of galaxies with a 'global' spiral structure and those with a 'stochastic' spiral, almost all barred galaxies are found to have density waves: the existence of a continuous and uniform spiral structure from the bar to the exterior, a spiral structure also at red wavelengths, and so on.

Table 6.1 lists the percentages of spiral galaxies that have a global spiral structure with two arms (G) compared with those with a stochastic spiral structure (S) according to the environment (isolated galaxies, binary galaxies, or groups of galaxies) and to their type (barred or not).

From these results one can deduce that tidal interaction with a companion is an efficient mechanism for exciting spiral waves.

Curved dust lanes, which are characteristic of SB galaxies, delineate the bar (Fig. 6.1); they do not pass through the centre of the galaxy but coincide with the leading (convex) edge of the bar either side of the centre. They are

150   6. Barred Galaxies

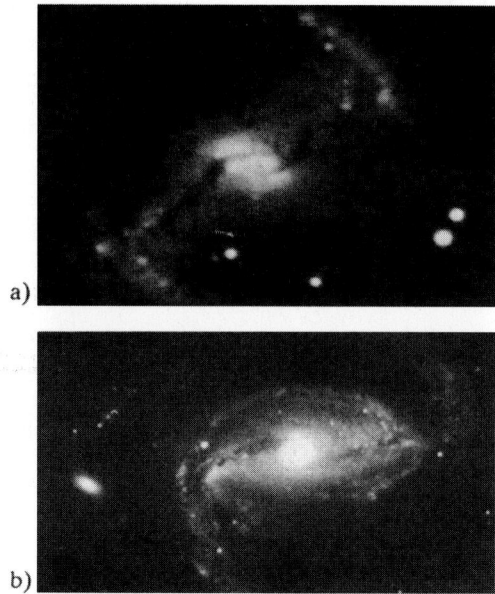

Fig. 6.1. SB barred galaxies have characteristic curved dust lanes along the bar and at the leading edge of the two arms of the bar. (**a**) NGC 5383 and (**b**) NGC 1097 are good examples. (From Sandage 1961)

Table 6.1. The Percentage of Galaxies with a (Global) Spiral Density Wave, according to their Environment. (From Elmegreen and Elmegreen 1983.)

Environment	Morphological Type	Number of Stochastic Spirals	Number of Global Spirals	Percentage of Global Spirals
Isolated	SA	15	7	32%
	SAB	7	16	70%
	SB	4	11	73%
Binary	SA	3	4	57%
	SAB	1	16	94%
	SB	1	11	92%
Grouped	SA	15	32	68%
	SAB	21	38	64%
	SB	12	45	79%
Total:		79	180	69%

(The error in the percentage of global spirals is estimated to be ±10%.)

the tracer of intense compression waves in the arms of the bar. The bar itself is composed of only old stars, its luminosity distribution being smooth and its colour red. The extremities of the bar and spiral arms, on the other hand, show some signs of star formation (H II regions and young stars).

The velocity field of a barred galaxy, which is very perturbed with respect to a velocity field of circular rotation, is also a characteristic that allows us to detect or confirm the presence of a weak bar. If the angles of projection on the plane of the sky are suitable, that is, if the galaxy is not seen face-on and the major and minor axes are not aligned with the bar, the isovelocity curves form an S shape instead of the classical spider pattern (Chap. 3). These systematic perturbations are due to very elliptical orbits, aligned along the bar as opposed to being, on average, circular.

As a result of these kinematic perturbations it was realized that our own Galaxy is probably barred. Towards the direction of the Galactic centre (longitude $= 0°$) expansion at $\pm 200\,\mathrm{km\,s^{-1}}$ has been observed in the H I gas at a wavelength of 21 cm. If the velocity field were circular, the gas would be observed with zero radial velocity. Owing to absorption effects in the radio-continuum emission spectrum of the Galactic centre, it has been possible to identify the gas moving at $200\,\mathrm{km\,s^{-1}}$ as being the nearest: this is the famous '3 kpc arm', whose apparent expansion was attributed for a long time to an explosion at the Galactic centre. In fact it is simply gas moving on elliptical orbits aligned with the bar. Besides the Magellanic clouds, which are irregular barred galaxies, a bar has also been identified in our nearest neighbouring galaxy, the Andromeda galaxy (M31).

Bars are biaxial or triaxial structures: we can consider them to be ellipsoids with two or three different axes. In the galactic plane their axial ratio is as high as 2.5 to 5, and in the perpendicular direction they are very flattened systems. It is possible that the two minor axes are similar and that the bar looks like a cigar (a prolate system). It is difficult to recognize a bar in a galaxy seen edge-on. However, there are central bulges that are seen edge-on, with a 'boxlike' or 'peanutlike' shape, which could be the sign of a bar seen at an inclination of 90° (see Sect. 6.2).

Bars do not have a very concentrated distribution of light: their surface brightness is almost constant along their major axis. In this respect these triaxial structures are very different from elliptical galaxies or the central bulges of spiral galaxies, for example.

6.2 The Theory of Bar Formation

Bars are also density waves. In fact they can be regarded as quasistationary waves, formed by the combination of trailing or leading wave packets. If there is no inner Lindblad resonance, short trailing waves are propagated towards the centre of the galaxy and leave again as short leading waves. These, by means of swing amplification, can return to the centre as short trailing waves. These successive reflections can create stationary waves if the amplification is strong enough. One conclusion is already obvious if linear theory is still valid: the bar must have a size equal to or less than the corotation radius, where the waves are reflected, and its speed must be higher than the maximum of

$\Omega - \kappa/2$; otherwise the existence of inner Lindblad resonances would cut the amplification cycle. These same conclusions can be obtained by considering the nature of orbits in a barred galaxy, which we shall now do.

6.2.1 Orbits in a Barred Galaxy

The bar creates a bisymmetric gravitational potential (the main term of its Fourier decomposition is $m = 2$), which rotates in the galaxy at an angular speed Ω_p. To reduce this to a time-independent potential, where test particles conserve their energy, we have to consider it in the rotating reference frame. In cylindrical coordinates (r, θ, z) the equivalent potential is then

$$\Phi_{\text{equiv}} = \Phi(r, \theta, z) - \Omega_p^2 r^2/2.$$

The energy of a particle in this reference frame is

$$E_J = v^2/2 + \Phi - \Omega_p^2 r^2/2.$$

This is an integral of the motion known as the Jacobian. The equation of motion in the rotating reference frame can be written as

$$\frac{d^2 \boldsymbol{r}}{dt^2} = -\boldsymbol{\nabla}\Phi - 2\boldsymbol{\Omega}_p \times \boldsymbol{v} + \Omega_p^2 \boldsymbol{r}$$

(the last two terms being respectively the Coriolis force and the centrifugal force). Taking the scalar product with $\dot{\boldsymbol{r}}$, we see that the Coriolis force does no work and that the centrifugal force derives from the potential $-\Omega_p^2 r^2/2$.

Notice that the Jacobian is written as a function of the energy in the fixed reference frame: $E_J = E - \Omega_p L_z$, L_z being the angular momentum. Obviously this is not a constant of the motion, since $\partial U/\partial \theta \neq 0$.

The appearance of the equivalent potential in the rotating reference frame is given by the equipotentials in Fig. 6.2. The five stationary points, L_1 to L_5, are called Lagrangian points. L_3 is the central minimum of the potential. Points L_4 and L_5 are maxima and L_1 and L_2 are saddle points, that is, maxima for the x direction and minima for the perpendicular direction. At all the Lagrangian points the gravitational force balances the centrifugal force, and a star can remain stationary there (corresponding to a circular orbit at an angular velocity Ω_p in the fixed reference frame). Only points L_4 and L_5 can be stable (according to the form of the potential), that is to say, a star can oscillate around these points at the epicyclic frequency κ. The equilibrium at L_1 and L_2 is always unstable. The four points L_1, L_2, L_4, and L_5 correspond to the corotation: although they are not necessarily at the same distance from the centre, they delimit a ring-shaped region that is called the corotation zone.

The orbits in a barred potential have been calculated in detail by Contopoulos and Papayannopoulos (1980). They examined a series of periodic orbits delimiting populations according to the energy (the Jacobian), and thus roughly according to the distance from the centre:

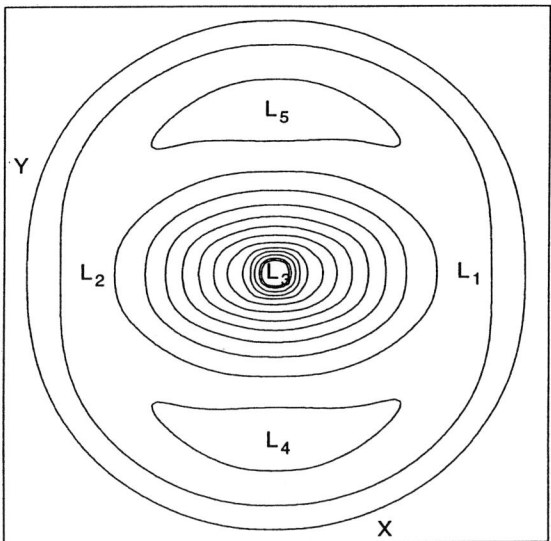

Fig. 6.2. Contours of the equivalent potential of a barred galaxy in the reference frame rotating at $\Omega_{\rm p}$: $\phi_{\rm equiv} = \phi(r,\theta) - \Omega^2 r^2/2$. The bar is parallel to the x axis. See the text for the definition of the Lagrangian points L_{1-4}

- Very close to the centre, at low energies, orbits are elongated parallel to the bar and form part of the so-called x_1 family (Fig. 6.3). There also exists an x_4 family of retrograde orbits (sparsely populated in spiral galaxies).
- Between the two inner Lindblad resonances (if they exist) we find the x_2 family of orbits elongated perpendicular to the bar, direct (in the sense of the rotation of the bar itself), and stable. Orbits of the same type but unstable also appear (the x_3 family). In the case of a very strong bar (with practically no axisymmetric component) the x_2 family no longer exists. The force of the bar necessary to eliminate the x_2 family depends on its angular velocity $\Omega_{\rm p}$: the lower this velocity, the stronger the bar must be.
- Between the second ILR and the corotation we again find orbits elongated parallel to the bar of the x_1 family, but they have undergone a change of shape at the transition; they have secondary lobes at their extremities. Orbits then become rounder, having more and more complex forms near to the corotation.
- At the corotation there exist families of periodic orbits around stable Lagrangian points L_4 and L_5 that do not revolve around the centre in the rotating reference frame.
- Far away from the corotation the angular velocity of the bar is now well above that of the stars. The perturbation potential is averaged in azimuth to obtain a quasiaxisymmetric potential. The orbits are then almost circu-

6. Barred Galaxies

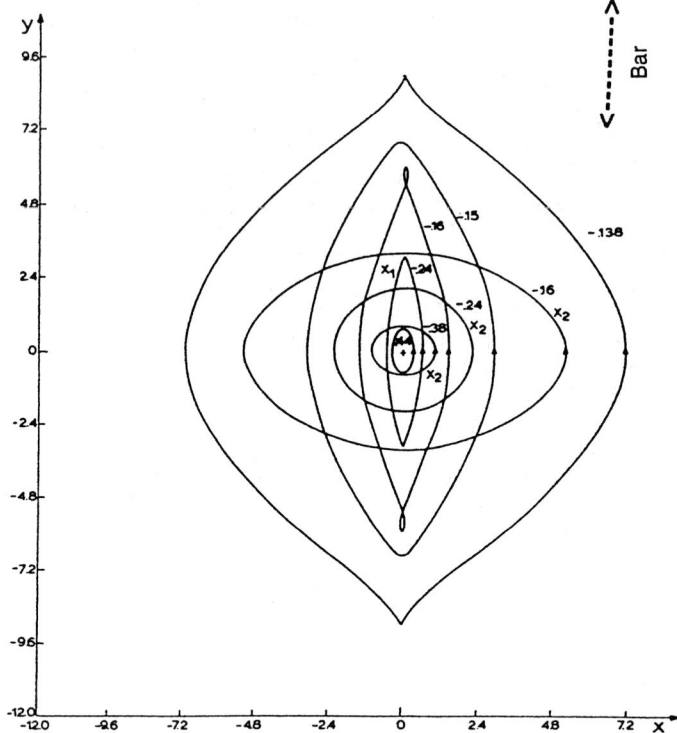

Fig. 6.3. Five periodic orbits from among the x_1 family and three from the x_2 family corresponding to different values of the Jacobian. The x_1 orbits are aligned with the vertical bar, while the x_2 orbits are perpendicular to it. The bar has an intermediate amplitude and rotates slowly, so as to allow the presence of inner Lindblad resonances. (From Contopoulos and Papayannopoulos 1980)

lar and there exist direct periodic orbits aligned with the bar and retrograde perpendicular orbits (the commonest in a galaxy beyond the corotation).

In the potential of a very strong bar the majority of orbits perpendicular to the bar between the corotation and the OLR become unstable. This is why this region often becomes depopulated to the benefit of a ring at the corotation (which is perhaps the origin of the θ form of some strong bars).

All these results can be summarized by the characteristic curves of the families of orbits; these are obtained by plotting, for example, the intersection of the orbits with the y axis as a function of their Jacobian (Fig. 6.4). The dashed curve represents the curve of zero velocity where $E_J = \Phi_{\mathrm{equiv}}$.

These families of orbits are sufficient to determine the nature of all the regular orbits of the galaxy: indeed most stars are trapped by stable closed orbits; they oscillate around the periodic orbit closest to the family. However, irregular orbits can exist, which are manifested in the surfaces of section as

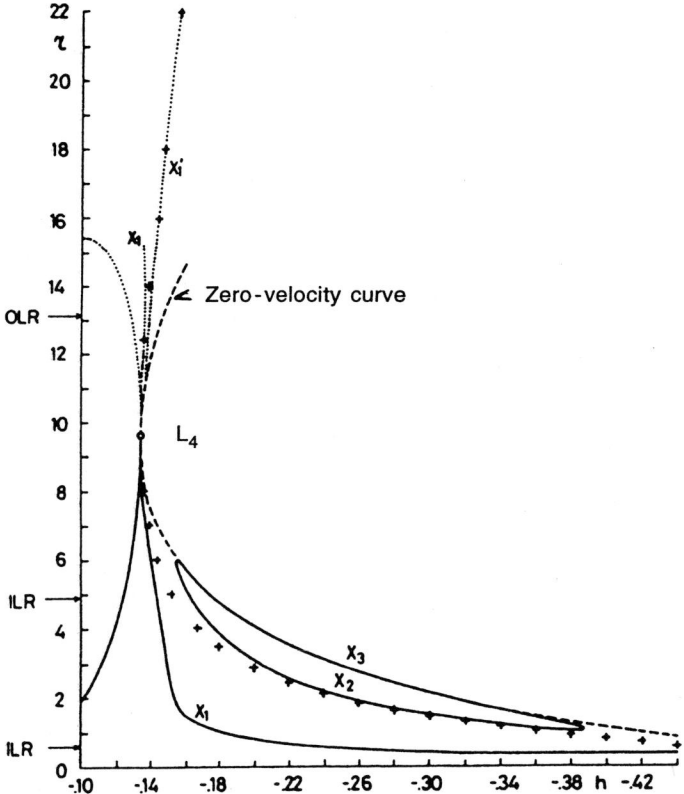

Fig. 6.4. Characteristic curves of periodic orbits for the slow bar of intermediate force corresponding to Fig. 6.3. The points of intersection of the orbits with the y axis perpendicular to the bar are represented as a function of the Jacobian E_J. (The solid line represents direct orbits, the dotted line, retrograde orbits.) The curve of zero velocity, which is used as a confining envelope for all the characteristics, is drawn as a dashed line. Crosses represent circular orbits in the axisymmetric potential unperturbed by the bar

a sea of random points, while the intersections of regular orbits are aligned on curves (suggesting the existence of a second integral in addition to the Jacobian). These irregular orbits, which are also said to be 'ergodic', appear as soon as the potential shows some important asymmetries (for example in the case of very strong bars). For realistic potentials the ergodic orbits are found mainly in the regions external to the corotation.

Which orbits are responsible for constructing the bar? Among the families given previously the orbits of the x_1 family aligned parallel to the bar spend most of their time in the bar's potential well and amplify it. But it can be seen that stable and direct orbits (those which interest us for the purposes of representing spiral galaxies) rotate by roughly 90° at each resonance: the x_2

156   6. Barred Galaxies

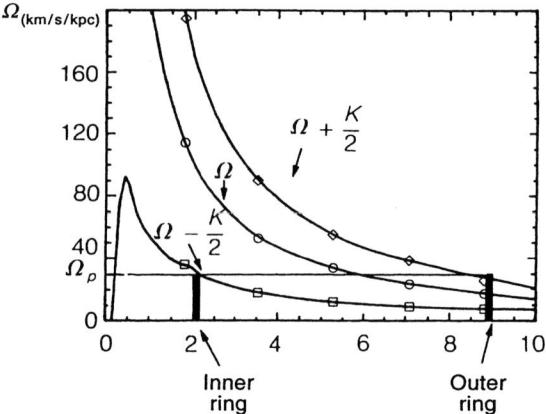

Fig. 6.5. Rotation curves $\Omega(r)$ and derived curves $\Omega - \kappa/2$ and $\Omega + \kappa/2$ in the barred galaxy NGC 4736, deduced from observations of ionized gas at the centre (the Hα line) and atomic H I gas (at 21 cm) in the outer regions. This galaxy has two well-contrasted rings at $R = 1.5$–2 kpc and $R = 9$ kpc. They correspond to the inner and outer resonances respectively. The function $\Omega(r)$ increases very rapidly when one approaches from the centre, as does $\Omega - \kappa/2$, and there is no doubt that an ILR exists

family, which is confined between the two ILRs and is oriented perpendicular to the bar, tends to weaken and even destroy it if the region considered is sufficiently large (Ω_p is small). We find here a result already obtained when considering the process of wave amplification: the presence of internal resonances destroys the bar. Beyond the corotation, stable orbits are also perpendicular to the bar and also ergodicity comes in, which is why a bar cannot be elongated beyond its own corotation. This result, which is also confirmed by the theory of density waves, is very valuable for determining the angular velocity Ω_p of a barred wave: all we have to do is measure Ω at the end of the bar on the observed rotation curve.

In many galaxies the velocity of the bar defined in this way generally allows the presence of two inner Lindblad resonances: the mass distribution is so concentrated that the rotation curve reaches a marked maximum at the centre, the bar thus having a velocity much less than $\Omega - \kappa/2$ (Fig. 6.5). Hence where the ILR is located there exists a ring of matter and sometimes a spiral structure inside the ring. This can be explained in terms of the periodic orbits determined above. In the region surrounding the ILR, orbits are no longer aligned with the bar; the bar is interrupted and a secondary system of much faster waves can develop, with its own resonances: corotation, an outer spiral, and an inner bar. This time this second bar does not have an ILR (see the photograph of the galaxy NGC 4314 in Fig. 6.6).

To summarize, in a barred galaxy, orbits forming the bar are all aligned with it, being butterfly-shaped at the centre, then becoming almost rectan-

6.2 The Theory of Bar Formation   157

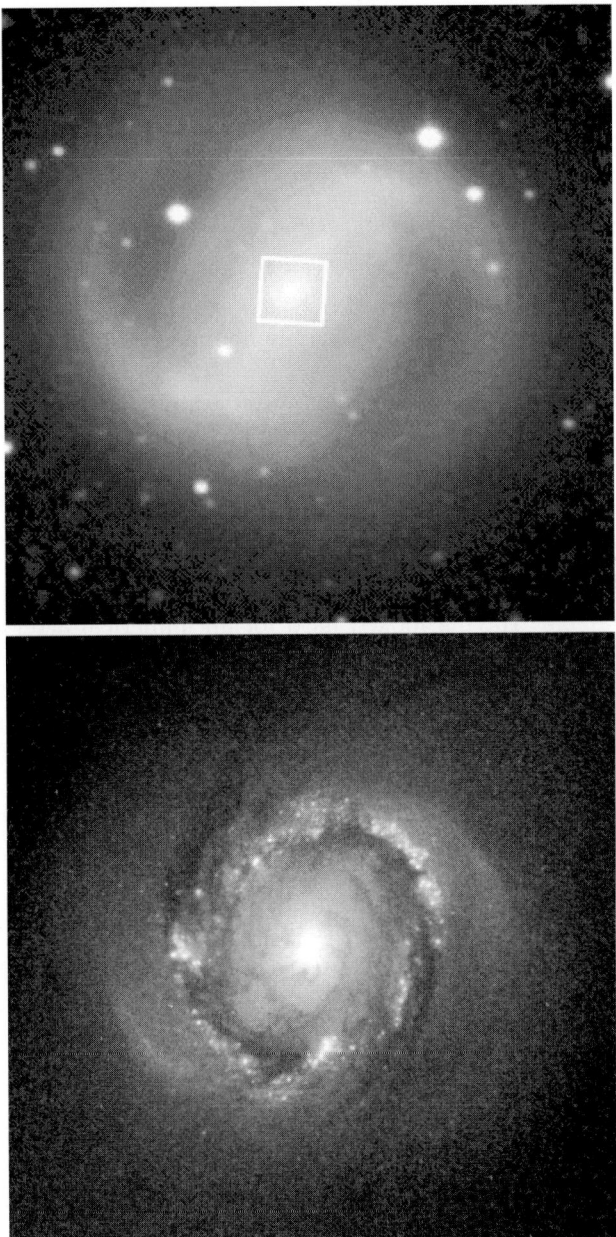

Fig. 6.6. Photographs of the barred galaxy NGC 4314. (*top*) A general view of the bar and the spiral arms (photograph from McDonald Observatory, Texas). (*bottom*) A closer view of the central parts (corresponding to the square in the first picture), with the Hubble Space Telescope. A nuclear ring can be discerned, within which a second independent spiral structure has developed (from Benedict et al. 1996)

gular, and finally rounding to ellipses between the ILR and the corotation. For an observer in a fixed reference frame the axis of the orbits is always aligned with the bar and rotates slowly with a velocity Ω_p. On the other hand according to the density-wave scheme, stars move at a much higher velocity and make several orbits during one revolution of the bar.

6.2.2 The N-Body Problem

Searching analytically for the modes and linear instabilities based on the theory of weak perturbations is useful but cannot give us complete information about the stability of a galaxy once highly nonlinear phenomena are involved (perturbation theory can then no longer be applied). This is why solution of the N-body problem by computer simulation is an indispensable and complementary tool; numerical calculations are behind most of the progress made in recent years in this field.

During the 1970s more and more simulations of spiral galaxies were made, in the hope of determining what the predominant spiral modes are and also what the mechanisms of amplification are. To the surprise of the researchers involved, the simulations would always terminate sooner or later by forming a stable stellar bar. Having looked hard for all numerical artefacts we must yield to the facts: the predominant instability of a flattened system of stars is not a spiral but a bar. Before considering all the consequences of this fundamental result, let us explain first how the gravitational interaction of the 10^{11} or so stars that form a galaxy can be accounted for numerically.

Numerical Methods. The direct summation of the forces acting on N particles by the $N-1$ neighbouring particles is very costly in terms of calculations, the time necessary to simulate the dynamical time increasing in proportion to N^2. Many improvements have been made by Aarseth (1979), in particular by treating each particle with its own time step: the periods generally vary from the centre to the edge of a system of stars. Nevertheless it is not possible to treat more than $N \approx 1000$ particles using such a method. Now the number of particles in an N-body simulation determines the spatial resolution, and this is therefore insufficient for a spiral galaxy.

As we explained in Sect. 5.1, a system of $N \approx 10^{11}$ stars can be considered collisionless and two-body collisions negligible. If we then simulate $N \approx 10^3$ particles interacting by means of the correct gravitational forces, the two-body relaxation time for two bodies comes a lot closer to the dynamical time: the potential due to the 1000 particles is too irregular, and there are far too many 'collisions' for the simulation of the system to be able to represent reality. The solution is to 'soften' Newton's law, by choosing, for example, a $1/(r^2+a^2)^{1/2}$ potential instead of $1/r$ (a is known as a 'softening' parameter). a is the distance below which simulations will be unable to give any information about the system: for example gravitational condensations are impossible at this scale. Thus the spatial resolution is limited by a, and the smaller N is, the more severe the limitation will be. To represent large-scale structures correctly, such as spiral waves, whose thickness is physically of the order of a kiloparsec, we must have a resolution of at least 300 pc, which requires N to be of the order of 10^4–10^5.

A method that allows us to treat such numbers makes use of *fast Fourier transforms (FFTs)* to solve the Poisson equation. Calculation of the potential $\Phi(r) = \iiint G\rho(r')\,\mathrm{d}r'/|r'-r|$ reduces to calculating the convolution of the density function by the $1/r$ law (or the chosen softened potential); in Fourier space the convolution product becomes a simple product, considerably reducing the difficulty of the calculations. Fourier transforms are taken in a discrete way over a two- or three-dimensional grid.

The use of Fourier series presupposes a periodic function, which implies that the galaxy will be represented by multiple images. To ensure that there are no interactions between the images, the Fourier transform is taken in each dimension on a grid twice as large as the grid used containing the galaxy (a grid 4 times larger in two dimensions and 8 times larger in three dimensions). If 'n' is the order of the series (the size of the grid), the calculation time necessary for the FFT increases as $n \log n$. As the number N of simulated particles is of the same order as the number of grid cells, the total calculation time increases roughly linearly with N. The calculation of forces and the advance in time being independent of each particle, the total time required also increases linearly with N. This is the enormous advantage compared with direct summation (where the time required is proportional to N^2), which allows the simulation of $N \approx 10^{5-6}$ bodies.

A variant of this method consists in replacing the Cartesian coordinates by cylindrical coordinates, which are well adapted to disc galaxies. The grid is then polar with cells of variable size from the centre to the edge, allowing a greater spatial resolution at the centre where the density is higher. Another advantage of this variant is that the Fourier transform along the azimuthal direction does not require a grid larger than the grid used for the galaxy, the periodicity in θ being natural.

These grid methods are well suited to the simulation of a single galaxy, but their disadvantage is not being able to treat particles once they have escaped from the grid during a strong perturbation (due to instabilities, a companion, and so on). The resolution of the N-body problem by means of multipolar developments avoids this stumbling block, while requiring a calculation time that increases linearly with N.

Another widely used method is the '*tree code*', which is particularly suited to clumpy systems, such as interacting galaxies. When the space filling factor is small, it is penalizing to compute the gravitational field throughout the entire volume of a large grid, since most of the cells are empty. It is then preferable to follow particles, as in Lagrangian methods, and to adapt the resolution which can be higher in populated regions. The tree code introduces approximations in the calculation of the potential, in order to keep the resolution high in dense clumps and to avoid constraining outer boundaries: the interaction between nearby particles is treated exactly, as in direct methods, but for the interaction between distant particles, only the first terms of the multipolar expansion are retained. To select the procedure, particles are first ordered in a hierarchical tree, which is composed of nested cells, the first one being the root containing all particles, which is divided into smaller and smaller cells, until the leaves are reached, which correspond to a single particle (Barnes and Hut 1986). The force on a given particle is computed by walking down the tree, beginning from the root. At each step, the size of the current cell is compared with its distance from the particle, the ratio determining the angle through which the particle sees the cell or cluster of distant particles. If this angle is smaller than a fixed tolerance parameter, then the force is computed with a truncated multipole expansion. If not, the descent down the tree is continued and the cell is subdivided, until the angle is small enough or a leaf is reached. Since in this procedure the size of the unresolved cells increases with the distance to the particle, the latter has a number of interactions $\propto \log(N)$ instead of N in the case of the direct sum: the cost of the simulation will then grow as $N \log(N)$ instead of N^2.

Note that the tree code is often combined with SPH ('Smoothed Particle Hydrodynamics') to treat a self-gravitating fluid. The two algorithms have a similar shape, a Lagrangian approach, with the highest possible resolution in dense regions, without grid limitations. In SPH, the gas is modelled as an ensemble of fluid elements, the size of which can vary with space and time, such that a few elements overlap at any point of the simulated volume. All physical quantities and forces are computed by averaging over the nearest neighbours. These are found using the hierarchical tree (Hernquist and Katz 1989).

Tests of Stability. *Test 1.* The first test attempted consists in ensuring the validity of the Toomre criterion for axisymmetric perturbations (a galaxy is stable if $Q = \sigma_r/\sigma_{r\min} > 1$. Hohl's experiment (1971), with 10^5 bodies on a two-dimensional 256×256 lattice, initially assumes a Kalnajs disc in rotational equilibrium with a surface density $\mu = (1 - r^2/a^2)^{1/2}$ and a potential $\Phi = \Omega_0^2 r^2/2$ for which the rotation is rigid: $v(r) = \Omega_0 r$. The initial radial velocity dispersion σ_r corresponds to $Q = 1$, and the azimuthal dispersion satisfies the relation $\sigma_\theta/\sigma_r = \kappa/2\Omega$ ($= 1$ here), deduced from the epicyclic theory (see the exercises in Chap. 5). As some of the kinetic energy corresponds to the disordered motion, the rotation speed is consequently less than the circular velocity at each point, according to the distribution function deduced from the Boltzmann equation. The advantage of the choice of a Kalnajs disc is that the equation can be solved analytically: the azimuthal speed of the particles still corresponds to rigid rotation but with a smaller angular velocity:

$$v_\theta = 0.8\Omega_0 r.$$

6.2 The Theory of Bar Formation

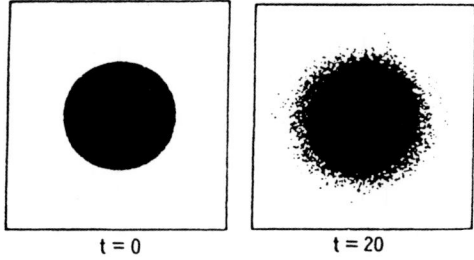

Fig. 6.7. Numerical simulations of the N-body problem, showing the evolution of a disc of stars in rigid rotation, $v(r) = \Omega_0 r$, with 10^5 particles. The unit of time is the period of rotation of the corresponding cold disc (itself in rigid rotation at Ω_0 with a period $2\pi/\Omega_0$). In this first step, the tangential forces are cancelled; only the radial forces calculated by the FFT method are considered: thus the simulation tests the Toomre stability criterion with respect to axisymmetric perturbations. With $Q = 1$ the disc is very stable with respect to these perturbations (From Hohl 1971)

No mass condensation occurs during this simulation, confirming the validity of the criterion (Figs. 6.7 and 6.8). To avoid confusing this problem with possible nonaxisymmetric perturbations Hohl kept only the radial forces and cancelled the tangential forces on the particles. Such simulations with $Q < 1$, on the other hand, give rise to fragmentation of the system into small aggregates.

Notice that to test these problems of stability there must already be a large number of particles, so as not to have to use too high a softening parameter a: indeed a brings an additional stability to the system, perturbations with a size $\lambda < a$ being unable to condense (the gravitational force tending towards zero at small distances). Even if the velocity dispersion is zero, the coefficient Q equivalent to such a cold system whose gravity is softened by a is $Q \approx 0.3\kappa(a/G\mu)^{1/2}$.

Test 2. If the above system, which was once relaxed, is now subjected to unconstrained tangential forces, it becomes globally unstable with respect to $m = 2$ perturbations (Fig. 6.8). The criterion $Q > 1$ is therefore not sufficient to guarantee the global stability of the galaxy. The $m = 2$ perturbation first has the form of a spiral, but this is only transient, lasting for just one galactic rotation. Then a barred $m = 2$ perturbation is set up, which is quasistationary and lasts for more than the Hubble time.

Although the result is not consistent with the expected quasistationary spiral wave, it nevertheless corresponds to an intense density wave which is strong and has a long lifetime. The barred wave rotates in the same direction as the stars but about two times slower. More precisely in the large number of experiments conducted with different initial conditions, the velocity Ω_p of the bar approximates the maximum of the curve $\Omega - \kappa/2$.

162   6. Barred Galaxies

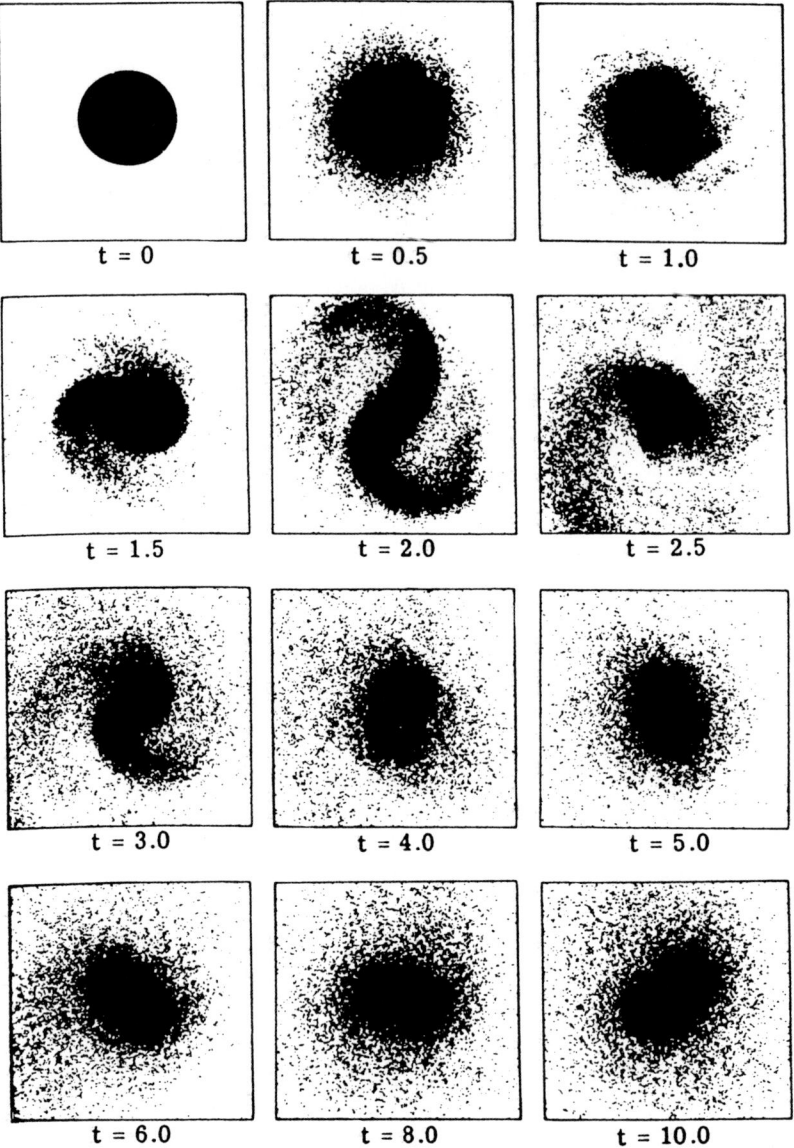

Fig. 6.8. Further numerical simulations of the N-body problem. In this second step, the tangential forces are taken into account: the same disc develops a characteristic bar-shaped instability, initially accompanied by a transient spiral structure. The bar persists until the end of the simulations. (From Hohl 1971)

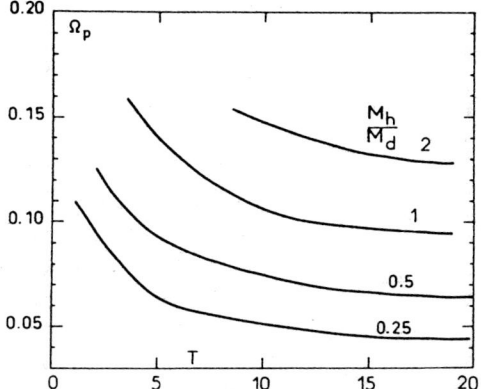

Fig. 6.9. Evolution (according to N-body simulations by Combes and Sanders 1981) of the angular velocity Ω_p of the bar as a function of time as it develops in intensity. The instability first affects the fastest central stars (with high Ω); more and more outer and slower stars are progressively trapped in its potential as the bar grows. The different curves correspond to initial discs that are to a greater or lesser extent unstable, by adding a variable massive halo ($M_\mathrm{h}/M_\mathrm{d}$ is the ratio of the halo mass to the disc mass)

Several theoretical predictions are here seen to be confirmed: the bar extends roughly up to the corotation, and there are few or no stars between the Lindblad internal resonances (these stars would weaken the bar). Furthermore during its evolution, even though the force of the bar always increases, the angular velocity Ω_p continuously decreases (Fig. 6.9). This phenomenon can be understood by noticing that the central particles, whose periods and dynamical times are all shorter, become unstable earlier and start to form a short bar before the outer particles join up with them. Now the velocity of the bar is equal to the rate of precession $\Omega - \kappa/2$ modified by the selfgravity, and this rate is also higher at the centre. As soon as the outer particles join the bar its angular velocity adapts to their lower rate of precession. Another way of viewing the phenomenon is that the bar ends at its corotation, and thus the more extended the bar is in the galaxy, the slower it is.

The role of the transient spiral in bar formation is very important. As we saw in Sect. 5.3.1, the spiral allows the transfer of angular momentum from the centre to the edge of the galaxy, thus decreasing the rate of precession. The barred wave, which has a negative angular-momentum density, can in this way be amplified.

Once a bar is quasi relaxed, in general it ends a little before its corotation, as observations and simulations seem to confirm. But its interaction with an almost spherical non rotating component, such as a bulge, and to a greater extent a dark matter halo, could slow down a bar rather quickly, as predicted through linear perturbative calculations by Weinberg (1985) and confirmed by N-body simulations (Debattista and Sellwood 1998). The mechanism is dynamical friction, which brakes a massive body moving at high speed in a sea of much smaller particles (see Chap. 7 for details). For a bar to be slowed down significantly in its lifetime, the dark halo would have to be very massive, however. The central part of the galaxy, where the bar rotates, would have to be dominated by dark matter, which is in general not the case, except for galaxies with very low surface brightness. Such a massive spherical component could then prevent the formation of a bar, explaining why slow bars are not frequently observed.

The Ostriker–Peebles Criterion. The instability described above explains the existence of bars in about two-thirds of galaxies splendidly. But what about the others? How are unbarred galaxies formed? What are the elements vital to the stability of a galaxy with respect to $m = 2$ perturbations, and can we determine a stability criterion that would play the role of the Toomre criterion for these perturbations?

An obvious element for stabilizing galactic discs is the velocity dispersion. But we need a value of $Q \approx 2$–4 to find the stability, which is well above the velocity dispersion observed in galaxies. Another factor that can be introduced is a massive quasispherical component known as the 'halo' and which would correspond to the invisible mass detected in galaxies by flat rotation curves very far from the centre. This massive component, which is stable in itself, because it derives its equilibrium from its velocity dispersion and not its rotation, should stabilize the disc, whose selfgravity would decrease. However, notice that *only the central part* of this massive component is important for stabilizing bars, which always appear at the centre of galaxies. Now the invisible mass of the 'haloes' is only obvious in the very outer parts of spiral galaxies. Within the optical limits of the galaxy the ratio of invisible mass to visible mass does not generally exceed 1 (see the corresponding simulations in Fig. 6.10).

Ostriker and Peebles (1973) tried to combine these two stability factors into just one criterion: a galaxy would be unstable if the percentage of kinetic energy in the ordered motion (rotation) were too high. The critical parameter chosen is the ratio $t = T_{\text{rot}}/W$; T_{rot} is the kinetic energy of the ordered motion and W is the potential energy. Denoting by T_{rand} the kinetic energy of random motion, we have, from the virial theorem,

$$T_{\text{rot}} + T_{\text{rand}} = -W/2.$$

This formulation has the advantage of taking into account the energy of the disorganized motion of the massive halo through the increase in the corresponding potential energy W. Ostriker and Peebles determined by means of numerical simulations that those systems with $t < t_{\text{crit}} = 0.14 \pm 0.02$ were stable with respect to bar formation. This criterion remains a very important empirical guide to judging the stability of a system.

However, it is possible that this criterion might be a sufficient but not a necessary condition for the stability. According to Toomre (1981), bars are

6.2 The Theory of Bar Formation

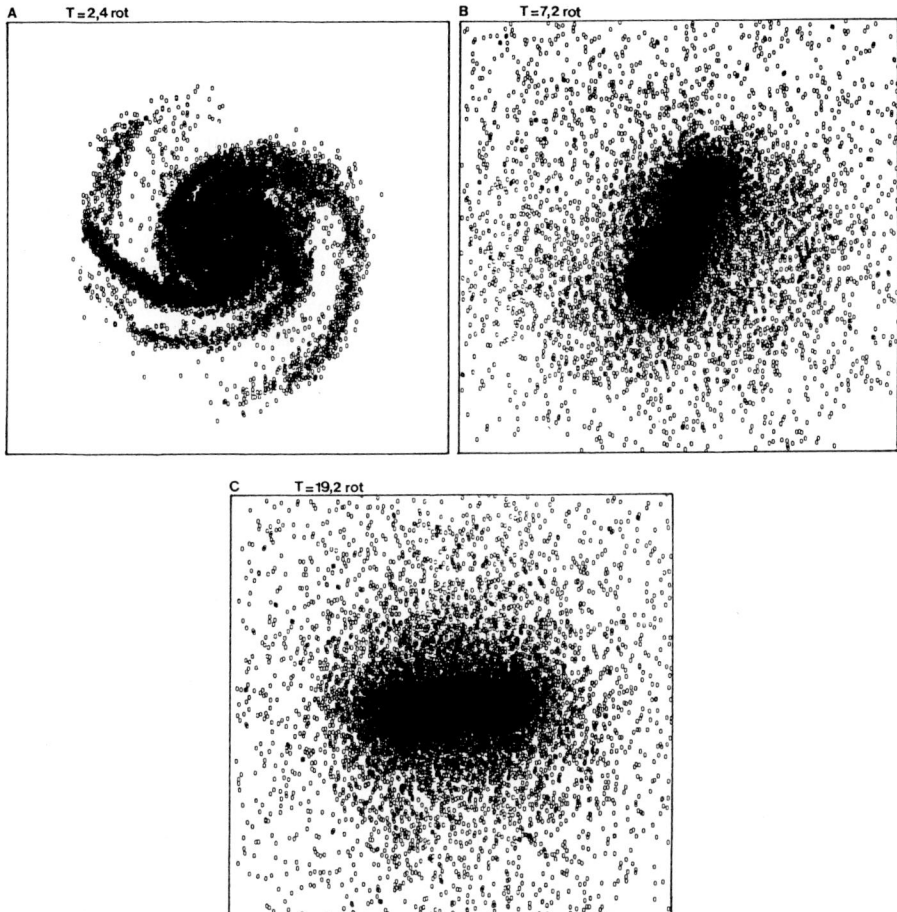

Fig. 6.10. Barlike instabilities, preceded by a transient spiral wave, in an initially cold disc in differential rotation, whose initial surface-density distribution corresponds to a Toomre disc ($\mu = \mu_0(1 + r^2/a^2)^{-3/2}$) (Toomre 1963). The simulation was performed by FFT in three dimensions (Combes and Sanders 1981). A massive spherical halo, whose mass within the optical radius is equal to that of the disc, stabilizes it but not sufficiently to avoid bar formation

density waves that owe their intensity to swing-amplification mechanisms. To prevent the instability we can imagine destroying the amplification, which is possible either by increasing the velocity dispersion (Q) or by varying $X = \lambda/\lambda_{\text{crit}} = \lambda\kappa^2/4\pi^2 G\mu$, that is, the mass contained in a massive spherical component (see Sect. 10.2.2). These two elements correspond exactly to Ostriker and Peebles's arguments: it turns out that the constraint $t < t_{\text{crit}}$ enables the efficiency of the swing to be suppressed. But there is another condition for the amplification to occur, and this is that new leading waves can move back out from the centre each time an amplified trailing wave is going to be propagated there. Now the existence of an internal Lindblad resonance prevents this return by damping down and destroying the trailing wave even before it arrives at the centre. Thus if the mass distribution is sufficiently concentrated towards the interior, then Ω and $\Omega - \kappa/2$ will be large enough. In the extreme model where the rotation velocity $v(r) =$ constant, for example, there has to be an ILR for every wave, and such a disc will be stable even if the swing amplification is efficient (even if $t > t_{\rm crit}$). This argument leaves some hope of being able to understand the existence of spiral galaxies that are not barred; for various reasons the mass distribution can be very concentrated, there being an important central bulge or contraction of the galactic core during the transfer of angular momentum by a transient spiral trailing wave (the passage of a companion) and so on.

6.2.3 Equilibrium Perpendicular to the Plane

The existence of a third degree of freedom, perpendicular to the plane of the galaxy, helps to stabilize the stellar disc. If the thickness of the disc is taken into account, the Toomre stability criterion is broadly unchanged, but the numerical factor defining the critical velocity dispersion $\sigma_{\rm crit}$ must be decreased. This is what numerical simulations confirm, making instabilities less violent for thick discs.

The appearance of the bar itself can also lead to thickening of the galactic plane in certain regions by a resonance phenomenon. The simulations in Fig. 6.11 show a barred galaxy seen edge-on for two possible lines of sight: parallel or perpendicular to the bar. When the bar is seen with its smallest cross-section face-on, it is square-shaped or box-shaped; in the perpendicular direction it has a 'peanut' shape. These two characteristic shapes are often observed in galaxies seen edge-on (and then classified as 'peculiar'). These shapes could therefore indicate the presence of a stellar bar in these galaxies. This hypothesis agrees with the observed frequency of box and peanut shapes in galaxies seen edge-on compared with the frequency of barred galaxies among disc galaxies.

It is difficult to check directly that a peanut-shaped galaxy is also a barred galaxy, since the two features are only seen from orthogonal viewing angles. Kinematical observations are, however, able to detect the peculiar signatures of the bar in edge-on galaxies (e.g. Kuijken and Merrifield 1995). Such spectroscopic work tends to confirm the bar mechanism in the formation of box- and peanut-shaped bulges.

Other hypotheses have been proposed to account for these structures, in particular the accretion of small companions or the fusion of two galaxies (see Chap. 7). The presence of profiles that are apparently not relaxed could be a sign of violent events in the relatively recent past.

How can these particular forms be generated by the appearance of bars? There is a resonance between the oscillatory motion of stars perpendicular

6.2 The Theory of Bar Formation

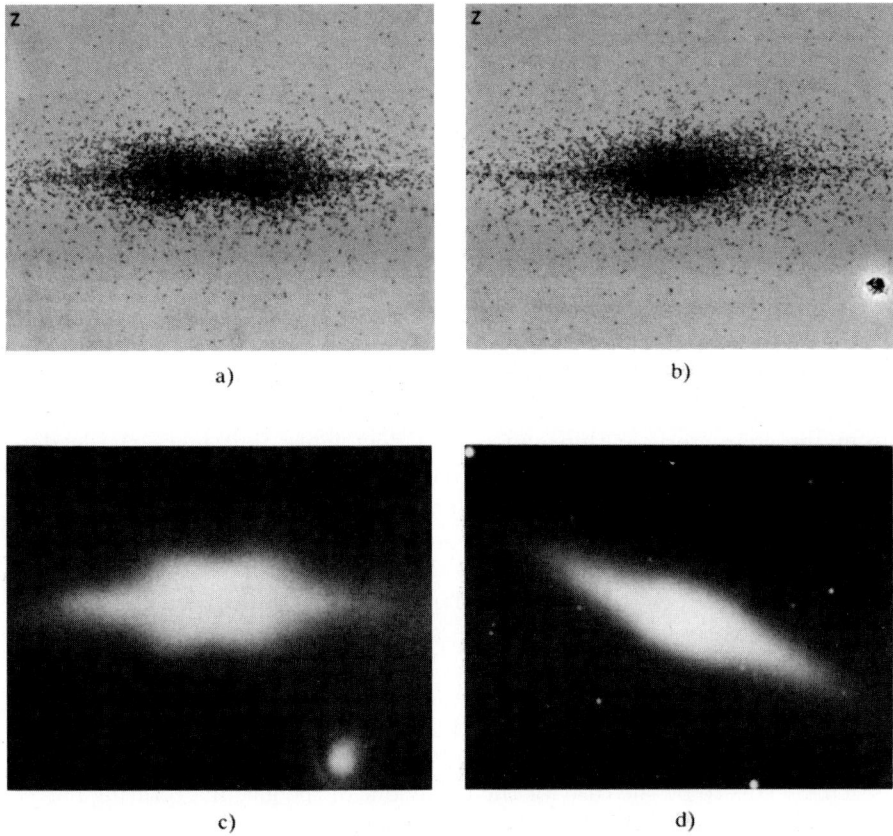

Fig. 6.11. The appearance of the bar can be accompanied by the formation of structures perpendicular to the plane of the disc. (**a**) and (**b**) show the galaxy simulated in Fig. 6.10 seen edge-on, the line of sight being respectively perpendicular and parallel to the bar. This thickening of the plane in certain areas is due to the resonance between the rotational motion of the bar with a velocity Ω_p and the oscillatory motion of stars perpendicular to the plane with a frequency ν_z. The resonance is analogous to the Lindblad resonance in the galactic plane and occurs when $\Omega_p = \Omega - \nu_z/2$. A certain number of galaxies seen edge-on are observed to have a peanut shape (**c**) or a box shape (**d**)

to the plane (with a frequency ν_z) and the rotational motion of the bar. This resonance is analogous to Lindblad resonances in the galactic plane. More precisely it can be shown that the ratio of ν_z to the epicyclic frequency κ is almost 1 at the centre of the galaxy. Given that the angular velocity Ω_p of the bar is always close to $\Omega - \kappa/2$ in this region (indicating the presence of an ILR), the relation $\Omega_p = \Omega - \nu_z/2$ also holds. Particles in this area will always cross through the plane at the same position relative to the bar, and the perturbative action of the bar will then be amplified there. The stars are therefore induced to rise high enough in their oscillations to create the 'peanutlike' profiles.

The phenomenon can be explained by the existence of periodic orbits in 3-dimensions (Combes et al. 1990). The thickening of discs might also be triggered by a bending or buckling instability (Raha et al. 1991). This instability occurs in cold and thin discs, when the ratio of vertical to horizontal velocity dispersion is sufficiently low (below 0.3). The bar can help in this instability, and therefore the resonant and bending-mode mechanisms are related.

6.3 The Response of the Gas to a Barred Stellar Potential

From the characteristic thin and curved dust lanes that mark the leading edges of the bar it can be claimed that the gas is subjected to significant compression waves inside the barred potential. But what would the stream lines of the gas throughout the galaxy, outside the bar, be like? Could the spiral originate in the gas? Remember that up until now the only permanent density waves obtained with the stellar component have been bars. Nevertheless once barred galaxies contain gas they are spirals. Only the barred lenticulars (SB0) and the less massive (Magellanic-type) irregulars are sometimes barred without having spiral waves. Is the presence of the gas, and therefore dissipative processes, needed in the formation of a quasipermanent spiral wave?

This seems to be the case for the behaviour of the gas in a rotating barred potential, as indicated in Fig. 6.12, which shows a simulation by Sanders and Huntley (1976). Contrary to intuition, no spiral has been introduced into the gravitational potential. Initially the gas is distributed homogenously in the disc. Its selfgravity is negligible.

6.3.1 Theory

To understand the phenomenon let us remind ourselves about stellar orbits in a barred potential, as described in Sect. 6.2.1. To sum up and simplify matters: periodic orbits are ellipses either parallel or perpendicular to the bar; their orientation rotates by 90° at each resonance. In particular the orbits are perpendicular to the bar between the two ILRs, if they exist (Fig. 6.13). Under the action of a rotating barred potential the gas clouds would tend to follow these orbits. But unlike stars they tend to collide, in particular when paths cross, when the orientation of orbits changes (during collisions gas clouds change orbits). These changes are no longer abrupt but progressive: the major axes of the ellipses will rotate gradually by 90° at each resonance. The result will resemble the orbit in Kalnajs kinematic spiral waves (Fig. 5.10). Hence the formation of spiral density waves in the gas is explained.

6.3 The Response of the Gas to a Barred Stellar Potential   169

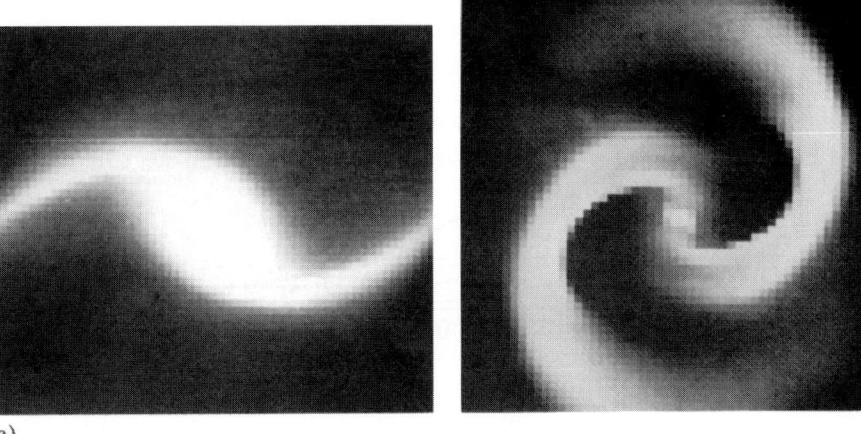

Fig. 6.12. A hydrodynamic simulation of the flow of gas in a rotating barlike potential. The bar is horizontal. The angular velocity Ω_p of the bar determines the geometry of the response by means of the number of resonances induced in the disc. In (**a**) the spiral is wide open and there are only two resonances (CR and OLR); in (**b**) there also exist inner resonances (that is why the response of the gas at the centre is perpendicular to the bar). (From Sanders and Huntley 1976)

Simulations confirm this interpretation: the spiral winds up by 90° at each resonance (Fig. 6.12). If no ILR exists, then stable periodic orbits are aligned with the bar from the centre to the corotation, and the response of the gas is barred all the way to the corotation; it then winds up by 180° (so we have a total of two resonances: CR and OLR). We have seen that in the case of a strong bar the orbits perpendicular to the bar between the two ILRs vanish. It can therefore be deduced in this case that whatever the rotation velocity of the bar is, the spiral begins at its ends and does not wind up by more than 180°. This is in fact what we observe in the strongly barred SB galaxies: wide-open spirals do not wind up more than half a turn.

This winding-up problem allows us to distinguish among spiral waves formed in the young component, those only created by the barred potential, and the selfgravity spiral density waves created by the tidal force of a companion: these latter can wind up for several turns (see NGC 5364).

Under what conditions can we obtain shock waves that could explain the presence of the famous dust lanes along the bars? According to the above arguments, strong collisions must occur and hence the beginnings of a change in the orientation of the gas stream lines: this can only take place if an ILR exists, implying a slow velocity Ω_p. Moreover if most of the gas clouds are close to the ILR, the frequency of the bar corresponds to their precession frequency $\Omega - \kappa/2$, and the response will be the stronger the more resonant it is. This is indeed what the simulations show: a weak perturbation is enough

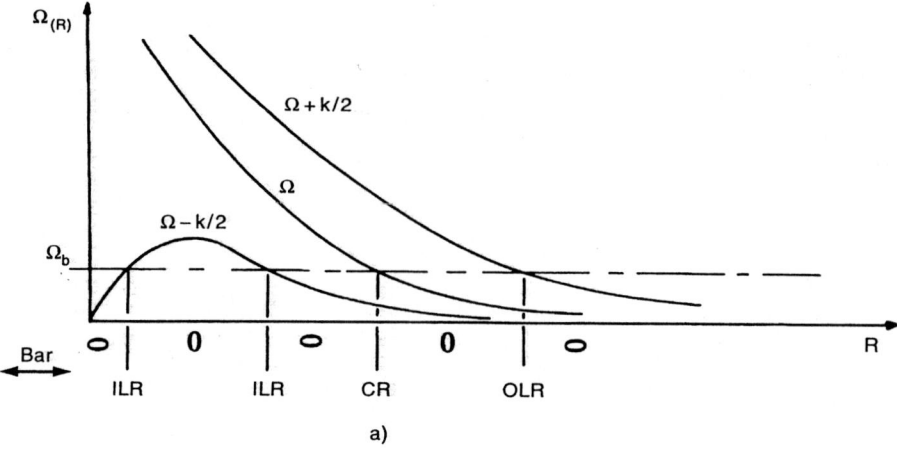

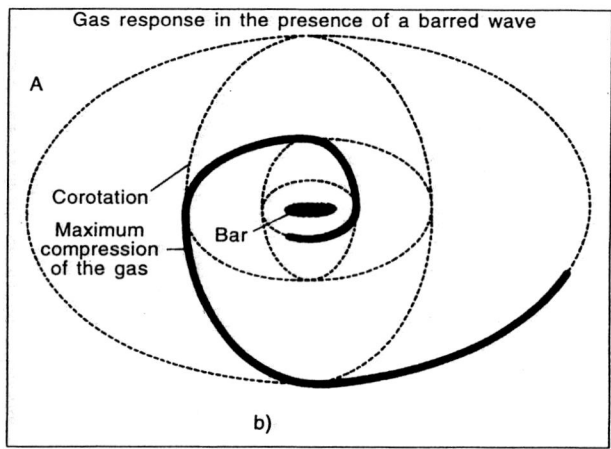

Fig. 6.13. Orbits of test particles in a rotating barred potential are elongated alternately along the bar and perpendicular to it. Put simply, the orientation changes by 90° at each Lindblad resonance (**a**). The gas tends to follow these orbits, but its viscosity (collisions between interstellar clouds) causes the stream lines to rotate progressively (and not violently) at 90° to the resonances. The points where the stream lines narrow form a spiral which turns at each resonance (**b**)

to form these shock waves as long as there is an internal resonance (Fig. 6.14, Athanassoula 1992). In these simulations the stream lines of the gas are particularly elongated. This illustrates well the large noncircular velocities observed in barred galaxies.

6.3 The Response of the Gas to a Barred Stellar Potential

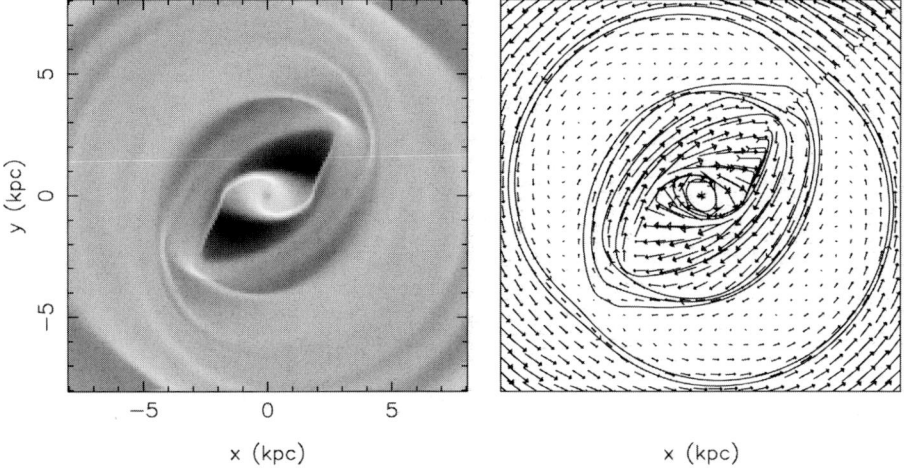

Fig. 6.14. Response of the gas to a bar potential. (*left panel*) The gas density in grey-scale. (*right panel*) The gas flow lines and the velocity vectors in the rotating frame of the bar. The bar potential is at 45° to the horizontal. (From Athanassoula, 1992)

6.3.2 Gas-Cloud Simulations

As in the case of the response of the gas to a spiral perturbation, the model of continuous interstellar gas is useful, but it does not correspond to the reality of the medium, which is composed essentially of dense clouds with a low volume filling factor (3%) immersed in a coronal medium too hot to develop shock waves. Moreover the degree of energy dissipation by collisions has great importance in determining the destiny of spiral waves in the gas. Indeed in the continuous-fluid models described above, the viscosity is introduced artificially in the numerical simulations. It is sufficient to consider the grid on which all the magnitudes of the fluid are calculated: in a given cell of the grid used, all the fluid elements coming from neighbouring cells with different velocities are averaged with respect to that cell; in other words the collisions of clouds would be perfectly inelastic on scales of the order of the cell (the spatial resolution of the simulations). It follows that in general the viscosity is excessive, and in the simulations mentioned above, the gas falls within a few rotations towards the centre of the galaxy, having lost its energy through viscosity. The lifetime of the spiral is therefore very short and depends on the spatial resolution used. Similarly the thickness of the shock zones is a function of this same resolution.

Simulations of interstellar clouds as ballistic particles subjected to dissipating collisions and periodically reacquiring energy in the processes of star formation enable the mean free path (m.f.p.) to be treated in a more realistic way. Indeed even if the number of particles simulated cannot equal the actual number of interstellar clouds, it is sufficient to assign to each particle an effective collisional cross-section higher than the real cross-section, so that the m.f.p. of the particles is equal to that of the clouds; in other words during one galactic revolution each particle will have been subjected to the same number of dissipating collisions as a real cloud. Nevertheless there must be enough particles for the artificial effective cross-section not to limit the spatial resolution beyond the m.f.p. With the physical data determined from observations of the molecular clouds, the m.f.p. is found to be of the order of 200 pc, and 10^4 particles are sufficient for the simulation to be accurate.

Simulations of clouds in a rotating barred potential allow us to study the formation of spiral waves as in a continuous medium and also to find the lifetime of these structures. Notice that if there are strictly speaking no shock waves, there nevertheless exist some very abrupt fronts along the bar and the arms (Fig. 6.15). The ensemble of clouds maintains its velocity dispersion, owing to the reinjection of energy during the dispersion of giant complexes by star formation. However, there is a secular evolution due to the transfer of angular momentum, which we shall now consider.

Torques Exerted by the Bar on the Gas. As the barred potential is not axisymmetric, it subjects the gas to tangential forces. If the gas moved on circular orbits, these forces would on average cancel along an orbit. This is also the case for elliptical orbits aligned with the major or minor axis of the bar. On the other hand as soon as the gas forms a spiral structure, the orbits are elongated but more inclined with respect to the axes of symmetry of the bar. A torque exerted by the bar on the gas results, as depicted in Fig. 6.16. The torque changes sign at each resonance, when the spiral winds up by 90°: between the second ILR and the corotation the torque is negative; beyond the corotation up to the OLR it is positive.

These torques tend to slowly depopulate the corotation region, so causing the gas to accumulate at the Lindblad resonances in the form of rings. Here the gas follows circular or slightly elliptical trajectories aligned along the bar and is no longer subjected to any torque. Schwartz's simulations (1981) (Fig. 6.17) clearly show the formation of an outer ring at the OLR after twenty or so galactic revolutions (half the Hubble time). The time taken to form these rings by a slow drifting of the gas is, of course, a function of the bar intensity, the rate of dissipation of the gas, and its rate of replacement by stellar ejections during the stars' lives (as stellar wind) or after their death (supernovae and so on). The gas trapped in stars does not take part in this slow drifting.

The inner ring at the ILR is formed much more quickly than the outer ring because the bar is much more intense between the ILR and the corotation and so produces stronger torques (Fig. 6.18). When there are two ILRs, the gas accumulates at the innermost resonance, which can be very close to the centre for a small value of Ω_p. This phenomenon might appear to be surprising, bearing in mind the sign of the torques exerted by the bar on

6.3 The Response of the Gas to a Barred Stellar Potential

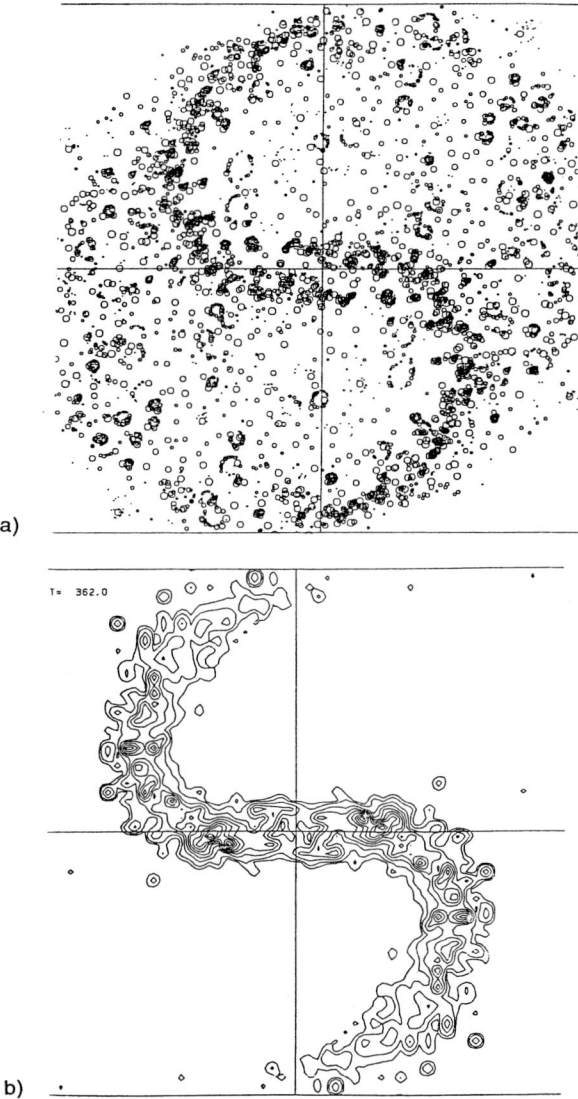

Fig. 6.15. A simulation of the ensemble of molecular clouds in the potential of a barred galaxy. The clouds collide with effective cross-sections that are functions of their respective sizes. The entire mass spectrum of the clouds is taken into account; at the massive end of the spectrum, giant molecular clouds (GMC) have a finite lifetime; they disperse by means of star formation, after some 10^7 years, and their mass is reinjected into the interstellar medium as small clouds with a velocity dispersion of the order of $12\,\mathrm{km\,s^{-1}}$. (**a**) Even though the mean free path of the clouds is of the same order of magnitude as the thickness of the spiral arms, they are well contrasted. The bar (which is horizontal here) rotates at an angular velocity Ω_p, which allows only two resonances (CR and OLR). (**b**) Contours of the energy dissipated in cloud collisions. (From Combes and Gérin 1985)

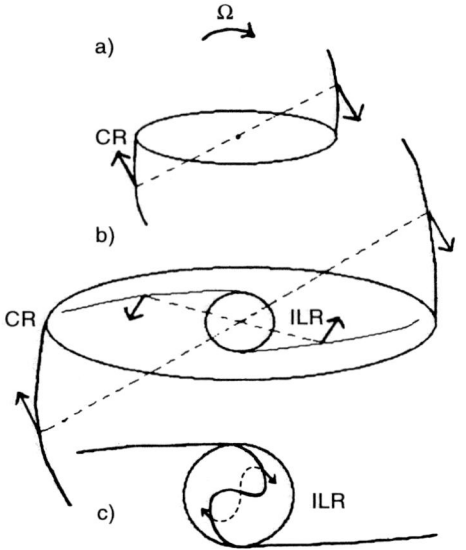

Fig. 6.16. A schematic representation of the direction of the tangential forces exerted by the stellar bar on the gas of the spiral arms. (**a**) Between the corotation and the OLR the gas acquires angular momentum and will drift to the exterior. (**b**) Between the CR and ILR the gas loses its angular momentum and falls towards the centre. (**c**) Inside the ILR, signs are changed once again. (Very near the centre the spiral can be leading (*dashed curve*) depending on the central mass concentration)

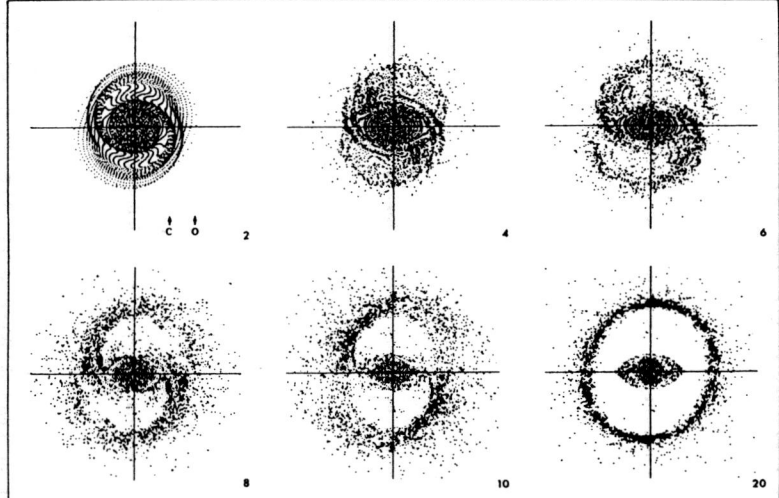

Fig. 6.17. The formation of an outer ring at the OLR by slow drifting of the gas owing to the couple exerted by the bar on the spiral arms. The unit of time is taken to be the period of rotation of the bar. (From Schwartz 1981)

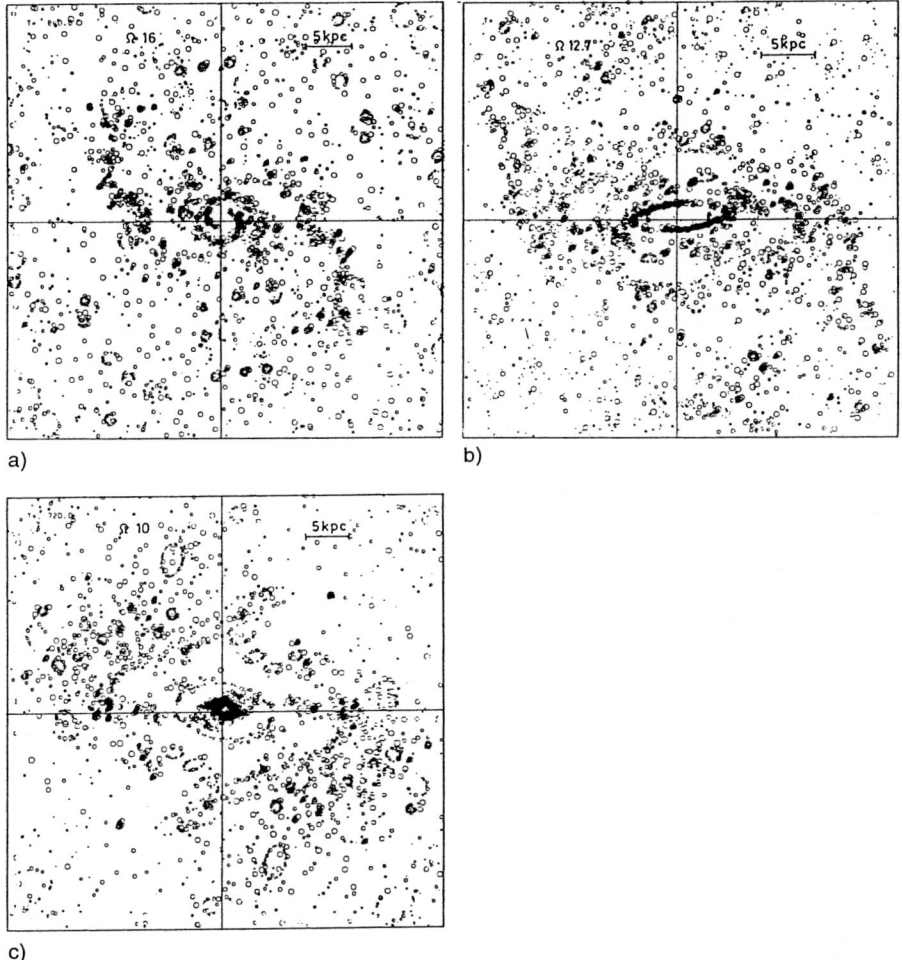

Fig. 6.18. The formation of an inner ring at the ILR by transfer of angular momentum from the gas to the bar. The angular velocity of the bar $\Omega_p = 16\,\mathrm{km\,s^{-1}\,kpc^{-1}}$ in (**a**), $12.7\,\mathrm{km\,s^{-1}\,kpc^{-1}}$ in (**b**), and $10\,\mathrm{km\,s^{-1}\,kpc^{-1}}$ in (**c**). There is an ILR in (**b**) and there are two in (**c**). (From Combes and Gérin 1985)

the gas (Fig. 6.16). In fact the spiral that is formed between the two ILRs is a mixture of trailing and leading waves. Very close to the inner ILR the leading orientation dominates. Therefore the distribution of the molecular gas, which is always maximal near the centre, favours the leading direction, and so the torque deduced from the stellar barred potential is negative (the torque changes sign with the direction of winding of the spiral).

Intermediate-sized rings are also formed at the corotation, corresponding to a higher-order, ultraharmonic resonance (UHR) (where $\Omega_p = \Omega - \kappa/4$).

These rings appear in particular when the barred potential has fourth-order harmonics.

Bars Within Bars. A strong bar is able to drive a large quantity of gas towards the centre of the galaxy. If the central accumulated mass is large enough (a few percent of the total mass), this modifies the rotation curve and increases considerably the precession frequency $\Omega - \kappa/2$. The orbital structure of the galaxy is then perturbed, and the bar can even be destroyed (Pfenniger and Norman 1990, Hasan et al. 1993). More generally, two inner Lindblad resonances and perpendicular (x_2) orbits appear, which weaken the bar at its centre. Given the high frequencies in the centre, together with a weakened bar, a secondary bar can decouple from the primary one, rotating much faster (Friedli and Martinet 1993). To avoid chaotic regions, there must be a common resonance between the two bars, and the most frequent case could be that the corotation of the embedded small bar corresponds to the ILR of the primary one: indeed, nuclear bars are observed inside the nuclear rings, that are generally attributed to the ILR (cf. Fig. 6.22 and the review by Buta and Combes 1996).

Comparison with Observations. Barred galaxies very often have inner, outer, or intermediate rings surrounding just the bar. The *outer* rings are very frequently pseudorings: these are spiral arms that separate from the bar and wind up around themselves to form rings (Fig. 6.19). They correspond to the rings obtained by transfer of the angular momentum from the bar to the spiral arms, and the fact that not all barred galaxies yet have an outer ring well separated from the bar and closed in on itself confirms that the time scale for formation is of the order of the age of the universe. The *intermediate* rings surrounding the bar, and from which the spiral arms extend, are fairly common (at least 30%) among strongly barred galaxies; their existence suggests the existence of harmonics of an order higher than $m = 2$ in the potential.

The *inner rings*, inside the bar, are also known as nuclear rings owing to their small size. Galaxies having this kind of ring generally have a rotation curve with a strong central gradient, implying very high values of $\Omega - \kappa/2$, which suggests the presence of ILRs if the corotation is really at the end of the visible bar. These rings are the scene of much star-formation activity, manifested by optically intense H II regions (hot spots) and by strong thermal and nonthermal radio-continuum emission delimiting the ring. The molecular-gas distribution is strongly concentrated in the ring, for example in NGC 1097 (Fig. 6.20), IC342, NGC 4321 and M51 (Fig. 6.21).

The formation of internal rings of such small radius (a few hundred parsecs, as in M82 or in the simulations of Fig. 6.18) suggests that the viscosity of the gas could drag it to the nucleus; this drawing of matter to the centre

6.3 The Response of the Gas to a Barred Stellar Potential 177

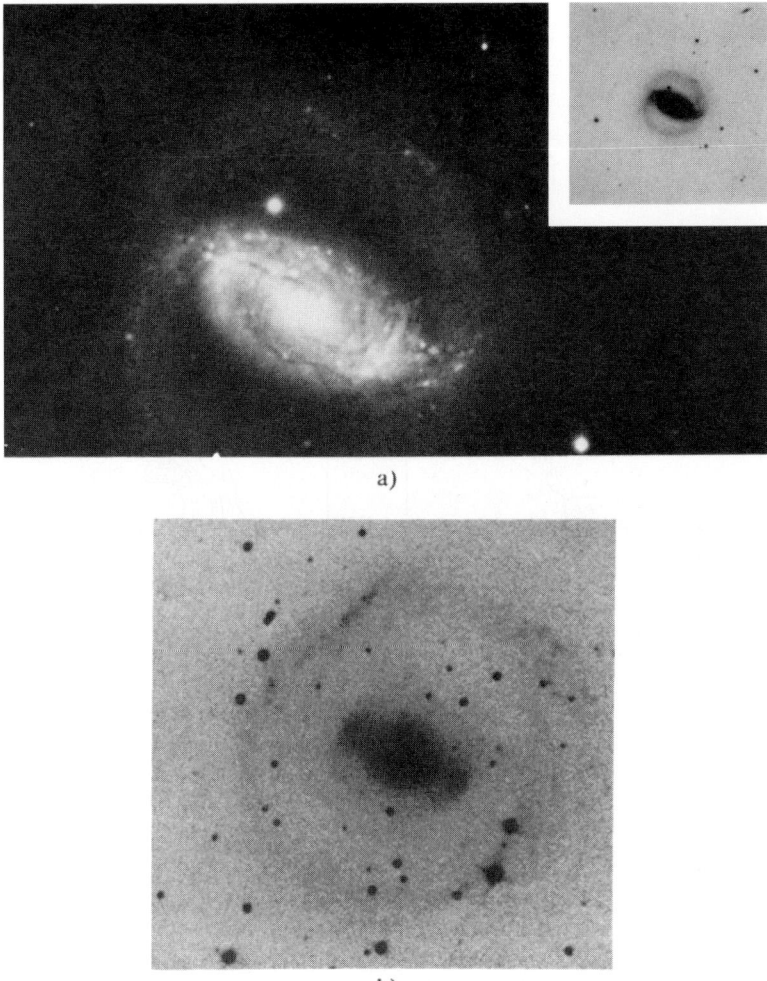

Fig. 6.19. Examples of galaxies with external pseudorings: (a) NGC 3504 and (b) NGC 2217. The ring is formed from two spiral arms that have not yet completely closed in on themselves. (From Sandage 1961)

itself, where there could be a supermassive object (especially a black hole), would be linked to activity in the nucleus. A definite correlation has been observed between the presence of a bar, particularly if it is accompanied by hot spots, and the activity of the nucleus (Seyfert nuclei and so on). More generally the existence of an increase in star formation at the centre of galaxies is relatively frequently correlated with the presence of an active nucleus: the mechanism behind the displacement of the gas towards the centre would help to feed the central 'monster'.

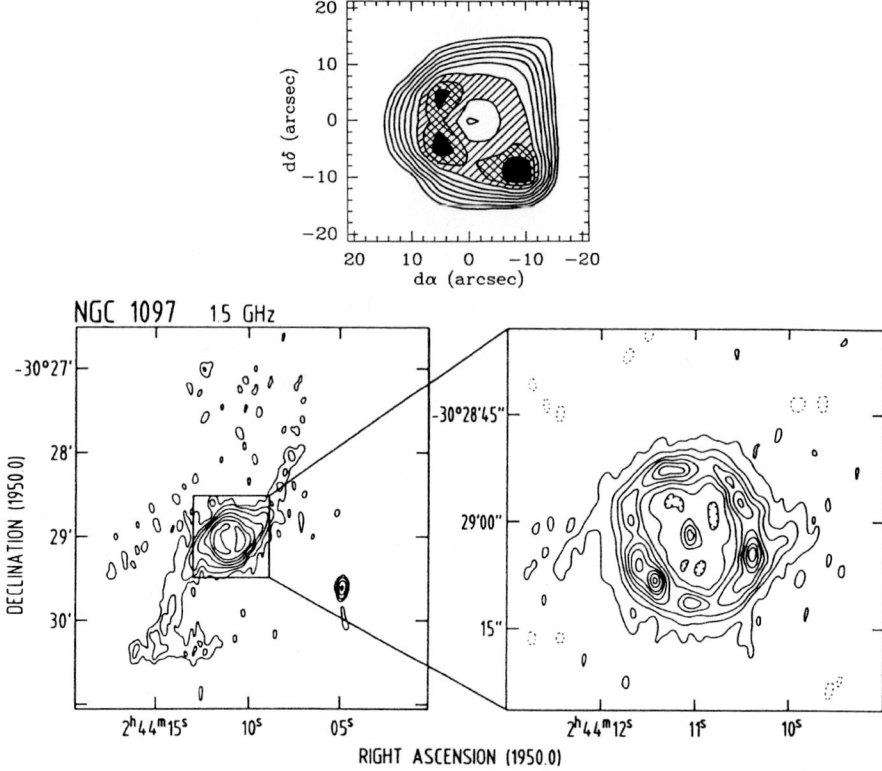

Fig. 6.20. A nuclear ring within the bar of NGC 1097. (**a**) The ring corresponds to a strong concentration of molecular clouds, observed by means of CO molecules by Gérin et al. (1988). (**b**) The ring also corresponds to a strong concentration of ionized gas, observed, as a radio continuum, at the VLA (Hummel et al. 1987). These observations are testimony to intense star-formation activity in the ring

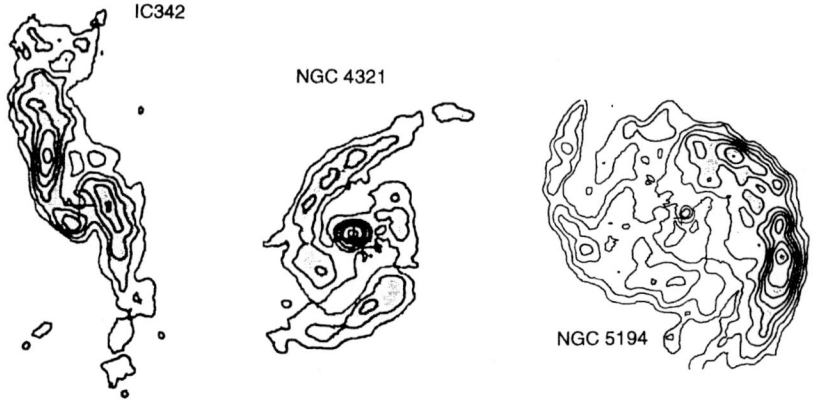

Fig. 6.21. Other examples of nuclear rings observed in the molecular component, through the CO(1-0) line: IC 342, NGC 4321, and M51 (from Sakamoto et al. 1999)

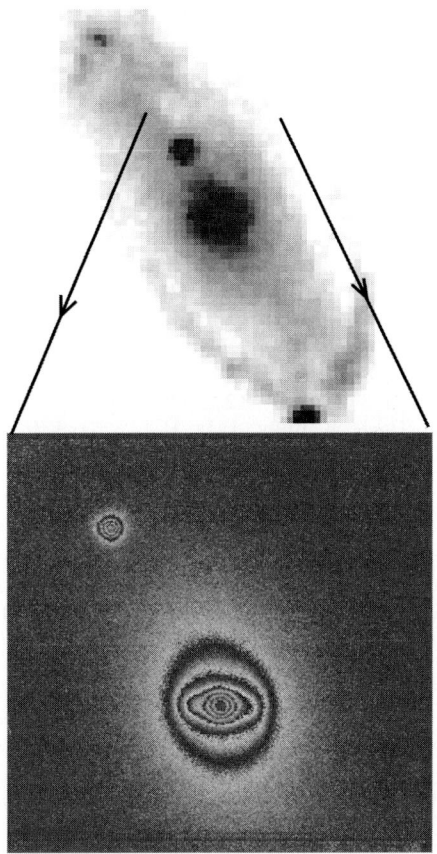

Fig. 6.22. Example of a secondary nested bar, inside the main bar and more precisely inside the nuclear ring corresponding to the ILR of the primary large bar, in NGC 5728. (*top*) A general view of the bar and the spiral arms (photograph from the DSS, Digital Sky Survey, size of the image 108"). (*bottom*) A close up view of the central parts, showing the nuclear bar, almost perpendicular to the main bar (size of the image, 36"). this image has been taken in the near-infrared, to be free from dust extinction, and to better trace the old stars, i.e. the main galactic potential (adaptive optic image taken at the Canada -France -Hawaii telescope, with a resolution of ~ 0.1", Combes et al. 2001)

Exercises

6.1 Simulations of galaxies that take selfgravity into account all use a softening parameter a: the $1/r$ gravitational potential between two particles is replaced by the softened potential

$$1/\sqrt{r^2 + a^2}$$

(see Sect. 6.2.2).

Hence show that a cold disc is stable with respect to small-scale axisymmetric perturbations if
$$a > 2\pi G\mu/\kappa^2 e,$$
where κ is the epicyclic frequency and μ is the surface density of the stellar disc.

References

Aarseth, S. J. (1979) in *Instabilities in Dynamical Systems*, edited by V. Szebehely (Reidel, Dordrecht), p. 69.
Athanassoula, E., (1992), *Mon. Not. Roy. Astron. Soc.* **259**, 345.
Barnes, J. E., Hut, P., (1986), *Nature* **324**, 446.
Benedict, F. G., Smith, B. J., Kenney, J. D. P. (1996) *Astron. J.* **111**, 1861.
Buta, R., Combes, F., (1996), *Fundam. of Cosmic Phys.* **17**, 95.
Combes, F., Debbasch, F., Friedli, D., Pfenniger, D. (1990), *Astron. Astrophys.* **233**, 82.
Combes, F., and Gérin, M. (1985) *Astron. Astrophys.* **150**, 327.
Combes, F., and Sanders, R. H. (1981) *Astron. Astrophys.* **96**, 164.
Contopoulos, G., and Papayannopoulos, T. (1980) *Astron. Astrophys.* **92**, 33.
Debattista, V.P., Sellwood, J.A., (1998), *Astrophys. J.* **493**, L5.
Elmegreen, B. G., and Elmegreen, D. M. (1983) *Astrophys. J.* **267**, 31.
Friedli, D., Martinet, L., (1993), *Astron. Astrophys.* **277**, 27.
Gérin, M., Nakai, N., and Combes, F. (1988) *Astron. Astrophys.* **203**, 44.
Hasan, H., Pfenniger, D., Norman, C., (1993), *Astrophys. J.* **409**, 91.
Hernquist, L., Katz, N., (1989), *Astrophys. J. Supp.* **70**, 419.
Hohl, F. (1971) *Astrophys. J.* **168**, 343.
Hummel, E., van der Hulst, J. M., and Keel, W. C. (1987) *Astron. Astrophys.* **172**, 32.
Kuijken, K., Merrifield, M.R., (1995), *Astrophys. J.* **443**, L13.
Ostriker, J. P., and Peebles, P. J. E. (1973) *Astrophys. J.* **186**, 467.
Pfenniger, D., Norman, C., (1990), *Astrophys. J.* **363**, 391.
Raha, N., Sellwood, J.A., James, R.A.,Kahn, F.D. (1991) *Nature* **352**, 411.
Sandage, A. R. (1961) *The Hubble Atlas of Galaxies* (Carnegie Institution of Washington, Washington DC).
Sakamoto K., Okumura S.K., Ishizuki S., Scoville N.Z., (1999), *Astrophys. J.* **525**, 691.
Sanders, R. H., and Huntley, J. M. (1976) *Astrophys. J.* **209**, 53.
Schwarz, M. P. (1963) *Astrophys. J.* **138**, 385.
Toomre, A. (1963) *Astrophys. J.* **138**, 385.
Toomre, A. (1981) in *The Structure and Evolution of Normal Galaxies*, edited by S. M. Fall and D. Lynden-Bell (Cambridge University Press, Cambridge).
Weinberg, M. D., (1985), *Mon. Not. Roy. Astron. Soc.* **213**, 451.

7. Interactions Between Galaxies

Gravitational interactions between galaxies are far more common than one might think, assuming that the space density of galaxies is uniform. The galaxies in effect group themselves in clusters, small groups, or pairs: far from being isolated systems they are formed and evolve in interaction with their environment and in particular with neighbouring galaxies.

The gravitational interactions produce giant tides in the discs of the galaxies, giving rise to spiral arms and provoking outbursts of star formation ('starbursts'). The morphology of the galaxies is therefore disturbed, even if the tides are not locally catastrophic: our Galaxy, for example, is interacting with the Magellanic clouds; it is reckoned that this interaction was at its maximum about one billion years ago.

Moreover interacting galaxies are slowed down in their relative orbit, and they end up merging together to form a single system, in general an elliptical system. Hence not all galaxies were created just after the big bang, by recombination of matter, but a number of them are still being formed today. Collisions between galaxies could therefore contribute to the activity at the nuclei of radio galaxies and quasars.

7.1 Galactic Tides

The nature of the interactions between galaxies has only recently been made clear; in the 1940s, however, the Swedish astronomer Erik Holmberg had already established the basis for the interpretation used today, but his pioneering work never made any impact. He predicted the giant tides developed in the interaction and the ultimate merger of the galaxies: he even made simulations of the phenomena on an analogue computer! Moveable lamps, whose intensity decreases with the square of the distance, placed on a table simulated the stars, whose gravitational interaction varies in an analogous way with respect to the distance. Two systems of 74 lamps thus simulated two interacting galaxies.

During the thirty years that followed, the theory only lost ground: astronomers were convinced that the long, very thin filaments observed around interacting galaxies (Fig. 7.1) could not be due to tidal interactions but resembled the channelling of matter by the flux tubes of the magnetic field.

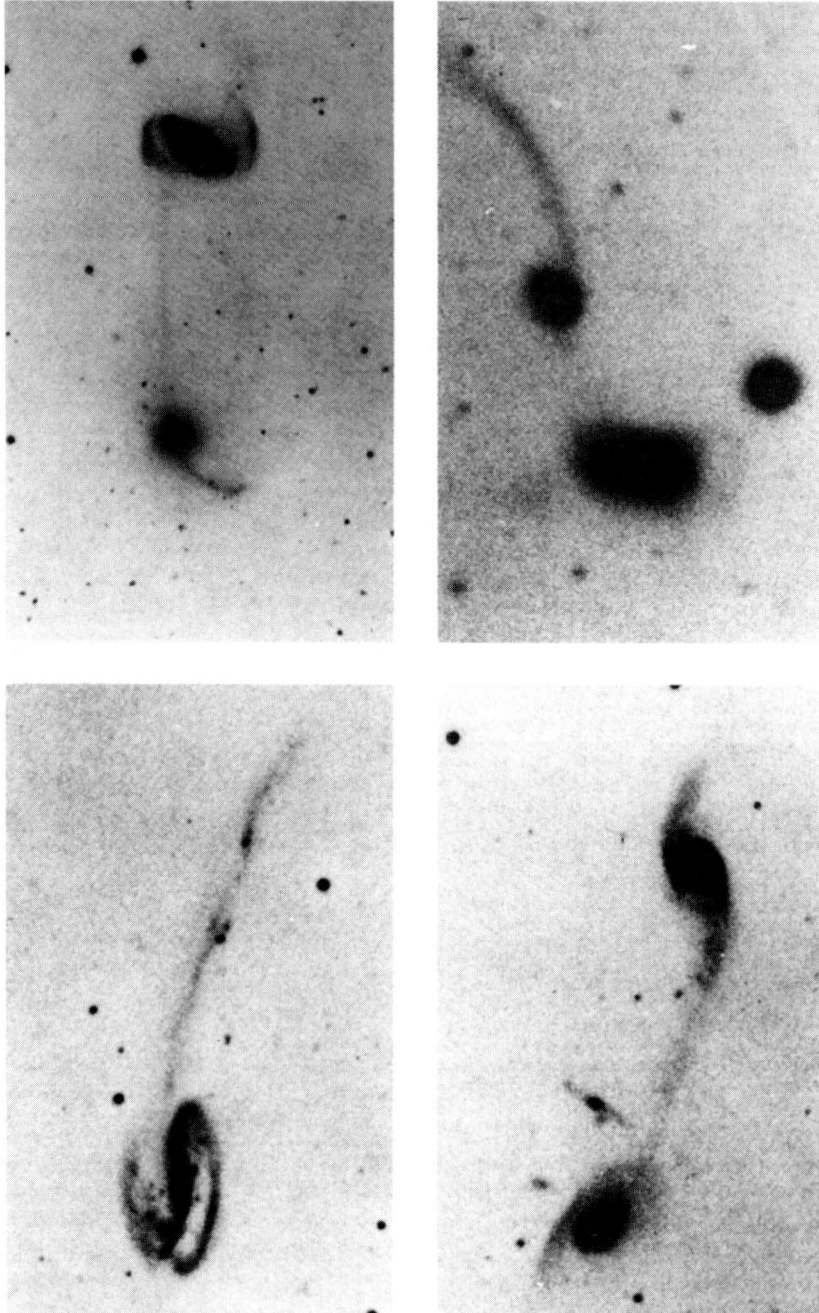

Fig. 7.1. The tidal interaction between galaxies manifests itself by the formation of very thin filaments stretching out from each galaxy. These are the filaments that confuse astronomers by suggesting the presence of magnetic fields. (From Arp 1966)

They therefore developed the theory of magnetohydrodynamics. The familiar tides observed in the Solar System, and terrestrial tides in particular, did not seem capable of accounting for such filamentary deformation; the terrestrial tides only act on solid materials and strongly cohesive liquids, whereas the galaxies are a loose grouping of stars and gases, without any cohesion except that due to selfgravity. After the discovery, in 1964, of quasars, galactic nuclei within which these very energetic phenomena occur, it was even thought that the centres of galaxies are the scenes of huge explosions that might be the origin of the filaments.

The famous article by the Toomre brothers in 1972 put an end to these distractions. With convincing numerical simulations they established that gravitational interaction with another galaxy could be the origin not only of spiral structures but also of the observed filamentary structures: gravity alone was responsible for the long and thin filaments, drawn from intergalactic space. Once again it is remarkable that ten years previously Pfeiderer and Siedentopf (1961, 1963) had made similar simulations, with the same conclusions, but had been almost ignored.

7.1.1 The Principles of Tidal Action

The tidal forces experienced by an object of diameter d in gravitational interaction with a mass M a distance D away correspond to the differential of the force of attraction of M: the parts closest to M are more attracted than those far away. The order of magnitude of the forces can be obtained by differentiating the gravitational force GM/D^2:

$$F_{\text{tide}} \approx GMd/D^3.$$

The tidal forces therefore decrease very rapidly with the separation of the two companion galaxies.

If the distance between two galaxies is greater than their individual radii, the main term in the tidal forces varies as $\cos 2\theta$ in the plane of the target galaxy (θ being the azimuth in the plane), a bisymmetric dependence analogous to a barred perturbation.

Consider the gravitational action of a companion galaxy of mass M situated at a distance D from a galaxy that we will refer to as the 'target'. In a first approximation, if the galaxies do not interpenetrate one another in the collision, we can assume that the distribution of the masses is almost spherical (like concentric central bulges of galaxies) and use Gauss's theorem to obtain the forces. This hypothesis neglects the mass of the disc, which is justified, as a first approximation. We start with the case where the orbit of the companion galaxy lies in the plane of the target galaxy (Fig. 7.2). The potential acting on a particle in the disc of the galaxy, located by its polar coordinates (r, θ) in its plane, is

$$V = -GM\left(r^2 + D^2 - 2rD\cos\theta\right)^{-1/2}.$$

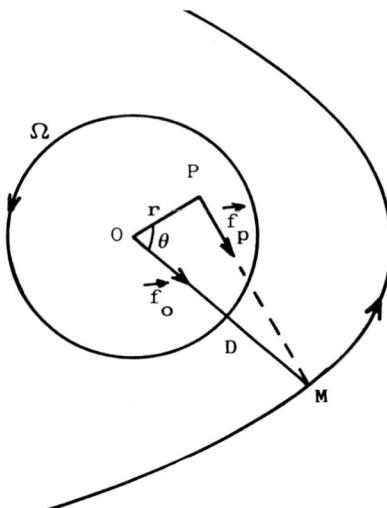

Fig. 7.2. The principle of tidal action. A companion galaxy M is assumed to move on a trajectory coplanar with the disc of the target galaxy (in the sense of the diagram). The tidal force experienced by the point P in the plane with polar coordinates (r, θ) is obtained by the subtraction of the force exerted on O (that is, the inertial force of the reference frame linked in translation to O) from the attraction that M exerts on P (along PM): $\boldsymbol{F}_{\text{tide}} = \boldsymbol{f}_P - \boldsymbol{f}_O$

On the other hand, since the galaxy is not stationary but moves with respect to the centre of gravity of the two galaxies (the target and companion), every particle is acted upon by the inertial force in the reference frame of the target, $-GM\boldsymbol{u}/D^2$, where $\boldsymbol{u}$ is a unit vector linking the two galactic centres. The corresponding potential is expressed by $GMD^{-2} r \cos\theta$.

The potential due to the tidal perturbation at every point of the target is expressed, in the reference frame of the target, by

$$V_{\text{tot}} = -GM \left(r^2 + D^2 - 2rD\cos\theta\right)^{-1/2} + GMD^{-2} r\cos\theta + \text{constant},$$

which gives, after expansion up to the second order in r/D,

$$V_{\text{tot}}(r,\theta) = -\frac{GM}{D}\left(1 + \frac{r}{D}\cos\theta + \frac{3r^2}{4D^2}\cos 2\theta + \frac{r^2}{4D^2} + \cdots\right) + \frac{GM}{D^2} r\cos\theta,$$

$$= \text{constant} - \frac{GM}{D}\frac{r^2}{D^2}\left(\frac{1}{4} + \frac{3}{4}\cos 2\theta\right) + O\left(\frac{r^3}{D^3}\right).$$

The term in $\cos\theta$ disappears in the potential, which becomes mainly bisymmetric. There therefore exist two poles of perturbation, which explains the formation of two spiral arms in the target galaxy.

This perturbation potential has the same azimuthal dependence as that of a stellar bar (Figs. 7.3 and 7.4). The only difference is the radial dependence:

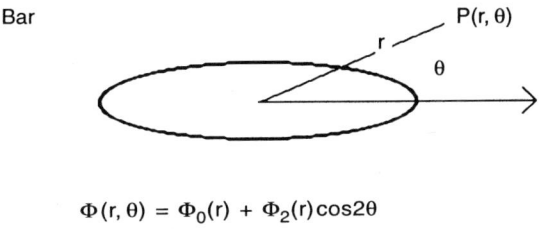

$$\Phi(r, \theta) = \Phi_0(r) + \Phi_2(r)\cos 2\theta$$

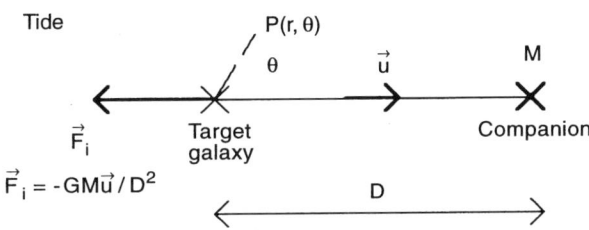

Fig. 7.3. A comparison of the potentials corresponding to a bar and a tidal interaction with a companion galaxy of mass M: the azimuthal dependence is the same as long as the bar is aligned parallel to the axis joining the two companion galaxies

the potential of a bar decreases in absolute value, starting from a certain radius $r_{\max}$ up to the edge of the galaxy, while the tidal potential does not stop increasing in r^2. The two perturbations rotate with an angular velocity Ω_p. Some results concerning the periodic orbits in such a rotating potential (Sect. 6.2.1) can therefore be generalized here.

In the case where the companion galaxy's orbit lies in an *inclined plane* at an angle i with respect to the plane of the target galaxy, the particles of the target at a distance r from the centre experience a force in the direction perpendicular to the plane (Fig. 7.5)

$$F_z = D\sin i\, GM\left[(D^2 + r^2 - 2rD\cos\theta\cos i)^{-3/2} - D^{-3}\right]$$
$$= \frac{3}{2}\frac{GM}{D^2}\frac{r}{D}\sin 2i \cos\theta.$$

The force here is calculated at the moment when the companion galaxy is at its highest point from the plane of the target. The tidal force is then of the order of GMr/D^3 (as in the plane), and its azimuthal dependence is no longer bisymmetric but contains the Fourier term $m = 1$. A passage of nonzero declination could thus excite modes of oscillation in the plane of order 1, which is observed very frequently in warps or S-shaped distortions (see Chaps. 2 and 3).

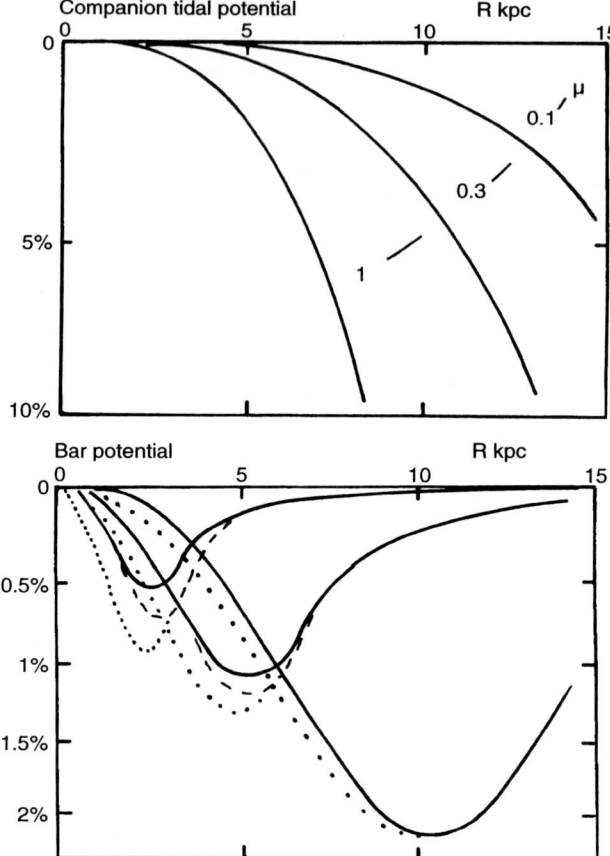

Fig. 7.4. Comparison of bar and tidal potentials: the radial dependence differs at large distances from the centre, where the influence of the bar weakens, whereas that of the companion galaxy does nothing but grow. (μ is the mass ratio of the companion)

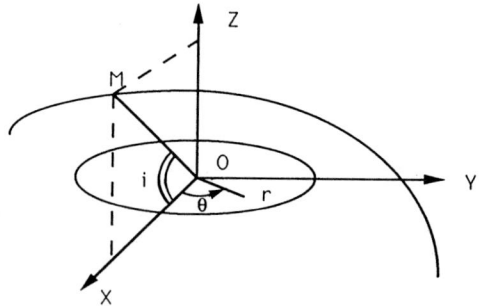

Fig. 7.5. The tidal force perpendicular to the plane: the orbital plane of the companion galaxy M now makes an angle i with the plane of the target galaxy; that is, the angle between xOy and yOM equals i

7.1.2 Numerical Simulations

The Three-Body Problem. In a gravitational encounter the tidal forces decrease with the distance as $1/D^3$; the interaction is especially important at the moment when the distance between the galaxies is a minimum: the particles of the target receive an impulse, and the acquired accelerations are transformed by the deformation, after the companion galaxy has passed. The principal effects are purely kinematic, which explains the success of the simple restricted three-body simulations, ignoring the selfgravity between the particles: all the matter of the target galaxy and the companion galaxy is represented by test particles, of negligible mass, which move independently. These particles are subject to the mean gravitational field of the two galaxies, whose relative orbits are calculated in the same way as in the two-body problem. Moreover we are justified in neglecting the selfgravity of the particles since the tidal perturbations modify especially the external parts of the galaxy, which are the least gravitationally bound.

Furthermore if the particles represent the stars as well as the gas and the dust, the approximation of three-body simulations is more justified for the gas, which represents 5 to 10% of the total mass of a spiral galaxy and whose selfgravity is more negligible. The gaseous component is far more perturbed by the tidal interactions than the stellar component, owing to its small velocity dispersion and its greater extension into the external regions (H I atomic gas). The formation of spiral density waves right to the centre even of perturbed galaxies is, however, a stellar phenomenon that relies essentially on selfgravity (see Sect. 5.3).

The kinematic spiral waves excited by the companion of M51 are well represented by a three-body simulation (Figs. 7.6 and 7.7). The deformations are triggered after the passage to perigalacticon; the two spiral arms generated are not symmetric and are distinguished by a bridge linking the two galaxies and a counterarm at the opposite side. Experiments show that the bridge has a short lifetime: the particles of the bridge are accumulated by one or other of the galaxies. On the other hand the counterarms can last up to several billion years before disappearing into intergalactic space.

Taking Account of Dissipation. The generation of the spiral waves in the gas by a $\cos 2\theta$ potential is due to the mechanism already described in Sect. 6.3. The tidal interaction can be very violent, however, and collisions induced between clouds in the spiral arms give rise to starbursts. By a comparative study of the colour–colour diagram of interacting galaxies and normal galaxies Larson and Tinsley (1978) put forward the idea that the tidal interaction considerably favours star-formation activity. Observations in the far infrared (FIR) by the *IRAS* satellite have largely confirmed their results: the vast majority of ultraluminous galaxies, with a luminosity $L_{\rm FIR} > 10^{11} L_\odot$, are interacting galaxies, and the most luminous ones correspond to galaxies that are now in the process of merging.

Fig. 7.6. Messier 51 and its small companion (NGC 5195, to the north) are the prototype system of interacting galaxies, where the spiral structure is triggered by the interaction

7.1 Galactic Tides 189

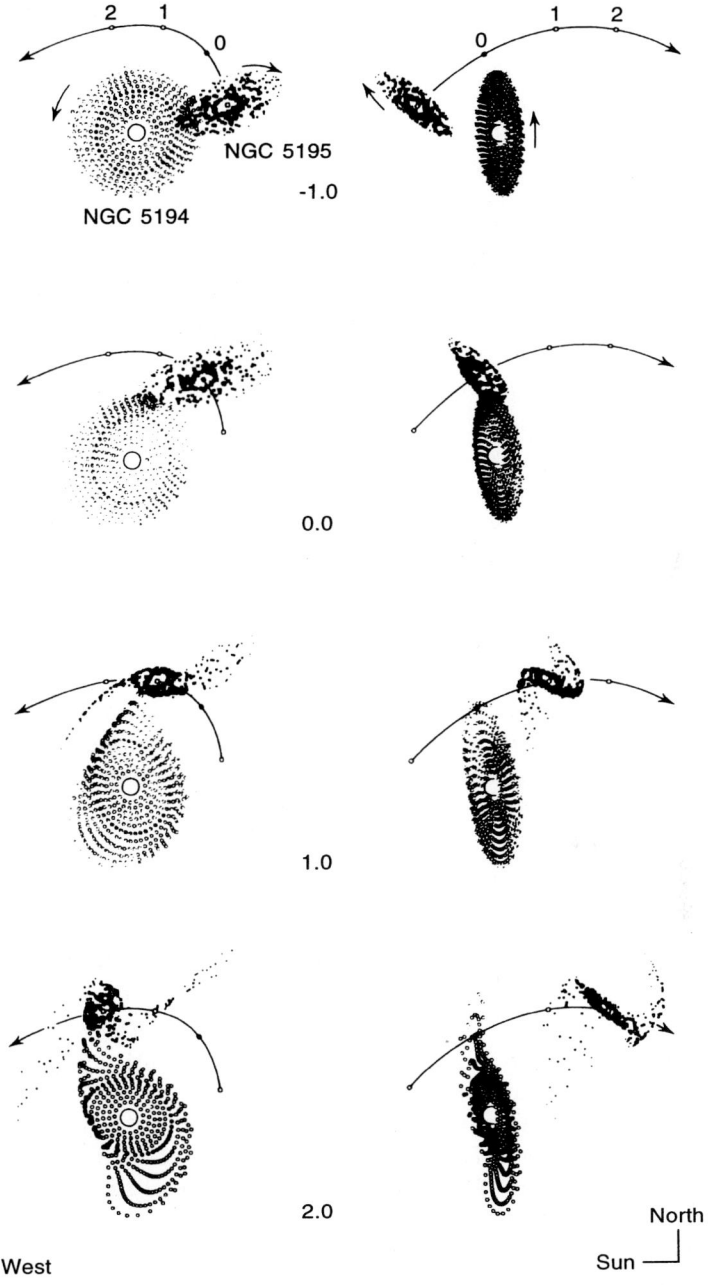

Fig. 7.7. A simulation using a restricted three-body model of the tidal interaction between M51 and its companion galaxy. The initial state consists of two homogeneous discs in rotational equilibrium. The time is measured in units of 10^8 years. Two spiral arms are formed in each galaxy, corresponding to the bipolar geometry of the tidal forces. (From Toomre and Toomre 1972)

Simulation of the ensemble of molecular clouds in a tidally interacting galaxy (Fig. 7.8) shows that the formation of spiral structure is accompanied by an increase in the number of collisions between the clouds and in the rate of formation of giant complexes, implying the activation of star formation. The spiral arms wind up from the inner Lindblad resonance (ILR), corresponding to the angular velocity Ω_p of the companion at the pericentre ($\Omega_p = V/D$, V and D being the velocity and distance of the companion at that instant). When the trajectory of the companion is almost circular and the perturbation is no longer an impulse but lasts for some rotations, the torque exerted by the companion on the spiral of gas has the same effect as a bar (Chap. 6): the gas, on losing its angular momentum, converges to a ring at the ILR (Fig. 7.8).

Introduction of the stellar component and its selfgravity could encourage the persistence of the perturbation, owing to the activation of the barred instability. Galactic discs are known to be very unstable with respect to the formation of bars, and the presence of a massive spherical halo in general does nothing but delay the instability. The passage of a companion galaxy and the $\cos 2\theta$ tidal force will contribute to destabilizing the disc and encouraging the formation of the bar, by carrying away angular momentum (the barred wave having a negative momentum density). The bar triggered in this way by the passage of the companion galaxy then has its action on the gaseous component prolonged. Observationally there exists a certain correlation between the presence of bars and companions: interacting spiral galaxies will yield a greater percentage of barred galaxies than will field galaxies; for example bars are twice as frequent in the galaxies at the heart of the Coma cluster than in the outer regions.

7.1.3 The Formation of Filaments and Ring Galaxies

Filaments. The symmetry of the tidal action, which contributes to the formation of two spiral arms in a perturbed disc, allows the formation of four arms if the two companions are disc galaxies. When the masses of the two interacting galaxies are equal or of the same order, the two internal spiral arms join up to form a bridge that disappears relatively quickly; the two external spiral arms are drawn into two 'antennae', which remain for one or two billion years, true relics of past interactions. Two famous examples are represented in Figs. 7.9–7.11 and 7.12–7.13: the Antennae (NGC 4038–4039) and the Mice (NGC 4676 A–B), respectively These two systems are quite close to merger, the last stage of the interaction. Simulations show that the antennae, whose particles reach escape velocity, disperse into intergalactic space soon afterwards. But at the ends of the antennae dense complexes of gas are formed, visible as a result of interferometric observations of atomic hydrogen (Figs. 7.11 and 7.13). These complexes also contain large quantities of ionized gas, suggesting an increase in star formation, probably triggered by the antennae encountering the intergalactic medium. Once detached from

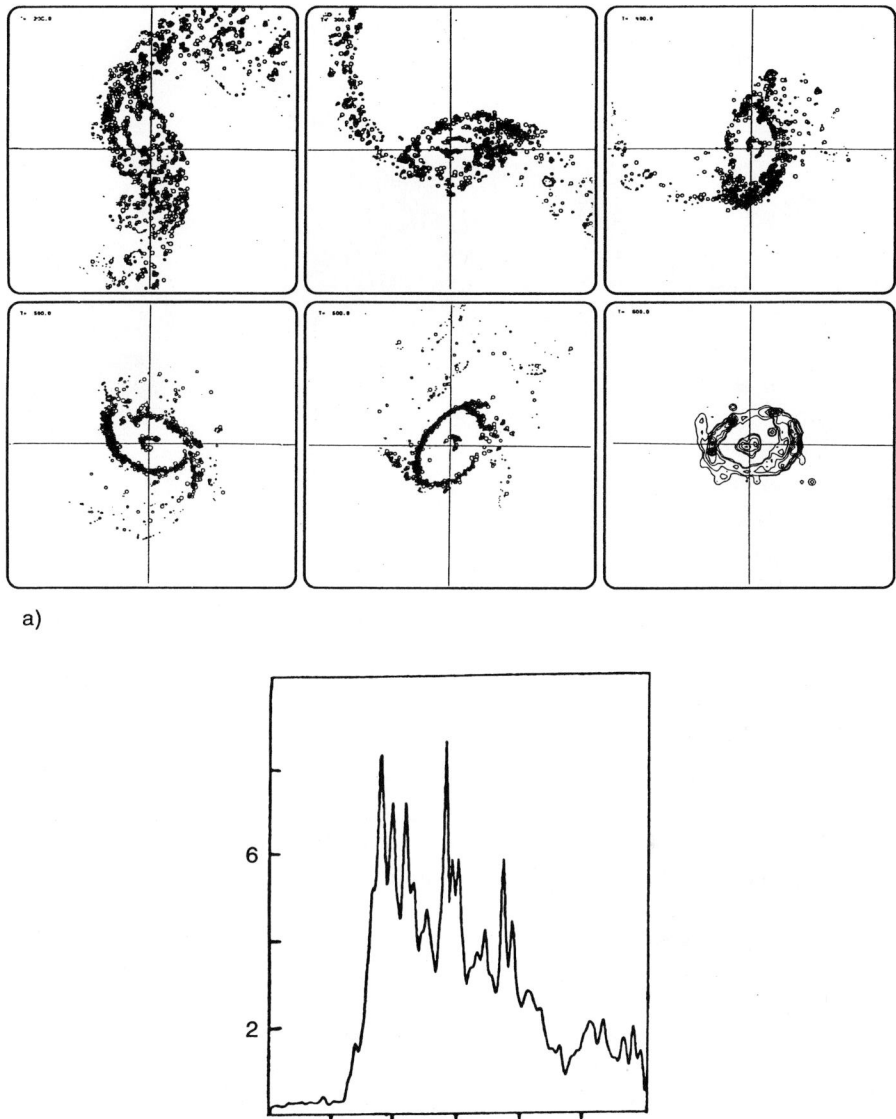

Fig. 7.8. (**a**) A simulation of the molecular-cloud ensemble in a tidally interacting galaxy with a companion of equal mass in a grazing parabolic passage. Dissipation of the gas is taken into account in the very inelastic collisions between the interstellar clouds. In time the clouds gather in a ring corresponding to the inner Lindblad resonance. The last 'snapshot' shows the contours of energy dissipated in the collisions. (From Combes et al. 1989.) (**b**) The evolution of the energy dissipated in the collisions as a function of time for the preceding simulation. The collisions are favoured by the tidal perturbation, which could explain the outburst of star formation

192   7. Interactions Between Galaxies

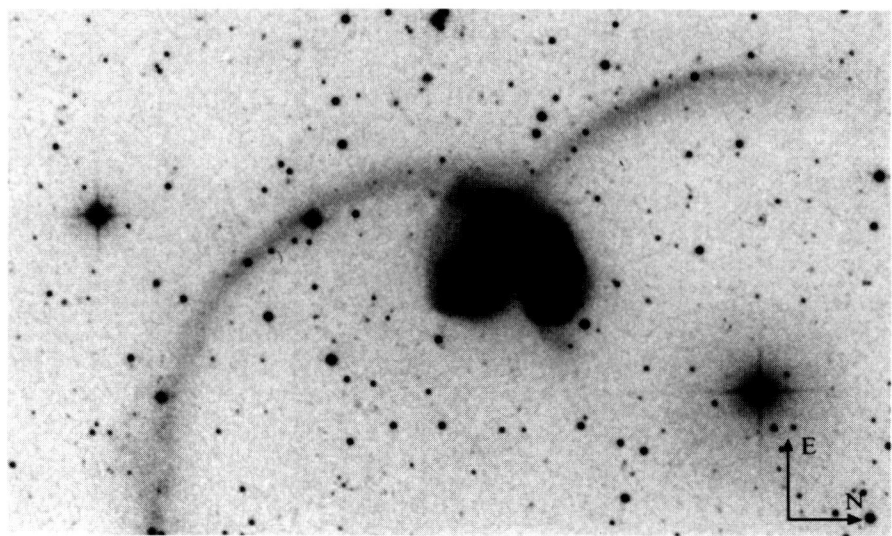

Fig. 7.9. The system of interacting galaxies NGC 4038–4039 known as the Antennae: A photograph from Arp's atlas (Arp 1966)

the mother galaxy these dense complexes can form compact dwarf galaxies, which can be said to be second generation.

'Coring' Galaxies. According to the value of the impact parameter of the collision, the spiral arms generated are open to a greater or lesser extent, until they close up in rings for a head-on collision, as seen in Fig. 7.14. This phenomenon, although rarer (the probability of small impact parameters b varying as b^2), is nevertheless observed in some 'ring' galaxies, such as the Cartwheel (Fig. 7.15). Most often the companion galaxy causing such a morphology is observed on the minor axis of the ring, confirming the hypothesis of a head-on collision. Sometimes the ring galaxy can completely lose its core in the collision. This core could correspond to one of the observed dwarf galaxies in the vicinity of the companion galaxy. When the two companion galaxies are disc galaxies, two rings are formed in an analogous manner (Fig. 7.16).

The rings, which are essentially kinematic waves, are well reproduced by test-particle simulations. The passage of a massive companion galaxy at the centre of the target galaxy momentarily attracts the particles to the centre; after the perturbation the particles return to the outside and start up oscillations driven by the force of the gravitational pull of the disc. These oscillations are produced with frequencies dependent on the initial radial position of the particles. A density wave slowly propagates to the outside, corresponding to where the particles accumulate (Fig. 7.16).

7.1 Galactic Tides   193

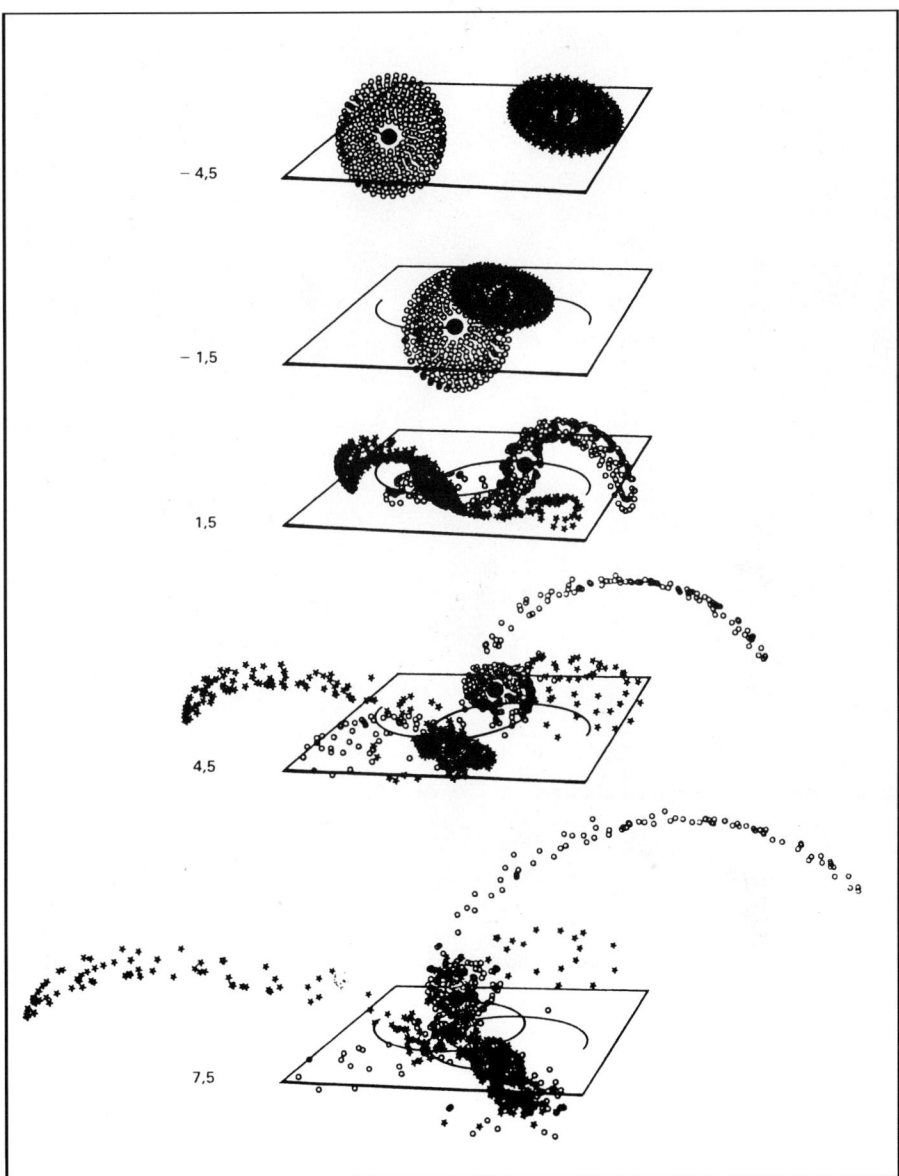

Fig. 7.10. The system of interacting galaxies NGC 4038–4039 (the Antennae): Toomre and Toomre's numerical simulation (1972)

194   7. Interactions Between Galaxies

Fig. 7.11. HI Contours (atomic hydrogen) superposed on an optical image of the Antennae, by Hibbard et al. (2001)

7.1 Galactic Tides 195

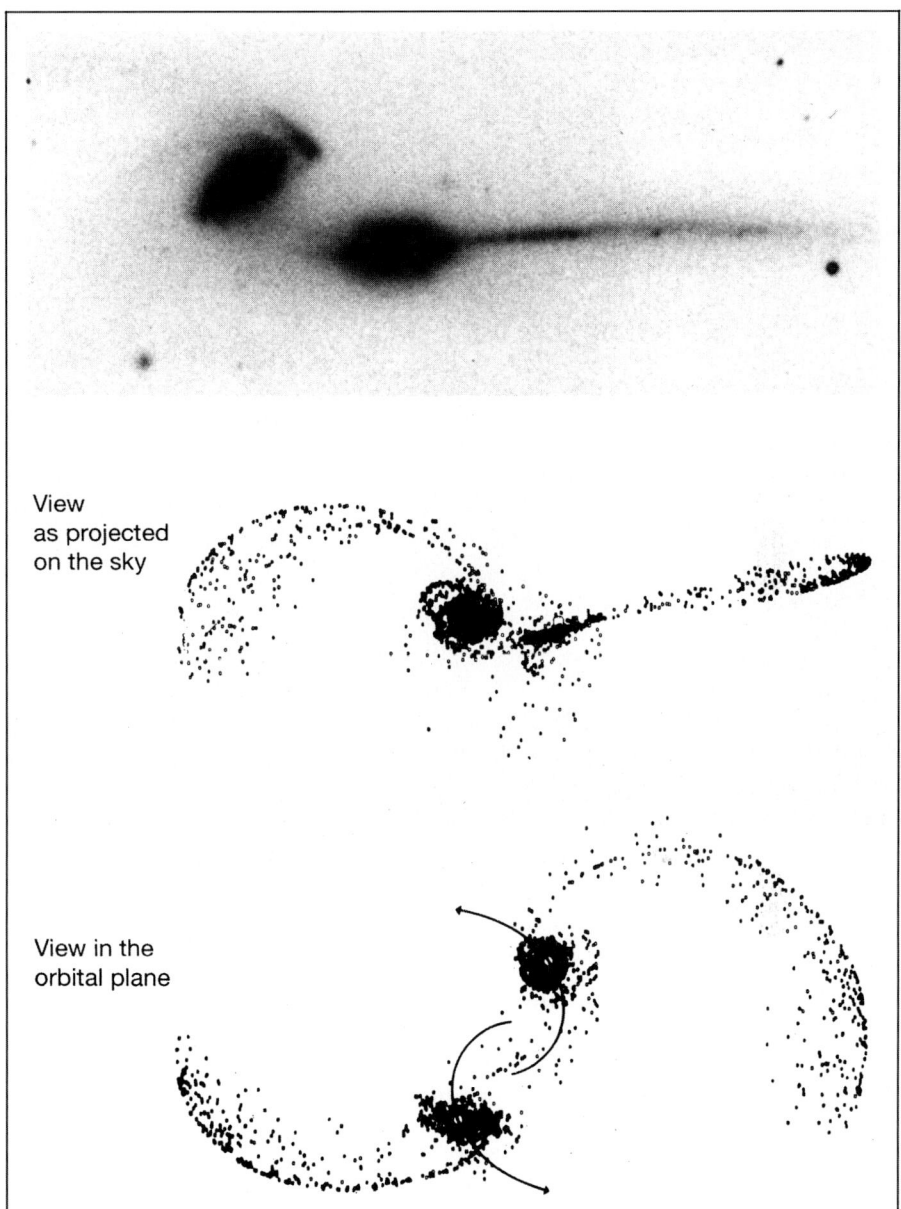

Fig. 7.12. The galaxies NGC 4676 A and B are well known as the Mice. This is a system of galaxies of equal mass and similar to the Antennae. (*top*) A photograph from Arp's atlas (Arp 1966). (*bottom*) A simulation taken from a video film by Dupraz and Combes (1986).

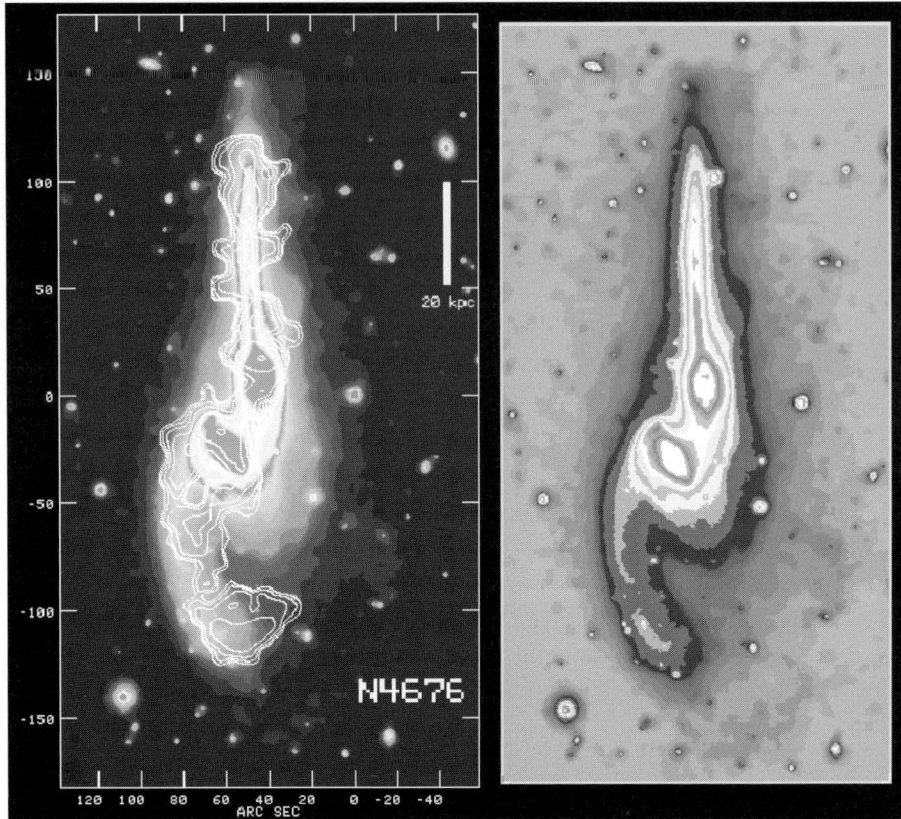

Fig. 7.13. An exposed R-band image of the Mice, NGC 4676 (*right*), and superposed on this image, the HI contours (*left*), From Hibbard and van Gorkom (1996)

7.2 Vertical Oscillations and Warps

The stars and the gas of a spiral galaxy do not inevitably remain confined to a plane, and motions in the perpendicular direction are frequent. We have already described such perturbations in our own Galaxy and in Andromeda (Chap. 2) and have remarked on the kinematic perturbations induced in the apparent velocity field of a galaxy (Chap. 3). Naturally such perturbations can be seen directly in galaxies seen edge-on (Fig. 7.17).

7.2.1 Differential Oscillations

The 'pancake'-shaped deformation (in which the disc rises at one side and falls in a symmetrical way at the other) can be described in terms of $\cos\theta$ as

7.2 Vertical Oscillations and Warps

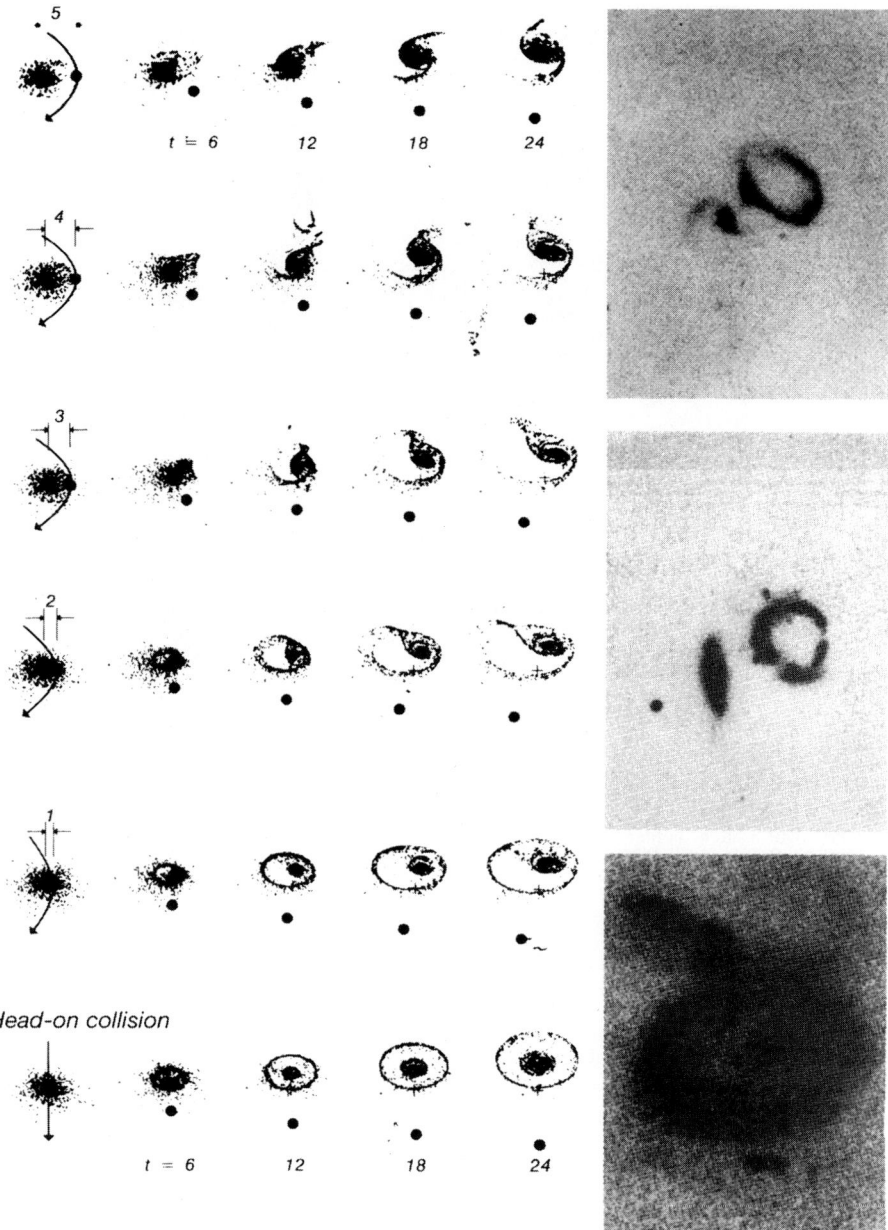

Fig. 7.14. When we go from the simple gravitational encounter between two galaxies to the head-on collision, the spiral arms formed by the tide transform into a ring; the unit of time here is 100 million years (simulation by Toomre 1977). The photographs on the right-hand side show some examples of ring galaxies in the sky (from Arp's atlas 1966)

198   7. Interactions Between Galaxies

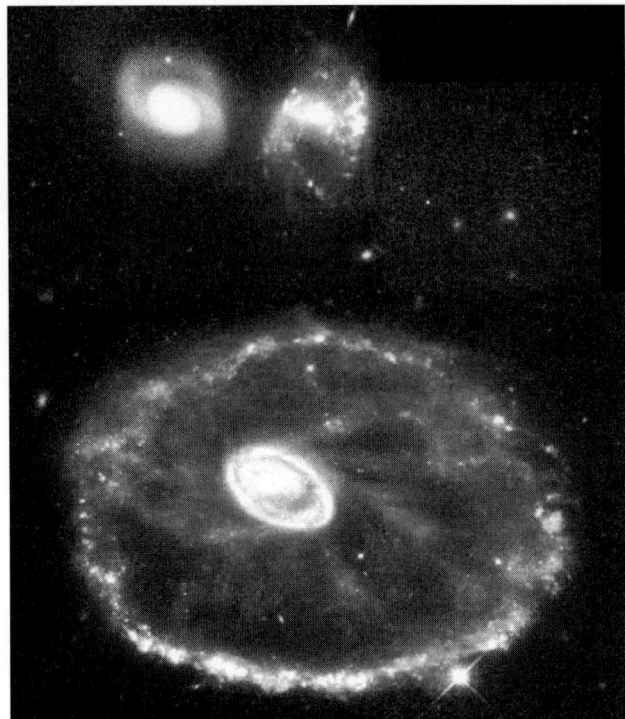

Fig. 7.15. The prototype of galaxies with a ring structure: the Cartwheel. This structure is due to a collision with a companion galaxy now situated close to the axis of the ring, may be one of the companions at the top right. (Photograph by the Hubble Space Telescope)

a function of azimuth in the disc:

$$z(r,\theta) = z_0(r)\cos m\theta,$$

with $m = 1$.

The deformation is of course not static: there is an oscillation either side of the plane of equilibrium, the restoring force being the gravitational attraction of the disc of the stars. This phenomenon immediately poses a problem, which is similar to the winding problem of the spiral arms in the plane.

If the disc could be regarded as a solid body, the phenomenon would have certain similarities with the oscillations of a coin which has been spun into the air: the instantaneous spin axis precesses around the symmetry axis (Fig. 7.18a). In this case the precession is rigid: the angular velocity of rotation and the frequency of vertical oscillation do not depend on the radius.

In a spiral galaxy the rate of precession is differential, the period of rotation and that of the oscillations in the z direction depending on the radius.

7.2 Vertical Oscillations and Warps

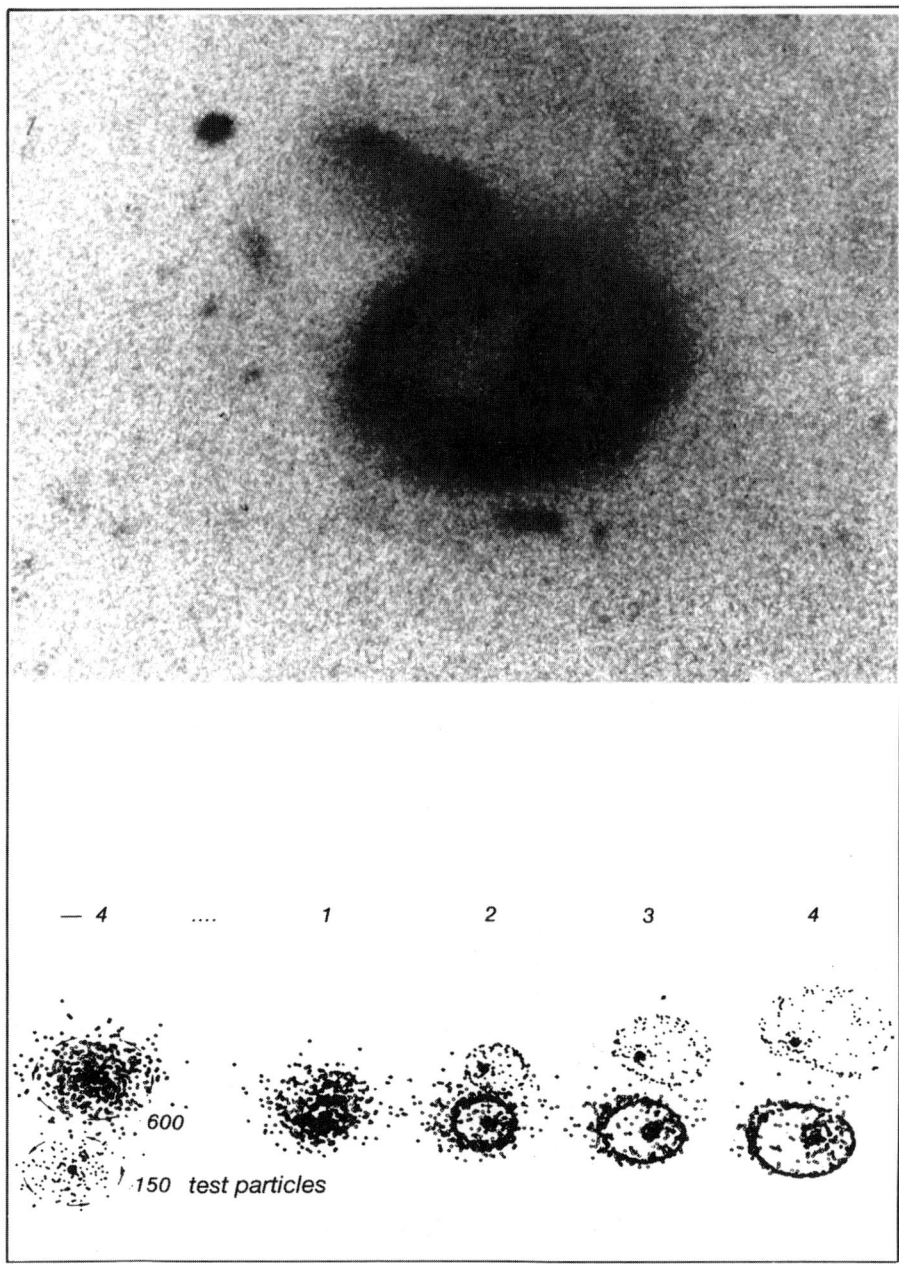

Fig. 7.16. In a head-on collision, when the two companion galaxies possess a disc, a ring can form symmetrically in the two systems. The system II Hz 4 is an example of these double rings. (Photograph and simulation from Lynds and Toomre 1976)

200   7. Interactions Between Galaxies

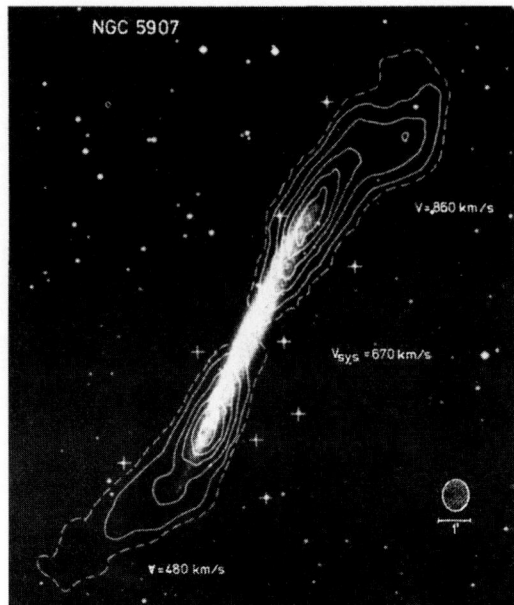

Fig. 7.17. In this spiral galaxy seen edge-on the deformation of the galactic plane into a pancake shape is clearly visible from the isophotes of atomic hydrogen observed by 21 cm interferometry. (From Sancisi 1976)

After several vertical oscillations the warp is damped by the formation of a disc like a corrugated sheet (a problem analogous to that of the spiral arms). How is the observation of pancake-shaped deformations in the majority of spiral galaxies to be explained?

7.2.2 Vertical Waves and the Dispersion Relation

To describe this phenomenon more quantitatively let us calculate the evolution in time of the height $z(r, \theta, t)$ of the galactic plane, which is assumed initially to be equal to $z_0(r) \cos \theta$. To simplify this, consider a particle with an almost circular trajectory in the plane with a radius $r = r_0$ and $\theta = \Omega(r_0)t + \theta_0$. To adapt the initial conditions let $z = z_0(r_0) \cos \theta_0$ at $t = 0$ for the particle under consideration. The oscillatory motion in the z direction has a frequency ν_z, giving $z(r_0, \theta, t) = z_0(r_0) \cos(\Omega t - \theta) \cos \nu_z t$.

This expression can be decomposed into two progressive waves:

$$z(r_0, \theta, t) = z_0(r_0) \big[\cos\big((\Omega - \nu_z)t - \theta\big) + \cos\big((\Omega + \nu_z)t - \theta\big)\big]/2.$$

These two waves have angular velocities equal to $\Omega_p = \Omega + \nu_z$ for the fast wave and $\Omega_p = \Omega - \nu_z$ for the slow wave. The latter is retrograde because Ω is

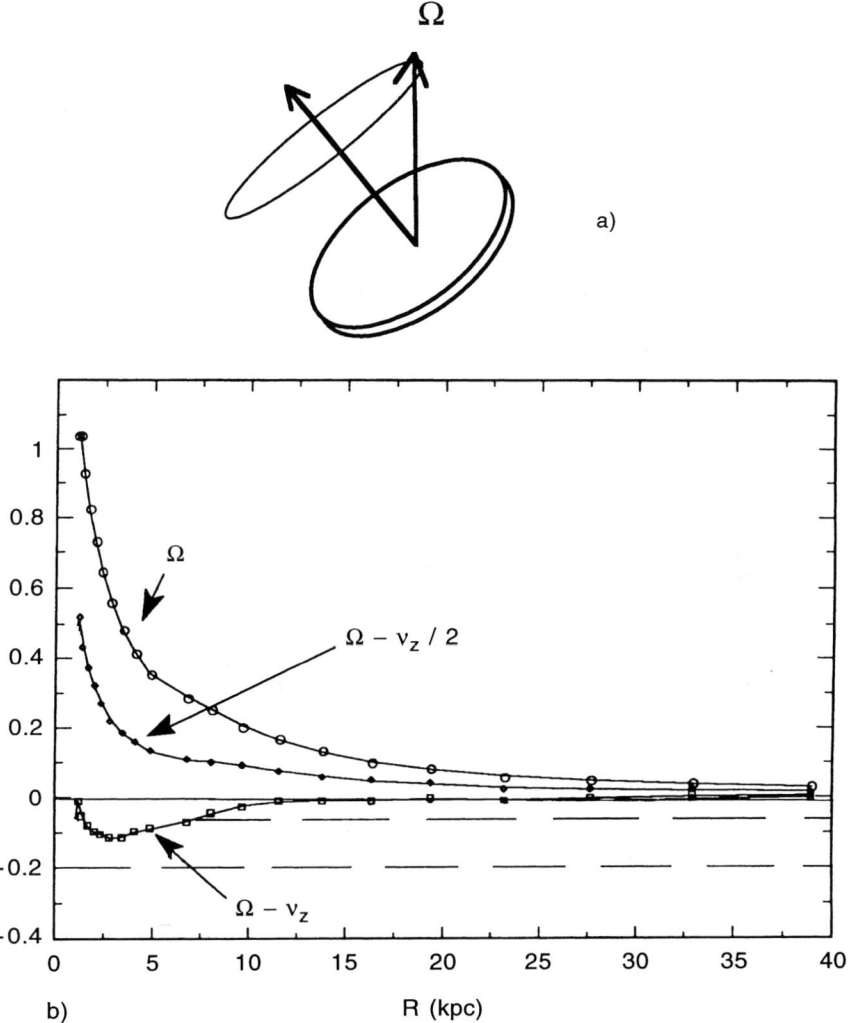

Fig. 7.18. (a) The precession of a coin when spun into the air. (b) The typical variation of $\Omega - \nu_z$ in a spiral galaxy. The vertical, retrograde WKB waves can only exist beyond the resonance $\Omega_p = \Omega - \nu_z$, but they can exist anywhere if there is no resonance

in general smaller than ν_z (itself of the same order as the epicyclic frequency κ) in spiral galaxies. As the winding problem is more critical for the fast wave, only the retrograde slow wave is considered.

The expression for $z(r_0, \theta, t)$ obtained above has the form of the crest of the wave $\theta(r,t) = [\Omega(r) - \nu_z(r)]t$. The wave can propagate without becoming too deformed if the rate of precession $(\Omega - \nu_z)$ does not depend on the radius. Would the introduction of selfgravity succeed in modifying the equivalent ν_z

and allow the vertical waves to propagate without deforming, in the same way that modifying κ allows the propagation of spiral waves? The calculations can be developed analytically using the WKB approximation and the dispersion relation for the waves deduced, in a similar way to that obtained by Lin and Shu (Sect. 5.2.2):

$$m^2 \left(\Omega - \Omega_p\right)^2 = \nu_z^2 + 2\pi G\mu|k|.$$

This is the dispersion relation obtained by Hunter and Toomre in 1969. Notice that here the selfgravity produces an increase in the equivalent frequency ν_z (an increase of the restoring force of the disc), whereas the opposite effect occurs for the epicyclic frequency κ, which decreases. As with the spiral waves the dispersion relation defines zones where vertical waves or warps could exist. Vertical resonances analogous to the Lindblad resonances are defined by

$$m\left(\Omega - \Omega_p\right) = \pm\nu_z.$$

A retrograde wave ($\Omega_p < 0$) with $m = 1$ can exist everywhere if there is no resonance $\Omega - \Omega_p = \nu_z$ but only beyond R_0 if there is a resonance at R_0 (Fig. 7.18b). Is the problem of the existence of warps then resolved? In fact, as with spiral density waves, the vertical wave packet has a group velocity and travels towards the edge of the disc in a few billion years, where it is absorbed. The warp of the disc cannot therefore be primordial, and it is necessary to find an excitation mechanism that has recently come into action to explain the observed deformations.

7.2.3 Tidal Interaction and Warps

While many galaxies with a distorted plane interact gravitationally with a nearby companion (this is the case, for example, for the Milky Way and the Magellanic clouds), some are observed to be relatively isolated; that is, they have not experienced the passage of a companion for a billion years. The problem of the persistence of the warps in these galaxies can be eased by postulating a spherical halo of invisible mass enveloping the galaxy.

The reason is that the influence of the restoring force of the disc becomes negligible in the presence of a massive halo; and in a spherical potential, where no direction is privileged, a ring of height $z = z_0 \cos\theta$ will have no tendency to precess. More quantitatively it can be shown that $\Omega = \nu_z$ in a spherical potential where $\Phi = \Phi(x^2 + y^2 + z^2)$.

The presence of invisible mass is suspected in spiral galaxies (owing to the flat rotation curve for atomic hydrogen far from the centre). But up until now nothing could be said about the spatial distribution of this invisible mass: it could just as well be confined to the disc. To resolve the problem of the persistence of the vertical waves it is necessary for this mass to be distributed spherically.

However, the problem of the persistence of the warping deformation is the most severe at the outskirts of the visible disc: there the hypothetical spherical dark halo is not sufficient to explain the persistence of warps, since its total mass is at most equal to that of the visible disc, according to the rotation curve. The solution could come from the way galaxies evolve and accrete mass. A galaxy accretes several times its initial mass in the Hubble time; this accretion is likely to be made through gas clouds with an angular momentum unrelated to that of the initial galaxy. The torques between the galaxy and its accreted elements will progressively align the various angular momenta in a common direction, which will become the new rotation axis of the galaxy. A spiral galaxy then changes its rotation direction several times in its life. The gaseous warps in the outer parts of galactic discs could only be a manifestation of this evolution.

Another way to explore the three-dimensional shape of dark haloes is provided by peculiar systems called polar ring galaxies (PRG; see Fig. 7.19). These systems are thought to be formed through accretion of gas during an interaction or merger event. The gas, through differential precession and dissipation, quickly settles in an equilibrium plane, which is in general the equatorial plane of the accreting object. In some rare cases, gas can be trapped in stable, nearly polar orbits, and form a PRG after star formation has processed. These systems make it possible to find out the rotational velocity in two perpendicular planes (the polar ring and the original galaxy), and therefore to access the three-dimensional shape of the potential. However, this method encounters intrinsic difficulties, since the polar ring itself is very massive and self-gravitating and considerably perturbs the original potential.

7.3 Dynamical Friction

Dynamical friction is the braking of a massive body P by the stars of a galaxy when P penetrates into this galaxy (or in a close encounter with P). To get an intuitive idea of this phenomenon we can represent the trail of stars that leave the body P behind, after it has deflected these stars. The increase in density behind P produces a supplementary force of attraction that slows it down (Fig. 7.20).

7.3.1 Estimating the Frictional Force

Calculation of the frictional force can of course only be made with the approximation of a uniform environment. The analytic formula was given by Chandrasekhar in 1943. To establish it we must first calculate the braking due to an individual star when it encounters a mass M.

Gravitational Deviation. Consider the gravitational encounter between the companion (of mass M) and the star (of mass m), their relative velocity being v_0. Figure 7.21a defines the impact parameter b and the angle Φ of deviation characterizing the encounter between the star, coming from infinity

Fig. 7.19. Optical image of NGC 4650A, the prototype of polar ring galaxies. The central nucleus is the lenticular galaxy, which happens to have accreted a large amount of gas, which has settled into a nearly polar disc a few Gyrs ago, according to the age of the stars that have formed in it (Photograph by the Hubble Space Telescope)

7.3 Dynamical Friction

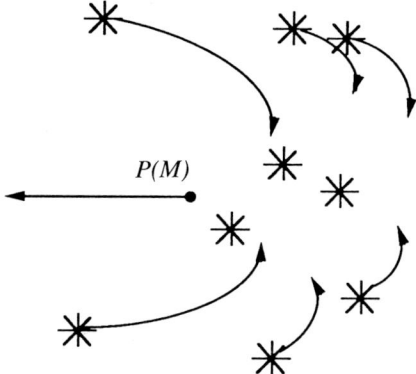

Fig. 7.20. The principle of braking by dynamical friction. The massive body P (of mass M) penetrates a distribution of stars, which are gravitationally deviated; these deviated stars accumulate in a trail behind P and their gravitational force slows P down

with a velocity v_0, and the companion, initially at rest. The calculation is easily done by using the reference system of the 'pseudoparticle': the centre of gravity (G) moves with a uniform velocity $mv_0/(m+M)$, and the pseudoparticle, of mass $\mu = mM/(m+M)$, has initial velocity v_0; its position is given by $\boldsymbol{r} = \boldsymbol{r}_m - \boldsymbol{r}_M$. As it is subject to a central force $\boldsymbol{f} = -GmM/r^2 \boldsymbol{u}$, its kinetic momentum is conserved: $L = L_z = bv_0 = r^2\dot\theta$, for all t. The trajectory is a hyperbola of the form

$$1/r = A\cos(\theta - \theta_0) + B. \tag{7.1}$$

The two constants A and B are obtained by writing down the equation for the conservation of energy:

$$E = \frac{\mu}{2}\left(\dot{r}^2 + \frac{L^2}{r^2}\right) - \frac{GmM}{r} = \frac{\mu v_0^2}{2} \quad \text{(at infinity)}. \tag{7.2}$$

The radial velocity is obtained by differentiating (7.1):

$$\begin{aligned}\dot{r} &= Ar^2\dot\theta\sin(\theta - \theta_0) \\ &= AL\sin(\theta - \theta_0),\end{aligned} \tag{7.3}$$

giving, on substitution into (7.2),

$$(\mu/2)\left(A^2L^2 + 2ABL^2\cos(\theta-\theta_0) + L^2B^2\right) - GmM\left(A\cos(\theta-\theta_0) + B\right) = \mu v_0^2/2.$$

As this equality is valid for all θ,

$$B = G(m+M)/L^2$$

and

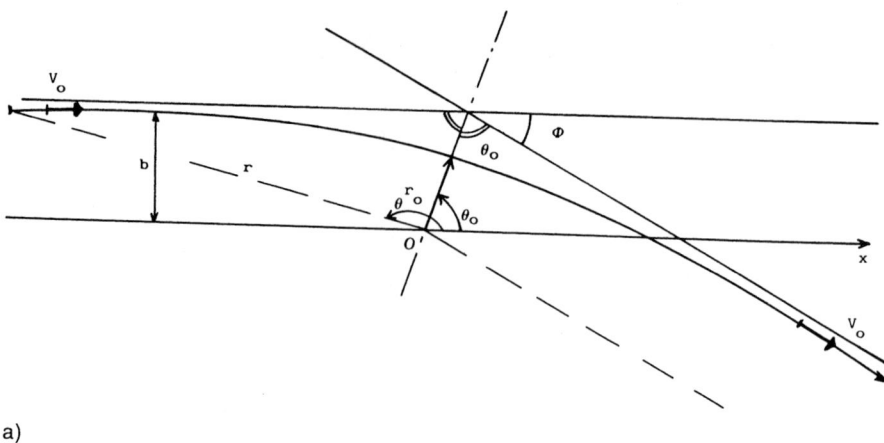

a)

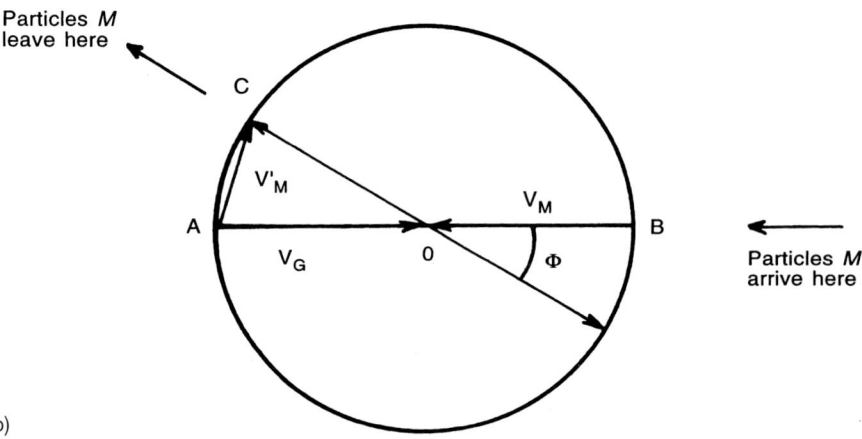

b)

Fig. 7.21. (a) The geometry of the gravitational encounter between two bodies of mass m and M. The diagram represents the trajectory of the 'pseudoparticle' of mass $\mu = mM/(m+M)$. Its polar coordinates (r, θ) are taken with respect to the Ox axis. The particle goes from $\theta = \pi$ to infinity with a velocity v_0 and impact parameter b. At the distance of minimum approach the coordinates are (r_0, θ_0). The corresponding axis OP is an axis of symmetry for the trajectory. After the encounter the particle moves away with the same velocity v_0 and an angle $\theta = -\Phi$ (the angle of deflection). (b) Back to the laboratory reference frame. The velocity of the centre of gravity is $\boldsymbol{v}_G = m\boldsymbol{v}_0/(m+M)$. In the reference frame of the centre of mass the two particles arrive in opposite directions with velocities $\boldsymbol{v}_m = M\boldsymbol{v}_0/(m+M)$, and $\boldsymbol{v}_M = -m\boldsymbol{v}_0/(m+M)$. As the collision is elastic, they each move off again with their velocity unchanged in magnitude but in another direction (Φ). The velocity of the body P will then be $\boldsymbol{OC}$ after the collision, in the centre-of-mass frame. In the fixed reference frame, $\boldsymbol{v}_{M'} = \boldsymbol{AO} + \boldsymbol{OC} = \boldsymbol{AC}$, where $v_{M'} = 2v_G \sin(\Phi/2) = \Delta v$, Δv corresponding to the perturbation of velocity experienced by particle M in the reference frame where it was initially stationary

$$A^2 = b^{-2} + G^2(m+M)^2/L^4.$$

By writing (7.3) and (7.1) at infinity, we get

$$-v_0 = A b v_0 \sin\theta_0$$

and

$$-A\cos\theta_0 + (m+M)G/b^2 v_0^2 = 0,$$

and eliminating A,

$$\tan\theta_0 = -\frac{b v_0^2}{(m+M)G}$$

and the deviation

$$\Phi = \pi - 2\theta_0,$$

where

$$\tan\frac{\Phi}{2} = -\frac{(m+M)G}{b v_0^2}$$

(the deviation Φ is thus not negligible, depending on the mass M of the companion galaxy P).

The velocity of the pseudoparticle is always v_0 at infinity, after the collision. To go back to the physical reference frame it is enough to add the velocity of the centre of gravity $\boldsymbol{v}_G = m v_0/(m+M)$, according to Fig. 7.21b: $2 v_G \sin(\Phi/2)$ is the final velocity of the companion galaxy, which is initially stationary. The perturbation experienced will be $\Delta v = 2 m v_0/(m+M)\sin(\Phi/2)$.

This perturbation of velocity is decomposed into two directions, parallel and perpendicular to the initial velocity $-v_0$ of the companion galaxy (assuming this time that the deviated stars are initially at rest). The parallel projection is $\Delta v_\parallel = \Delta v \sin(\Phi/2) = 2 m v_0/(m+M)\sin^2(\Phi/2)$. This perturbation is always of the opposite sign to $-v_0$; the companion galaxy P is therefore always slowed down. In the perpendicular direction, on the other hand, all the perturbations due to the uniform distribution of stars (deviations Φ and $-\Phi$ being equiprobable) average to zero.

Calculating the Braking Force. Suppose that the companion galaxy P is surrounded by a uniform and infinite sea of stars with a density $f(\boldsymbol{v}_m)$ in phase space. The rate of encounters with stars having an impact parameter between b and $b+db$ and a velocity in the element $d\boldsymbol{v}_m$ is

$$2\pi b\, db\, v_0 f(\boldsymbol{v}_m)\, d\boldsymbol{v}_m.$$

The change in the total velocity due to all these accumulated encounters is

$$\frac{d\boldsymbol{v}_M}{dt} = \int \boldsymbol{v}_0 f(\boldsymbol{v}_m)\, d\boldsymbol{v}_m \int_{b_{\min}}^{b_{\max}} \frac{2 m v_0}{m+M}\left(1 + \frac{b^2 v_0^4}{G^2(m+M)^2}\right)^{-1} 2\pi b\, db;$$

$b_{\max}$ is the largest possible impact parameter, being the radius of the galaxy, for example. Integrating over b is easy:

$$\frac{d\boldsymbol{v}_M}{dt} = 2\pi \ln\left(1 + \Lambda^2\right) G^2 m(m+M) \int f(\boldsymbol{v}_m) \frac{d\boldsymbol{v}_M(\boldsymbol{v}_m - \boldsymbol{v}_M)}{|\boldsymbol{v}_m - \boldsymbol{v}_M|^3},$$

where $\Lambda = b_{\max} v_0^2 / G(m+M)$ is of the order of 1 for galaxies.

The frictional force therefore has the relative-velocity direction

$$\boldsymbol{v}_0 = \boldsymbol{v}_m - \boldsymbol{v}_M;$$

its modulus is inversely proportional to this velocity (to v_0^{-2}). Integration over all possible velocities $\boldsymbol{v}_m$ of the stars of the target system is easily carried out, noting that formally the equation reduces to integrating at a point R_M the gravitational force due to all the masses at r_m of a spherical distribution:

$$\boldsymbol{f}_{\text{grav}} = \rho \int \frac{d\boldsymbol{r}_m(\boldsymbol{r}_m - \boldsymbol{R}_M)}{|\boldsymbol{r}_m - \boldsymbol{R}_M|^3}$$

(this is supposing of course that the distribution of velocities $\boldsymbol{v}_m$ is isotropic, which corresponds to the hypothesis of a uniform density ρ).

In this integration, according to Gauss's theorem only the mass contained within the radius R_M contributes to the gravitational force: after we change variables to velocities, only particles with a velocity less than v_m contribute to the frictional force:

$$\frac{d\boldsymbol{v}_M}{dt} = -\frac{\boldsymbol{v}_M}{v_M^3} 16\pi^2 (\ln \Lambda) G^2 m(m+M) \int f(\boldsymbol{v}_m) v_m^2 \, d\boldsymbol{v}_m.$$

When the intruder is massive ($M \gg m$) and also slow enough for v_m to be small, the above formula can be further simplified by taking $f(v_m) \approx f(0)$:

$$\frac{d\boldsymbol{v}_M}{dt} \approx -\boldsymbol{v}_M \frac{16\pi^2}{3} (\ln \Lambda) G^2 mM f(0).$$

The force is therefore proportional only to v_m, like the force corresponding to a viscous friction. Note in the expression for the force that the mass of the individual stars (or of the deflected bodies) does not play a part; only the density ($mf(0)$) of the target environment is relevant. In addition the frictional force is very dependent on the mass M: as the braking acceleration is proportional to M, the force is proportional to M^2.

The Limitations of Chandrasekhar's Formula. Chandrasekhar's formula is the only analytic expression available for estimating the frictional force. Nevertheless the strongest approximation, that of a uniform and homogeneous environment, can be seriously questioned, especially if the trajectory of the companion galaxy leads it to cross the core of a galaxy, where the density varies over many orders of magnitude. In addition the effects of selfgravity in response to the stars in the target environment can modify the braking. On the other hand the deformations experienced by the companion galaxy (which is taken to be rigid in Chandrasekhar's simple formula) are also an important factor in the braking.

When a target galaxy is an elliptical galaxy, where the dispersion of the velocities of the stars is high, the selfgravitational response of the stars of the target can be modelled by a multipolar development (in essence a quadrupolar deformation). When the target is the disc of a spiral galaxy, resonance effects have to be considered (there being more marked effects for a direct trajectory than for a retrograde one): the deformations can be much more important, and the braking is therefore greatly underestimated by Chandrasekhar's formula.

Figure 7.22a shows, for an initially circular orbit of the companion galaxy, the type of trajectory obtained by integrating Chandrasekhar's formula. The time taken for the rapid decay (towards the centre) from the start of the integration is 2×10^9 years. A comparison with an N-body simulation of the same problem is given in Fig. 7.22b.

7.3.2 The Criteria for Merger

The Merger of Two Elliptical Galaxies. Can we determine the physical parameters necessary (and sufficient) for the encounter between two galaxies to end in a merger? Consider for simplicity two spherical galaxies of the same mass M whose equilibrium is entirely due to the velocity dispersion (with no rotation). The fate of an encounter between these two galaxies can be examined in terms of the orbital energy $E = v^2/2$ and the angular momentum $L = bv$ of the pseudoparticle of mass $\mu = M/2$. For systems that are not initially bound there exists a velocity $v_{\max}$ (corresponding to an energy $E_{\max}$) beyond which the merger will never happen, even for $L = 0$. On the other hand below $E_{\max}$ there exists a critical $L_{\max}$ permitting merger. For bound systems it is known that there will always be merger at some point, but it is interesting to know which systems will have merged in the Hubble time. In Fig. 7.23 are shown the dimensionless energy $e = E/\langle v^2 \rangle$ and the angular momentum $l = L/r_c \sigma$, which define the zones of merger in a time less than the age of the universe.

The Merger of Two Spiral Galaxies. It is a simple matter to recognize the merger of two spiral galaxies owing to the two antennae that spread out into intergalactic space, for up to many billions of years after the interaction. Figure 7.24 shows the prototype of a new galaxy formed by the merger of two spirals, NGC 7252. In this object the two nuclei of the spiral galaxies have already merged. What happened during the merger? The stars of each galaxy are thrown into a gravitational potential that does not correspond at all to

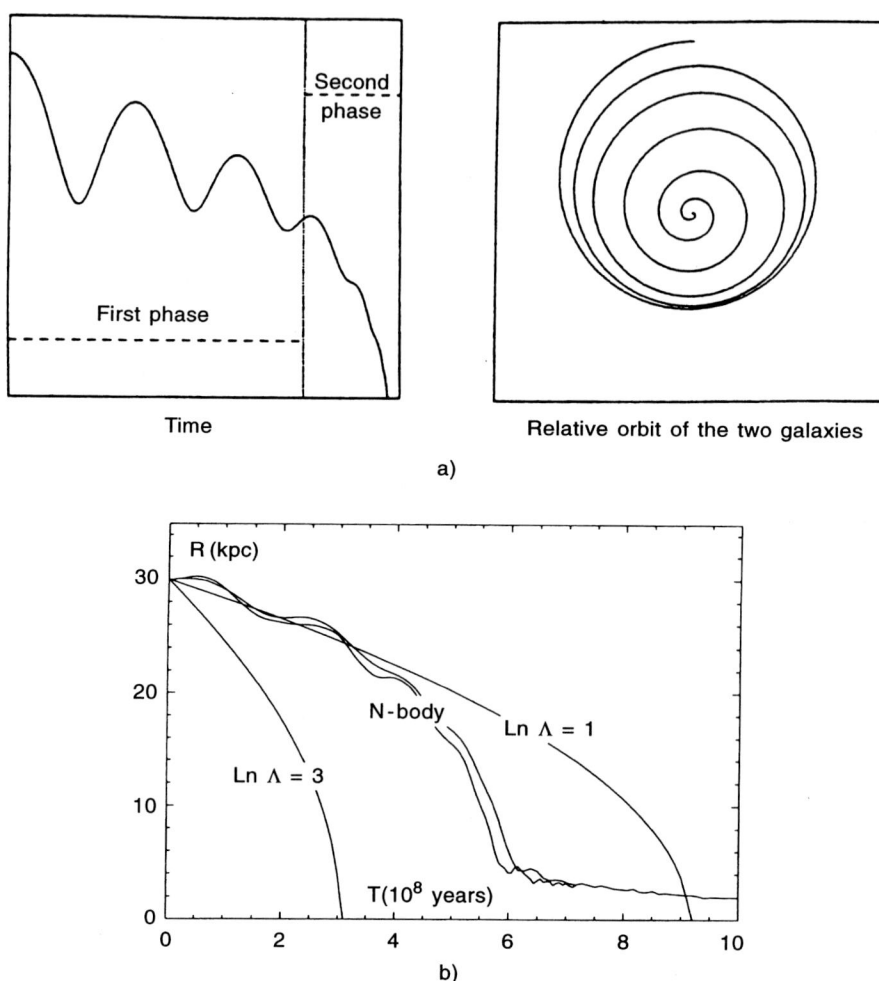

Fig. 7.22. (a) An example of the relative trajectory between two colliding galaxies, calculated for a ratio of mass of 1/10, by using Chandrasekhar's formula to describe the dynamical friction. (b) Comparison of the results obtained from Chandrasekhar's formula and from N-body simulations for the evolution of the distance between the two galaxies during their merger

an equilibrium position: there is a violent relaxation in some crossing time of the system and the particles are repositioned to occupy the new phase space that is now available. The final system is a mass distribution that resembles a cut-off isothermal ensemble, very closely resembling an elliptical galaxy (Chap. 4). The light profile observed in the galaxy NGC 7252 corresponds exactly to the $r^{1/4}$ profile of the ellipticals.

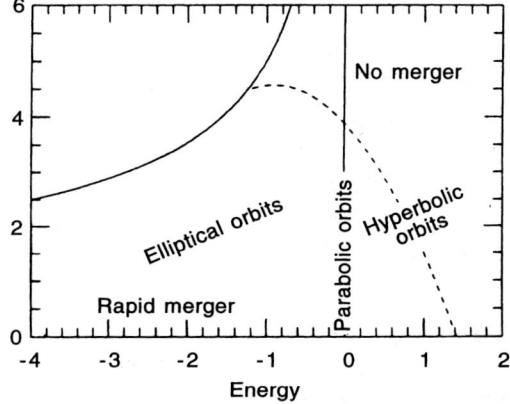

Fig. 7.23. A diagram representing the possibilities for the merger of two spherical galaxies as a function of the energy e and the initial angular momentum l of their relative orbit, in reduced coordinates; the possible orbits are grouped to the lower left of the solid curve, which represents those that are circular. In principle all the bound orbits (elliptical types) end in merger but not necessarily within the Hubble time; the dashed line precisely defines the domain where merger occurs within the Hubble time. (From Binney and Tremaine 1987)

7.4 Shells Around Elliptical Galaxies

In 1961 Halton Arp observed some very fine structures like rings encircling certain elliptical galaxies, but it was mainly the Australian astronomers Malin and Carter who, in 1983, revealed the phenomenon by observing 137 shell galaxies (Fig. 7.25). The shells are very thin and regular luminous filaments in the form of circular arcs whose centre is the galaxy. These luminous shells in fact correspond to the projection on the plane of the sky of small portions of spheres formed by stars: if the structures were two dimensional, like rings in a plane, for example, they would very frequently appear elliptical in projection, and not always circular.

The discovery of the shells is very recent: their luminosity is indeed so weak that on an original photograph they are completely drowned out by the brightness of the elliptical galaxy. The evidence of their existence is a tribute to the development of new image-processing techniques. The technique of *spatial filtering* employed by Malin and Carter is one of them: it involves producing a 'blurred' negative of the original photograph (by defocusing the optical equipment, for example); the superposition of this blurred negative on the original removes everything that varies slowly in the image, that is to say, the low spatial frequencies. The very delicate and contrasted structures (among them the shells) that correspond to large spatial frequencies remain: this is the so-called unsharp-masking technique.

It is also possible to reveal these fine structures by replacing the photographic plate by a linear receiver, the CCD camera; the filtering techniques

212   7. Interactions Between Galaxies

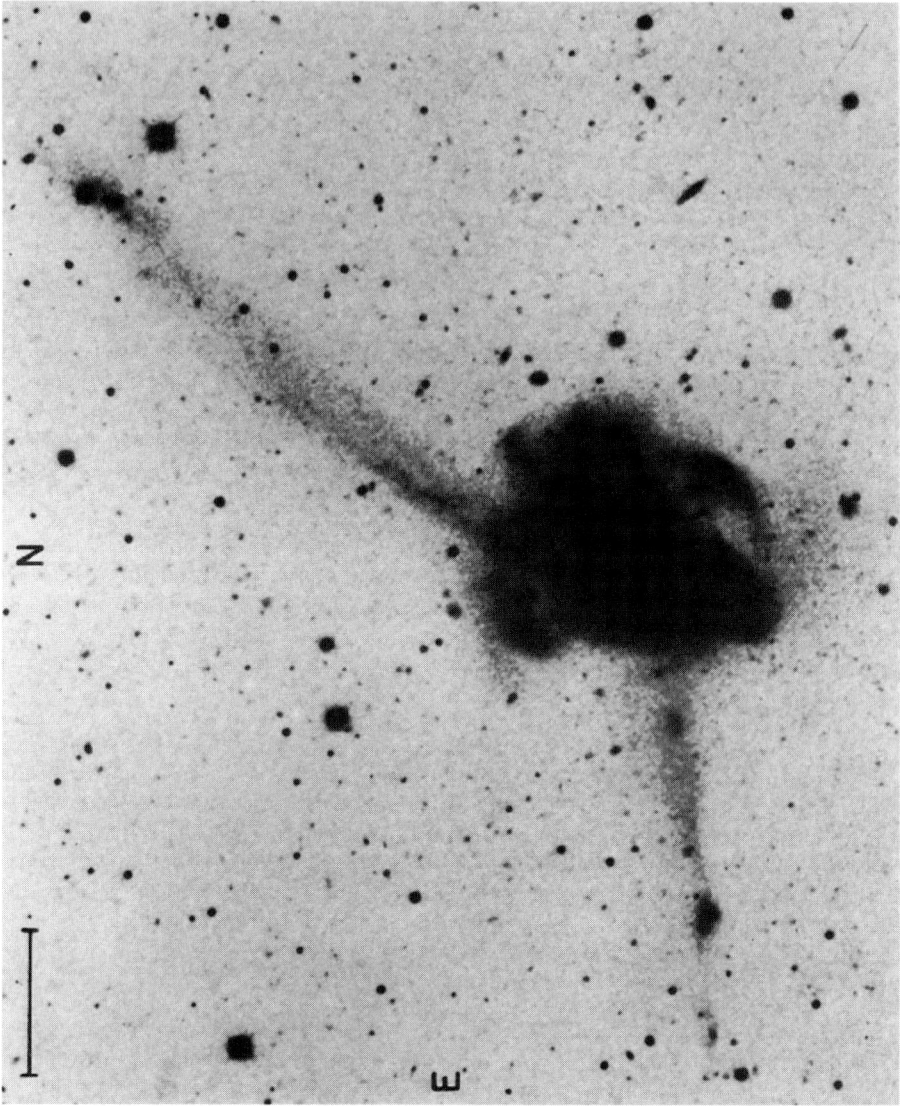

Fig. 7.24. The prototype of a system of galaxies in the process of merging, NGC 7252; the two filaments coming from this curious object, out of a shapeless mixture of galactic nuclei, are the vestiges of the tidal interaction that caused the merger. (From Schweizer 1982)

employed are thus digital (removal of the $r^{1/4}$ luminosity profile of the elliptical galaxy and so on) and give an equivalent result. The first statistics carried out on such observations obtained up to the present day show that close to 20% of isolated elliptical galaxies have shells; in clusters the shells are

7.4 Shells Around Elliptical Galaxies

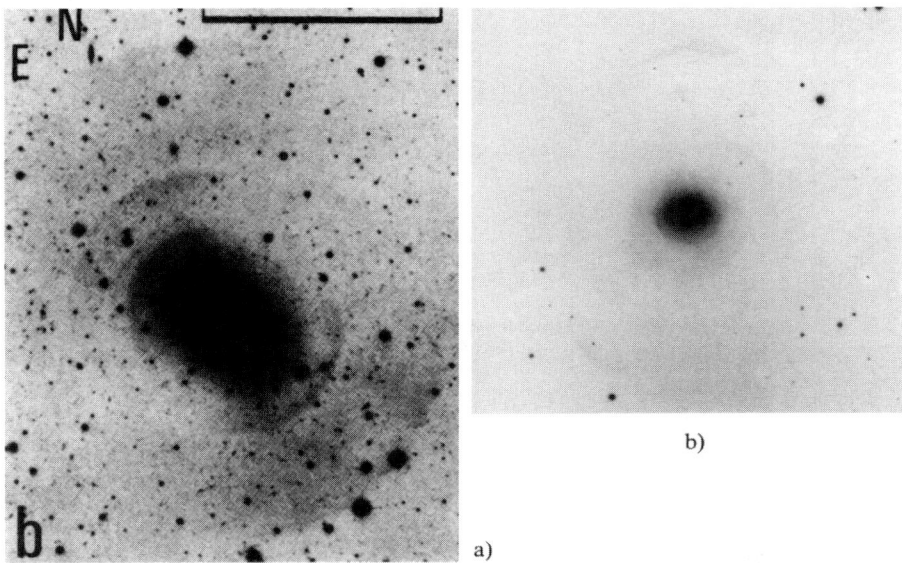

Fig. 7.25. Two examples of shell galaxies. (**a**) The galaxy NGC 3923 has the appearance of a dark oblong stain, encircled by a system of shells aligned with its major axis and positioned like 'parentheses'. Only the internal shells are visible on this photograph of the centre of the galaxy. Twenty-five shells have been counted in this galaxy, up to ten galactic radii across. (From Malin and Carter 1983.) (**b**) The galaxy NGC 474 is circular, and the concentric shells that encircle it this time have no preferred direction. (From Arp 1966)

destroyed by interactions with the other galaxies. The shells are sometimes observed very far from the centre of the galaxy (ten galactic radii, or 120 kpc away) and they can number up to 25 around a single galaxy.

7.4.1 The Shell Formation Mechanism

How do the shells form? The most plausible hypothesis was put forward by François Schweizer, in 1983. After noticing that the structures are frequently associated with distortions (spiral arms and filaments of material ejected to great distances) that characterize the collision and the merger of two spiral galaxies (Fig. 7.26), he suggested that a small spiral galaxy impinging onto a giant elliptical galaxy lies behind the formation of the shells. The elliptical galaxy, being very massive, will hardly be disturbed in such a collision, whereas the small galaxy, totally engulfed, will lose its identity: its stars, scattered in the gravitational field of the elliptical galaxy, therefore form the shells.

Simulations, using the restricted three-body model (see Sect. 7.1.2), made by Peter Quinn in 1984 confirmed this scenario: a head-on collision between a large elliptical galaxy ($M \approx 10^{12} M_\odot$) and a small spiral companion galaxy

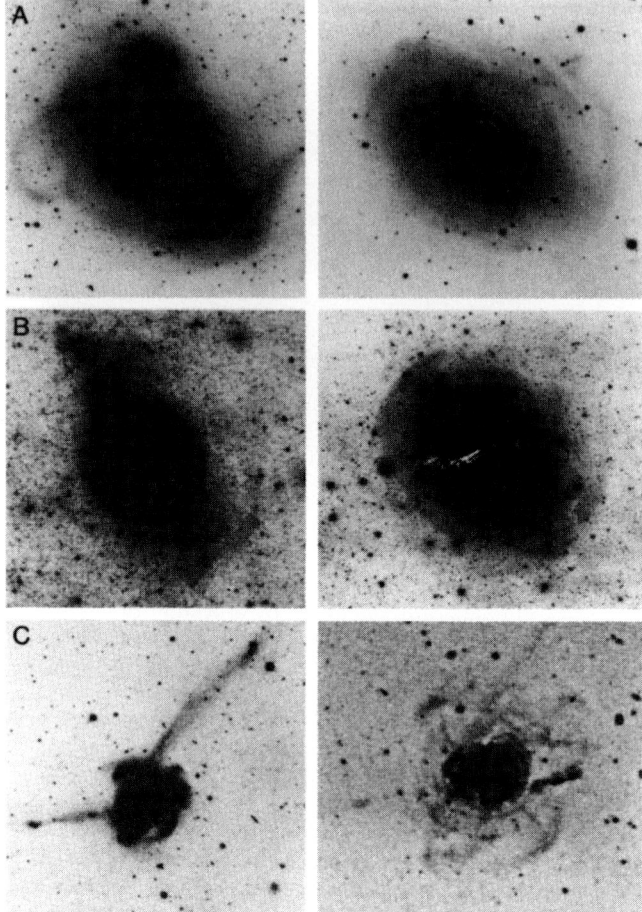

Fig. 7.26. Three systems of merging galaxies, (**a**) Fornax, (**b**) Centaurus, (**c**) NGC 7252, seen in direct photography, on the left, and after filtering out of low spatial frequencies, on the right. This processing increases the contrast of the shells that are present in the three systems, suggesting a relationship between the interaction of the galaxies and the shells. (From Schweizer 1986)

roughly a hundred times less massive ends in the formation of these shells. What happens after the collision? Once the small galaxy has been completely dissociated by the tidal forces of the giant galaxy, the stars that composed it oscillate independently in the gravitational field of the elliptical galaxy. Their trajectories are quasiradial, since the impact parameter of the collision is zero. At the extremities of these trajectories (apocentres) the stars slow down, to retrace their orbit: they therefore spend most of the time in the neighbourhood of this position rather than in the rest of their orbits, where the velocity is greater. It is these points of accumulation that form the shells. If shells exist at various distances from the centre, this is due to the spreading-out of the apocentres of the stars: they do not all have the same energy, and some can reach greater distances from the centre than others. The distribution of the particles is not continuous because the stars not having the same periods do not oscillate in phase and do not reach their apocentres (the shells) at the same time.

To illustrate the mechanism better, consider a system of stars in one dimension whose energies are spread between the values $E_{\min}$ and $E_{\max}$. The corresponding periods of oscillation in the potential well of the giant elliptical galaxy lie between $T_{\min}$ and $T_{\max}$, and the apocentres (maximum elongations of the oscillation) lie between $a_{\min}$ and $a_{\max}$; the particles having a lower energy oscillate more rapidly with a smaller elongation. Figure 7.27α represents the system of stars at the start of the collision by a line of points in the x–T diagram, where x is the distance to the centre of each star and T is the period. As the gravitational field of the central elliptical galaxy is independent of time, the energy of each star is conserved, as is its period T. One star in the x–T diagram then describes a segment parallel to the Ox axis from $-a$ to a (a is the maximum elongation corresponding to T).

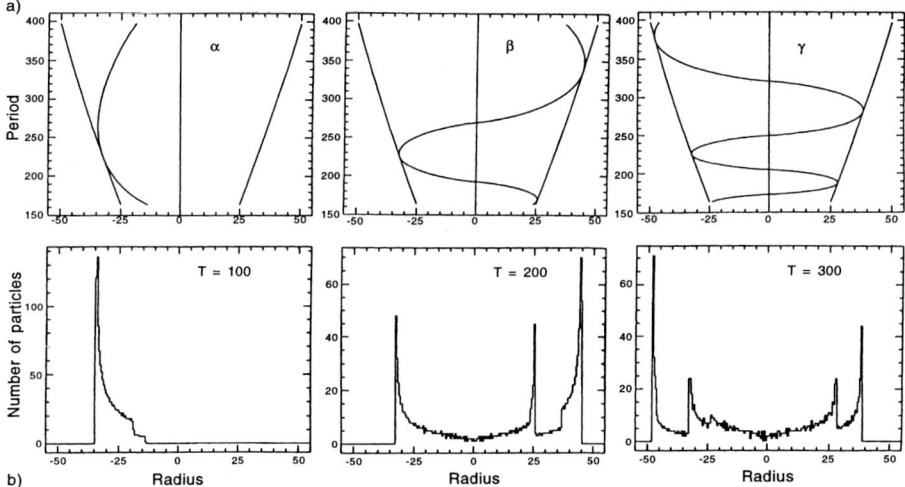

Fig. 7.27. The shell formation mechanism. (**a**) The diagram represents a system of stars in one dimension oscillating in the gravitational potential corresponding to an elliptical galaxy, centred at $x = 0$. Each star is represented by a point in the x–T diagram, x being the position with respect to the central elliptic and T the constant period of oscillation. The trajectory of each star in the diagram is therefore a right segment parallel to the Ox axis and bounded by $x = \pm a$, the maximum elongation of the oscillation. The line of points represents the stars of the system spiralling as the shells form. (**b**) The x–N diagram, where N is the number of stars at the position x, illustrates in a remarkable way the presence of shells, which are points of accumulation of the stars at their apocentres

The stars at the bottom of the diagram, which have a smaller period T, carry out many more oscillations than the stars at the top of the diagram in the same time, and the line of points will spiral (Fig. 7.27β and γ). This is known as *wrapping in phase space*. It is easy to pick out the shells in the x–N diagram (N is the number of stars at the position x) as the points of accumulation of the stars, at which time they retrace their orbit in their oscillations. It can be seen that the number of shells rises relatively quickly with time, because more form at the centre than disappear towards the outside. When the spacing between shells becomes of the order of their width, the system of shells gets mixed up, which allows us to define their lifetimes.

Having reached its apocentre and remained there for a certain time, each star moves farther away. The stars that constitute a shell are therefore replaced at each instant by others, which reach their apocentres a little later and whose period T is then slightly greater. But a greater period corresponds to a greater elongation a: the distance to the centre of the shell increases uniformly with time up to the point where the shell disappears towards the exterior. Each shell is thus a *density wave* that moves in a purely radial manner towards the outside, without any rotational movement around the elliptical galaxy, like the waves produced by dropping a stone onto a surface of water.

7.4.2 Sampling the Gravitational Potential of an Elliptical Galaxy

The results of simulations carried out by Peter Quinn (Fig. 7.28) agree well with the observations. In particular the theory predicts the formation of shells alternating in radius: the shells selected according to their distance to the centre lie alternately to the left and to the right of the central elliptical galaxy. This allows us to deduce important properties concerning the form and distribution of the mass of the elliptical galaxy. Assuming a given mass distribution we find that the model calculates the exact position of the shells at each instant: comparison with a system of observed shells would then enable us to deduce the mass profile. Since elliptical galaxies do not contain atomic gas very far from the centre, as the spiral galaxies do, it is not easy to determine, for example, if there exists a halo of invisible mass around elliptical galaxies. Could the shells help us on this point? In fact the formation of shells by the accretion of a small companion galaxy means that dynamical friction plays a role; the exact position of the shells thus depends critically on the trajectory of the companion galaxy before the merger, the ratio of the masses, and so on. There are so many free parameters of the model that a direct determination of the mass distribution is impossible; it would appear that the presence of invisible mass is not necessary to explain the position of the observed shells (Dupraz and Combes 1987).

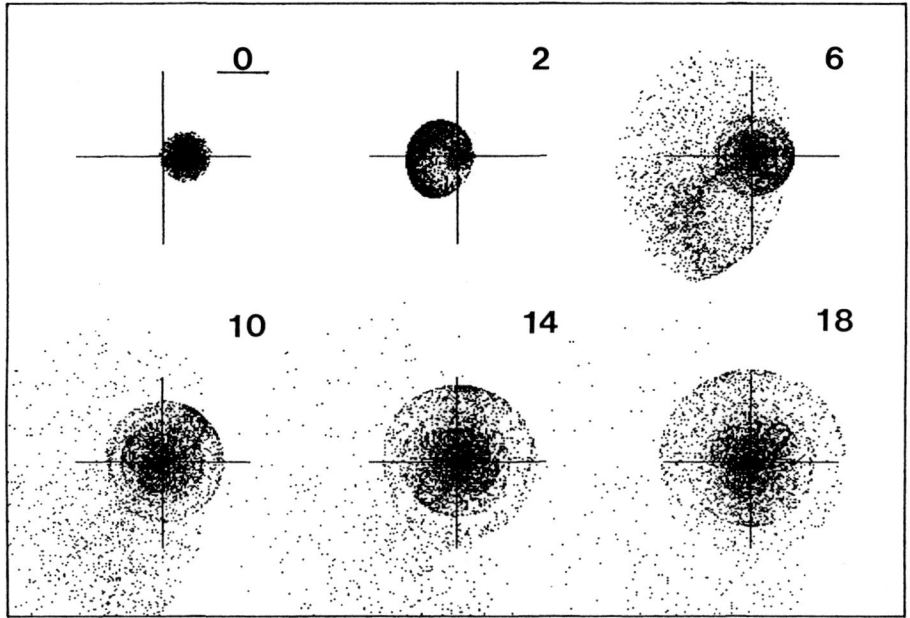

Fig. 7.28. A simulation using a model of test particles of the formation of shells around a spherical galaxy, by Quinn (1984). The stars of the companion galaxy arrive from the right and are scattered in the potential of the central galaxy

7.4.3 The Three-Dimensional Shape of Elliptical Galaxies

The shells are formed because the stars of the former companion galaxy move along very elongated radial trajectories and thus practically in one direction. If the central galaxy (and its gravitational field) were spherical, all the directions would be equivalent and the shells would be portions of spheres having no particular orientation and whose position and size depend only on the initial conditions (for example the angle of arrival and the energy). On the other hand in the case where the central galaxy is elongated (a 'cigar' or 'prolate' galaxy) the trajectories of the stars of the companion galaxy are trapped along the length of a privileged axis: the symmetry axis. Even if it is flattened (a 'pancake' or 'oblate' galaxy), the stars are confined to the neighbourhood of the plane of symmetry. We therefore expect that the geometry of the system of shells varies according to that of the giant elliptical galaxy.

Numerical simulations have been carried out (Dupraz and Combes 1986) corresponding to the effect of a small galaxy on an elliptical galaxy of successively prolate and oblate form (Figs. 7.29 and 7.30): the results obtained confirm the qualitative approach presented below. In the prolate case the shells are portions of spheres aligned with the major axis of the galaxy (as polar caps); these are seen as parentheses encircling the galaxy, once the line

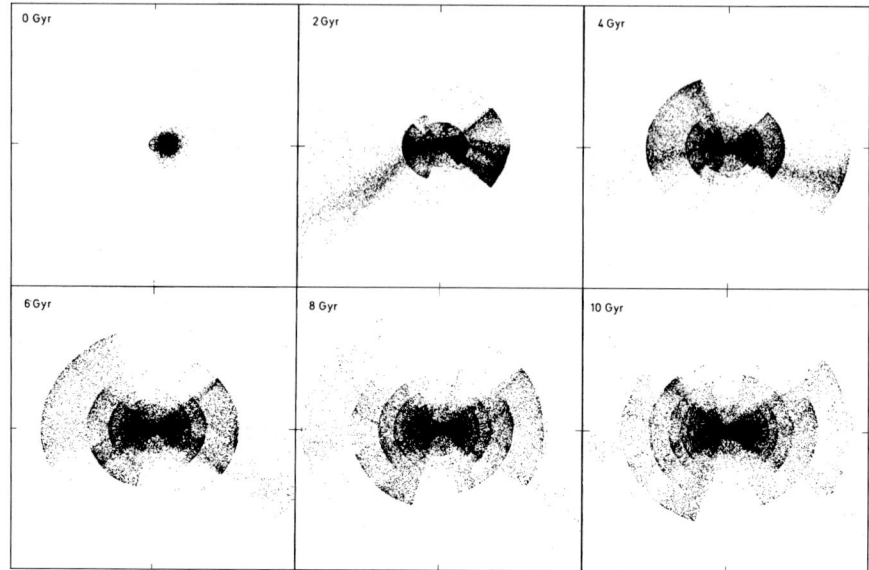

Fig. 7.29. A simulation of the accretion of a small companion galaxy by a massive, nonspherical elliptical galaxy. In this first case the elliptical galaxy is elongated (prolate) parallel to the horizontal axis: the shells are therefore aligned with the major axis of the galaxy. (From Dupraz and Combes 1986)

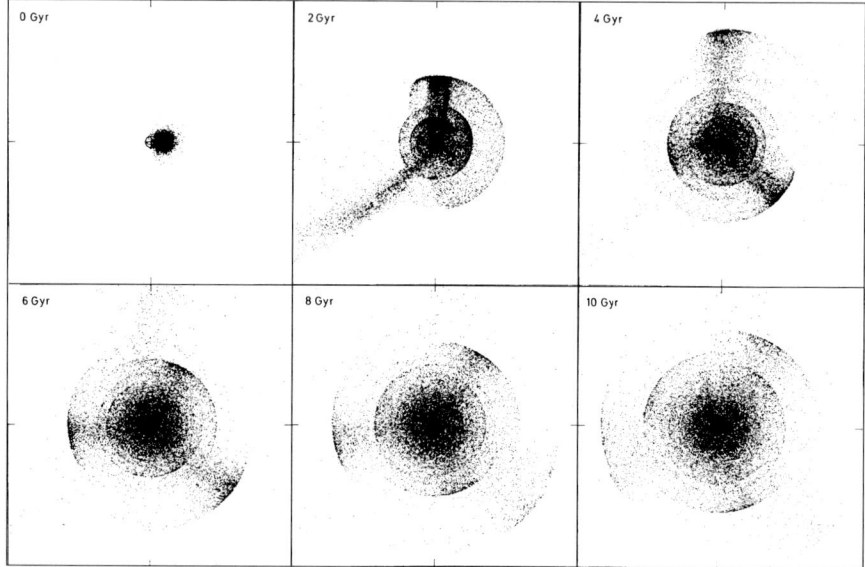

Fig. 7.30. A simulation of the accretion of a small companion galaxy by a massive, nonspherical elliptical galaxy. In this second case the elliptical galaxy is flattened (oblate), and the line of sight is perpendicular to the equatorial plane: the shells spiral randomly around the whole galaxy. (From Dupraz and Combes 1986)

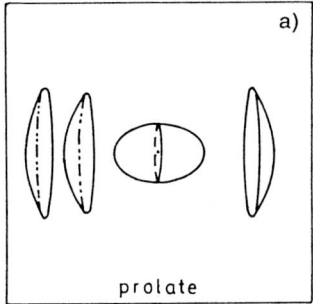 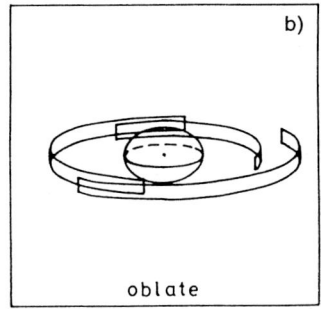

Fig. 7.31. A representation of the three-dimensional shape of the shells, according to the shape of the central galaxy. (**a**) In the 'cigar' case the shells are sections of spheres aligned with the apparent major axis of the galaxy; they are visible when the galaxy is observed in profile. (**b**) In the 'pancake' case the shells are confined to the neighbourhood of the equatorial plane of the galaxy and display no privileged direction; they appear only when the galaxy is seen face-on

of sight no longer coincides with the major axis (see Fig. 7.31). In the oblate case the shells are arranged in the neighbourhood of the plane of symmetry of the galaxy, as equatorial bands that are very thin but almost entirely encircle the galaxy, with no privileged direction; the shells are then only visible if the galaxy appears circular in projection on the plane of the sky.

It is thus possible to determine the three-dimensional form of the elliptical galaxies represented in Fig. 7.25: the model suggests, for example, that NGC 3923 has an elongated (prolate) form and NGC 474 has a flattened (oblate) form. The first statistics carried out with the systems of shells already observed show that about 30% of the shell systems come under the aligned geometry (that of NGC 3923) and 30% under the 'scattered' geometry (that of NGC 474), the other systems being difficult to classify because they are too irregular or contain too small a number of shells. These results confirm the coexistence of prolate and oblate objects at the heart of elliptical galaxies, a result already suggested by the observation of dust lanes. It appears that this is the only thing that we can be certain of and that formation models for elliptical galaxies must account for.

7.5 The Formation of Ellipticals. Conclusions

The very large percentage of galaxies observed that have shells (20%) suggests a very significant rate of merger between galaxies. Many favourable conditions must come together for the accretion of a small companion galaxy by a giant galaxy to produce visible shells. We can estimate the average number of companion galaxies devoured by each galaxy in the sky to be 4 or 5. Now the merger of two spiral galaxies of comparable mass can result in an elliptical galaxy (see Sect. 7.3). *Can the merger of two galaxies be the origin of all the ellipticals?*

The existence of two principal types of galaxy, ellipticals (spheroidals) and spirals (discs), is an old problem in extragalactic astronomy. According to the traditional interpretation, in which the galaxies were formed just after the big bang, the stars of the spheroidal systems form very quickly from the protogalactic cloud, before there is complete dissipation of the gaseous component. The stars of the discs, on the other hand, appear more slowly from the clouds of gas that have had time to contract and to flatten into discs by rotation. The origin of this two-speed star formation is poorly understood.

Furthermore it is possible to estimate the number of mergers of galaxies since the beginning of the universe. Basing our considerations on systems in the process of merging, with vestiges of the interaction (NGC 7252), and knowing that this transition period lasts roughly one to two billion years, we can extrapolate the number of galaxies that have merged since the big bang to be 10–20%. This percentage corresponds to quite close to the proportion of elliptical galaxies in the universe!

However, many arguments have been raised against this theory of the formation of ellipticals, notably the abundance of globular clusters, which is very great in ellipticals and cannot be explained by the merger of two spirals, where these clusters are far less abundant; moreover the fusion of two less concentrated systems of stars cannot lead to the very strong concentration of the ellipticals; finally dwarf elliptical galaxies could not be formed by the same mechanism.

These arguments can nevertheless be discussed. The formation of elliptical galaxies would not only occur through the merger of galaxies of comparable mass: if account is taken of small companion galaxies, the formation of the ellipticals would be more progressive. The merger of two gaseous and nonstellar systems would allow the formation of concentrated stellar systems by dissipation. On the other hand, observations made by the *IRAS* satellite, in 1984, have revealed that the brightest galaxies in the sky in the infrared are systems in the process of merging and interacting galaxies. Now the far-infrared flux is directly dependent on the rate of star formation (emission being by the dust heated by the new stars). Enormous outbursts of star formation are therefore present in these merging galaxies, which could explain the great abundance of globular clusters in elliptical galaxies. The many supernovae and the powerful stellar winds generated in the course of such starbursts could also drive out the gaseous component carried along by the spirals, which explains its absence in the final elliptical galaxy.

One sufficiently convincing argument due to Alar Toomre remains: if the merging of galaxies is not the origin of the elliptical galaxies, how are we to explain the traces of the merger whose prototype is NGC 7252? These mysterious objects must represent at least about 15% of the galaxies in the universe!

7.6 Cosmological Implications

Most galaxies are embedded in massive extended dark haloes. The merger of two galaxies involves the merging of the haloes first. In two-galaxy merger simulations, individual dark haloes have been included around each galaxy, and their role has been shown to be very important: it is to receive the relative orbital angular momentum of the merging discs and to accelerate the merger of the visible galaxies (Barnes 1988). In cosmological scenarios of hierarchical galaxy formation (see Chap. 12), small dark haloes virialize first and progressively merge to form larger and larger haloes, the latter being the root of complex merging trees. The merger of the visible systems does not follow at the same speed, and it is likely that interacting galaxies already have a common halo. This might help us to understand the high efficiency of visible galaxy mergers.

The number of interactions and mergers experienced by a galaxy during its life time can put constraints on the different models of the universe, and in particular on the values of its density. In standard Cold Dark Matter (CDM) models, with a critical density ($\Omega = 1$, see Chap. 13), the mass accreted by galaxies during the last Gyrs is still a large fraction of their mass today, while the accretion declines faster in open models ($\Omega < 1$). Observations of present galaxies reveal that most of them possess very thin stellar discs, while satellite accretion is very efficient in disc thickening. The statistics on galaxy disc thickness today can then serve to constrain the rate of mergers and determine models of the universe (Toth and Ostriker 1992).

Exercises

7.1 With the help of Chandrasekhar's formula for the deceleration due to friction,

$$\gamma = \frac{dv}{dt} = \frac{4\pi G^2 M \rho(<v)}{v^2} \ln \Lambda,$$

where M is the mass of the body that is braked (the satellite) and $\rho\,(<v)$ is the density of the braking medium corresponding to stars with a velocity less than v, find the trajectory of the satellite M in a spherical galaxy with an isothermal distribution and velocity dispersion σ. The density at a radius r is written as $\rho(r) = \sigma^2/2\pi G r^2$ for $r < r_{\max}$ and $\rho(r) = 0$ for $r > r_{\max}$.

Given that the velocity distribution in the isothermal sphere is Maxwellian, show that

$$\rho(<v) = \rho(r) \left[\mathrm{erf}\left(v/\sigma\sqrt{2}\right) - \left(v/\sigma\sqrt{2}\right) \mathrm{erf}'\left(v/\sigma\sqrt{2}\right) \right],$$

where erf is the error function and erf$'$ is its derivative.

Suppose that the satellite is launched on a circular trajectory and that its orbit is almost circular at each instant. Give the value of its velocity as a function of σ. Note that this velocity is constant. Calculate the potential energy of the satellite as a function of r.

Thus integrate the equation for the friction and show that the position $r(t)$ of the satellite satisfies

$$r^2(t) = r^2(0) - 0.6 GM \ln \Lambda/\sigma t.$$

7.2 In a plane defined in terms of its polar coordinates (r, θ) consider an axisymmetric gravitational potential $V(r)$. For elongated periodic trajectories with an energy E, what is the relation linking the period of oscillation T at maximum elongation a, $T(a, E)$?

Show that in the case of an isochronal potential (Hénon 1959)

$$\Phi(r) = -GM / \left[b + \left(r^2 + b^2 \right)^{1/2} \right]$$

(where b is a characteristic size), the period depends only on the energy:

$$T(E) = 2\pi GM(-2E)^{-3/2}.$$

In this isochronal potential assume that stars with an energy lying between $E_{\min}$ and $E_{\max}$ are thrown in by a companion at the time $t = 0$. Show that the number of shells (see Sect. 7.4) increases in time as

$$N(t) = \frac{t}{2\pi GM} \left[(-2E_{\min})^{3/2} - (-2E_{\max})^{3/2} \right].$$

References

Arp, H. (1966) *Atlas of Peculiar Galaxies* (California Institute of Technology, Pasadena CA).
Barnes, J.E., (1988), *Astrophys. J.* **331**, 699.
Binney, J., and Tremaine, S. (1987) *Galactic Dynamics* (Princeton University Press, Princeton NJ).
Chandrasekhar, S. (1943) *Astrophys. J.* **97**, 255.
Combes, F., Dupraz, C., and Gérin, M. (1989) in *Dynamics of Interacting Galaxies*, edited by R. Wielen (Springer, Berlin, Heidelberg), p. 205.
Dupraz, C., and Combes, F. (1986) *Astron. Astrophys.* **166**, 53.
Dupraz, C., and Combes, F. (1987) *Astron. Astrophys.* **185**, L1.
Hénon, M. (1959) *Ann. d'Astrophys.* **22**, 491.
Hibbard, J. E., van Gorkom, J. H. (1996) *Astron. J.* **111**, 655.
Hibbard, J. E., van der Hulst J.M., Barnes J. (2001) *Astron. J.* in prep.
Holmberg, E. (1941) *Astrophys. J.* **94**, 385.
Hunter, C., and Toomre, A. (1969) *Astrophys. J.* **155**, 747.
Larson, R. B., and Tinsley, B. M. (1978) *Astrophys. J.* **219**, 46.

Lin, C. C., and Shu, F. H. (1964) *Astrophys. J.* **140**, 646.
Lynds, B., and Toomre, A. (1976) *Astrophys. J.* **209**, 382.
Malin, D. F., and Carter, D. (1983) *Astrophys. J.* **274**, 534.
Pfleiderer, J., and Seidentopf, H. (1961) *Z. Astrophys.* **51**, 201.
Pfleiderer, J., and Seidentopf, H. (1961) *Z. Astrophys.* **58**, 12.
Quinn, P. J. (1984) *Astrophys. J.* **279**, 596.
Sancisi, R. (1976) *Astron. Astrophys.* **53**, 159.
Schweizer, F. (1982) *Astrophys J.* **252**, 455.
Schweizer, F. (1983) in *Internal Kinematics and Dynamics of Galaxies*, edited by E. Athanassoula (IAU Symposium 100, Kluwer, Dordrecht), p. 319.
Schweizer, F. (1986) *Science* **231**, 227.
Toth, G., Ostriker, J.P., (1992), *Astrophys. J.* **389**, 5.
Toomre, A. (1977) in *The Evolution of Galaxies and Stellar Populations: Conference at Yale University, May 19–21, 1977*, edited by R. B. Larson and B. M. Tinsley (Cambridge University Press, Cambridge).
Toomre, A., and Toomre, J. (1972) *Astrophys. J.* **178**, 623.
Van der Hulst, J. M. (1978) *Astron. Astrophys.* **71**, 131.

8. Extragalactic Radio Sources

Twelve years after the publication of the first results obtained by Jansky in 1932, which marked the beginning of radio astronomy, Reber observed a radio-emission maximum in the constellation Cygnus. It was the first detection of Cygnus A, one of the brightest of all extragalactic sources in the radio region. These observations, made at low frequencies (160 MHz) and with the aid of small antennae, gave only a rough estimate of the source position. It was necessary to await the construction of interferometers (in Australia by Bolton and Stanley around 1948 and simultaneously in Britain by Ryle and Smith; the resolution obtained was of the order of 10') before the radio emission could be better localized and possibly associated with a known object.

The extragalactic nature of at least some of the sources was established in 1949 when Bolton, Stanley, and Slee were able to identify Virgo A and Centaurus A (so named because they were the first radio sources detected in the constellations Virgo and Centaurus), two of the brightest sources in the sky, with the nearby galaxies M87 and NGC 5128. Later on, Baade and Minkowski (1954) showed that Cygnus A is also associated with a galaxy, this time distinctly fainter ($m \approx 15$) and more distant (redshifted by $z = 0.06$[1]). This result suggested that radio astronomy was capable of revealing distant objects in the universe, possibly undetectable in other spectral ranges. Note that such searches for the optical object associated with radio sources led to the discovery of quasars (see Chap. 9).

When large interferometers such as the VLA came into service more than a decade ago, it became possible to obtain very detailed maps of the radio emission of extended sources, since the resolution attained is of the order of 1" (at about 1 GHz), whereas the Cygnus A lobes, for instance, extend to more than 1'. Likewise, for compact radio sources, intercontinental interferometry (VLBI, or very long-baseline interferometry) allows details on the scale of

---

[1] Recall that z is defined by $z = (\lambda_{obs} - \lambda_{lab})/\lambda_{lab}$, where λ_{obs} and λ_{lab} are respectively the wavelength as observed and as measured in the laboratory. In cosmology z is an important parameter because it is measurable; the other main quantities allowing one to specify the chosen cosmological model are the rate of expansion $H_0 = \dot{R}_0/R_0$ and the deceleration parameter $q_0 = -\ddot{R}_0 R_0/\dot{R}_0^2$, where R is the scale parameter and the index 0 refers to the present epoch. For more details on these points, refer to Chaps. 11 and 13.

milliarcseconds to be revealed. Adding to that the possibility of observing at various frequencies and determining the polarization of the radiation, one can see that the radio domain is characterized by a great wealth of information. The variety of observed morphologies constitutes a real challenge to theoreticians, and at present the interpretation of all the data is still uncertain. That is why, in what follows, we shall describe mainly information provided by observations. First the main characteristics of phenomena involving the plasma responsible for the radio emission will be given, and then the various types of morphology encountered will be described. For each of the three basic elements, the extended lobes, the jets, and the compact nuclear components, we shall specify the spectral and polarization properties. Finally some elements of radio-source modelling will be presented; in this last part we shall deliberately leave aside all discussion about the ultimate central object, whose nature and properties are covered in Chap. 10.

8.1 Physical Processes

In a region where a magnetic field is present, rapidly moving charged particles experience the Lorentz force; they follow a helical path and, thus accelerated, emit electromagnetic waves known as synchrotron radiation. In contrast with the 'free–free' emission process, by which the radiation originates from a population of charges with a Maxwellian velocity distribution, this mechanism is said to be 'nonthermal'. Owing to their tiny mass, the electrons undergo a more intense acceleration and largely dominate the emitted flux (the same is true for thermal radiation where interactions between the charges themselves are the source of the acceleration). This has an immediate consequence: the observed radio brightness provides no information about the energy carried by massive particles (protons and nuclei). By analogy with the cosmic radiation observed in our Galaxy, the energy of massive particles may be 100 times that of electrons, but this factor is very uncertain.

8.1.1 Radiation Emitted by an Ensemble of Relativistic Electrons

We shall first consider the case of a single electron, with energy E and a Lorentz factor γ ($\gamma = 1/\sqrt{1-v^2/c^2}$), where v is the velocity. When $v \approx c$, the radiation emitted is concentrated in a narrow cone (with an angle at the vertex of the order of $1/\gamma$): a 'properly located' observer would then see a series of very short pulses. It can be shown that the corresponding radiation comprises a narrow interval of frequencies around the value

$$\nu_{\max} = 0.069 \gamma^2 \omega_L \sin \psi,$$

where ω_L is the Larmor frequency ($\omega_L = eB/mc$) and ψ is the angle between the line of sight and the direction of the magnetic field $\boldsymbol{B}$. If the velocity distribution is isotropic, the average over ψ is given by the practical expression

$$\langle \nu \rangle = 5E^2 B,$$

with $\langle \nu \rangle$ expressed in MHz, E in GeV, and B in µG. For sufficiently broad energy distributions $N(E)$ there exists an almost one-to-one relation between the observed frequency and the energy. This simply means the following: a measured spectrum $F_\nu = k_1 \nu^{-\alpha}$ corresponds to a distribution $N(E)$ also satisfying a power law, $N(E) = k_2 E^{-\lambda}$, the spectral index α being related to λ by $\lambda = 2\alpha + 1$. We have implicitly assumed that the medium is optically thin. Under the above conditions one can show that the opacity $\tau(s)$ over a distance s varies as $s\nu^{-(\lambda+4)/2}$ and that at low frequencies (the plasma then becoming optically thick) the spectrum received is of the form

$$F_\nu = a\nu^{2.5}.$$

To obtain the luminosity from the observed flux one generally considers the emission to be isotropic; it is important to bear in mind that this hypothesis may be quite wrong in certain cases, in particular when the emitting plasma is moving relativistically in the direction of the observer (the luminosity would then be greatly overestimated).

8.1.2 The Internal Energy of a Gas

The observed radio flux does not allow an unambiguous determination of the total energy of the electrons; nevertheless we can obtain a lower limit, reached when there is equipartition between the magnetic energy and that of the relativistic particles. In this case the electron energy density is

$$U_{\min} = 9.3 \times 10^{-2} B^2$$

(in erg cm^{-3} with B in gauss).

By making several assumptions concerning the spectrum of the source, the direction of B, and the fraction of energy associated with the electrons, we can estimate U and B. The order of magnitude obtained is several µG for extended sources.

8.1.3 Energy Losses

We shall first examine what the evolution of a radio source would be if the emitting gas underwent a simple adiabatic expansion after being ejected by the active nucleus. Arguments relating to the conservation of magnetic flux lead to the following result: if the cloud of gas increases in size by a factor β, the radio flux will decrease by a factor $\beta^{4\alpha+2}$. However, sources having comparable luminosities are observed with sizes ranging from 1 to 100 kpc. For a typical value, $\alpha = 0.75$, we see that it is impossible to imagine these different objects as members of the same sequence having undergone adiabatic evolution. This therefore implies a permanent reinjection of energy during the expansion.

Because they radiate, the energy of the relativistic electrons progressively decreases. This is not the only source of energy loss, since they also undergo inverse Compton scattering with the photons associated with the 2.7 K cosmic background radiation or, in the most compact sources, with the photons produced there by synchrotron radiation. There is therefore a transfer of energy from the electrons to the photons (which move from the radio domain to the X-ray domain). As the variation ΔE is proportional to E^2, for the losses by synchrotron radiation as well as for those due to the inverse Compton effect, these processes should result in a modification of the distribution $N(E)$ and the appearance of a high-frequency cutoff, if the plasma were not steadily fed by energetic electrons. The existence of inverse Compton scattering implies the presence of X-ray emission associated with radio sources. This can be observed in several cases (Centaurus A and Cygnus A), which allows an independent estimate of B to be obtained. Indeed the ratio of radio and X-ray luminosities depends very sensitively on this parameter ($F_R/F_X \propto B^{1+\alpha}$), and its determination does not depend on any particular hypothesis. The problem with this method is more an observational one: the sensitivity and the spatial resolution of X-ray measurements are much less than those attained at radio wavelengths, which confines the application of the method to the brightest and most extended sources. In practice one obtains only a lower limit for B, since a thermal contribution to the measured flux F_X cannot be excluded (only the inverse Compton component of the X-ray flux enters into the estimate of B). The values thus obtained are also of the order of $1\,\mu G$.

8.1.4 Polarization of Radiation. Faraday Rotation

Theory predicts that when the magnetic field acting throughout the plasma is uniform, the synchrotron radiation emitted is strongly polarized perpendicular to the direction of B. The degree of polarization may reach 75% for $\alpha = 1$; in practice, however, the direction of B may vary along the line of sight, and the observed value is distinctly less than 1. Another complication arises if one attempts to interpret the direction of polarization, since this depends on the frequency, owing to Faraday rotation. When a linearly polarized wave travels through ionized gas (the emitting medium itself or foreground gas), the direction of the E vector turns through an angle $\Delta\theta$, given in degrees by

$$\Delta\theta = 4.64 \times 10^6 n_t B_\parallel L \lambda^2,$$

where n_t (in cm^{-3}) is the density of thermal electrons, $B_\parallel$ (in G) the longitudinal component of B (n_t and $B_\parallel$ are assumed here to be uniform), L the distance travelled (in kpc), and λ the wavelength (in cm). Measurements at various frequencies allow the determination of $\Delta\theta(\lambda)$ and give an estimate of the product $n_t B_\parallel$.

8.2 The Various Types of Radio Source and Associated Optical Objects

8.2.1 Compact Sources. Extended Sources

In general one distinguishes between two large classes of radio sources according to their apparent angular size: the sources are called 'compact' or 'extended' depending on whether they are less than or greater than $1''$ in size. This criterion may appear arbitrary because it has nothing to do with the intrinsic size of the objects. In fact there is a pronounced difference between the two classes of radio sources: extended sources are generally quite a bit larger than $1''$ and extend beyond the limits of the associated galaxy observable in the visible (linear dimension $\geq 10\,\mathrm{kpc}$). Centaurus A, shown in Fig. 8.1, is an example. On the other hand, compact sources are far smaller, always less than several tens of parsecs across. This distinction in size corresponds to other differences concerning the spectra and variability: compact sources are characterized by spectral indices α of the order of 0–0.3 and a strong variability (a factor of two or three on a time scale of several months), whereas the emission of extended sources remains constant in time and exhibits a steeper spectrum ($\alpha \approx 0.7$–1.2). Furthermore an estimate of the total energy contained in the emitting gas leads to quite distinct values, typically 10^{58}–10^{61} ergs for extended radio sources, compared with only 10^{52}–10^{58} ergs for compact sources. Note that it is possible to encounter the two types of component either separately or together in the same source.

8.2.2 The Optical Identification of Radio Sources

So far the great majority of the thousands of extragalactic radio sources detected have not been identified optically. The search for a corresponding visible object is difficult since it requires (1) the position of the radio source to be known with great accuracy ($\approx 1''$) and possibly a detailed knowledge of the morphology if the emission is extended, and (2) deep optical-imaging observations. The correlation between the morphology of a radio source and that of the corresponding optical object, where it has been possible to detect it, is not very strong; nevertheless some general trends may be noted:

- The sources dominated (near 1 GHz) by an extended component are frequently associated with elliptical galaxies, the latter thus being called 'radio galaxies'. Optically these are similar to ordinary ellipticals; sometimes nuclear activity manifests itself through the presence of optical or X-ray emission lines.
- When the compact component predominates, the associated object may possibly be of the BL Lacertae type, the active nucleus of a Seyfert galaxy,

230   8. Extragalactic Radio Sources

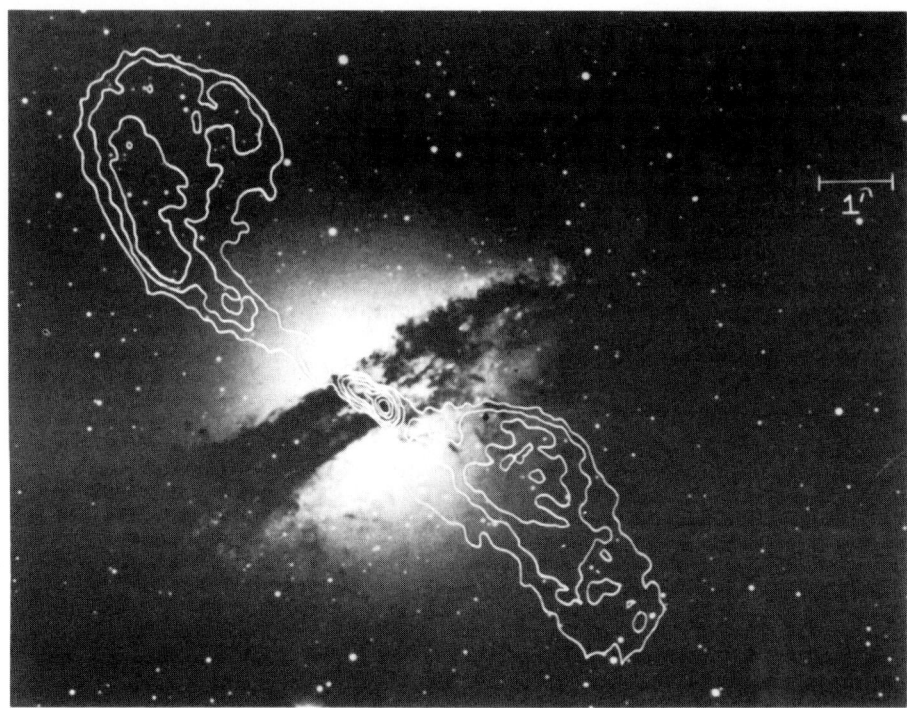

Fig. 8.1. The radio emission at 5 GHz for the source Centaurus A superimposed on the optical image (taken in the B band) of the associated galaxy, NGC 5128. (From Burns et al. 1983)

or simply (for nearby sources of low luminosity) the central region of an ordinary spiral galaxy.
- In the case where the contribution of the extended lobes is comparable to that of the compact source, the associated object is often a quasar (this would be true up to about 1 GHz; at higher frequencies the compact component always dominates). In brief (see Chap. 9 for details), quasars are characterized by the presence of strong and broad optical emission lines. Seyfert galaxies are similar to quasars in many respects but they are much less luminous; these objects are therefore detected only in nearby regions of the universe. Lastly BL Lacertae objects, whose luminosity may well reach that of some high-redshift quasars, show much weaker emission lines.

There are, however, many exceptions; for instance there are quasars for which the emission is exclusively of the compact or extended type. As we shall see later on, radio emission is just one of the forms of activity of some galactic nuclei. The absence of a relation between the type of radio source and the nature of the associated optical object is reflected in the fact that 'nuclear activity' is a phenomenon as such and that the very energetic processes that take place are relatively independent of the environment, that is, of the global properties of the galaxy. This justifies the study of radio sources in their own right.

8.3 Extended Sources

8.3.1 Observed Morphologies

Extended Edge-Brightened Double Sources. This is the type most commonly encountered: two elongated radio lobes (the length to width ratio being $L/l > 4$) are observed on either side of a central object (an elliptical galaxy or a QSO).

Cygnus A is the prototype of this class (Figs. 8.3c and 8.6); 3C 284 (the 284th source in the 3rd Cambridge catalogue) shows a similar morphology (Fig. 8.2). The strong emission observed towards the outer parts is often due to the existence of hot spots located there. The size of these structures is of the order of kiloparsecs, whereas the global size can exceed 1 Mpc. A compact component is often associated with the nucleus of the underlying galaxy. The structure of these radio sources shows remarkable features suggestive of formation mechanisms:

- alignment of hot spots with the optical object or the central component;
- global symmetry, in terms of both position and flux.

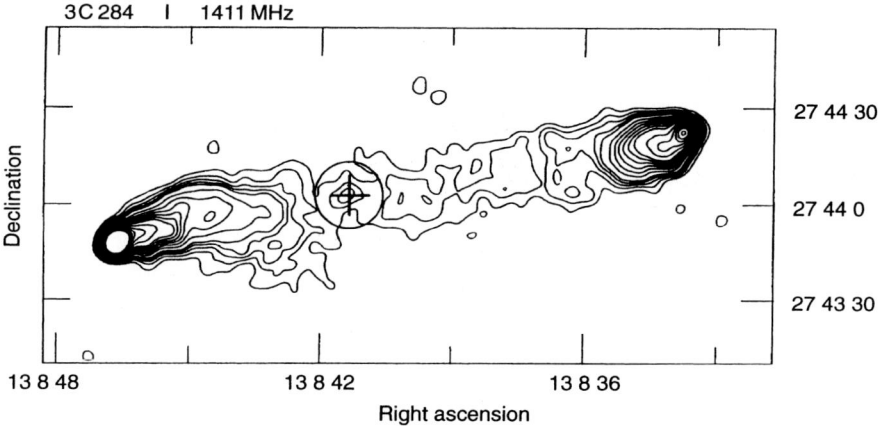

Fig. 8.2. A map of the 1.4 GHz emission from the source 3C 284. The circle, centred on the associated galaxy (shown by a cross), has a radius of 25 kpc. (From Leahy and Williams 1984)

One can, for example, study for these sources the distribution of ratios L_1/L_2 and F_1/F_2, where L and F are respectively the separation from the optical object and the flux of one of the components. The histograms show a pronounced maximum around 1 with a relatively small dispersion. It is likely that biases play a role here. Thus, one can imagine that identification would appear more likely if the optical object were located exactly between the two lobes; likewise, it is possible that observations with a low dynamic range disadvantage those sources with one lobe clearly weaker than the other. Nevertheless the high degree of symmetry is real because it is confirmed by recent more sensitive observations. This property strongly suggests the existence of ejection in two opposite directions along a well-defined axis.

Extended Edge-Darkened Double Sources. Unlike in the previous case, a slow decrease in flux with distance from the central object is observed; the external limit of the lobes is less well defined. The closest radio galaxy ($D = 5\,\text{Mpc}$), Centaurus A, belongs to this class (Fig. 8.1). For this category of sources, observations at high resolution show, in general, a compact component connected to the lobes by narrow jets.

Wide Double Sources. These are characterized by a length-to-width ratio of the order of 2. Virgo A, associated with the brightest galaxy (E type) in the Virgo cluster, M87, is a radio source of this type (Fig. 8.3b).

Sources with Two Tails. This morphological type is similar to that observed for double sources with a dark external edge. The difference concerns the angle between the directions of the two lobes, which now differs from 180°. The sources 3C 465 and 3C 83.1B (Fig. 8.3a) belong to this class.

Sources with a Single Narrow Tail. The structure here is entirely asymmetrical. It is composed of a bright head coinciding with the nucleus of an elliptical galaxy and a diffuse elongated tail. However, the head–tail system appears to be split; this suggests that these sources might be of the previous type but with a very small angle (an example, IC 310, is shown in Fig. 8.3a).

8.3.2 Morphological Classification

To better understand the great variety of observed structures, astronomers tried to bring together the various morphologies and look for some continuity with respect to some parameters. For example we noticed earlier on that when two elongated components are present, the angle between them can lie between 0 and 180°. Can we construct a sequence according to the value of this angle and thereby find what factor is responsible for the curvature defined by the two lobes: is it the age of the sources, the energy contained, or the interaction with an external medium? In answering such questions, we hope to understand which phenomena are responsible for the evolution of a given parameter along a sequence.

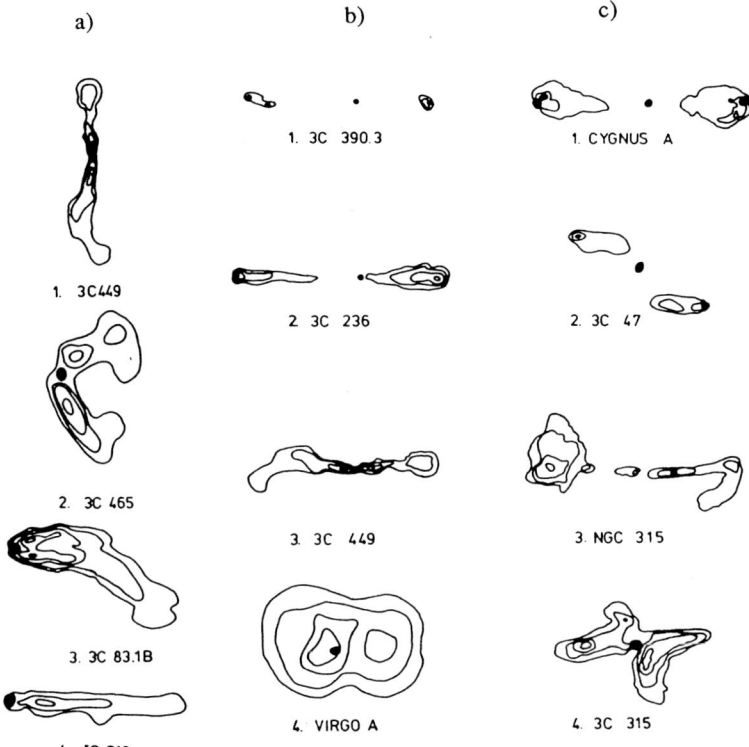

Fig. 8.3. Classification of observed morphologies for extended sources, according to: (**a**) the curvature, (**b**) the radial brightness gradient, and (**c**) the type of symmetry. (From Miley 1980)

In a classification according to curvature, the factor considered is the angle α defined by the two components, which is zero in the limiting case of sources with a single narrow tail. Figure 8.3a illustrates the observed gradation as a function of α. It appears that the morphology of sources for which $\alpha < 180°$ is satisfactorily interpreted by interaction with an external medium. Indeed the most distinctly curved sources are found in clusters of galaxies rich in intergalactic gas (this is the case in particular for 3C 83.1B, which belongs to the Perseus cluster). Furthermore, small values of α correspond, on average, to galaxies that are less massive and as a consequence move at higher velocities with respect to intergalactic gas.

A classification according to the relative brightness of the outer edge shows that this parameter is strongly correlated with the luminosity of the sources: when $L(\nu = 178\,\mathrm{MHz}) \geq 10^{26}\,\mathrm{W\,Hz}^{-1}$ the source almost always has a bright edge. This result clearly indicates that the most luminous sources are associated with the most violent and recent ejections.

For those sources with a global central symmetry a classification according to the shape outlined by the radio emission can be made (Fig. 8.3c). S or Z shapes are sometimes observed; here again it seems that these morphologies are related to an interaction with the surrounding medium. Thus it was discovered that the quasar 3C 275.1 ($z = 0.55$) is located at the centre of a cluster of galaxies and that it is surrounded by a large, rotating gaseous cloud (with a diameter of 100 kpc). It is also possible that this type of distortion (with reference to very linear double radio sources) is due to a variation in time of the direction of gas ejection.

From the variety of morphologies described above it appears that we can, following Fanaroff and Riley, extract two broad classes according to both the global appearance of the distribution of the emission (in the sense of the brightness increasing in an inward or outward direction) and the power radiated at 1.4 GHz, $P_{1.4}$ (these two parameters being strongly correlated). 'FR I-type' objects are characterized by a dark outer edge and low luminosity, $P_{1.4} \leq 10^{25}$ W Hz^{-1} (an example is 3C 449), whereas for sources belonging to the 'FR II' class, whose prototype is Cygnus A, the external edge is clearly defined by the presence of hot spots, and $P_{1.4} \geq 10^{25}$ W Hz^{-1}. This separation into two groups very probably corresponds to important intrinsic differences and undoubtedly to distinct mechanisms for the production and evolution of the sources.

8.3.3 The Intrinsic Size of Radio Sources

The size of extended sources spans more than two orders of magnitude, from several tens of kpc to 4 Mpc in the case of 3C 236 (a radio galaxy located at 1.7 Gpc). Detailed results are necessarily limited to edge-brightened double sources, the only ones for which it is easy to define a size independent on observational limitations (this is obviously not the case for sources with a dark outer edge). The majority have a size between 150 and 300 kpc; values greater than 1 Mpc are rare (less than 5% of all sources). Moreover the size is correlated with luminosity. A similar trend is observed for double sources in general and probably means that the brightest sources are associated with the most energetic ejections.

8.3.4 The Spectrum and Polarization of Extended Lobes

As indicated above, extended sources are characterized by a spectrum that decreases rapidly at high frequencies ($F_\nu = \nu^{-\alpha}$ with $\alpha \approx 1$). In fact the large size of the emitting regions relative to that of the instrumental beam often allows us to obtain detailed properties of the substructures; these are considered separately below.

Diffuse Emission. This provides by far the most important contribution to the total flux; it is therefore characterized by the same spectral index as given previously ($0.7 \leq \alpha \leq 1.2$). More interesting is the possible existence of a systematic gradient across the source. Note that for several sources with a dark outer edge (doubles or those with a single tail) α increases with distance to the optical object. This suggests that the proportion of very energetic electrons decreases with distance, and therefore that the plasma ejected undergoes important energy losses. However, the evolution of the gas is not determined only by losses due to synchrotron radiation or inverse Compton effect. If this were the case, the energy density would have to decrease much more rapidly with distance from the centre: it is therefore necessary that the electrons are reaccelerated in situ as they move along.

The diffuse emission is in general strongly polarized (at a level of approximately 50% near 1 GHz). This indicates that the magnetic field has a relatively constant direction along the line of sight; otherwise, degrees of polarization close to the maximum value could not be observed. This order at large scales is also apparent in the distribution observed for directions of polarization over the emitting regions. Figure 8.4 illustrates the results obtained for the source 3C 282 with the Westerbork interferometer. Determination of the directions of polarization at two different frequencies (here 1.4 and 5 GHz) has allowed Faraday-rotation effects to be removed. As can be noted, the direction of the projected magnetic field is essentially orthoradial.

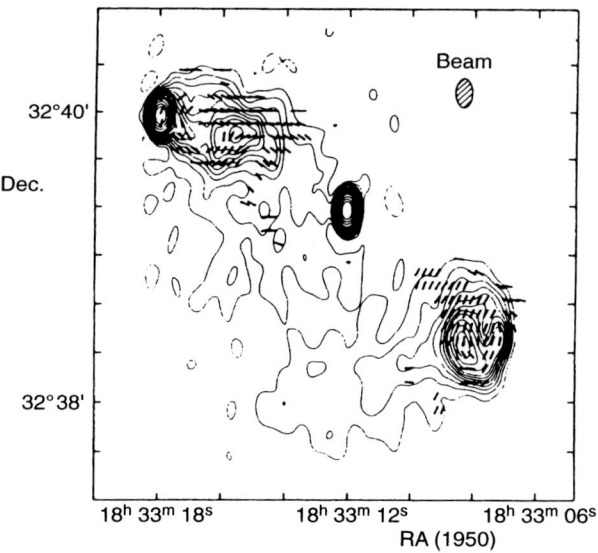

Fig. 8.4. The distribution of the projected magnetic-field direction superimposed onto the map of 3C 282 obtained at $\lambda = 6$ cm. The size (at half maximum) of the instrumental beam is indicated at the upper right. (From Strom et al. 1978)

236   8. Extragalactic Radio Sources

Hot Spots. These structures are present in sources with a bright edge, such as Cygnus A, 3C 284, and 3C 282. Interferometric observations now allow us to obtain detailed information about their properties. Originally a different technique was used: it consists in analysing the temporal variations of the radiation received, which, before reaching the observer, passes through the interplanetary medium filled by the solar wind. At low frequencies (this method was used at $\nu \approx 80$ MHz at Cambridge) refraction associated with fluctuations of the electron density becomes significant and induces (interplanetary) scintillation analogous to that observed when looking at a star through the Earth's atmosphere. If the source is purely diffuse, scintillations of the various parts are decorrelated and compensate one another; on the other hand if a sufficiently small and bright point is present, temporal fluctuations are seen (similar reasoning explains why the stars but not the planets scintillate at night). In practice the resolution obtained with this method is of the order of $0.1''$ at 100 MHz, which is excellent; on the other hand the information obtained is very sketchy compared with that contained in a full brightness map.

The linear size of the hot spots is of the order of 1 kpc. Their spectrum is close to that determined for diffuse emission. Sometimes it is slightly flatter (that is, α is smaller). The global degree of polarization is less than that of diffuse emission; this is natural, since the direction of $\boldsymbol{B}$ varies rapidly close to the emission peak.

In several cases, such as 3C 33 and Pictor A, optical emission (or even X ray emission for Pictor A) associated with radio peaks has been detected. The shape of the spectrum and the high degree of polarization observed ($> 30\%$) indicates, as in the radio domain, a synchrotron origin. The particular interest of these difficult measurements is related to the lifetime of the electrons responsible for this emission, which decreases as the frequency increases. For instance $\tau \leq 10^2$ years for X-ray emission, a value much less than the time necessary for propagation at the velocity c from the nucleus. This provides a direct confirmation of the necessity for in situ acceleration of the electrons.

8.4 Radio Jets

This type of structure is often observed either on a large scale in extended diffuse emission or on a smaller scale in the immediate neighbourhood of the compact central source. Jets are clearly seen, for example, in the sources 3C 449 and NGC 315, shown in Fig. 8.3. Examination of the observed morphologies for extended sources strongly suggests that the plasma responsible for the emission of the lobes has been ejected from the nucleus of the associated galaxy. Jets are then naturally thought to trace the path followed by the ejected material. In practice about a hundred jets have so far been identified. They are found in all types of radio source, whatever their luminosity, size, or morphology. A few are found in Seyfert galaxies; the structure of the radio source associated with the centre of our own Galaxy, Sgr A, is similar to that of some extragalactic jets. Jets then appear to be strongly correlated with the presence of activity in galactic nuclei.

8.4.1 Symmetry, Shape, and Size

Studies of the morphology of extended sources reveal remarkable symmetry properties, in particular for objects of the Cygnus A type. It is interesting to see whether similar properties are observed for jets. The answer depends on the scale considered. In the immediate neighbourhood of the compact component the jets are, in general, strongly asymmetric and often show up on just one side. At greater distances ($d \geq 1\,\mathrm{kpc}$) the brightness distribution becomes gradually more symmetric. This property is nicely illustrated by the example of M84 (Fig. 8.5). Furthermore the jets appear on average more symmetric in FR I-type sources (such as 3C 449). On the other hand for those like Cygnus A (FR II types) the jet is often detected on one side only, even at large scales. For this source, maps of the 6 cm emission have been obtained at the VLA. The high dynamical range attained (that is, the ratio of the maximum brightness observed to the minimum detected – here 3000) has for the first time revealed a very narrow and tenuous jet linking the compact source to one of the hot spots in the western lobe (Fig. 8.6). The characteristics of associated jets are thus consistent with the distinction between FR I- and FR II-type objects, which is based mainly on the morphology of the emission from the lobes.

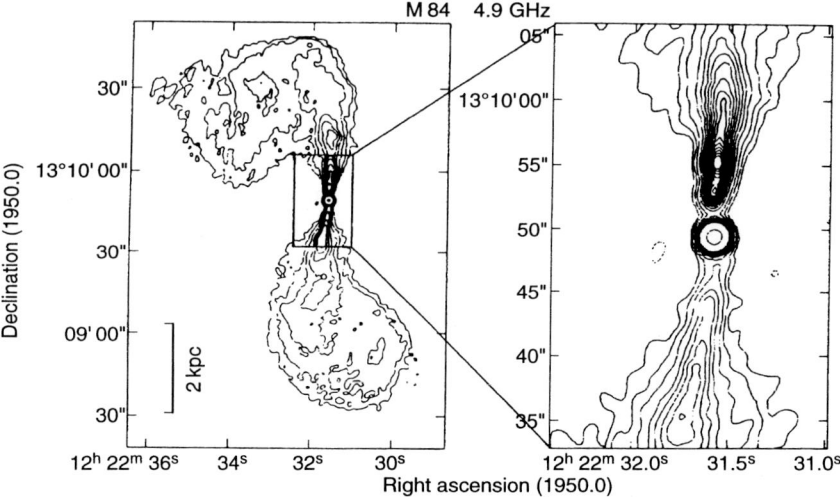

Fig. 8.5. A 4.9 GHz map of the radio galaxy M84, taken with the VLA. The view on the right (an enlargement of the central region) shows the asymmetry of the jets emanating from the compact component. (From Bridle and Perley 1984)

238   8. Extragalactic Radio Sources

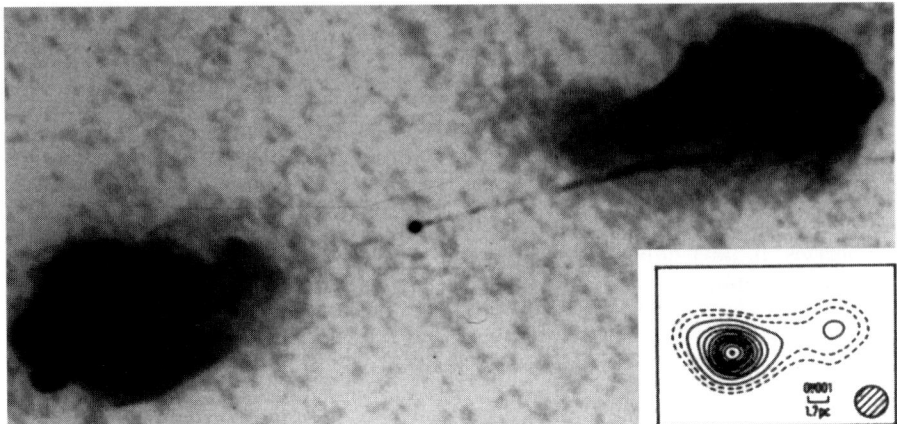

Fig. 8.6. A photographic representation of the 6 cm emission from Cygnus A observed at the VLA. The resolution is 0.4″. The box at lower right shows the VLBI map obtained at the same wavelength for the compact component. (From Perley et al. 1984 and Kellerman et al. 1981)

The direction defined by the jets identified in compact sources (in general from VLBI observations) is almost always in good agreement with the direction of the jets observed on larger scales in diffuse emission or with that of the axis joining the hot spots. This is the case for Cygnus A in particular. Its compact component has been resolved using the VLBI network (Fig. 8.6) and appears to be elongated in the direction of the north-west lobe with an extent of a few milliarcseconds. This remarkable property shows that the process by which the transport of matter is collimated preserves the long-term stability of the ejection axis.

The jets frequently have an elongated shape. However, many deviations exist. In sources with two tails the jets are clearly curved and thus seem, like all the diffuse emission, strongly influenced by the interaction with the external medium. Notable curvatures are also often encountered in sources dominated by their central component. Sometimes this deviation from linearity takes the form of a 'wiggle' about the mean direction. Various processes can be invoked to explain such effects; for example it is possible that the radio galaxy is interacting with other neighbouring galaxies and is therefore in orbital motion. Alternatively the flow of material may be subject to Kelvin–Helmholtz instabilities at the interface with the surrounding medium. Furthermore in sources dominated by their central component the size of the jets is on average smaller. In fact only the projected size can be measured, and it is quite plausible that for these sources the angle between the line of sight and the direction of the jet is systematically small. This may explain why the apparent size and the presence of a strong compact component seem to be correlated, because in such a configuration the emission towards the observer can be enhanced if the emitting plasma is moving relativistically.

Moreover this interpretation would also explain the more pronounced curvature observed for these sources; when the angle between the line of sight and the source axis is small, a given lateral deviation of the flow corresponds to a shift that is much more apparent in projection.

8.4.2 The Spectrum and Polarization

The spectral index measured for radio emission from jets lies in the range 0.5–0.8; its value is relatively independent of position, because the gradient of α is always small. When significant variations are detected, α increases towards the edge; therefore the jet gradually contains fewer energetic electrons. Notice that this weak evolution of α is consistent with the spectra of the jets and lobes having a similar index.

Jets are characterized by high degrees of polarization, reaching 50%, which indicates the existence of a well-ordered magnetic-field structure on the scale of the jet itself. Put very simply, two types of distribution for the apparent magnetic field $\boldsymbol{B}_a$ are encountered: one where $\boldsymbol{B}_a$ is perpendicular to the direction of the jet, which is observed mainly among the sources of the FR I class, and the other where $\boldsymbol{B}_a$ is parallel (FR II sources). It is very likely that the magnetic field plays an important role in channelling and stabilizing the flow of matter, but this role remains difficult to clarify, because the observed distributions are in general more complex than described above. Furthermore the effect of Faraday rotation due to a thermal plasma (either associated with the jet or located in the foreground) is difficult to estimate; indeed it is strongest at low frequencies – precisely where it is almost impossible to attain the spatial resolution required to separate the jets from the diffuse emission. A still more fundamental limitation is that the knowledge of the projected field, $\boldsymbol{B}_a$, does not allow the three-dimensional configuration of $\boldsymbol{B}$ to be obtained.

8.4.3 Lateral and Longitudinal Variation of the Radio Emission

The variation with distance D to the active nucleus of the jet radius R can give valuable information about the flow of matter, its velocity, and the confinement mechanism. It has been possible to obtain this information for the nearest sources (radio galaxies) and for several quasars whose jets are resolved. For most of them the brightness is maximal at the centre and not at the edges (Fig. 8.5); this clearly shows that radio emission does not result from the interaction with the medium surrounding the flow but from the jet itself.

One may also wonder whether the transport is 'free' or constrained by an external medium. In the case of a free flow – which assumes that $p_i \gg p_e$, where p_i is the pressure inside the jet (including both that of the particles and that of the magnetic field proportional to B^2) and p_e is the external pressure – the expansion rate of the jet becomes constant:

240   8. Extragalactic Radio Sources

$$\frac{\mathrm{d}R}{\mathrm{d}D} = \frac{v_\mathrm{s}}{v_\mathrm{j}},$$

v_s denoting the sound velocity in the plasma and v_j the flow velocity. Moreover the maps give the angles subtended by R and D, θ_R and θ_D, and consequently

$$\frac{\mathrm{d}R}{\mathrm{d}D} = \cos i \frac{\mathrm{d}\theta_R}{\mathrm{d}\theta_D},$$

incorporating the angle i of the jet with the plane of the sky. Assuming i to be uniform along the jet, the absence of external constraints would imply that $\mathrm{d}\theta_R/\mathrm{d}\theta_D$ is constant.

In reality the results differ appreciably for the least luminous radio galaxies and for the quasars. In the first case, illustrated by the example of NGC 315 in Fig. 8.7, the broadening of the jet is small near the nucleus ($\mathrm{d}\theta_R/\mathrm{d}\theta_D \leq 0.05$). It then opens out just beyond ($D \approx$ several kpc) and likewise at its end ($D \geq 100\,\mathrm{kpc}$), where it merges with the lobe; an intermediate zone of recollimation ($\mathrm{d}\theta_R/\mathrm{d}\theta_D \leq 0$) is sometimes observed (Fig. 8.7). The most powerful sources of the FR II type are characterized by an almost zero, or very small, expansion, which in general is not compatible with a free evolution of the flow over the entire jet. If it is nevertheless assumed that on the portions for which the expansion rate is constant the jet is free of any lateral constraint, the small values of $\mathrm{d}\theta_R/\mathrm{d}\theta_D$ obtained indicate that the flow is highly supersonic ($M \geq 10$, M denoting the Mach number). The small transverse size of the jets in FR II sources is comparable to that of hot spots; these would then be, as assumed in the majority of models, the point where jets impact onto the intergalactic medium.

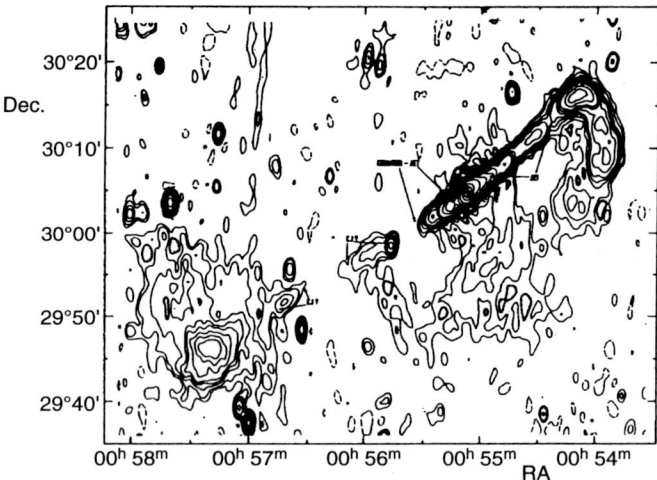

Fig. 8.7. A VLA map of the radio source associated with NGC 315 obtained by Bridle at 610 MHz. The cross indicates the position of NGC 315. (From Willis et al. 1981)

The variation in brightness along the jet, $I(D)$, can also yield very interesting information. Indeed assuming magnetic flux conservation and equipartition between the energy of the particles and the field $\boldsymbol{B}$ along the jet, we get a variation of the type

$$I(D) = kR^{-4.1}v_j^{-1.9}$$

for $\alpha = 0.65$ (v_j is the velocity of the flow). In reality $I(D)$ differs from the above expression at several points. On the one hand the decrease of I with D is less steep than predicted; in the neighbourhood of a compact source, I is often observed to increase with D. On the other hand the variation is far from being monotonic and has many local maxima (Figs. 8.6 and 8.7). The real conditions in which the flow proceeds are then probably much more complex; we can imagine in particular that oblique shocks as well as turbulence on a large scale (vortices of a size roughly equal to R_j) will be present.

8.4.4 Optical and X-ray Emission Associated with Radio Jets

The existence of a jet in visible light (as a continuum) emerging from the galaxy M87 was noted by Curtis as long ago as 1918. After the intense radio source Virgo A had been associated with M87 and mapped in detail, a remarkable similarity appeared in the structures of optical and radio emission. As can be seen in Fig. 8.8, the radio and optical maxima coincide exactly.

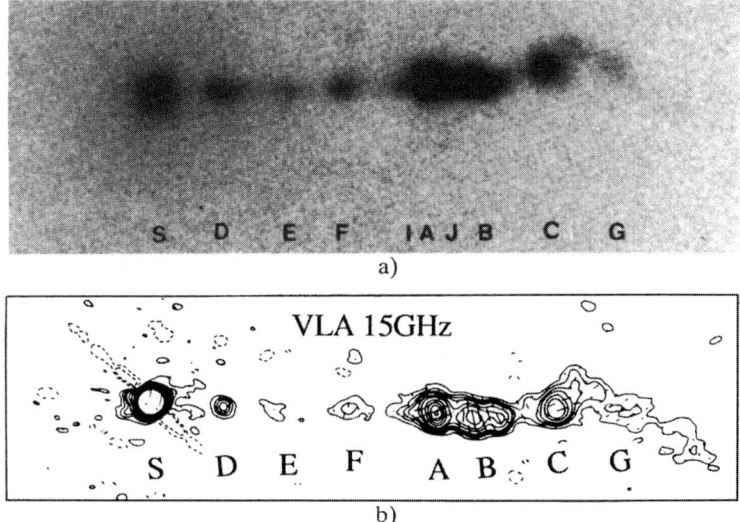

Fig. 8.8. Optical emission (**a**) and radio emission at $\lambda = 2\,\text{cm}$ (**b**) of the jet in Virgo A. (From Lelièvre et al. 1984 and Charlesworth and Spencer 1982)

The optical, infrared, and radio spectra of the knots are very closely connected, which suggests a common origin for these emissions. An additional argument in favour of synchrotron radiation is the high degree of polarization ($\approx 20\%$) measured in the visible. Several other optical jets are known, such as that of the nearby quasar 3C 273; they show, in general, the same properties as those of the jet of M87.

In the visible region Seyfert 2-type emission lines (Chap. 9) have been detected in several cases. This emission is localized preferentially near the end of the radio jet, and, more precisely, close to the bright knots or to regions showing a strong curvature. It probably results from the interaction of the jet with interstellar clouds of the associated galaxy, the gas being highly ionized by photoionization or heating due to shock waves.

8.5 Compact Sources

Compact radio sources can be clearly distinguished from the structures described above by several of their properties. First, their angular size is less than 1 arcsecond and they generally remain unresolved by interferometers such as the VLA. Only VLBI observations are able to provide data concerning their morphology. Unlike extended components, these sources are also very often variable, on time scales of the order of months or years. Finally their emission is characterized by a flatter spectrum (that is, a smaller value of α), with possibly a low-frequency cutoff (a so-called inverted spectrum).

Compact sources always coincide with the active optical nucleus when it has been identified. In particular no such component has ever been found in diffuse radio lobes. Thus compact radio emission appears as one signature of active nuclei themselves; its study may clarify what processes occur around the central object owing to the power of interferometric techniques in this frequency range. Whereas sources dominated by their extended component are preferentially associated with radio galaxies, compact sources are in general identified optically with a BL Lac object or a quasar; in the latter case associated extended emission is sometimes observed. In fact the properties of compact sources do not depend much on the nature of the optical object that hosts them.

For those sources having both compact and diffuse components the fraction of the flux coming from the nucleus varies over a large domain, from a few per cent to 100%, at about 1 GHz. Compact sources of low luminosity, but with similar characteristics to those described above, are also encountered at the centre of Seyfert galaxies, or even at the centre of 'normal' galaxies like our own.

8.5.1 The Radio Spectrum

Unlike extended emission, for which the spectral flux density typically varies as $\nu^{0.8}$, compact sources often have a flat ($\alpha \approx 0$) or inverted ($\alpha < 0$ at low frequencies) spectrum. Sometimes several local maxima are present, suggesting a more complex variation. The example of the source PKS 0735+178, which is associated with a BL Lac object, is shown in Fig. 8.9.

Recall that a homogeneous synchrotron source is optically thin at high frequencies and thick at lower frequencies (see Sect. 8.1.1). The maximum in between these two domains appears at a frequency ν_{max} determined by the value of B, the electron energy distribution, and the size of the source. If the medium is expanding, ν_{max} gradually shifts towards lower frequencies. Thus extended emission is naturally attributed to a thin medium whereas compact sources with an inverted spectrum are those for which ν_{max} falls in the frequency range observed.

The sources whose spectrum remains flat over several frequency decades seem to contradict this simple picture. In fact for these sources small-amplitude variations can often be detected in the spectrum. Furthermore VLBI observations have shown the existence of several substructures, each

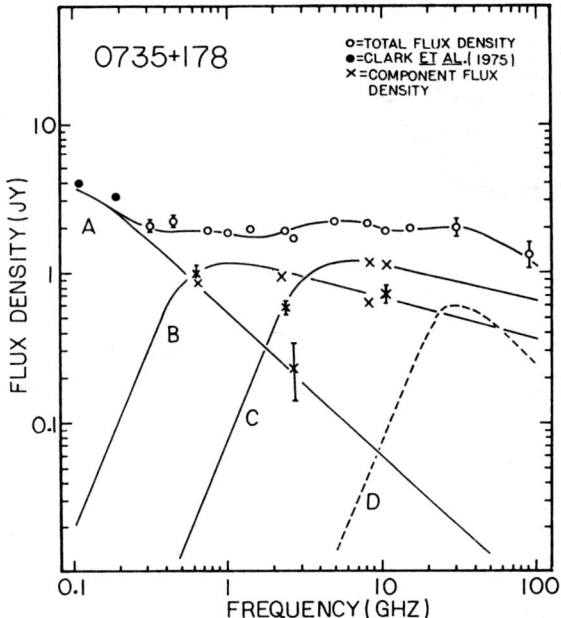

Fig. 8.9. The spectrum of the source 0735+178 between 0.1 and 100 GHz; the total flux is indicated (by solid and open circles) as well as that of the three components (A, B, and C) resolved by VLBI (shown by crosses). The fourth component (D), which is required to account for the total flux (continuous line) is hypothetical. (From Cotton et al. 1980)

having a spectrum and $\nu_{\max}$ value of its own. This is the case in particular for the source PKS 0735+178, whose spectrum can be represented as the sum of the four 'homogeneous-synchrotron-type' spectra. It is likely that a smooth systematic variation of the magnetic field and of the density of electrons (or of their energy distribution) in the emitting medium is present; this also tends to broaden the predicted spectrum.

The radio spectrum of compact sources is generally consistent with the extrapolation of that obtained in the near infrared and the visible. This is an additional indication in favour of synchrotron emission in these latter spectral ranges. The data obtained recently in the mid-infrared and far infrared (10–100 µm) by the *IRAS* satellite are also in agreement with this view. At present, only the submillimetre domain remains largely unexplored (observations are difficult owing to the large atmospheric opacity). However, it is at these wavelengths that the break in the spectrum is expected for many active nuclei having intense emission at $\lambda < 100$ µm but a weak or undetected emission at low frequencies (several GHz). Forthcoming experiments in this spectral range, carried out using balloons or satellites, should then bring new and very significant results in the near future.

8.5.2 Variability

Most compact radio sources are variable. This makes it difficult to obtain data concerning the spectrum of the emitted radiation such as that described above, since only simultaneous measurements at various frequencies are unambiguous.

Two main types of variation are observed: either a slow increase or decrease of the flux over several years, or sporadic bursts on the scale of a year. The flux observed at $\lambda = 2$ cm and $\lambda = 11$ cm from the radio galaxy 3C 120 ($z = 0.033$) is shown in Fig. 8.10. At $\lambda = 2$ cm three bursts with an amplitude $\Delta F/F \approx 2$–3 can be clearly seen; the same sequence is observed at 11 cm but with a smaller amplitude and a time delay of about seven months. This result is typical. It corresponds well to the behaviour expected from an initially opaque and expanding medium recently fed with high-energy particles. The source gradually becomes transparent at increasing wavelengths; hence the observed time delay.

In other cases, and in particular for BL Lac objects, which show both the most abrupt and pronounced variations, bursts appear simultaneously and with comparable amplitudes at various frequencies. The most plausible origin is the injection of new relativistic electrons, the decline then following naturally from the expansion of the plasma or radiative losses.

In general, variations of active nuclei in visible light are uncorrelated with respect to those observed at radio frequencies. However, there are several objects that have undergone spectacular variations simultaneously in these two domains.

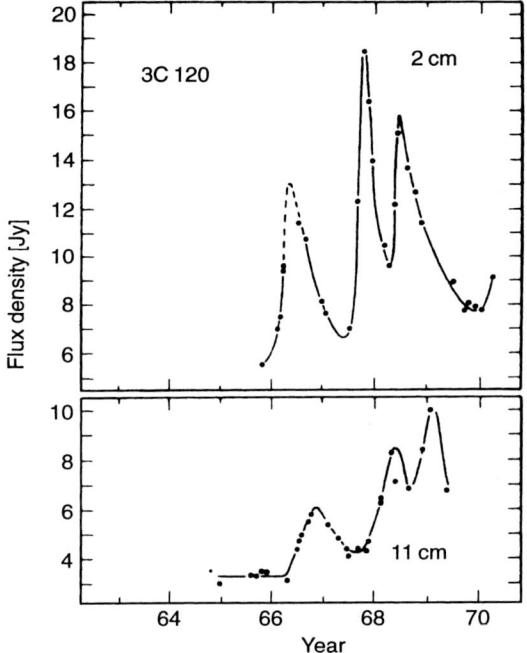

Fig. 8.10. Variation of the flux emitted by 3C 120 at 2 cm and 11 cm between 1965 and 1970. (From Verschuur and Kellerman 1974)

The data concerning radio-flux variations give valuable information not only about the energy injection mechanism but also on the structure of the sources. Indeed the size l of the source (or at least the part responsible for the observed burst) and the characteristic time scale of variations Δt_{var} must necessarily satisfy $\Delta t_{\text{var}} > l/c$, which provides an upper limit for l. Furthermore a lower limit can be obtained by noting that if the source is extremely compact, the density of radio photons can be sufficiently large for inverse-Compton-type collisions with relativistic electrons playing an important role, corresponding to the existence of associated X-ray emission. It can be shown that this phenomenon implies an upper limit for the brightness temperature of the radio emission: $T_b < 10^{12}$ K. In practice no X-ray emission associated with a radio burst has been detected, which leads to a lower limit for l that greatly exceeds the value $c\,\Delta t_{\text{var}}$.

Another argument is directly related to the above paradox; it involves the absence of interstellar scintillation of compact sources. If their size were really of the order of $c\,\Delta t_{\text{var}}$, we would expect to observe, as we do for pulsars, variations of the radio flux due to the presence of inhomogeneities in the ionized phase of the interstellar medium. Such fluctuations are generally not detected (except very recently for some sources located at low galactic lat-

246   8. Extragalactic Radio Sources

itude). A contradiction then results, which, as we shall see in what follows, can be resolved by accepting that part of the emitting plasma is moving relativistically towards the observer.

8.5.3 Morphology, Changes of Structure, and Superluminal Velocities

Apart from the analysis of radio variations, the only information available on size and morphology comes from VLBI observations. In practice the number of antennae is limited (to four or five, in general) and their location on the earth is far from being optimal. The information gathered is consequently only sketchy compared with that obtained (on the arcsecond scale) with interferometers such as the VLA; moreover only the most intense source can be studied with this technique.

The structure of compact sources does not seem to depend on the presence of extended emission. When extended lobes or jets are associated, we can look for a relation between the large and small structures observed. In general a direct link appears; this is the case, for example, for the double source 3C 111, shown in Fig. 8.11. Is the compact source also composed of two components aligned on the axis defined by the diffuse emission? Sometimes an asymmetric structure with a nucleus–jet morphology can be observed. Again the direction clearly coincides with that defined by the extended component (see Fig. 8.6 for the case of Cygnus A, discussed above). This correlation of structures on quite different scales (their ratio reaches 3×10^5 for 3C 111) reveals an

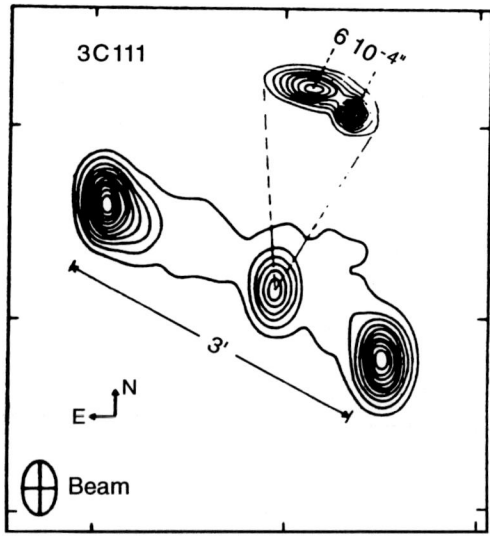

Fig. 8.11. A radio map of the source 3C 111 at small (VLBI) and large scales. (From Strittmatter 1979)

important feature of the central object: the permanence of the ejection axis over periods greater than 10^8 years.

The substructures that appear in nucleus–jet-type objects are often self-absorbed at low frequencies (≈ 1 GHz) and appear more clearly at higher frequencies ($\nu \geq 5$ GHz). The brightness maxima observed reach temperatures of about 10^{11}–10^{12} K as the theory predicts. Moreover the frequent detection of X-ray emission is a direct indication that inverse Compton processes are at work.

One of the most remarkable results obtained from VLBI observations is that in some compact sources changes in position of substructures are detectable over intervals as short as a year. Figure 8.12 shows the emission at 2.8 cm of the quasar 3C 273 at three different epochs. One component that moves away from the nucleus with an apparent angular velocity of 2.2×10^{-3} arcseconds per year can be clearly seen. This corresponds to a projected linear velocity close to $10(H_0/50)^{-1}c$, given the redshift, $z = 0.158$, of 3C 273 ($q_0 = 0$). The source is then said to be 'superluminal'. At first this phenomenon was observed only between two distinct epochs and interpreted in terms of relative variations of the flux emitted by adjacent fixed substructures. When the evolution could be followed for a larger time interval, the reality of this displacement became indisputable. Up to now this phenomenon has been

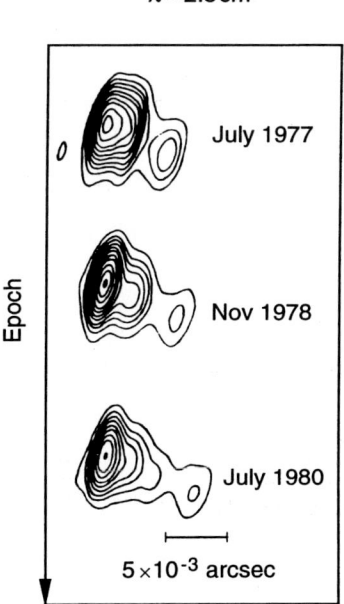

Fig. 8.12. Variations in the structure of 3C 273. (From Pearson et al. 1981)

observed in a dozen objects. Notice that there also exist sources for which no relative motion is observed (for example $V < 10^4\,\mathrm{km\,s^{-1}}$ in 3C 84).

Many theories have been proposed to account for superluminal motion. They must explain the main characteristics of the phenomenon, which can be summarized as follows:

- only outward motions have been observed;
- if the data are extrapolated back, the time of zero separation can be associated with an outburst of the source;
- the direction of motion remains constant for components associated with several different outbursts, but the velocity can vary;
- the apparent linear velocity of expansion can reach $45c$.

One mechanism, proposed by Lynden-Bell, was suggested by the theory of luminous echoes developed at the beginning of the century. A light wave front corresponding to the outburst propagates outwards and illuminates the matter, which responds to this excitation by an increase in radio emission. If the matter were to be found in the plane of the sky, the apparent velocity observed would be $2c$; if, on the other hand it is confined in an elongated structure making a small angle θ with the line of sight, the apparent velocity will equal $2c/\sin\theta$, so long as the redshift remains smaller than 1.

The most common interpretation accepted today assumes that it is the matter itself that moves relativistically in a direction close to the line of sight. In this case, the apparent transverse velocity is

$$v_\mathrm{t} = v\sin\theta \bigg/ \left(1 - \frac{v}{c}\cos\theta\right)$$

for the component approaching the observer (v is the velocity of the matter). Indeed consider the photons emitted by the moving source at two epochs separated by Δt. In projection the distance covered is $v\,\Delta t\sin\theta$, but, owing to the finite velocity of light, the corresponding time interval as measured by the observer becomes $\Delta t[1 - (v/c)\cos\theta]$. The above expression for v_t has a maximum when $\sin\theta = 1/\gamma = (1-\beta^2)^{1/2}$ (with $\beta = v/c$); it is most pronounced when β is close to unity ($v_{\mathrm{t\,max}} = \beta\gamma c$). This model introduces an additional parameter, β, that characterizes the motion of the ejected gas. There are several important differences with respect to the first scenario:

- Owing to its relativistic motion, the matter no longer radiates isotropically for an external observer. The radiation emitted is concentrated in a cone of size $\Delta\theta \approx 1/\gamma$ around the direction of motion. The apparent 'amplification' is given by the factor $(1-\beta\cos\theta)^{-\alpha-3}\gamma^{-3}$, which reduces to γ^{-3} for $\sin\theta = 1/\gamma$ and $\alpha \approx 0$.
- The real variations in the rest frame of the source occur on time scales larger by a factor γ compared with the observed time scales ($\Delta t_\mathrm{source} = \gamma\,\Delta t_\mathrm{var}$).

It must then be possible to understand the origin of the superluminal motions and simultaneously solve the paradox raised by the small size of the sources inferred from the observed variations. One difficulty remains: to explain velocities greater than $10c$ the angle θ must necessarily be very small, a situation that is extremely unlikely. However, a significant fraction of compact sources show superluminal motions. This argument no longer holds if the observed flux is strongly affected by relativistic beaming; this phenomenon induces an obvious observational bias since it greatly favours those sources for which $\theta \ll 1$ and $\gamma \gg 1$. Therefore only compact sources satisfying these criteria would be detected.

The above model seems to be consistent with observations. It explains in particular the strong apparent curvature of jets associated with compact sources (a projection effect if $\theta \approx 0$) and the large number of one-sided jets (in the direction opposite to that of the detected jet the emission is much less intense since the emitting material is moving away). Nevertheless several points remain to be clarified:

- The values of β and γ which enable us to explain the observed transverse velocities lead to a very large flux difference between the component approaching the observer and the central core (a ratio of 100 for $\gamma \approx 4.6$). This difference is even greater if the second component observed is ejected backwards at the same velocity v, since it is then *attenuated* by a factor γ^{-3} (the ratio reaching 10^4 for $\gamma \approx 4.6$).
- What is the mechanism responsible for the ejection of the matter at $v \approx c$?

Some predictions of this model can be tested. For example high-sensitivity searches for compact sources should lead to the detection of those sources with a lower ejection velocity ($\beta \ll 1$) or an ejection axis $\theta \neq 0$. A systematic study of the properties of superluminal sources (apparent luminosity, variability, and so on) and a better estimate of the incidence of such motions among all compact sources detected are also essential.

8.6 Radio-Source Modelling

From the data described above, a general picture emerges, even if the models proposed still have difficulties to account for the observed phenomena in detail. The active nucleus as it appears at optical wavelengths or at high radio frequencies ejects material, probably at relativistic velocities in some cases. The transport of this gas takes place along a well-defined axis and is possibly apparent in the form of a narrow jet. Finally the jet feeds the extended lobes with high-energy particles. Much effort has been devoted in the past twenty years to develop a model of radio sources that accounts for the phenomena described above. We shall just mention here its main features and a few points of particular interest.

The study of the morphology of extended sources leads us to distinguish two large classes of objects, those of type FR I (such as 3C 449) and type FR II (such as Cygnus A). What parameters are responsible for this dichotomy? As the luminosity of FR II-type sources is systematically larger, it seems that the difference is really intrinsic to the active nucleus and not due to properties involving the external medium (for example its density). The same is likely to be true for the different appearance of the jets associated with each type. Thus FR II-type sources would be those for which the ejection of the plasma is sufficiently energetic for the flow to remain collimated up to the point where it impacts onto the intergalactic medium (hot spots). The question then should involve the properties of the active nucleus itself.

One can imagine that, depending on the 'regime' of ejection, the interaction with the external medium will be quite different. The appearance of jets, and in particular the radial variation of the spectral index, strongly suggests that for FR II-type sources the Mach number characterizing the flow is large; the hot spots correspond to regions of strong shocks and the 'rebound' of the plasma towards the centre leads to the formation of lobes. On the other hand for sources such as 3C 449 the lower velocity of the material in the jet would allow a smooth connection with the external medium without any discontinuity. Numerical simulations have been conducted to check whether jets with high Mach numbers can effectively develop and maintain their coherence in spite of instabilities that may appear at their ends (broadening out and spread of the plasma) or at the edges (Kelvin–Helmholtz instabilities). One result is shown in Fig. 8.13. The jet can be seen clearly together with a shell separating it from the external medium and discontinuities associated with shock waves. These studies confirm the interpretation of sources similar to Cygnus A initially proposed by Blandford and Rees in 1974. Furthermore they allow us to investigate what types of substructure will appear depending on the initial conditions (for example bright nodules). Comparison with the data nonetheless remains a delicate matter because the observed radio emission is not a *direct* tracer of the matter (only relativistic electrons associated with a magnetic field show up).

Fig. 8.13. A numerical simulation of a jet with a Mach number equal to 6. Contours represent lines of constant projected density. At each point of the grid the direction and modulus of the velocity are represented by a short segment. Oblique shocks along the jet as well as Kelvin–Helmholtz surface intabilities can be seen. (From Norman et al. 1982)

8.6 Radio-Source Modelling

Radio sources located in clusters of galaxies are of particular interest, since the parameters characterizing the properties of the external medium are relatively well defined thanks to X-ray observations of intracluster gas ($n \approx 10^{-3}\,\mathrm{cm}^{-3}$, $T \approx 10^7\,\mathrm{K}$). We have already mentioned the peculiar morphologies of radio sources found in such an environment. It is possible to account for their properties in detail, as in the case of NGC 1265 (Fig. 8.14). In these models the appearance of the jet results from the motion of the galaxy in the intracluster gas, and only two parameters need to be determined: the ratio of the velocity in the jet, V_j, to the velocity of the galaxy, V_g, and the angle between the jet and the direction of $\boldsymbol{V_g}$. For NGC 1265 and other similar cases one gets $V_j \approx 10^4\,\mathrm{km\,s^{-1}}$, a value much less than c.

The radio sources inside clusters are also interesting for the modelling of lobes. Indeed the relatively high density of the external medium imposes an important confinement, and pressure equilibrium should be quickly reached. Then losses related to expansion are negligible and the plasma evolution is due to radiative losses only (synchrotron and Compton). This particular situation has remarkable consequences for the spectrum, which is characterized by breaks separating domains where $\alpha \approx 0.8$, $\alpha \approx 1.3$, and $\alpha \approx 2.1$, from lower to higher frequencies.

Such behaviour is well understood (the Kardashev effect): since the most energetic electrons have a shorter lifetime, each frequency domain allows the study of a given epoch in the history of the radio source (the earlier, the lower the frequency). The first transition, $\alpha = 0.8$–1.3, corresponds to the epoch when the spectrum is no longer determined only by the initial distribution of the ejected electrons (corresponding to $\alpha \approx 0.8$) but also by radiative losses. The second transition is related to a break in the activity of the nucleus: the

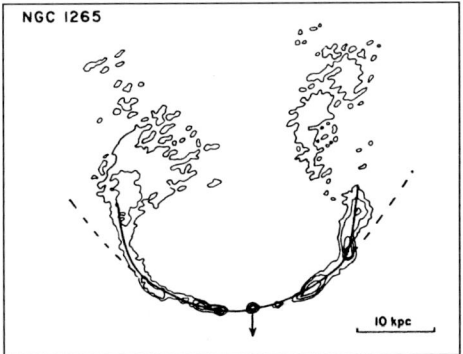

Fig. 8.14. A model for the radio source NGC 1265 located in the Perseus cluster. The isophotes trace the emission at 5 GHz while the solid line represents the fit obtained for the velocity indicated and an ejection angle equal to 6°. The break observed along the jet is probably associated with the transition between the interstellar medium of the galaxy and the intracluster medium. (From Baan and McKee 1985)

plasma then quickly loses very energetic electrons and radiates less at high frequencies.

Regarding diffuse emission in general, an important point has not yet been discussed: the typical spectral index $\alpha = 0.8$ corresponds to an electron energy distribution $\mathrm{d}N/\mathrm{d}E \propto E^{-2.6}$, which diverges when $E \to 0$ for both the number and total energy of the particles. Then there must exist a low-frequency cutoff in the spectrum; its position is directly linked to the lower bound of the energy distribution and consequently to the total energy content. This cutoff has been detected at about 200 MHz in Cygnus A.

In previous sections we mentioned that compact sources generally have a flat or inverted spectrum. This rule, however, has exceptions: indeed compact radio sources with $\alpha \approx 0.7$ to 1 exist. The systematic study of these objects, undertaken only recently, should enable us to understand the origin of their small size and to decide, for instance, whether they are very young sources in which activity has recently begun or galaxies whose particularly dense interstellar medium prevents the formation of the classical lobes. It is natural to search first for the general characteristics of radio sources from which unifying schemes can be proposed, but it is also important to study in detail the atypical cases. This is true in particular for the morphology of extended sources. At the beginning of this chapter we stressed the remarkable symmetry and collinearity properties of most sources. However, for extended sources too, many exceptions to the rule are observed; the source 3C 196 (Fig. 8.15) is a remarkable example. The peculiar structure and orientation of the north-east lobe could, in this case, be explained by the presence of a dense cloud on which the jet impinges.

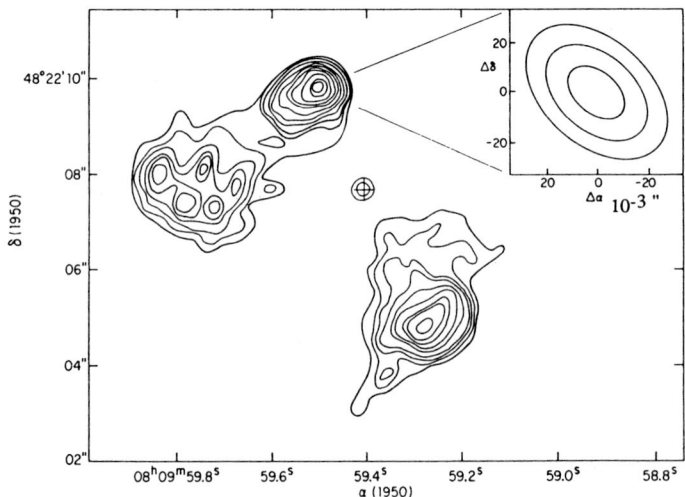

Fig. 8.15. A 5 GHz map of 3C 196. In the upper right corner is indicated the structure of the hot spot measured by VLBI. (From Brown et al. 1986)

8.7 Radio-Source Counts. Evolution

Detailed observation of individual radio sources gives information on the nature of these objects and on the physical processes involved. Statistical studies can also be very useful to describe their spatial distribution throughout the universe. As a significant fraction of radio sources are associated with very distant objects ($z \approx 1$ or more), it should also be possible, therefore, to investigate the evolution with cosmic time of phenomena connected with the radio activity. When the first studies of this type were undertaken, astronomers thought that it would be possible to constrain cosmological models (in particular the value of q_0) by regarding radio sources as standard candles. It quickly turned out that the evolution in time of radio sources is so strong that purely 'geometric' effects are, as we shall show later on, totally masked. However, the angular distribution of sources is not affected by these effects: this then enables us to test the isotropy of the universe on a large scale (for other, stricter constraints see Chap. 13). In fact the analysis of the distribution on the sky of radio sources is entirely compatible with the assumption of isotropy. The existing catalogues set an upper limit of the order of 3% on large-scale variation of the surface density of sources. The latter takes the following value:

$$\frac{\mathrm{d}N}{\mathrm{d}\sigma}(F_{1.4\,\mathrm{GHz}} > 0.24\,\mathrm{Jy}) = 0.5\,(\mathrm{degree})^{-2}.$$

On the whole sky ($\approx 4 \times 10^4\,(\mathrm{degree})^2$) the number of sources available for such an analysis would be of the order of 10^4 (recall that $1\,\mathrm{Jy} = 10^{-26}\,\mathrm{W\,m^{-2}\,Hz^{-1}}$).

8.7.1 Expected Counts in the Absence of Evolution

Let us consider now the information contained in the flux distribution of observed sources. Two functions can be introduced. The first simply gives the number of sources whose flux F' at a given frequency is greater than a certain value F ($N(F' \geq F)$ or integrated counts). This can be determined directly from a *homogeneous* map of a significant fraction of the sky. Likewise we introduce the associated differential distribution $\mathrm{d}N/\mathrm{d}F$.

First, we assume that we inhabit a Euclidean universe and consider only sources having a luminosity lying between L and $L + \mathrm{d}L$ (at the frequency considered). For these objects the flux distribution is directly related to the 'distribution in depth', since there is a one-to-one correspondence between flux and distance. $N_\mathrm{E}(F' \geq F)$ is thus written simply as

$$\mathrm{d}N_\mathrm{E}(F' \geq F) = \frac{4\pi}{3}\,\mathrm{d}n(L)\left(\frac{L}{4\pi F}\right)^{3/2},$$

$\mathrm{d}n(L)$ denoting the volume density of sources with luminosity L. In practice the spread in luminosity is important and we get

$$N_{\rm E}(F' > F) = \frac{(4\pi)^{-1/2}}{3} F^{-3/2} \int L^{3/2}\, {\rm d}n(L).$$

The important result is the dependence on $F^{-3/2}$ or on $F^{-5/2}$ for the differential distribution ${\rm d}N_{\rm E}/{\rm d}F$.

Changes introduced by considering a more realistic model of the universe have to be taken into account, since many sources contributing to $N_{\rm E}$ have a substantial redshift. The variation of the ratio $({\rm d}N/{\rm d}F)/({\rm d}N_{\rm E}/{\rm d}F)$ with redshift z in standard models has the following form for $q_0 = 1/2$:

$$\frac{{\rm d}N}{{\rm d}F} \Big/ \frac{{\rm d}N_{\rm E}}{{\rm d}F} = \frac{(1+z)^{-(3/2)(1+\alpha)}}{\left[(1+\alpha)(1+z)^{1/2} - \alpha\right]},$$

where α denotes the spectral index ($\alpha \approx 0.8$ for many sources). It is easy to show that for increasing values of z (and thus decreasing values of F at a given luminosity), the above ratio decreases monotonically from 1 (for $z = 0$). This indicates that if the observed counts are represented by the relation

$$N_{\rm obs}(F' > F) = kF^{-\beta},$$

we should have $\beta < 1.5$. This remains true for any q_0 value.

8.7.2 Observed Counts and Their Consequences

The results obtained at frequencies of 408 MHz, 1.4, 2.7, and 5.0 GHz are shown in Fig. 8.16. The variations with F of ${\rm d}N/{\rm d}N_{\rm E}$ are all of the same type. For fluxes greater than about 1 Jy, ${\rm d}N/{\rm d}N_{\rm E}$ decreases with F, which then corresponds to $\beta > 1.5$, contrary to predictions. At 408 MHz, for example, one gets $\beta \approx 1.8$ for $F > 2$ Jy and $\beta \approx 0.8$ at small fluxes ($F < 0.1$ Jy). This is an unambiguous indication that sources are not uniformly distributed in distance. The behaviour of ${\rm d}N_{\rm E}/{\rm d}F$ at high fluxes implies a strong cosmological evolution of very luminous sources, in the sense that they were more numerous in the past. In fact many questions are raised by these results:

- Is the evolution similar for different types of source (compact and extended)? Does it depend on the luminosity?
- At what epoch (at what value of z) did radio activity first appear?

As one might expect, it is not easy to obtain a definite answer to these questions. The information contained in the counts is global and clearly, quite different scenarios allow the observations to be accounted for. Additional information is needed to better constrain models. For instance, it has been possible to conduct systematic searches for optical identifications, which gives the redshift and thus the luminosity of the objects.

The aim of evolutionary models is to determine the numerical density of the sources $n(z, L)$ as a function of the redshift z and the luminosity L of the

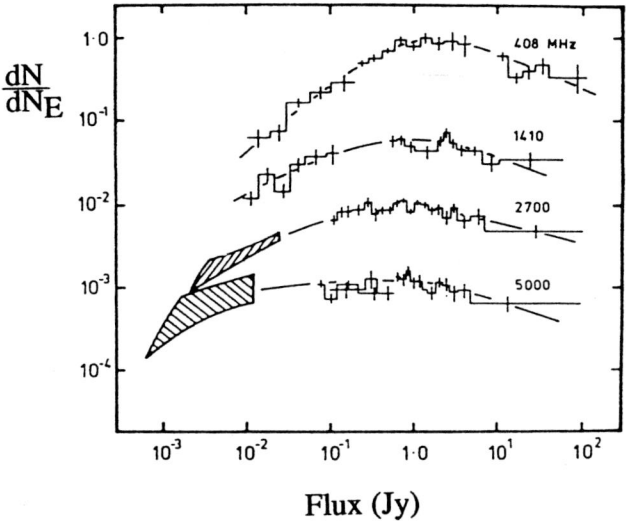

Fig. 8.16. Differential counts for several frequencies. (From Longair 1978)

sources. In general a distinction is also made between those sources with a flat spectrum (compact) and a steep spectrum (extended). The results indicate for these two types of sources a similar evolution, although more pronounced for radio sources with $\alpha \approx 0.8$–1.0. The number density (per unit comoving volume) of the most luminous objects – and of these only – increases rapidly with z and probably shows a cutoff at around $z \approx 4$. The fact that we end up with similar conclusions for extended and compact sources indicates that the governing factor of the evolution is intrinsic to the sources and does not involve the conditions imposed by the external medium. For example inverse Compton collisions with cosmic background photons whose energy density was larger in the past ($n \approx (1+z)^3$) might possibly weaken the radio emission of high-redshift compact sources. However, it seems more likely that the efficiency with which the host galaxy feeds the nucleus is the governing factor.

8.7.3 The Size–Redshift Relation. Size Evolution

Recall that in standard cosmological models an object of projected dimension d and redshift z is seen at an angle θ given by

$$\theta = \frac{H_0 q_0^2}{c} \frac{(1+z)^2}{\left[q_0 z + (q_0 - 1)\left(\sqrt{1+2q_0 z} - 1\right)\right]} d.$$

If objects were known whose size could be regarded as a constant, it would be possible to determine q_0 by studying the function $\theta(z)$. It has been suggested that radio sources of the Cygnus A type may be used for this purpose. This is worth mentioning since it clearly illustrates how difficult it is to separate purely cosmological effects from evolutionary effects.

Edge-brightened extended sources have a maximum size that is well defined and largely independent of frequency and instrumental spatial resolution. Moreover they are frequently associated with quasars, which enables their redshift to be measured. They are then well suited for such studies. For a given z the spread in values of θ is large; this can be attributed to projection effects to which should probably be added an intrinsic dispersion. It is thus rather the maximum value $\theta_{\max}(z)$ that has to be considered. This varies approximately as z^{-1}, in contradiction with the above relation, which depends on the q_0 value (for example with $q_0 = 0$ we can infer that $\theta(z = 0.1)/\theta(z = 1.0) = 4.3$ instead of the observed value of about 10). Friedmann models (see Chap. 13) lead to a slower decrease of θ with z than that given by the data. How should this be interpreted? Two assumptions have been proposed:

- there is an anticorrelation between the luminosity and the linear dimension (d is smaller for large L); since at large z only the most luminous sources are detected, they appear smaller;
- the size of radio sources does evolve with cosmic time and sources were indeed smaller in the past.

It is not easy to separate the two effects; it seems that the first effect plays some role but is not large enough to explain the relation $\theta \propto z^{-1}$. There should then be a real evolution of the radio-source size. This is especially likely beyond $z = 2.5$, where counts also imply a strong evolution. Which process governs the size evolution remains to be understood.

Exercises

8.1 For the Cygnus A (Fig. 8.6; $z = 0.057$) and 3C 111 sources (Fig. 8.11; $z = 0.0485$) give a lower limit for the time interval during which the ejection axis remained constant.

8.2 Establish the relation giving the *apparent* transverse velocity when the source is moving rapidly towards the observer in a direction close to the line of sight (see Sect. 8.5.3):

$$v_t = v \sin\theta \left/ \left(1 - \frac{v}{c}\cos\theta\right)\right. .$$

Give as a function of v the value of θ that renders v_t maximal. What happens to v_t if the source is moving away from the observer? For a symmetric ejection (with the same velocity v and angle θ for the components moving away from or approaching the observer) what is the apparent separation velocity of the two components?

8.3 Establish the expression for dN_E and $N_E(F' \geq F)$ introduced in Sect. 8.7. Adapt the reasoning to the case where (in a Euclidean universe) the luminosity function $n(L, D)$ varies with the distance D to the observer.

8.4 Give the relation $\theta(d)$ from Sect. 8.8.3 in a simpler form valid when $q_0 = 0$. Show that in this case, beyond some value of z, θ increases with z for a given linear size d. Establish that the nondetection of extended radio sources at large z ($z > 2.5$ to 3) can in no way be due to geometric effects.

References

Baan, W. A., and McKee, M. R. (1985) *Astron. Astrophys.* **143**, 136.
Blandford, R. D., and Rees, M. J. (1974) *Mon. Not. Roy. Astron. Soc.* **169**, 395.
Bridle, A. H., and Perley, R. A. (1984) *Ann. Rev. Astron. Astophys.* **22**, 319.
Brown, R. L., Broderick, J. J., and Mitchell, K. J. (1986) *Astrophys. J.* **306**, 107.
Burns, J. O., Feigelson, E. D., and Schreier, E. J. (1983) *Astrophys. J.* **273**, 128.
Charlesworth, M., and Spencer, R. E. (1982) *Mon. Not. Roy. Astron. Soc.* **200**, 933.
Cotton, W. D., et al. (1980) *Astrophys. J.* **238**, L123.
Ferrari, A. (1998) *Ann. Rev. Astron. Astophys.* **36**, 539.
Gunn, J. E., Longair, M. S., and Rees, M. J. (1978) *Observational Cosmology* (Saas-Fee Advanced Course 8, Geneva Observatory, Geneva).
Kellerman, K. I., and Pauliny-Toth, I. I. K (1981) *Annu. Rev. Astron. Astrophys.* **19**, 373.
Kellerman, K. I., et al. (1981) *Astron. Astrophys.* **97**, L1.
Leahy, J. P., and Williams, A. G. (1984) *Mon. Not. Roy. Astron. Soc.* **210**, 929.
Lelièvre, G., Nieto, J. L., Horville, D., Renard, L., and Servan, B. (1984) *Astron. Astrophys.* **138**, 49.
Longair, M. S. (1978) in *Observational Cosmology* (Saas-Fee Advanced Course 8, Geneva Observatory, Geneva).
Miley, G. (1980) *Annu. Rev. Astron. Astrophys.* **18**, 165.
Norman, M. L., Smarr, L., Winkler, K. H. A., and Smith, M. D. (1982) *Astron. Astrophys.* **113**, 285.
Pacini, F., Ryter, C., and Strittmatter, P. A. (1979) *Extragalactic High Energy Astrophysics* (Saas-Fee Advanced Course 9, Geneva Observatory, Geneva).
Pearson, T. J. et al. (1981) *Nature* **290**, 365.
Perley, R. A., Dreher, J. W., and Cowan, J. J. (1984) *Astrophys. J.* **285**, L35.
Strittmatter, P. A. (1979) in *Extragalactic High Energy Physics* (Saas-Fee Advanced Course 9, Geneva Observatory, Geneva).
Strom, R. G., Willis, A. G., and Wilson, A. S. (1978) *Astron. Astrophys.* **68**, 367.
Verschuur, G. L., and Kellermann, K. I. (1988) *Galactic and Extragalactic Radio Astronomy* (Springer, Berlin, Heidelberg).
Willis, A. G., Strom, R. G., Bridle, A. H., and Fomalont, E. B. (1981) *Astron. Astrophys.* **95**, 250.
Zensus, J. A. (1997) *Ann. Rev. Astron. Astrophys.* **35**, 607.

9. Quasars and Other Active Nuclei

Radio astronomy led to the discovery of the first quasars. As long ago as 1960 several radio sources in the 3C catalogue had been noticed for their remarkably small angular size and were therefore particularly suitable for searches for the associated optical object. Optical exposures of the 3C 48 field indicated an object with a stellar appearance; its spectrum showed very strong emission lines that at first could not be identified. In 1962 Hazard, Mackey, and Shimmins, using the Parkes telescope, succeeded in locating the source 3C 273 with an excellent accuracy (better than $1''$), thanks to a lunar occultation. Analysis of the light-curve profile at the beginning and end of the occultation showed, moreover, the existence of two components, A and B, separated by $20''$; the second, 3C 273B, coincides exactly with a stellarlike object ($m_V \approx 13$) whose spectrum also turned out to have very strong emission lines. Schmidt discovered that these were in fact hydrogen lines redshifted by an amount $z = \Delta\lambda/\lambda_0 = 0.158$. Hence it was realized that if 3C 273 obeys the Hubble law, it is at an extremely large distance and has an enormous intrinsic luminosity (of the order of $10^{47}\,\mathrm{ergs\,s^{-1}}$).

The difficulty of imagining what processes could account for such peculiar properties led some astronomers to question the nature of the redshift. There followed a lively debate, which to a large extent is now closed. Indeed many arguments favour a cosmological origin. In what follows we shall take this as accepted; at the end of the chapter we shall return to this question, in the light of all the observational data available.

The definition generally adopted for quasars or QSOs (quasistellar objects) is the following: an object with a stellar image whose spectrum shows strong emission lines significantly redshifted and with an absolute B-band magnitude less than -23 ($M_B < -23$). The latter criterion relating to the intrinsic luminosity allows us to establish a dividing line with other objects that are less luminous but nonetheless have several common properties: Seyfert galaxies. Many observations have shown that it is always the central region of a galaxy (which is otherwise relatively normal) that is responsible for the unusual properties characterizing these objects, whence the expression 'active galactic nuclei'.

Since 1963 the study of active galactic nuclei has seen many developments. Some have been investigations of 'the active galactic nuclei phenomenon' as such (their description and modelling). It is this aspect that we shall describe first of all by giving the main properties of the continuous and line radiation observed over the whole electromagnetic spectrum; quasars, which in a way represent the archetypal active galactic nuclei (the underlying galaxy having a negligible contribution), will then be discussed, and later on we shall indicate the distinctive features of other classes of AGN (Seyfert galaxies, BL Lac objects, and so on). Other developments, on the other hand, seek to consider QSOs as a population. Systematic counts have been made to specify the large-scale depth distribution (related to the evolution of the AGN phenomenon with cosmological time) or small-scale distribution (clustering). We shall then describe the various methods currently used to identify quasars, the results obtained, and their implications. We shall also describe the remarkable phenomenon of gravitational lensing and its potential interest for cosmology or the study of galactic mass; in fact active nuclei thus enable us to probe the intergalactic medium (as do the absorption-line systems described in Chap. 10). Finally, after having described the AGN in their appropriate environment (the type of the surrounding galaxy, the possible existence of a cluster), we give the main constraints imposed by observations on models describing the central engine and possible scenarios considered plausible.

9.1 Emission from the Nucleus

In this section we shall review the principal characteristics of the continuous and line emission of quasars: the shape of the spectrum, the profile and intensity of the lines, variability, polarization, and so on, with the aim of identifying the physical processes responsible for the observed radiation.

9.1.1 Continuous Emission

Quasars are remarkable in that the energy they radiate is – roughly speaking – spread out uniformly over the whole of the spectrum accessible to observation (from $\nu \approx 1\,\text{GHz}$ to γ rays). Indeed the global variation of the flux F_ν (given, for example, in $\text{erg}\,\text{cm}^{-2}\,\text{s}^{-1}\,\text{Hz}^{-1}$) with the frequency ν is approximately represented by a power law of the type $F_\nu = k\nu^{-\alpha}$, where the spectral index α is close to 1 (Fig. 9.1). Thus the energy radiated around the frequency ν (for example in a band of width $\Delta\nu = \nu/2$), proportional to νF_ν, is independent of ν. This is, of course, only a first approximation that needs to be made more precise by examining the various spectral domains.

The Radio Domain. As we have seen above, the first quasars were discovered by their radio emission; however, these objects are identified on the basis of purely optical criteria and it turns out that the large majority of QSOs do not have easily detectable radio emission (at $\nu \approx$ a few GHz). More

9.1 Emission from the Nucleus

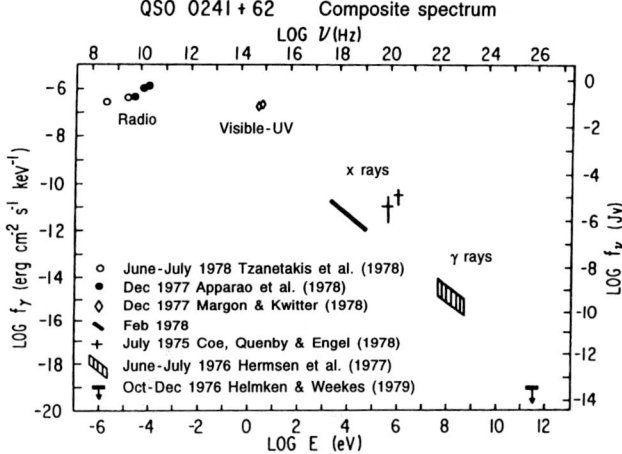

Fig. 9.1. The spectrum of the quasar 0241+622 ($z = 0.044$). Beyond 10^{13} Hz, F_ν varies approximately as ν^{-1}. (From Worall et al. 1980)

precisely the fraction of quasars discovered optically having a measurable radio emission at 5 GHz ($((\nu F_\nu)_{5\,\text{GHz}}/(\nu F_\nu)_{2500\,\text{Å}}) \geq 10^3$) is of the order of 10%. As radio-loud and radio-quiet quasars have otherwise very similar properties, we can wonder whether the presence of radio emission is an essential aspect of the AGN phenomenon, or, on the contrary, just a secondary characteristic (associated, for example, with a different environment for the active nucleus).

When it is present, the radio emission generally comes from a compact source with a relatively flat spectrum with an index $\alpha \approx 0.5$, sometimes associated with more extended components (see Chap. 8). Recent studies of high-redshift quasars show a complete absence of extended sources with a steep spectrum at $z > 2.5$. This evolution might be explained by the presence, at these corresponding epochs, of an intergalactic medium having a much higher density (this quantity should vary at least as $(1+z)^3$) and preventing the formation of the large radio lobes seen at small z.

The Infrared Domain. From $\lambda \approx 100\,\mu\text{m}$ up to the visible, the whole spectrum follows a power law with $\alpha = 1$ to 1.2. In fact near-infrared data obtained from the ground added to that from the *IRAS* satellite show an excess of emission between $2\,\mu\text{m}$ and $1\,\text{mm}$. This region of the spectrum contains about 3% of the total energy emitted. The origin of the radiation in this domain is still not exactly known; nevertheless it seems clear that a significant fraction (if not all) of the emission is thermal in nature. It would come from the dust located around the active nucleus at distances from about $10\,\text{pc}$ to about $1\,\text{kpc}$, the ultraviolet and visible radiation from the central source constituting the heating source of the dust grains.

IRAS observations have also shown a great similarity between the radio-loud and radio-quiet quasars' spectra between $\lambda = 10$ and $100\,\mu\text{m}$. This

262   9. Quasars and Other Active Nuclei

indicates that for the latter the cutoff occurs in the submillimetre domain. Measurements in this still largely unexplored part of the spectrum should enable us to determine why emission is absent at low frequencies: is there a break in the electron energy distribution, absorption by a thermal plasma located in the foreground, or synchrotron selfabsorption?

The Optical and Ultraviolet Domains. To compare the properties of QSOs at different redshifts, we need to consider the wavelengths in the reference frame of the source (rest-frame wavelengths). In the visible the continuum shows a spectrum that is clearly flatter: between $\lambda = 5000$ Å and $\lambda = 1000$ Å, $\alpha \approx 0.5$. For adjacent domains this corresponds to a bump centred at about $\lambda = 2500$ Å (Fig. 9.2). In models that assume the existence of a massive black hole it is natural to interpret this excess as thermal emission from the accretion disc that 'feeds' the central object. Indeed it is expected that the differential rotation combined with the viscosity raises the gas to temperatures of the order of 10^4 K, which corresponds to the position of the observed maximum (in reality a relatively broad range of temperatures is probably present owing to the radial extent of the disc). A more systematic study of the characteristics of this bump (such as its variability, its relation to other properties of the quasar, and so on) should allow us to better understand its origin.

At $\lambda < 912$ Å the continuum is rarely absorbed; this shows that, in general, the gas responsible for the emission lines does not hide the central source of the continuum radiation. In the far ultraviolet (down to approximately

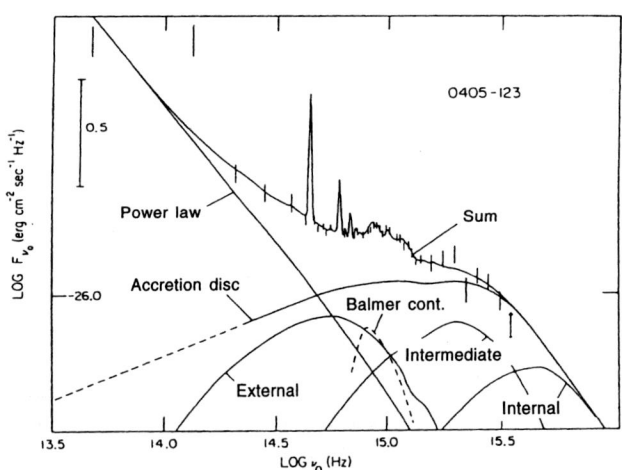

Fig. 9.2. The spectrum of the QSO 0405−123 ($z = 0.57$) in the near infrared, the visible, and the ultraviolet. The fit involves three components (corresponding to three regions at various temperatures) and the Balmer continuum. (From Malkan 1983)

10 Å) observations are very difficult and we only have indirect indications; for example the detection of the emission lines of Fe X requires the presence of photons at $\lambda < 53$ Å ($E > 235$ eV, the ionization potential of Fe IX). In a similar way one can verify that the continuum observed in the UV is consistent with the ionization necessary to account for the hydrogen emission lines (Lyα, in particular). One can then infer that the observable continuum is not notably affected by dust absorption.

The X-ray Domain. X-ray data exist in the range 0.5–100 keV and mostly concern nearby quasars. Between 2 keV and 100 keV the spectrum (F_ν) follows a power law with an index $\alpha \approx 0.7$ (for $E < 2$ keV, the spectrum shows a break due to absorption by heavy elements). To join this up with the UV and X-ray ranges, we must take an index $\alpha \approx 1$ to 2 in the intermediate region.

The origin of the X-ray emission remains uncertain. Regarding nonthermal processes, besides synchrotron emission originating from very energetic electrons, which probably plays a negligible role, account must be taken of inverse Compton radiation. This involves collisions between relativistic electrons or electrons belonging to a thermal component at a very high temperature and low-energy photons (at radio wavelengths, for example).

The X-ray data for quasars are essential because they give us information about regions very close to the 'central engine'. They also involve the interpretation of the diffuse isotropic X-ray background (a thermal spectrum with $T \approx 10^7$ K). Indeed it seems that the cumulative emission of quasars is not large enough to explain this radiation. However, these estimates do not take into account the properties of QSOs at large z and thus of high luminosity (their flux being undetectable by the *Einstein* satellite). Future observations will allow us to better study the emission of these objects (their evolution, the ratio L_X/L_{visible}) and perhaps solve the enigma posed by the X-ray background radiation.

9.1.2 Emission Lines

Emission lines are an essential feature of the 'active nucleus' phenomenon. Indeed objects whose luminosity differ by several orders of magnitude have approximately the same ratio of energy emitted in lines as in the continuum. Figure 9.3 represents a composite spectrum of a quasar showing lines commonly detected. We can distinguish two main categories: the broad, permitted lines and the clearly narrower, forbidden lines. Analysis of the profiles, of relative intensities, gives information about the kinematics of the emitting gas and other important parameters, such as the density or mass.

Permitted Lines. The lines concerned are principally those of hydrogen (Lyα and the Balmer series Hα, Hβ, and Hγ), C IV (the doublet at 1548 and 1550 Å), and Mg II (the doublet at 2796 and 2803 Å). In a first approximation

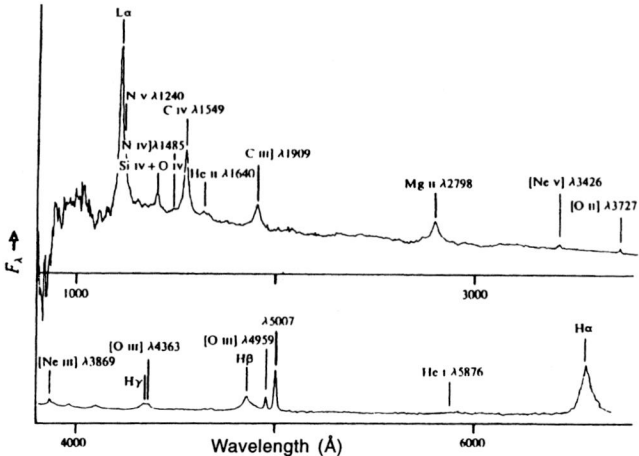

Fig. 9.3. The composite rest-frame spectrum of a quasar, showing the strongest emission lines. (From Baldwin 1977)

the profile appears as the sum of two components: a broad one characterized by $\Delta V = c\Delta\lambda_{1/2}/\lambda = 10^3$ to $10^4\,\mathrm{km\,s^{-1}}$, where $\Delta\lambda_{1/2}$ is the width at half maximum, and a narrow one with $\Delta V = 300$ to $100\,\mathrm{km\,s^{-1}}$. This second component is clearly better defined by the forbidden lines since it appears alone (it then allows us to precisely determine the redshift). The profile of the broad components is sometimes strongly asymmetric and often differs from one transition to another. As one can imagine, these data do not determine the kinematics in a unique way; nevertheless detailed modelling favours a radial outward flow rather than a rotating disc.

The distinction between broad and narrow lines leads to the definition of two different regions in which these lines are produced: the BLR (or broad-line region) and NLR (narrow-line region). In the first the density is too high for forbidden lines to appear; the absence of a broad component for the doublet [O III]$\lambda\lambda 4959, 5007$ implies an electron density of $n_e > 10^8\,\mathrm{cm^{-3}}$; on the other hand the semiforbidden line C III]$\lambda 1909$ is present where $n_e < 10^{10}\,\mathrm{cm^{-3}}$. The density is thus around $n_e \approx 10^9\,\mathrm{cm^{-3}}$.

The emission lines of quasars are to a large extent common to H II regions. They are characteristic of a photoionized medium at approximately 10^4 K. This coincidence for regions under such different conditions is not a matter of chance. The line emission (allowed and forbidden since the excitation is at least partly collisional) plays the role of a very efficient thermostat: if $T \ll 10^4$ K, the heating prevails over the cooling, and the opposite occurs if $T \gg 10^4$ K. Thus it is possible to model the BLR gas in a similar way to that of H II regions. Here we shall give only some indications concerning the intensity ratio of the hydrogen lines. The equation expressing the stationarity of the population of a level i ($i \neq 1$) is written

9.1 Emission from the Nucleus

$$n_+ n_\mathrm{e} \langle v\sigma_{\infty,i}(v)\rangle_v + \sum_{j=i+1}^{\infty} n_j A_{ji} = n_i \sum_{j=2}^{i-1} A_{ij},$$

where n_+ is the proton density, n_i the density of H atoms in the level i, v the relative velocity of electrons and protons, $\sigma_{\infty,i}$ the effective recombination cross-section in the state i, A_{ij} (assuming that the gas is optically thick for photons of the Lyman series: this is why the level $j = 1$ does not appear). For QSOs with low z we can simply compare the flux in the lines Hα ($3 \to 2$) and Hβ ($4 \to 2$). The Balmer ratio, or decrement, is written

$$\frac{F(\mathrm{H}\alpha)}{F(\mathrm{H}\beta)} = \frac{n_3\, A_{3\,2}\, h\nu_\alpha}{n_4\, A_{4\,2}\, h\nu_\beta}$$

and depends only on n_3/n_4. This latter parameter can be obtained by solving all the equations of the type written above, from which a theoretical value $F(\mathrm{H}\alpha)/F(\mathrm{H}\beta) \approx 3$ is obtained. The observed ratio is clearly higher. A possible explanation is selective extinction caused by dust; however, the same discrepancy exists for lines of the Paschen and Brackett series (in the near infrared), which should be much less sensitive to this phenomenon. An alternative interpretation assumes that collisions contribute to increase the population of levels $i > 1$ and also that the medium is thick even for lines of the Balmer series (other terms must then be introduced into the equation for n_i). The column density of hydrogen for this must reach $N \approx 10^{23}\,\mathrm{cm}^{-2}$.

The ionization equilibrium is governed by the equation

$$n_+ n_\mathrm{e} \langle v\sigma_\mathrm{r}(v)\rangle = n_\mathrm{H} \int_{\nu_0}^{\infty} 4\pi \frac{J(\nu)}{h\nu} a(\nu)\, d\nu,$$

where σ_r is the effective recombination cross-section, ν_0 the frequency at the Lyman limit, $J(\nu)$ the mean intensity of the radiation, and $a(\nu)$ the absorption coefficient. The fraction of ionized gas, n_+/n_H, is proportional (for a given T) to $\Gamma = n_\mathrm{ph}/n_\mathrm{e}$, where n_ph is the number density of ionizing photons, whence the term 'ionization parameter' used to designate Γ. n_ph varies as LR^{-2}, where R is the distance to the source and L is the luminosity. As the models (via the ratio C III]λ1909/C IVλ1549, in particular) allow us to determine $\Gamma \approx 10^{-2}$, one can derive an estimate for R of the order of a parsec (for $L_\mathrm{ionizing} \approx 10^{47}\,\mathrm{erg\,s}^{-1}$). The size deduced from the ratio N/n is clearly smaller, being around 10^{-6} pc. This is one indication, among others, that the BLR is not a homogeneous region but is formed out of small dense clouds occupying a small fraction of the volume.

The flux in the Lyα line allows us to estimate what fraction of the ionizing radiation is really absorbed by the gas. Indeed each ionization is associated (in a steady state) with the emission of a Lyα photon. We thus find that only 10% of the ionizing continuum observed is absorbed by the gas; this value then represents the order of magnitude of the surface covering factor of the BLR clouds.

Forbidden Lines. Forbidden lines correspond to transitions for which the probabilities A_{ij} are much smaller (for example $A_{ij} \approx 10^{-2}\,\text{s}^{-1}$ compared to approximately $10^8\,\text{s}^{-1}$ for permitted transitions). Under laboratory conditions collisional deexcitation is much more likely than radiative deexcitation, whence the impossibility of observing such transitions in the laboratory and the term 'forbidden' (conventionally the species responsible is written between brackets; for example [OIII]λ4959). The forbidden lines observed in the spectrum of QSOs do not come from ion–electron recombination but are excited by collisions (this is also the case for nearly all the permitted lines, Mg IIλ2798 and C IVλ1549, for example). For a two-level atom, the equation determining the ratio of populations in states 1 and 2 is written

$$n_1 n_e \langle v\sigma_{1\,2}(v)\rangle = n_2 A_{2\,1} + n_2 n_e \langle v\sigma_{2\,1}(v)\rangle,$$

where $\sigma_{1\,2}$ and $\sigma_{2\,1}$ are cross-sections for excitation and deexcitation by collisions with electrons. We can see that if n_e exceeds the critical density $n_c = A_{1\,2}/\langle v\sigma_{2\,1}(v)\rangle$, deexcitation is caused mainly by collisions (laboratory conditions); the opposite case ($n < n_c$) is, on the other hand, favourable to the observation of the line. Constraints on the density in the BLR mentioned above come from this analysis. For the NLR, which emits both permitted and forbidden lines, we end up with $n_e \approx 10^5\,\text{cm}^{-3}$.

In some nearby active nuclei, similar to QSOs but with a smaller luminosity (of Seyfert 1 type; see Sect. 9.6.1), it has been possible to resolve the NLR spatially. Its size is at least of the order of 1 kpc. As for the BLR the gas is grouped into clouds that occupy a small fraction of space and are optically thick at $\lambda < 912\,\text{Å}$. The various lines observed (from O I to Fe X) show a great variety of ionization states. The mass is estimated to be of the order of $10^6\,M_\odot$ (compared to less than $10^4\,M_\odot$ for the BLR).

The models of the BLR and NLR lead also to an estimate of the relative abundance of heavy elements of the order of solar abundances. This is a remarkable result since the look-back time for the more distant quasars ($z > 4$) is about 80% of the age of the universe. The matter observed at this time must have undergone an active phase of stellar nucleosynthesis.

9.1.3 Variability and Polarization

The majority of quasars have a significant degree of polarization, especially in the visible ($P > 1\%$). Moderate variability ($\Delta I/I \approx 20\text{--}30\%$) is common but generally occurs only on time scales of one year or more. In fact these results are strongly influenced by observational constraints (difficulty of obtaining a sample covering various time scales due to the limited duration of observing runs, uneven meteorological conditions etc). Systematic observation of 3C 273 has revealed variations in the near infrared of 100% over $\Delta t \approx 1$ day ($\Delta L/\Delta t \approx 10^7\,L_\odot\,\text{s}^{-1}$), which implies that the emitting region has a size of the order of light-days at most (if the matter is not moving relativistically towards the observer). Flux variations generally consist of brief and recurrent outbursts following quiet phases during which the characteristic time of variation is of the order of months (in the visible and UV). At present no periodic variation has been clearly observed.

A small fraction of quasars (several per cent) have, by contrast, outbursts with a high amplitude ($\Delta I/I \approx 5$) and at the same time a far greater degree of optical polarization, reaching 20%. The preferential direction for $\boldsymbol{E}$ also varies on the scale of several days and without any clear correlation with the intensity fluctuations. OVVs (optically violent variables), blazars, or HPQs (highly polarized QSOs) fall within this category.

The variability and polarization of lines have both been studied. Figure 9.4 shows the C IVλ1550, C III]λ1909, and Mg IIλ2800 lines of the Seyfert 1 galaxy NGC 4151, at times when the continuum was weak and then strong. There are important variations of the C IVλ1550 and Mg IIλ2800 lines which reflect those of the continuum with delays of about 13 and 60 days respectively (the CIII line does not vary). These durations probably correspond to the time required by the light emanating from the central continuum source to reach BLR clouds and give a direct size estimate. Very little data exists for

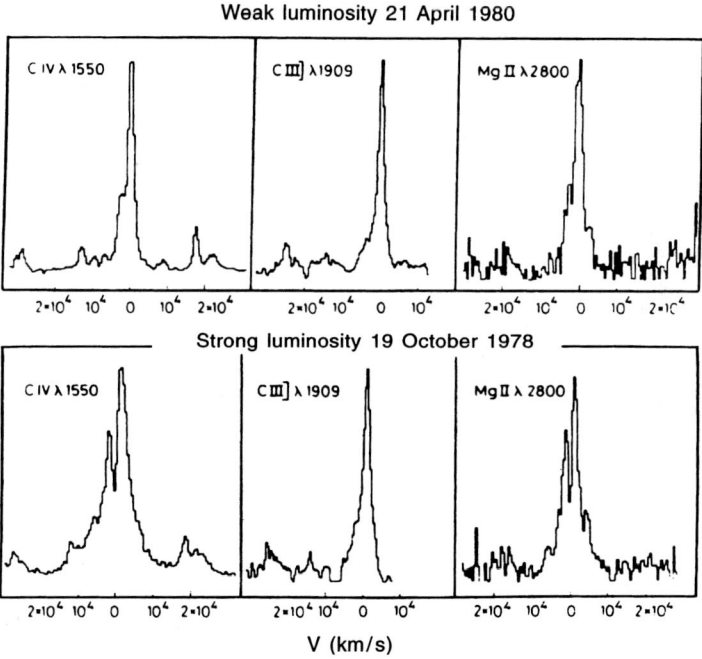

Fig. 9.4. The profile of three emission lines from the nucleus of the Seyfert 1 galaxy NGC 4151 at two different epochs. When the continuum emission is strong (October 1978), an additional component appears in the C IV and Mg II lines. (From Ulrich et al. 1984)

quasars (which are brighter) but we suspect that sizes increase with intrinsic luminosity. The emission lines also appear to be polarized (with the same field direction as the continuum). The polarization mechanism remains uncertain: it might be scattering by dust grains or by electrons. A more systematic study of the dependence of this phenomenon with respect to wavelength should help to identify the process involved and give valuable information on the geometry of the region; indeed these mechanisms involve the relative position of the continuum source, the scattering material, and the observer, which are different for the BLR and the NLR and may also vary according to the degree of ionization.

9.2 Systematic Quasar Searches

The various characteristics presented in the preceding section enable us – possibly after cross-checking – to identify the quasars present in a given region of the sky. At present, catalogues contain more than 10 000 such objects; only a small fraction of them have been studied in detail. Current surveys mainly aim at measuring the surface density of quasars in well-defined fields or the identification of objects with specific properties (QSOs with high z, with broad absorption lines or pairs of quasars).

9.2.1 The Radio Domain

Instruments such as the VLA allow mapping of extended fields with a very high sensitivity ($F_\nu \leq 0.1$ mJy) at $\nu \approx 1.4$ to 5 GHz. The optical identification of sources detected in this way is, in general, tricky or impossible (the objects are too weak for spectroscopy or the fields are empty). In addition in many cases a radio galaxy or an ordinary galaxy is found. The quasars that are found are preferentially associated with compact sources ($\Delta\theta < 1''$) with a flat spectrum; unfortunately they only represent a small fraction of all field quasars, since it is known that only 10% of them emit at radio frequencies.

9.2.2 The Visible Domain

Two main methods are used. The first consists in selecting possible candidates on the basis of criteria involving the colour or possibly the variability. Indeed the typical spectrum has the form $F_\nu \propto \nu^{-1}$, that is, one very different from the majority of stellar spectra (QSOs are clearly bluer). This results in the fact that in a colour–colour diagram ($U - B$, $B - V$, or $B - R$, $R - I$, for instance) quasars occupy a very well-defined region (Fig. 9.5). In fact some white dwarfs also lie in this area and it is always necessary to get a spectrum to verify that the characteristic broad emission lines are present (which at the same time gives the redshift). To take account of the variability, wide-field

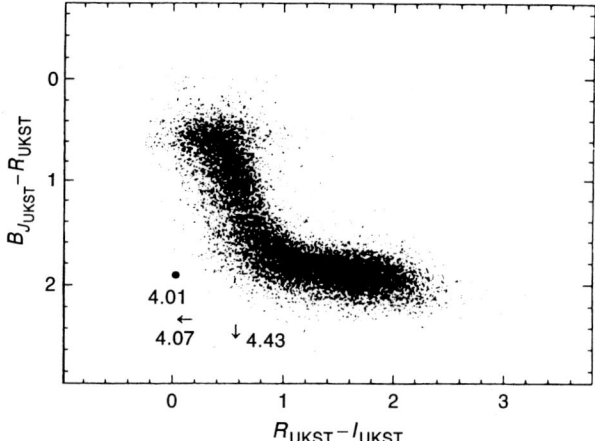

Fig. 9.5. A colour–colour diagram for objects with $m_R < 20$ in a field centred at $\alpha = 0^{\rm h}53^{\rm m}$, $\delta = -28°03'$. The positions of three QSOs with z greater than 4 are shown. $\leftarrow$: not detected in I; $\downarrow$: not detected in B. (From Warren et al. 1987)

images can be obtained through a given filter (U or B) at regular intervals; digitization of the plates then enables candidate quasars whose magnitudes fluctuate significantly to be automatically identified.

The second method consists in inserting a weakly dispersive element into the light path, an objective prism or transmission grating giving a resolution $R \approx 100$ Å. On the image of the field given by the telescope a small spectrum is recorded for each object (this is known as 'slitless spectroscopy') allowing those quasars having strong emission lines to be easily picked out (Fig. 9.6). If at least two lines appear, the redshift can be estimated without it being nec-

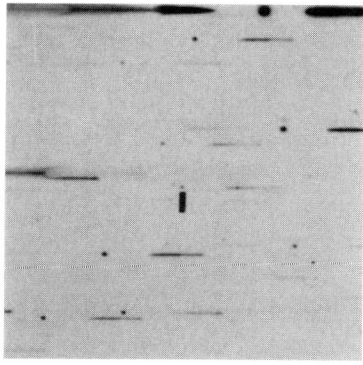

Fig. 9.6. A plate illustrating the use of slitless spectroscopy. Each point corresponds to an image of order 0; the first-order spectrum is to the right of the latter image. The object indicated at the centre is a quasar of magnitude $m_B = 18.5$ and redshift $z = 2.17$. (From Hoag and Smith 1977)

essary to take another spectrum. Nevertheless because the light is dispersed, this method does not allow the selection of objects as faint as the procedure based on colours (the difference in sensitivity is about 2 magnitudes).

It is important to recognize that each of these methods is affected by systematic biases, making it very difficult to obtain complete samples. Selection by *UBV* colours turns out to be inefficient when z becomes large enough for the Lyα line to fall within the B band; indeed this strongly modifies the $U - B$ colour and makes the object appear redder than it is in reality. To find QSOs with large z other filters must be used (for example R and I). With this method one of the few QSOs with $z > 4$ has been identified, 0051–279 at $z = 4.43$. Likewise the use of variability criteria gives a much stronger weighting to objects which fluctuate strongly (OVVs and blazars). Finally the use of objective prisms only highlights those QSOs with clearly contrasted emission lines. The recent use of Kodak IIIa-J and IIIa-F emulsions has enabled the Lyα line to be detected up to 5200 Å and 6800 Å respectively (which corresponds to a limit for z of 3.3 and 4.6).

9.2.3 Selection Based on X-ray and Infrared Emission

The *Einstein* satellite has observed in detail some fields and revealed many point sources which turned out to be QSOs. The optical properties of these objects (the continuum and lines) are very similar to those of quasars selected by the above methods, but generally these QSOs have relatively small z. This result is consistent with the correlation between X-ray and optical luminosities, $L_X/L_{opt} \propto L_{opt}^{-0.3}$: the QSOs observed at large z have a large optical luminosity and a relatively weak X-ray luminosity, often lower than the limiting flux detectable with current instruments.

We mention finally that some quasars have been discovered by the *IRAS* satellite on the basis of their emission in the mid-infrared (10–100 µm). Their spectrum shows weak optical emission, probably owing to significant extinction due to dust situated around the nucleus.

In the future many instruments will allow existing catalogues to be completed as a result of improved sensitivity. The extension of searches into very different regions of the spectrum (IR: *ISO* and *SIRTF*; UV and visible: the *Hubble Space Telescope*; X-ray: *Chandra, XMM-Newton*) is an important way of avoiding the biases affecting each method.

9.3 The Spatial Distribution of Quasars

Before describing how quasars are distributed in projection on the sky, on the one hand, and in depth (that is, in z), on the other, we first give the order of magnitude of the number of objects detected. Optical methods – by far the most effective – lead to the detection of about 30 QSOs brighter than $m_B = 20$ per square degree. A plate obtained with a Schmidt telescope covers a field greater than $10\,\text{deg}^2$, and then more than 300 quasars can be identified. Over the whole sky ($\approx 4 \times 10^4\,\text{deg}^2$) there are therefore quite a considerable number of potentially observable quasars.

9.3.1 The Distribution Projected on the Sky

Just as it is very instructive to see whether galaxies are distributed at random in space or whether they appear preferentially grouped, so it is interesting to see what the case is for quasars. We can investigate this by using the two-point correlation function $\epsilon(\theta)$ defined by

$$\sigma(\theta) = \bar{\sigma}\bigl(1 + \epsilon(\theta)\bigr),$$

where $\sigma(\theta)$ is the surface density of QSOs at an angular distance θ from a given QSO and $\bar{\sigma}$ is the mean density. Recent determinations of ϵ show for $z < 1.5$ the existence of clustering, giving $\epsilon > 0$, for (comoving) scales $r \leq 10\,\text{Mpc}$. It is possible that gravitationally lensed QSOs (see Sect. 9.4) not recognized as such are in part responsible for this effect.

We can also look for a correlation between the distribution of QSOs and that of galaxies close by on the sky. This type of study, initially motivated by noncosmological interpretations of the redshift, has seen a renewal of interest in quite a different context. Indeed this is a way to investigate the possible apparent increase of σ in the neighbourhood of a galaxy owing to gravitational amplification caused by the galaxy's mass distribution. In fact estimates of σ do not show a significant excess in the neighbourhood of galaxies. These determinations are performed either by conducting a detailed survey of all QSOs located around a given galaxy or by considering the QSO–galaxy correlation function for well-defined samples.

9.3.2 The Luminosity Function of Quasars

The quantity $\phi(L \text{ or } M)$ represents the number of objects per unit comoving volume whose luminosity lies between L and $L + \mathrm{d}L$ (or equivalently whose absolute magnitude $M = -2.5 \log L + \text{constant}$ lies between M and $M + \mathrm{d}M$). L and M can be monochromatic or bolometric quantities.

We also define the integrated function $\phi(L \text{ or } M)$ (QSOs such that $L' > L$ or $M' < M$). The principal motivation for determining Φ is the study of a possible cosmological evolution: were QSOs more numerous and more luminous in the past? An answer to this question would represent a valuable constraint for models of active nuclei.

Of course, only part of the luminosity function is accessible, the objects that are too weak falling beyond the detection limit. At best we can hope to gather a complete sample up to a given limit, for example $m_B < 20$. This is already difficult because the properties of QSOs show a large intrinsic scatter and some objects, according to the selection method used, may be missed owing to their peculiar characteristics. Thus there exist QSOs that are particularly red ($\alpha > 3$ in the visible); others have very weak emission lines; and so on. The precise definition of a 'completeness' limit, even for a sample of objects with well-defined characteristics, is no longer a simple affair; for example galactic extinction should be subtracted (typically a few 0.1 mag at galactic latitudes $b > 30°$), which is tricky owing to the very inhomogeneous character of the interstellar medium. In principle this limit can be determined from the counts themselves, the catalogue no longer being complete where ϕ stops varying as $F^{-3/2}$ (see Sect. 8.7.1), but such a criterion assumes a uniform distribution of objects without evolution in a Euclidean universe. Another method consists in using the so-called $V/V_{\max}$ test, whose principle is as follows. For each QSO from a sample characterized by a limit $F_{\min}$, we know the flux F_i (or m_i) and its redshift z_i. An object with a flux F_i could have been detected up to a redshift $z_{i\max}$ such that

$$F_{\min}d_L^2(z_{i\max}) = F_i d_L^2(z_i),$$

where d_L is the luminosity distance calculated in a given cosmological model. Let V_i and $V_{i\max}$ be the comoving volumes of the spheres $z < z_i$ and $z < z_{i\max}$; if the sample is complete we then have $\langle V_i/V_{i\max}\rangle = 0.5$ (in the absence of evolution). If $F_{\min}$ has been chosen too small, then $\langle V_i/V_{i\max}\rangle < 0.5$. It is important to notice that, whatever the cosmological model adopted (in particular whatever q_0), $\langle V_i/V_{i\max}\rangle > 0.5$ can only be obtained if the population evolves (for $F_{\min}$ large and no evolution $\langle V_i/V_{i\max}\rangle = 0.5$).

9.3.3 Quasar Evolution

Figure 9.7 shows an example of the quasar luminosity function estimated in different redshift intervals.

The distribution of Seyfert 1 nuclei is in good agreement with that of QSOs with small z. The nuclei of Seyfert 1 galaxies are thus not a population distinct from QSOs; they are simply members with smaller luminosities. Surveys made at small z have interesting consequences regarding the frequency with which quasars appear. Indeed one can estimate that among the galaxies capable of hosting a QSO-type or Seyfert 1-type nucleus at their centres, that is, spirals of normal size, several per cent of them are active.

In Fig. 9.7 it clearly appears that, at constant M, Φ increases with z. The $V/V_{\max}$ test confirms this result: all samples give $V/V_{\max} > 0.5$. From the local distribution ($z \approx 0$) to that at $z \approx 2.2$ the ratio is of the order of 10^4. This evolution can be due either to an increase of L with a constant number of objects (a horizontal shift of the curve towards the right when z increases) or to an increase in n at constant L (a vertical shift). In reality this latter possibility is not very likely, because there would not then be enough galaxies available to put all the quasars in (unless dwarf galaxies are considered, but

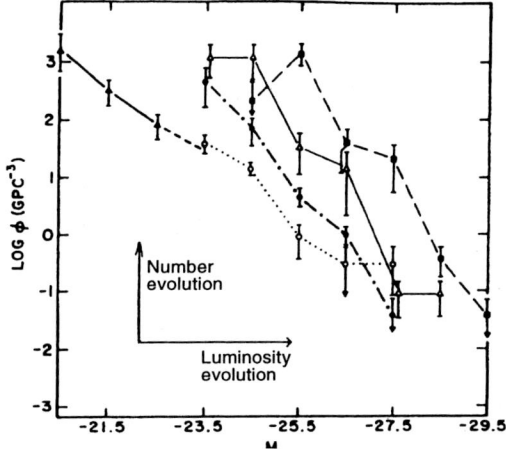

Fig. 9.7. The quasar luminosity function in various redshift intervals: ○ = 0.1–0.3; ● = 0.3–0.8; △ = 0.8–1.2; ■ = 1.2–1.5; ▲ = Seyfert galaxies ($z \approx 0$). (From Green 1985)

no active nuclei are seen in them at $z \approx 0$). On the other hand, it is possible to account for the observed evolution purely by a variation of L. Formally one can write

$$L(z) = (1+z)^{\alpha} L(z=0)$$

with $\alpha \approx 3.7$. This does not necessarily mean that the luminosity of the same quasars varies continuously following this law; recall that for $q_0 = 0$ the lookback time is $\Delta t(z) = z/[(1+z)H_0]$ giving $\frac{2}{3}H_0$ at $z = 2$. It is also possible that the activity is intermittent on short time scales ($\ll 1/H_0$) and that in the past, for some unknown reasons, more powerful quasars appeared.

At large z ($z > 3$–3.5), Φ is clearly smaller than would be expected from the evolution law obtained between $z = 0$ and 2.5. The possibility of an average extinction increasing with z caused by dust associated with intervening galaxies has been proposed to explain this 'cutoff' at large z (see Sect. 10.2.3), but this appears unlikely since radio-selected QSO samples, which do not suffer from such biases, display a similar behavior.

The existence of strong quasar evolution is an incontrovertible fact, in spite of the inherent uncertainties in the composition of the samples (biases in z) or the analysis of the results (the presence of extinction and the cosmological model used). We shall see later that it is not easy for models of active nuclei to account for such an evolution in luminosity. X-ray samples also yield some information on this question. We have already mentioned that L_X/L_{opt} decreases with L_{opt} (as $L_{opt}^{-0.3}$); it is thus natural to find a weaker evolution in X rays. As for optical samples, a variation of L alone can explain the data, with, for instance,

$$L(z) = e^{\Delta t(z)/\tau_X} L(z=0),$$

with $\tau_X \approx \frac{1}{5}H_0$ (adopting this form for optical samples one would obtain $\tau_{opt} \approx \frac{1}{6}H_0$). This result mainly concerns objects with $z < 1$, which contains the majority of quasars detected in X rays.

9.4 Gravitational Lenses

The phenomenon of gravitational deflection of light rays predicted by the theory of general relativity was observed for the first time in 1919 by Eddington during an eclipse of the Sun. In agreement with the theory the light of a star situated at the edge of the solar disc was deflected by 1.7″. Zwicky, as early on as 1937, proposed investigating this effect for galaxies, but only in 1979 could a clear example be discovered: the twin quasar 0957+561. Since then, many observers have systematically searched for other similar cases, with only moderate success. Gravitationally lensed quasars – a situation where clearly separated images appear ($\Delta\theta >$ a few arcseconds) – are thus rare. Conversely it is probable that, more frequently, the presence of a massive object along the line of sight (a galaxy or cluster playing the role of a 'lens') causes amplification of the flux without inducing multiple images (an effect that is impossible to recognize as such). This area of investigation is then related to the study of quasars themselves. Indeed all perturbations of the apparent flux (amplification, but also possibly attenuation) will affect the observed samples – which are always flux-limited – and might create important biases. Thus the interpretation of counts (that is, the study of evolution; see Sect. 9.3.3) will be distorted if quasars with large z are more often amplified than nearby quasars (the probability of finding a potentially deflecting object being greater). Likewise the number of absorption systems will be wrong (being overestimated) if the QSOs observed spectroscopically, selected from among the brightest objects, are chosen preferentially when they are amplified by an intervening galaxy (we would thus certainly have at least one system; see Sect. 10.2.1). It is then essential to be able to estimate how our vision of the distant universe is altered by the phenomenon of gravitational deflection.

In what follows we shall simply summarize the main characteristics of the phenomenon and describe a few observations.

9.4.1 Characteristics of the Phenomenon

The parameters involved are defined in Fig. 9.8. The deviation θ, due to a point mass, is $4R_S/b$, where R_S is the Schwarzschild radius and b the impact parameter ($R_S = 2GM/c^2$, M being the mass of the deflector). On the other hand θ also satisfies (with $\alpha = b/D_G$)

$$\theta = (\alpha - \beta)\frac{D_Q}{D_{GQ}}.$$

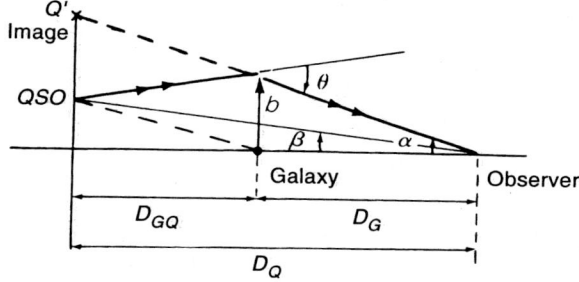

Fig. 9.8. Gravitational deflection of light rays emanating from the quasar Q by a galaxy G; the observer O sees the quasar as Q′ separated from the deflector G by an angle α

For a three-dimensional mass distribution, the phenomenon is similar when α is sufficiently large but very different for $\alpha \to 0$. Indeed the mass M to be considered is only that contained within the cylinder of radius β if the distribution shows spherical symmetry, and M thus tends towards 0 with α. Figure 9.9 shows how we can obtain by graphical means the number and the position of the images according to the values of β, D_Q, and D_{GQ} (we assume here Euclidean space). For $\beta = 0$ a ring is observed; this particular case put aside, the number of images is always odd and is 1 as soon as β is sufficiently large. According to their position, the images show various fluxes (they can be amplified or attenuated with respect to the original image). In practice for a mass of $10^{12}\,M_\odot$ placed halfway to a QSO at $z = 1$ we obtain at best a separation of a few arcseconds. It is therefore difficult to detect gravitationally lensed quasars.

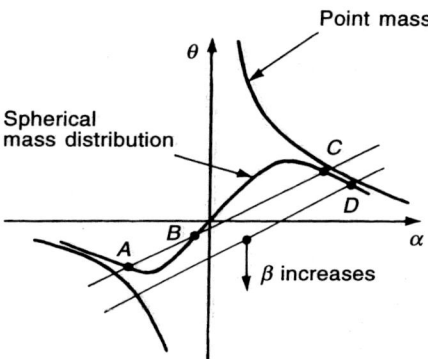

Fig. 9.9. The position of images according to the value of β (see Fig. 9.8) for a point mass or a spherically symmetric distribution. When O, G, and Q are no longer aligned (when β is large) we go from three images (A, B, C) to one (D)

276   9. Quasars and Other Active Nuclei

The various signatures of the phenomenon are the following: (a) the existence of multiple images with an identical spectrum, (b) the presence of a galaxy or cluster along the line of sight, and (c) the correlation between the temporal flux variations of the various images. Criterion (a) might seem sufficient; in fact the tendency of QSOs to cluster on small scales and the uncertainties in the determination of the spectra (often one of the images is relatively faint) lead to ambiguities. The second criterion is often difficult to use, especially if the deflector is distant ($z > 0.5$). As for the third, it requires long-term monitoring and has been used for a few objects only. The principle of the method is the following: let $F_1(t)$ and $F_2(t)$ be the fluxes of two images of the same QSO. The optical trajectories being different (image 2 being delayed by $\Delta t = \tau$, for instance), the variations of the source (if any!) are not detected simultaneously; the cross-correlation $\langle F_1(t-T)F_2(t)\rangle$ then has a maximum for $T = \tau$. The difficulty lies in obtaining well-sampled series of measurements over sufficiently long time intervals (τ being of the order of a year). Figure 9.10 shows measurements accumulated over 7 years for the twin quasar 0957+561. As can be seen, the possibility of detecting a correlation for $\tau = 415$ days largely depends on the fact that it has been possible to make measurements in the two intervals 5700–5800 and 6100–6200. Indeed it is only during these periods that it is possible to recognize on each of the light curves a well-defined feature (a decrease in flux of about 0.25 mag in about 100 days).

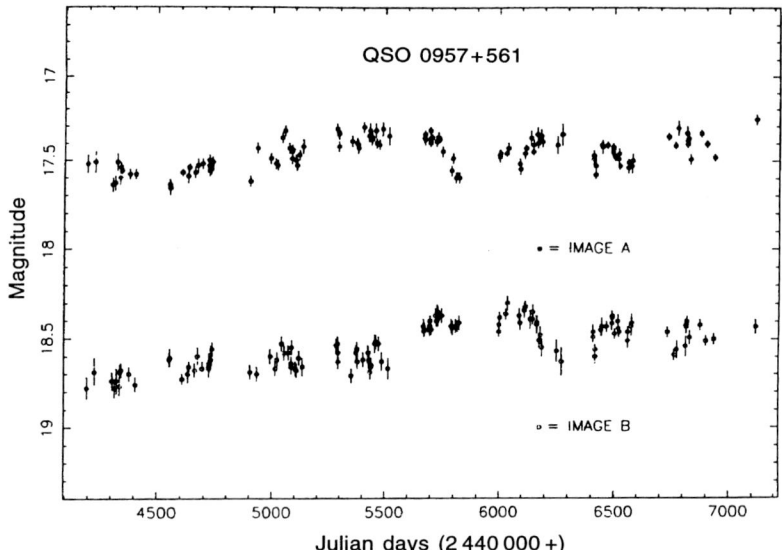

Fig. 9.10. Light curves for images A and B of 0957+561 in the B band. (From Vanderriest et al. 1989)

9.4.2 An Example: The Gravitationally Lensed Quasar 0957+561

Figure 9.11 shows optical and radio images of the twin quasar 0957+561. In the visible, two components, A and B, appear separated by 6″ and both have a magnitude close to 17.5. The spectra of A and B are identical and lead to a redshift of 1.41. One year after the discovery the lensing galaxy was identified; it is a giant elliptical with a redshift of 0.39 located just north of image B. It happens that 0957+561 is a powerful radio source (≈ 0.5 Jy at $\lambda = 20$ cm) so that the phenomenon can be observed in the radio range too (the radio 'photons' follow the same trajectory as the optical ones: just as the mechanical acceleration in a given gravitational field does not depend on the mass, the geodesics are identical whatever the frequency). The advantage of radio observations is the high spatial resolution accessible thanks to the VLA or VLBI. On the map obtained at $\lambda = 6$ cm, the complex structure of the source can be seen: the core (A), which coincides with the optical image A, the jet, and the lobes (C and E). The second image B only shows the compact component: this is due to the fact that for C, D, and E the alignment with the deflector is less good. Finally the lensing galaxy (G) is clearly visible. The

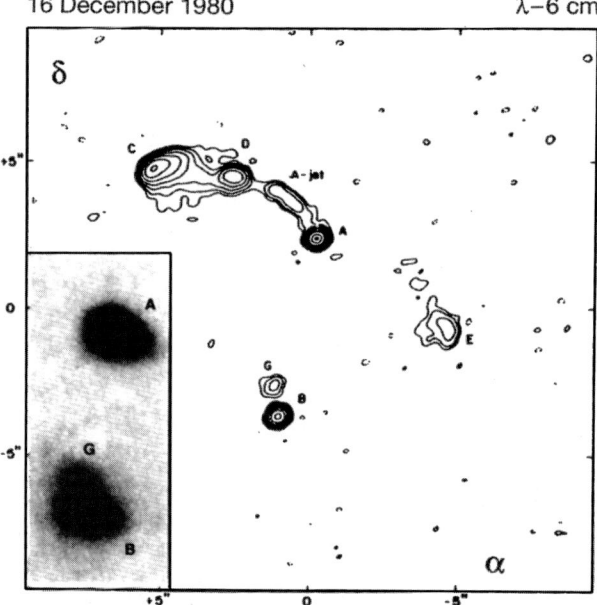

Fig. 9.11. A radio map of the double quasar 0957+561 obtained with the VLA at $\lambda = 6$ cm. One can distinguish the two images of the nucleus (A and B) as well as the extended structure associated with A (the jet, C, D, and E). The box at the lower left shows the image obtained in the visible (with a blue filter); the image of the lensing galaxy ($z = 0.39$) is partially blended with B. (From Greenfield et al. 1985 and Stockton 1980)

ratio of the radio spectra of A and B is independent of λ between $\lambda = 2\,\mathrm{cm}$ and $\lambda = 21\,\mathrm{cm}$ and is identical to the ratio observed in the visible. A study of the polarization at $\lambda = 6$, 18, and 20 cm leads to different values for the Faraday rotation towards A and B; there is no contradiction because this can be explained by the distinct trajectories followed through the galaxy G.

We have seen that an odd number of images is expected: where, then, is the third? VLBI observations have shown that it is weak and found very close to B.

It is possible to construct a detailed model, using the constraints imposed by the data. One conclusion is that the galaxy G alone cannot account for the phenomenon. Indeed G belongs to a cluster that itself contributes to the deflection. Unfortunately the large number of parameters does not allow us to construct a unique model (the situation is still less favourable for other known lensed quasars). One can show, however, that the masses required for G and for the cluster are much greater than those observed (assuming a 'standard' mass–luminosity ratio): these observations thus give a direct indication in favour of the existence of 'dark' matter.

There are also arc-shaped structures (with an angular size of the order of $10''$) discovered in 1986 in several clusters of galaxies. Spectroscopic observations have shown that the arc observed in the cluster A370 ($z = 0.374$) is in fact an image formed by the cluster itself of a spiral galaxy at $z = 0.7244$) (Fig. 9.12). This is another remarkable consequence of gravitational lensing.

Giant arcs recently turned out to be not so rare, and many have been dicovered towards rich clusters. Furthermore in some cases many smaller and fainter arclets have been detected (these are images of normal background galaxies too). These observations offer promising opportunities since (1) modelling of the arcs and arclets is a powerful means of probing the cluster dark

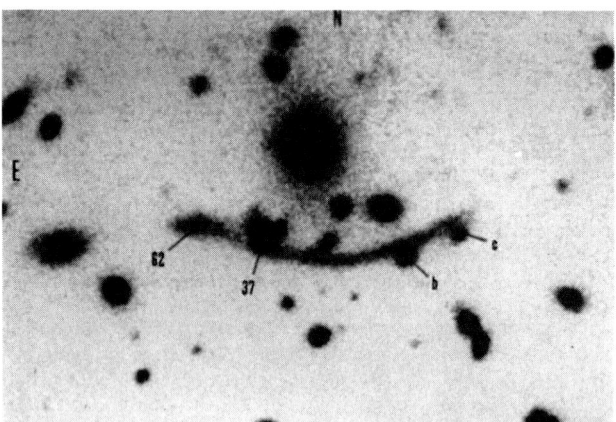

Fig. 9.12. A CCD image of the arc in the cluster A370 obtained at CFHT. (From Soucail et al. 1988)

matter distribution, and (2) clusters thus act as 'gravitational telescopes', which enable us to study the properties of distant normal galaxies that would, in the absence of amplification, remain undetectable.

9.5 The Quasar Environment. The Nature of the Redshifts

At small redshifts ($z \leq 0.5$) many observations have shown that quasars reside at the centre of some galaxies. It is instructive, in order to understand the quasars themselves, to determine what these host galaxies' properties are (their morphological type, size, and gas content, the presence of interacting galaxies, and so on). In this way one can expect to understand why the QSOs appear only at the centres of galaxies and not elsewhere (what the formation mechanisms are) and what their relation is to the surrounding medium. As the detection of galaxies associated with quasars is important in establishing the cosmological interpretation of the redshift, we shall summarize the various arguments for and against this interpretation.

9.5.1 The Host Galaxies of Quasars

Many quasars have been observed in order to study the nebulosity that surrounds them. In many cases it has been possible to obtain the spectrum of this nebulosity: it shows the usual emission lines of ionized gas at the redshift of the quasar. In general the intensity of these lines is such that it implies a spiral galaxy. More quasars have been studied using imaging alone. The use of CCD detectors provides a numerical image, which enables the unresolved central component to be subtracted out (the QSO image), the point spread function being determined from stellar images. This enables the colours, morphology, size, and brightness profile of the host galaxy to be determined. By fitting these parameters to those expected for a spiral or elliptical galaxy, one can determine the properties of the associated nebulosity.

The results of these observations indicate that the nebulosity has all the characteristics of a galaxy that is relatively 'normal' in size and nebulosity (our Galaxy or M31, for instance). Although there should be no marked differences between nebulosities associated with QSOs selected by different methods (optical, X ray, or radio), the host galaxies of radio-loud quasars are in most cases ellipticals and are on average about one magnitude brighter than those of radio-quiet quasars. There also appears to be a correlation between the luminosity of the host galaxy and that of the quasar ($L_{\text{gal}} \approx kL_{\text{QSO}}$).

The example given in Fig. 9.13 illustrates another distinctive property of QSOs: the frequent presence of a nearby companion, sometimes with signs of interaction with the host galaxy. This is consistent with the systematic association of QSOs with clusters or groups of galaxies (where collisions are favoured).

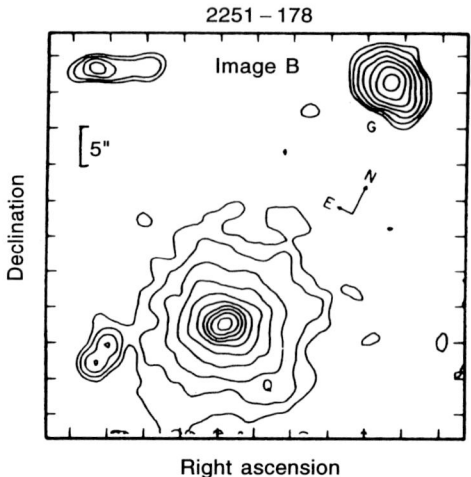

Fig. 9.13. A B image of the quasar 2251−178 ($z = 0.068$). The nebulosity associated with the QSO (Q) as well as a galaxy (G) approximately $40''$ to the north can be clearly seen. (From Hutchings et al. 1984)

9.5.2 The Nature of the Redshifts

The cosmological interpretation of large redshifts was proposed when the first QSOs were discovered. According to this frame, quasars just follow the Hubble flow and their large redshift simply means a large distance. The implications for the properties of quasars (in particular that a high luminosity is emitted by a very small volume) very quickly suggested alternative explanations. We could imagine, for instance, that the redshift has a gravitational origin; indeed photons emitted at the surface of a spherical body of radius R and mass M experience a redshift given by

$$z = \frac{\Delta\lambda}{\lambda_0} = \frac{1}{\sqrt{1 - (2GM/Rc^2)}} - 1.$$

The radius R, or more generally the distance at which the photons are emitted, is intimately linked to z ($z \propto R^{-1}$ for $z \ll 1$). The data concerning permitted and forbidden emission lines (see Sect. 9.1.2) impose an important radial size limit on the emitting regions that is incompatible both with the emission-line widths and with a unique value of z for the two types of line. In the same way, the assumption that the QSOs are local objects with a very rapid proper motion fails because we should then also observe blueshifts as well as redshifts, which is not the case. We could also assume that an unknown physical process is responsible for the phenomenon, but this approach does not provide any framework for the interpretation of the data. The most natural and simplest possible theory is to be preferred, so long as it does not

9.5 The Quasar Environment. The Nature of the Redshifts

not explicitly contradict any data. What then are the facts that, according to some astronomers, call into question the cosmological interpretation? They are:

- the existence of apparent superluminal velocities and brightness temperatures exceeding the Compton limit in compact radio sources: these phenomena can be explained by relativistic aberration;
- the existence of remarkable configurations (the proximity in the sky of objects with different z) which would be highly improbable if the objects were at their 'cosmological distance'.

An example of the latter is shown in Fig. 9.14. In fact the probability of such associations is indeed low, but the argument is questionable because it is based on an a posteriori calculation. To illustrate this point, imagine a series of five throws of a die; the configuration observed had a probability of $(\frac{1}{6})^5 \approx 10^{-4}$ of appearing but nevertheless is seen! The correct approach consists in defining beforehand the hypothesis to be tested, then testing the facts by observing a sufficiently large sample (notice that in our case of a die, to determine whether the results observed are consistent with randomness, we need at least 10^6 series of five throws!). Furthermore it is possible that the effects of gravitational amplification favour the detection of such associations (the closest object playing the role of a lens with respect to the others).

Conversely we shall now summarize the points that, in our opinion, confirm the cosmological interpretation:

- QSOs are situated at the centres of 'normal' galaxies with the same z; in one case it is even thought that a supernova explosion in the underlying galaxy has been detected;
- there is continuity between the QSO population and that of Seyfert 1 galaxies (see Sect. 9.3.3), which themselves obviously follow the Hubble law;

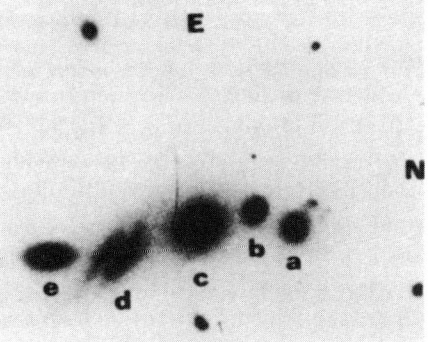

Fig. 9.14. The group of galaxies VV172. Although the systemic velocities of a, c, d, and e are around $16\,000\,\mathrm{km\,s^{-1}}$, $V(\mathrm{b}) = 36\,900\,\mathrm{km\,s^{-1}}$. (From Sulentic and Lorre 1983)

- at small z, where the cosmological effects (both 'geometric' and evolutionary) are negligible, $V/V_{\max}$ tests give $\langle V/V_{\max}\rangle \approx 0.5$;
- absorption-line systems are never found for z_a clearly larger than $z_{\rm QSO}$; in the same way in a pair of quasars Q_1–Q_2 (at redshifts z_1 and z_2, with $z_1 < z_2$) one sometimes finds absorption at $z = z_1$ in the spectrum of Q_2 (owing to the galaxy associated with Q_1) but never the converse.

Argument (d) shows in a very direct way that, at the very least, z is a monotonic function of the distance.

In summary, it seems to us that the cosmological interpretation is now well supported by the available observational facts and clearly represents more than a simple working hypothesis.

9.6 Other Classes of Active Nuclei

So far we have only discussed the properties of quasars. This population already contains a large variety of objects, and several subclasses have been mentioned according to the level of radio emission (radio-loud or radio-quiet QSOs), the amplitude of continuum variations (OVVs), and the existence of broad absorption lines (see Sect. 10.3). We now define the distinctive features of the other main types of active nucleus. In the absence of a comprehensive model, this classification is necessarily empirical: it implicitly gives more weight to those properties that are easily observable – which are not necessarily the most relevant.

9.6.1 Seyfert Galaxies

Seyfert galaxies are spiral galaxies whose nucleus is particularly bright; moreover they emit intense lines, which implies the existence of a nonstellar ionizing continuum. One can distinguish two main types: Seyfert 1 and Seyfert 2 galaxies.

The nuclei of Seyfert 1 galaxies are very similar to quasars. Approximately the same line profiles (broad permitted lines, narrow forbidden lines) and the same spectral distribution of continuum radiation are found. Their radio emission is systematically small. They are distinguished from QSOs by a lower luminosity, comparable to or less than the luminosity of the underlying galaxy. The prototype of this class is NGC 4151.

On the other hand Seyfert 2 nuclei (NGC 1068 being a representative example) do not have broad lines (both the permitted and forbidden lines are narrow). Furthermore the continuum is much steeper ($\alpha \geq 1.5$ from the near infrared to the ultraviolet) and X-ray emission is generally much weaker. There exist, however, a subclass of objects, the NELGs (narrow-emission-line galaxies), whose X-ray emission is comparable to that of Seyfert 1 galaxies;

the allowed lines in fact have a broad weak component, and the strong inclination of these galaxies suggests that they might be very obscured Seyfert 1 galaxies. In a more general way it is possible that the differences between Seyfert 1 and 2 galaxies are due in part to strong absorption (of the broad lines and of the visible and ultraviolet continuum) by dust. One recent observation confirms this: observed in polarized light the nucleus of NGC 1068 has characteristics of Seyfert 1 galaxies. The photons emitted would be scattered by the free electrons encountered at the surface of an opaque torus masking the nucleus.

The nuclei of Seyfert 2 galaxies are more intense radio sources than Seyfert 1 galaxies; their morphology is reminiscent of that of extended double radio sources but on a much smaller scale (of the order of 1 kpc).

9.6.2 BL Lac Objects

The characteristic property of BL Lac objects is the absence of emission lines. Aside from this they are very similar to strongly variable QSOs (OVVs): their radio morphology is that of a compact radio source, the degree of polarization is high ($P > 5\%$), the time scale of variations is very short (a few days), and the UV–visible spectrum is steep ($\alpha > 1$). The shape of the whole spectrum is difficult to observe owing to the very rapid variations (requiring simultaneous observations). On a scale of several years, very large-amplitude variations ($\Delta m = 5$, or a factor of 100) can be observed; when the object is weak, the spectrum is steeper and one then often observes the usual emission lines of QSOs (but they are much weaker). BL Lac objects appear at the centre of elliptical galaxies; this fact added to spectacular variations of the flux, but also of the direction of polarization, suggests that BL Lac objects may be radio galaxies viewed along the direction of ejection and whose emission is altered by relativistic boosting (the amplified continuum masking the lines and time scales being reduced).

9.6.3 Radio Galaxies

Aside from the radio activity (a compact source, jets, and lobes; see Chap. 8), broad or narrow emission lines in the optical and UV domains are sometimes detected. This feature is more common among FR II-type sources (symmetric and very luminous double radio sources). The region producing the narrow lines coincides, in general, with the starting point of the radio jet.

9.6.4 Modelling Active Galactic Nuclei

To summarize: a suitable model should answer four major questions, as follows:

- What is the nature of the central object and the mechanism responsible for the energy emitted?
- How can the wide diversity of active nuclei (the range of radio, optical, and X-ray luminosities observed; the specific character of Seyfert 2 galaxies; BL Lac objects) be explained?
- What is the strong cosmological evolution of the QSOs due to?
- What relation is there between normal galaxies and galaxies with active nuclei?

Needless to say, we are at present far from being able to give a firm answer to these questions. Most of the radiation is emitted by a volume so small that to observe it in detail is out of the question. Theoreticians must then often resort to indirect arguments that are rarely free from ambiguity; for example sizes can be inferred from the time scales of variations, but this estimate can be wrong if the emitting matter is moving rapidly towards the observer (Δt is reduced by relativistic aberration). One can therefore hardly envisage looking for *the* proof of a theory; at best we can try to gather all the arguments in its favour into a coherent picture and extract predictions that can be verified by observation.

Another difficulty resides in the weight to be given to each of the various types of data available. Some properties, although pronounced, may just be related to the subtle details of one process, whereas others, which are more important, may not have been noticed because of observational difficulties. In this respect recognizing the common nature of QSOs and the nuclei of Seyfert 1 galaxies is without doubt a major step because the small redshift of these latter objects allows a much more detailed investigation.

Before coming to the theory that involves the presence of a supermassive black hole, accepted at present by the majority of astronomers, we mention two other scenarios proposed just after the discovery of QSOs. One of them assumes the presence of a very dense cluster of stars: the frequent collisions led to the formation of massive stars, and then supernova explosions. This model has difficulties in accounting for (a) the existence of a preferred direction (defined by the radio jets), and (b) the amplitude of the observed variations, because the energy associated with an outburst is greater than 10^{51} ergs, the energy radiated by a single supernova explosion. An argument in favour of a unique central object is the constancy of the axis of ejection over $\tau \geq 10^7$ years manifested by the linear form of the radio jets. The second scenario assumes that this object is a supermassive star ($M > 10^3 \, M_\odot$), possibly in rapid rotation (a spinar), or supported by the pressure of a strong magnetic field. This type of model remains plausible although it is difficult to establish the stability of such objects and to explain their formation.

On the other hand it seems that the evolution of the central regions of a galaxy naturally leads to the formation of a massive black hole (one reason being the very many collisions between the stars). A black hole cannot, however, be detected in a direct manner. It is only the effects on the surrounding regions that may be detectable. In particular a very massive compact object determines a quasi-Keplerian kinematics for objects (stars or gas clouds) that surround it. This characteristic signature has been recognized in our Galaxy (by observing the 12.8 μm line of Ne II around the galactic centre) and in several other nearby galaxies, such as M81 and M87. In our Galaxy, the analysis of stellar dynamics close to the center (proper motions and radial velocities) leads to a mass estimate of $\approx 3 10^6 M_\odot$ for the central object. In NGC 4258, VLBI observations of OH masers give $\approx 4 10^6 M_\odot$.

One can also estimate the mass required to account for the observed luminosities ($L \approx 10^{44}$ to $10^{48}\,\mathrm{erg\,s^{-1}}$) by introducing the Eddington luminosity. In a steady state and with spherical symmetry the gravitational attraction (acting mainly on protons) is at least equal to the repulsive force (radiation pressure) exerted by the photons on the electrons owing to Thomson scattering (the matter is ionized, but protons and electrons evidently remain 'coupled'). This condition may be written as

$$\frac{GMm_\mathrm{p}}{r^2} \geq \frac{\langle \sigma_\mathrm{T} L \rangle_\nu}{4\pi r^2},$$

where m_p is the mass of the proton, σ_T is the effective Thomson cross-section (the mean $\langle\ \rangle_\nu$ is taken over the whole spectrum), and r is the distance to the centre. In order of magnitude this leads to

$$L < L_\mathrm{E} \approx 10^{38}(M/M_\odot)\,\mathrm{erg\,s^{-1}}.$$

If L were greater than the Eddington luminosity L_E, the radiation of the source would expel the matter of which it is formed (the 'supercritical' case). It is thus necessary to invoke masses of the order of $10^8\,M_\odot$. Recall that the Schwarzschild radius R_S, beyond which no particle can escape from the black hole (the escape velocity being equal to c), is written

$$R_\mathrm{S} = \frac{2GM_\mathrm{BH}}{c^2} \approx 10^{13}(M_\mathrm{BH}/M_\odot)\,\mathrm{pc} \approx 10^{-5}\,\mathrm{pc}$$

for a black-hole mass $M_\mathrm{BH} = 10^8\,M_\odot$. The most realistic mechanism by which energy can be produced is the transformation of gravitational potential energy into radiation. For example by bringing a mass m from infinity to $5R_\mathrm{S}$, we get $mc^2/10$, which is a significant fraction of the total energy; under these conditions, the accretion of $1\,M_\odot\,\mathrm{year}^{-1}$ allows the production of $6 \times 10^{45}\,\mathrm{erg\,s^{-1}}$; it is thus a very efficient process. It remains to be clarified by what process and with what efficiency ρ this energy can be transformed into radiation. Many scenarios have been proposed, in particular those assuming the existence of an accretion disc. Thin discs (with a thickness much less than the radius of the inner edge) can be studied in detail, but an essential parameter which determines ρ remains unknown: the viscosity. Its actual value is probably set by turbulent or magnetic processes. It is likely that the accretion structure has the geometry of a torus (which has a higher stability than a thin disc), which moreover allows us to explain the collimation of matter and radiation being ejected (on the scale of the black hole, the inner edge of the torus defines a quasicylindrical channel). The accretion can also occur isotropically (a radial flow with spherical symmetry); this mechanism leads to at best $\rho \approx 0.1$ when the turbulent or magnetic dissipation is strong.

It is not possible at present to choose between these various possibilities; so far as the black hole itself is concerned, we cannot say whether it is static (a Schwarzschild black hole) or rotating (a Kerr black hole). We are then led to study the various objects involved in the models (stability, the physical processes taking place, the observable properties, and so on); Figure 9.15 represents the main building blocks of the 'standard model'. It is still a very crude picture because the various subsystems presumably interact strongly

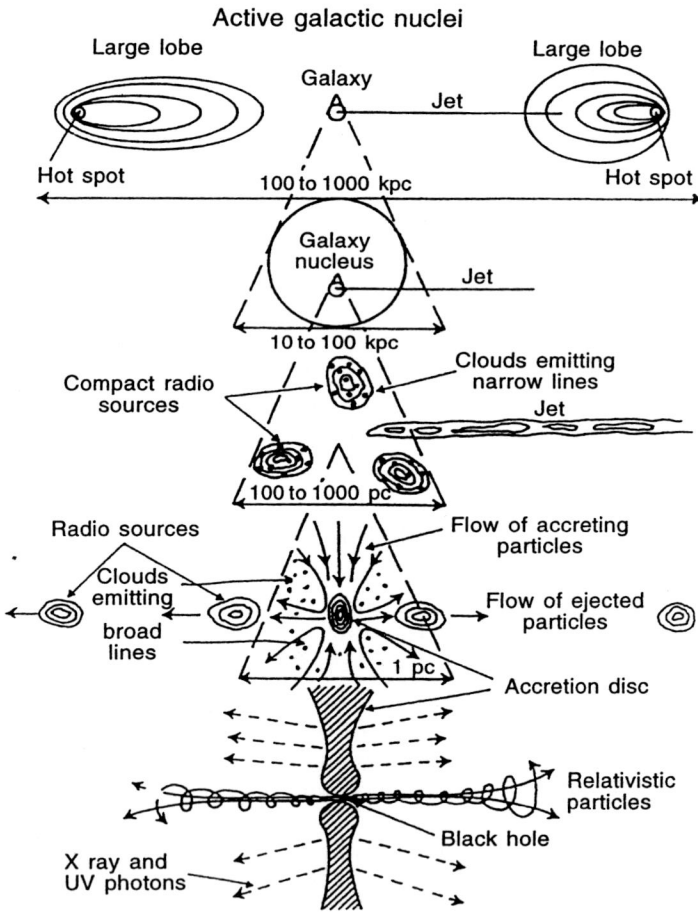

Fig. 9.15. The standard galactic-nuclei model viewed at various scales. (From Collin 1987)

with each other (implying exchanges of matter between the disc and the BLR, for example). We expect, however, that observations will allow the identification of each of the elements by some clear signature: these could then be included in a more global model taking account of the mutual interactions.

Exercises

9.1 For a QSO with a magnitude $m_B = 20$ ($\lambda_B \approx 4400$ Å) and a spectral index $\alpha = 1$ assumed constant from the visible to radio frequencies, calculate in Jy the value of F_ν at 100 µm and at 6 cm ($m_B = 0$ corresponds to $F = 6.6 \times 10^{-12}$ W cm^{-2} µm^{-1} and 1 Jy $= 10^{-26}$ W m^{-2} Hz^{-1}). Discuss these results knowing that we can detect several times 0.1 Jy at 100 µm (with the *IRAS* satellite) and a few mJy at 6 cm.

9.2 What will the time scale be for the variability associated with the circular rotation of matter situated at a distance $D = 5R_S$ (R_S is the Schwarzschild radius) of a giant black hole of mass $M = 10^8 \, M_\odot$?

9.3 Show that the deflection produced by a spherically symmetric mass distribution on a light ray characterized by an impact parameter r is determined only by the mass contained in a cylinder of radius r (proceed as in classical mechanics by attributing to photons a fictitious mass $m = E/c^2$, which, to within a factor of 2, gives the correct result). Derive the curve shown in Fig. 9.9 for a sphere of constant density with a radius $R = 10$ kpc and a mass $M = 10^{10} \, M_\odot$.

References

Baldwin, J. E. (1977) in *Active Galactic Nuclei*, edited by C. Hazard and S. Mitton (Cambridge University Press, Cambridge), p. 51.
Blandford, R. D., Netzer, H., and Woltjer, L. (1990) *Active Galactic Nuclei* (Saas-Fee Advanced Course 20, Springer, Berlin, Heidelberg).
Collin, S. (1987) in *L'Activité dans les Galaxies*, edited by G. Stasinska (Les Editions de Physique, Les Ullis), p. 3.
Green, R. F. (1985) in *Quasars*, edited by G. Swarup and V. K. Kapahi (Reidel, Dordrecht), p. 429.
Greenfield, P. E., Roberts, D. H., and Burke, B. F.(1985) *Astrophys. J.* **293**, 370.
Hazard, C., and Mitton, S. (1979) *Active Galactic Nuclei* (Cambridge University Press, Cambridge).
Hoag, A. A., and Smith, M. G. (1977) *Astrophys. J.* **217**, 362.
Hutchings, J. B., Crampton, D., Campbell, B., Duncan, D., and Glendenning, B. (1984) *Astrophys. J. Suppl.* **55**, 319.
Malkan, M. A. (1983) *Astrophys. J.* **268**, 582.
Osterbrock, D. E. and Miller, J. S. (editors) (1989) *Active Galactic Nuclei* (IAU Symposium 134, Kluwer, Dordrecht).

Peacock J.A. (1999) *Cosmological Physics* (Cambridge University Press, Cambridge).
Soucail, G., Mellier, Y., Fort, B., Mathez, G., and Cailloux, M. (1988) *Astron. Astrophys.* **191**, L19.
Stockton, A. (1980) *Astrophys. J.* **242**, L141.
Sullentic, J. W., and Lorre, J. J. (1983) *Astron. Astrophys.* **120**, 36.
Swarup, G, and Kapahi, V. K. (editors) (1986) *Quasars* (IAU Symposium 119, Reidel, Dordrecht).
Ulrich, M. H. et al. (1984) *Mon. Not. Roy. Ast. Soc.* **206**, 221.
Ulrich, M. H., Maraschi, L., and Urry, C. M. (1997) *Annu. Rev. Astron. Astrophys.* **35**, 445.
Vanderriest, C. et al. (1989) *Astron. Astrophys.* **215**, 1.
Warren, S. J., Hewett, P. C., Osmer, P. S., and Irwin, M. J. (1987) *Nature* **330**, 453.
Weedman, D. (1986) *Quasar Astronomy* (Cambridge University Press, Cambridge).
Worall, D. M., Boldt, E. A., Holt, S. S, and Serlemitsos, P. J. (1980) *Astrophys. J.* **240**, 421.

10. Quasar Absorption-Line Systems

Three years after the discovery of quasars (1963) the presence of narrow absorption lines in the spectrum of one of them, 3C 191 (redshift $z_e = 1.955$) was observed for the first time. It is remarkable that 9 clearly significant lines associated with the same redshift $z_a = 1.9470$ could be unambiguously identified. The spectrum of another quasar, 1159+123, obtained at the 10m Keck telescope is presented in Fig. 10.1 and illustrates how much information is contained in such an exposure.

The majority of the observed absorption lines are narrow (we discuss the so-called broad lines later) and in general remain unresolved at a spectral resolution $\Delta\lambda \geq 1\,\text{Å}$. This characteristic facilitates their identification and the determination of the associated redshift.

From a historical point of view it is undoubtedly significant that the first absorption-line system was discovered at a redshift close to that of the quasar

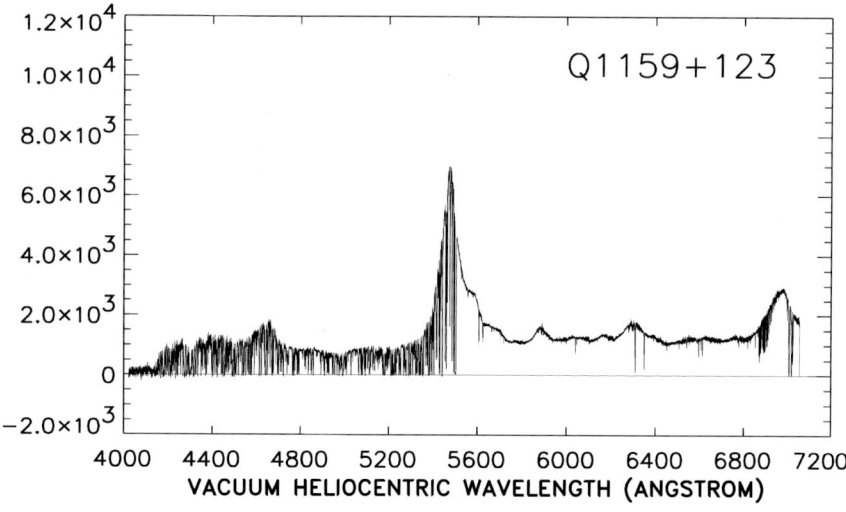

Fig. 10.1. The spectrum of the quasar 1159+123, obtained at the Keck telescope. The most prominent emission line near 5500 Å is Lyα at the emission redshift $z_e = 3.502$. Shortward of it, the Lyα forest can be seen while the few absorption lines longward of 5500 Å are from metals. (From Songaila 1998)

itself (3C 191) even though it quickly turned out that this is not the case in general. This discovery focused attention on those models involving matter associated with the quasar (possibly ejected by it, since a small difference between the emission and absorption redshifts, z_e and z_a, can easily be attributed to relative motion of the gas). However, subsequent observations gradually supported the idea that the majority of absorption line systems are due to intervening objects lying along the line of sight. Spectroscopic studies of quasars thus turned out to be a very powerful means of probing diffuse gas lying at the outskirts of galaxies or in the intergalactic medium.

10.1 General Remarks

10.1.1 The Information Contained in Line Profiles

As most of the data concerning absorption systems is derived from the analysis of spectra, we recall here some basic related definitions and properties.

The profile of an absorption line centred on λ_0 can be written as

$$I(\lambda) = I_0 e^{-\tau(\lambda)},$$

where $\tau(\lambda)$ is the opacity (τ_0 denotes the opacity at the line centre) and I_0 is the intensity of the adjacent continuum. Several parameters are involved in $\tau(\lambda)$: the oscillator strength f and the natural width characterize the transition and are in principle given by laboratory measurements or possibly theoretical calculations. On the other hand the amount of absorbing matter, or more precisely the *column density*, $N = \int n\,dl$ (n being the volume density), of the species considered, together with its velocity distribution characterizes the intervening medium: this is what we attempt to determine from the data. In most cases, the natural width does not matter, and the width of the line (as seen with a very high-resolution spectrometer), $\Delta\lambda_L$ (estimated at half maximum, for example), depends on both N and the velocity distribution. If a single component is present, the velocity distribution is often represented by a Gaussian distribution:

$$N'(v)\,dv = \frac{N}{\sqrt{2\pi}\,\sigma_v} e^{-(v-v_0)^2/2\sigma_v^2}\,dv,$$

where v_0 is the mean velocity and σ_v the standard deviation.

The profile is then a Voigt profile. In practice, observations give the convolution of the instrumental profile (full width at half maximum: $\Delta\lambda_I$) and the intrinsic profile of the line. If both are assumed to be Gaussians, the resulting profile is itself Gaussian and has a width

$$\Delta\lambda_{\mathrm{obs}} = \sqrt{\Delta\lambda_L^2 + \Delta\lambda_I^2}.$$

There are therefore two cases to be considered:

- $\Delta\lambda_\mathrm{obs}$ significantly exceeds $\Delta\lambda_\mathrm{I}$: the line is resolved. The spectrum thus contains information about the intrinsic profile that can be extracted by fitting the observed profile to a theoretical profile convolved with the instrumental response (whence N and σ_v).
- To within errors, $\Delta\lambda_\mathrm{obs} = \Delta\lambda_\mathrm{I}$. The only meaningful value provided by the data, *independent of the instrumental resolution*, is the equivalent width defined by

$$W = \int_\mathrm{line} \frac{I_0 - I(\lambda)}{I_0}\,\mathrm{d}\lambda.$$

W represents the area of the line if the adjacent continuum is normalized to 1. To determine σ_v and N, measurement of W for at least one other line of the same species and involving the same lower energy level is required. The principle of the method is as follows: it can be shown that $W/\lambda_0\sigma_v$ is a 'universal' function of τ_0 (given by the 'curve of growth'). The expression for the opacity τ_0 is

$$\tau_0 = 1.06 \times 10^{-15} \frac{N\,[\mathrm{cm}^{-2}]\,f\lambda_0\,[\mathrm{\AA}]}{\sigma_v\,[\mathrm{km\,s}^{-1}]}.$$

- In a $\log(\lambda_0 f)$–$\log(W/\lambda_0)$ diagram we can fit the data to the curve of growth and this gives σ_v and N. If only two lines are available (the so-called doublet method), the result is uncertain because it is impossible to check that only one velocity component is present. W is a strongly nonlinear function of N and σ_v; a component with a very small velocity dispersion but containing much material will contribute very little to W and can remain completely unnoticed (in such a case N will be considerably underestimated).

A particularly simple case should be considered: that of an optically thin line ($\tau \ll 1$). In this limit W becomes independent of σ_v and N is written

$$N\,[\mathrm{cm}^{-2}] = 1.13 \times 10^{21}\frac{W\,[\mathrm{\AA}]}{\lambda_0^2\,[\mathrm{\AA}]\,f}.$$

Another important situation in which W no longer depends on σ_v is when N is very large. In this limit, due to natural broadening, the opacity is significant far away from the line centre and the line profile is 'damped'. W (and the whole line profile) is determined solely by N and increases as $\sqrt{N}$. Lyman α lines are often observed in this regime in QSO spectra, and if resolved, the full profile can be used to get an accurate estimate of N(HI).

For a redshifted line the equivalent width, as well as the central wavelength, is multiplied by $1+z$:

$$\lambda_\mathrm{obs} = (1+z)\lambda_\mathrm{r}$$

and

$$W_\mathrm{obs} = (1+z)W_\mathrm{r},$$

where λ_r and W_r characterize the line in the rest frame of the absorbing material. The analysis outlined above which gives σ_v and N evidently requires λ_r and W_r to be used instead of λ_obs and W_obs.

10.1.2 The Identification of Absorption-Line Systems

In the laboratory the wavelength at which a line appears allows the transition concerned to be recognized without ambiguity. This is no longer the case for quasar spectra owing to the large spectral shifts observed. Nevertheless, as these shifts introduce a multiplicative factor, the ratio of two wavelengths $\lambda_{\mathrm{r}1}$ and $\lambda_{\mathrm{r}2}$ is preserved, and it is this property that allows lines to be identified. In practice, the doublets of C IV (1548 Å and 1551 Å) and of Mg II (2796 Å and 2803 Å) played an important role: these lines are among the strongest observed and the small wavelength difference between the two lines guarantees that both fall in the observed range. It is then easy to recognize them and measure the redshift. Next, other lines can be searched for at the same redshift: together they form a set, the *system* of lines, which has its origin in a single physical object.

We give in Table 10.1 the lines most commonly detected.

Table 10.1. Major Absorption Lines

Species	λ_0 [Å]	Species	λ_0 [Å]
H	1215.7	Al III	1862.8
N V	1238.8	Fe II	2344.2
N V	1242.8	Fe II	2374.5
Si II	1260.4	Fe II	2382.8
Si II	1304.4	Fe II	2586.6
O I	1304.9	Fe II	2600.2
C II	1334.5	Mg II	2796.3
Si IV	1393.8	Mg II	2803.5
Si IV	1402.8	Mg I	2853.0
Si II	1526.7	Ca II	3934.8
C IV	1548.2	Ca II	3969.6
C IV	1550.8	Na I	5891.6
Al II	1670.8	Na I	5897.6
Al III	1854.7		

This list is determined by:

- the existence of strong transitions in the domains considered (note the richness of the UV domain, where resonance lines of many species appear);
- the relative abundance of elements in the universe;
- the ionization state of these elements.

Occasionally, identification may be impossible, for instance when a single line from a given system appears in the spectral range observed. By improving the signal-to-noise ratio or extending the spectral interval it is in general possible to detect other lines and determine the absorption redshift.

10.1.3 The Empirical Classification of Absorption-Line Systems

Three classification criteria are generally considered:

- the width of the lines;
- the difference between the emission-line redshift z_e of the quasar and the redshift of the system of absorption lines z_a;
- the presence of lines from metals (in practise, elements other than H).

The systems were originally divided into four categories:

- *Type A:* very broad lines always appearing close to and on the blue side of the corresponding emission lines. These systems are very probably associated with gas ejected by the quasar. The ejection velocity can therefore be evaluated using
$$V = c\frac{R^2 - 1}{R^2 + 1}$$
with
$$R = \frac{1+z_e}{1+z_a}.$$
Values reaching $V = 65\,000\,\mathrm{km\,s^{-1}}$ are observed, the width of the lines corresponding to $\Delta V \approx$ several $10^4\,\mathrm{km\,s^{-1}}$.
- *Types B and C:* narrow-line systems containing metals, either at about the quasar redshift (type B) with $(z_e - z_a)/(1 + z_e) < 0.01$ (relative velocity less than $3000\,\mathrm{km\,s^{-1}}$) or unrelated to the quasar (type C).
- *Type D:* 'Lyman α' systems in which no metal lines are apparent (other weaker lines of the Lyman series may possibly be detected). These systems appear in large numbers on the blue side of the quasar's Lyman α emission line and constitute the 'Lyman α forest'. High S/N observations performed with 10m class telescopes revealed later that metals are in fact present and that the apparent absence of metal lines associated to low column density Lyman α absorbers was due to an observational limitation.

Figure 10.2 shows several systems of these various types observed in the spectrum of the quasar 1246−057.

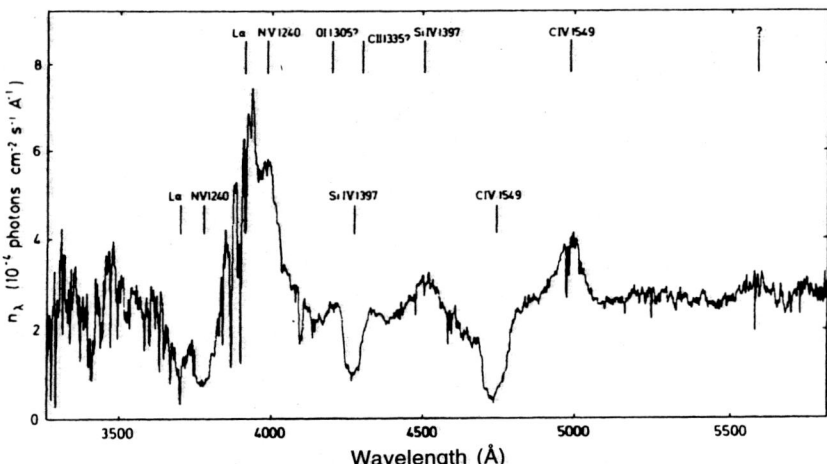

Fig. 10.2. The spectrum of the quasar 1246−057. One can see emission lines ($z_e = 2.22$), a broad-line system at $z_a = 2.05$, several narrow metal lines longward of the Lyα emission, and the Lyman α forest at $\lambda < 3900$ Å. (From Boksenberg et al. 1978)

10.2 Narrow-Metal-Line Systems

Narrow-metal-line systems are of types B and C, as defined in Sect. 10.1.3. We shall mainly discuss the properties of systems with $z_a \ll z_e$ (C) – by far the most numerous – before moving on to the features of group B. The information available about the lines is of two types. By analysing the profiles we can determine properties of the medium such as the velocity dispersion, the column densities of the various ions, the degree of ionization, and so on. It is also possible to simply count the systems and derive constraints on the number, size, clustering, and hence nature of the absorbing objects. We now come to this point.

10.2.1 The Redshift Distribution of the C IV and Mg II Systems and Its Implications

Most of the quasar spectra have been obtained in the blue owing to observational constraints (more effective detectors, less severe contamination by lines from the background sky). The C IV doublet enables us to explore the universe in the interval $1.1 < z < 2.5$ for observations between 3300 and 5500 Å, while that of Mg II only enables us to reach $0.18 < z < 0.96$. These data have to be considered separately for two reasons. Firstly they involve distinct elements and a priori there is no simple correspondence between the lines of Mg II and C IV. Secondly the redshift domains involved are different and evolutionary effects (on the degree of ionization, the relative abundance of the elements, and so on) may be important.

In a Friedmann universe with a zero cosmological constant (see Chap. 13) the number density of the systems, dN/dz, is given by

$$\frac{dN}{dz} = \frac{c}{H_0} n_0 \sigma_0 \frac{1+z}{\sqrt{1+2q_0 z}}, \qquad (10.1)$$

where n_0 is the volume density of the absorbing objects and σ_0 is their effective cross-section projected onto the sky at the current epoch ($z = 0$).

Before this relation is compared with the data a sufficiently large sample must be collected. As only a few C IV doublets are detected in one spectrum, the observation of a large number of quasars is required. After obtaining a well-defined sample, it must be ensured that the signal-to-noise (S/N) ratio is sufficient for all doublets satisfying a criterion such that $W_r(1548\,\text{Å} \text{ and } 1550\,\text{Å}) > W_{\min}$ can be detected. Finally quasars for which information is already available must be excluded in order to avoid biases towards too large a number of systems. Only a few surveys satisfying all these constraints have been carried out.

Analysis of the results obtained shows that the redshift distribution of the systems is compatible with that expected from objects distributed at random in space. It can be shown, for example, that the number of lines of sight for which 0, 1, 2, ... systems are detected obeys a Poisson distribution. This indicates that the majority of systems are not associated with quasars (other arguments point in the same direction: when $(z_e - z_a)/(1 + z_e) \ll 1$, the ejection hypothesis would imply a velocity V_{ej} close to c incompatible with the velocity dispersion – several $100\,\text{km}\,\text{s}^{-1}$ at most – inferred from the line width).

On the other hand, by studying the two-point correlation function of the C IV doublets, a significant excess is noted for splittings corresponding to $\Delta V_r \le 600\,\text{km}\,\text{s}^{-1}$ in the reference frame of the absorbing material. This is most likely the signature of galaxies. Following this idea it is possible to interpret the absolute value of dN/dz: $dN/dz \approx 2$ at $z = 2$ for the C IV doublets satisfying $W_r(1548\,\text{Å}) > 0.3\,\text{Å}$ and $W_r(1550\,\text{Å}) > 0.3\,\text{Å}$. Indeed, the number of systems is directly related to the product $n\sigma$ (cf. (10.1)), where n(galaxies) is known approximately. To obtain a realistic estimate, it is necessary to take account of the large variety of luminosities and sizes of galaxies. In this analysis the radius of the C IV halo is assumed to vary with the luminosity L as the size of the visible image measured by the Holmberg radius R_H, for which $R_H \propto L^{5/12}$. Next (10.1) is integrated over the luminosity function. We thus end up with an effective cross-section of the C IV clouds such that $R_{z=0}(\text{C IV}) \approx 4 R_H$ for lines such as $W_r > 0.3\,\text{Å}$ ($R_0(\text{C IV})$ is even larger if weaker systems are considered). For 'standard' galaxies this implies that $R_0(\text{C IV}) \ge 40(100/H_0)\,\text{kpc}$; these dimensions, although large, are comparable to the size of galaxies. We thus obtain an additional argument in favour of galactic haloes.

If the data allowed us to obtain a good fit to (10.1), we might be tempted to constrain q_0. Unfortunately, evolutionary effects may have a significant impact on the observed variation of dN/dz. Indeed, if recent C IV data are fitted to a $(1+z)^\gamma$ law (by introducing an evolutionary factor $(1+z)^\alpha$ into (10.1) with $\gamma = 1 + \alpha$ for $q_0 = 0$), we end up with $\gamma \approx -1$ (between $z \approx 1.3$ and 3.4), which implies a strong negative evolution ($\alpha \approx -2$). This might be due to a slight deficiency of carbon in galaxies with $z \geq 3$, but other factors, such as the ionization state, are also subject to evolution and could play a role.

Mg II doublets are on average rarer and a similar statistical analysis can be performed. At $z \approx 1$, dN/dz is only 0.5, which, interpreted as indicated above, corresponds to a size $R_0(\text{Mg II}) \approx 2R_\text{H}$. Similarly to the C IV doublets, the redshifts appear to be distributed at random. The variation of dN/dz (between $z \approx 0.2$ and $z \approx 2$) seems compatible with an absence of evolution.

We shall see in what follows that, owing to their relatively small redshift, Mg II systems enable us to directly pick out the galaxies responsible for the metal lines and therefore confirm the above statistical arguments.

10.2.2 Physical Properties of the Absorbing Gas

We are interested here in parameters that characterize the absorbing clouds, such as the velocity dispersion, temperature, density, size, and so on. Their determination, based on a detailed study of the line profiles, allows a comparison with the properties of the gas observed in our Galaxy or in nearby galaxies (QSO–galaxy pairs).

The Velocity Dispersion. In general at high resolution ($\approx 10\text{–}20\,\text{km}\,\text{s}^{-1}$ in the rest frame of the absorbing gas) the lines split into multiple components. Figure 10.3 shows a portion of the spectrum of PKS 0215+015 (a BL Lac object of redshift z_e=1.72) containing the C IV doublet from a system at $z_\text{a} = 1.549$; for each of the two lines (1548.2 and 1550.8 Å) 6 components are clearly present. The velocity dispersion inside individual clouds can be measured, since the corresponding components are partially resolved; they are of the order of $7\,\text{km}\,\text{s}^{-1}$.

For other systems values as small as $3.5\,\text{km}\,\text{s}^{-1}$ are measured. If the velocity distribution is dominated by thermal motions of the C IV ions, this implies a temperature of 1.8×10^4 K ($m\sigma_v^2 = kT$, where σ_v^2 is the variance along the line of sight). This is a significant result because a plasma in an equilibrium state under the effect of collisions still contains very little carbon in the form of C IV at this temperature (C IV/$\text{C}_\text{total} \ll 10^{-3}$). As the column densities measured for these clouds indicate that a significant fraction of the carbon is in the form of C IV, these ions are produced not by collisions but by photoionization. The temperature mentioned above is then just an upper limit. Another characteristic of these multiple systems is the large range of velocities covered by the various components, $300\,\text{km}\,\text{s}^{-1}$ in

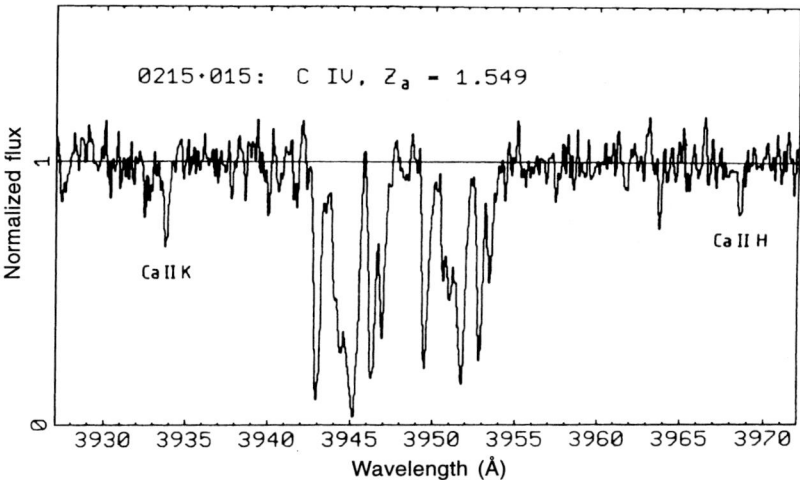

Fig. 10.3. A high-resolution spectrum ($20\,\mathrm{km\,s^{-1}}$) of 0215+015 (a BL Lac object, $z_e = 1.72$). Six components are detected for each line of the C IV doublet at $z_a = 1.549$. Furthermore the Ca II doublet from gas in our own Galaxy can be seen. (From Pettini et al. 1983)

the case of Fig. 10.3 ($\Delta\lambda \approx 4\,\text{Å}$). Total dispersions in excess of $500\,\mathrm{km\,s^{-1}}$ involving more than 10 components have been observed. Such ranges could correspond to the velocity dispersion of galaxies in a cluster; however, it is very unlikely that a line of sight crossing the central regions of a cluster as rich as the nearby Perseus and Coma clusters simultaneously intersects such a large number of galaxies. Systems of the above type are relatively common (the spectrum of 0215+015 additionally contains a second system with 9 components at $z_a = 1.649$) and their high degree of multiplicity is still to be understood.

The Ionization State. By comparing the equivalent widths of the various lines – in particular from the doublets of Mg II and C IV – it is possible to estimate the ionization degree of the absorbing gas. Beyond $z = 1.2$, where these two species are both observable in the visible, three types of systems can be distinguished. Those with a high degree of ionization show the C IV doublet but not that of Mg II. The two doublets are detected in systems with an intermediate degree of ionization. Finally the existence of a third category characterized by a small degree of ionization (Mg II is present, C IV is not) has also been found. This observed sequence in ionization states of the gas seems to be associated with a monotonic variation of the amount of material along the line of sight. Indeed, the first systems do not show the discontinuity at the Lyman limit ($\lambda = (1+z) \times 912\,\text{Å}$), which indicates a column density $N(\mathrm{H\,I})$ less than $2 \times 10^{17}\,\mathrm{cm^{-2}}$. On the other hand this discontinuity is observed for systems of mixed ionization levels.

It is interesting to compare the characteristics mentioned above with the properties of local gas at high galactic latitude. The *IUE* satellite has enabled us to obtain UV spectra of bright stars belonging to the halo of our own Galaxy or to the Magellanic clouds and thus probe the diffuse gas surrounding our own Galaxy. The results obtained correspond to the properties of systems with low or intermediate degrees of ionization. It is then natural to believe that the latter are due to the presence of an intervening galaxy. The origin of systems with a high degree of ionization is less clear; the scenario could be the same but with the line of sight crossing the most peripheral region of the halo, which is strongly ionized by the UV background radiation (due essentially to all quasars). Note that the significance of such a comparison with our own Galaxy may be limited: at $z > 1$ cosmological evolution may be strong, in particular because of the much greater background ionizing radiation.

The Size of the Haloes. The size of the absorbing haloes can be determined in a more direct way than stated in Sect. 10.2.1 when two adjacent lines of sight with a small angular separation can be investigated. Several groups (mainly pairs) of quasars have been listed and can be used to this end if they are bright enough. For pairs with a small separation ($\leq 1'$) the spectrum of the most distant QSOs often shows absorption at the redshift of the other quasar owing to the halo of the host galaxy. Other systems that possibly appear are not in general common to the two spectra when the linear separation of the two lines of sight exceeds 100 kpc. Nevertheless several remarkable exceptions exist. Thus, in the pair 0307−195A,B ($\Delta\theta = 58''$ on the sky between A and B), a common system is detected at $z = 2.034$, although the separation is $760(H_0/50)^{-1}$ kpc at this redshift; the difference $\Delta z_a = z_a(A) - z_a(B)$ measured is low and corresponds to $\Delta V \approx 300 \text{ km s}^{-1}$. The lines can be due either to a single giant halo or to two distinct galaxies belonging to the same cluster. Other such cases have been found which strongly suggest that studies of correlated absorptions along several adjacent lines of sight can provide useful information on the large scale distribution of the gas.

The multiple images of a quasar (due to gravitational lensing) can be used in the same way to obtain a lower bound on the size of the haloes. In this case the angular separations are much smaller ($\Delta\theta$ is of the order of a few arcseconds), and the systems are generally common. Occasionally line-intensity differences between the two lines of sight reveal the existence of inhomogeneities or the presence of substructures in the absorbing gas. The interpretation of the results concerning gravitationally lensed quasars is not straightforward because the correspondence between the transverse dimension at z_a and $\Delta\theta$ is not known precisely (the position and characteristics of the deflector are also involved). The sizes inferred using these methods do not contradict those quoted previously. The discovery of pairs with $\Delta\theta \approx$ several $10''$ would allow more precise constraints to be obtained.

To conclude this section let us mention that some information may be obtained concerning other important parameters such as the fraction of heavy elements or density. The determination of metal abundances is not an easy task. First, the lines are frequently saturated (that is, on the logarithmic part of the curve of growth), which renders the calculation of column densities N very uncertain if the resolution is not sufficient to separate the various components possibly present. The situation becomes more favourable when N is sufficiently large for the opacity in the wings of the lines to be large compared with unity; these are the damped lines quoted previously. Strong Lyα lines of atomic hydrogen belong to this category and for $N(\mathrm{H\,I}) > 2 \times 10^{20}\,\mathrm{cm}^{-2}$ we have

$$N(\mathrm{H\,I}) = 10^{20} \left(W_\mathrm{r}\,[\text{Å}]/7.3\right)^2\,\mathrm{cm}^{-2}.$$

Unfortunately for most of the elements, the spectra give access only to some ionization states (H I but not H II for H; C II and C IV but not C I or C III for C; and so on). The ionization of the absorbing gas must then be modelled to estimate the contribution of unobserved ionization states. The metal abundances obtained are generally lower than solar abundances (the ratio being about 0.1 to 1; abundances higher than solar are occasionnally observed for systems with $z_\mathrm{a} \approx z_\mathrm{e}$).

Constraints on the electron density n_e are given by fine-structure lines. Thus, C II has two transitions at 1345.5 Å and 1335.7 Å, the first originating from the ground state, the second from a level 8 meV above the ground state. When n_e is sufficiently large, the upper level is populated by collisions with electrons, and the line at 1335.7 Å can be observed. In the majority of cases where C IIλ1334 is present the second line at 1335.7 Å is not detected, which implies an upper limit on n_e, $n_\mathrm{e} \leq 1\,\mathrm{cm}^{-3}$.

10.2.3 Damped Lyman α Systems

Prior to systematic surveys of these systems several cases where $N(\mathrm{H\,I}) > 5 \times 10^{20}\,\mathrm{cm}^{-2}$ were discovered by chance. Such column densities roughly correspond to those expected from an intervening galactic disc. In this case, the large amount of matter present along the line of sight should allow the detection of other components usually associated with discs (e.g. molecules, dust grains) and the study of these up to early epochs; hence the special interest of these systems.

Detection: Lyman α and 21 cm Absorption. We have already mentioned that the presence of a discontinuity at the Lyman limit ($\lambda_\mathrm{r} = 912$ Å) indicates a column density $N(\mathrm{H\,I})$ greater than $2 \times 10^{17}\,\mathrm{cm}^{-2}$. Two methods can be used to detect systems for which $N(\mathrm{H\,I})$ exceeds $10^{20}\,\mathrm{cm}^{-2}$: the search for damped Lyman α lines and, for radio-loud quasars, the search for absorption from atomic hydrogen at $\lambda_\mathrm{r} = 21$ cm. The first method has been employed with success; the large H I column density results in a large equivalent width and a broad profile easily recognizable in a low-resolution spectrum (≈ 10 Å). Candidate damped Lyman α lines are next reobserved at a higher resolution (a few Å) to discard those due to a chance coincidence of several narrow Lyman α lines. Systems thus detected at $z \approx 2$ systematically show the lines of low-ionization species (O I, C II). These systems are five times more numerous than expected from known galactic discs. For nearby galaxies ($z \approx 0$) the radius beyond which $N(\text{H I})$ becomes less than 2×10^{20} cm^{-2} is $R(\text{H I}) \approx 1.5 R_\text{H}$, where R_H is the Holmberg radius (the size of the optical image, as defined for a limiting brightness of 26.5 mag arcseconds^{-2}); at $z \approx 2$ the size of the discs would therefore be larger, with $R(\text{H I}) \approx 3.5 R_\text{H}$, in the assumption of a constant comoving galaxy number density.

Searches in the radio range for absorption from the H I line at 1420 MHz ($\lambda_\text{r} = 21$ cm) is very difficult. Indeed the band covered by radio receivers is quite narrow ($\Delta \nu / \nu \ll$ several times 10^{-2}): one observation then only allows the exploration of a very small redshift interval. Nevertheless, the first detection was obtained in 1973 at $z_\text{a} = 0.692$ in the spectrum of the quasar 3C 286 ($z_\text{e} = 0.849$) without prior knowledge about the presence of optical absorption at that redshift (optical lines were detected afterwards). About ten '21 cm absorptions' have been found at cosmological redshifts. Figure 10.4 shows the complex feature observed at $z = 0.524$ in the spectrum of the BL Lac object 0235+164. These observations reveal preferentially 'cold' gas; indeed the opacity τ integrated over the line is proportional to $N(\text{H I})$ and inversely proportional to the spin temperature of the gas T_s. By combining the measurements of $\int \tau \, d\nu$ and $N(\text{H I})$ (derived from the equivalent width

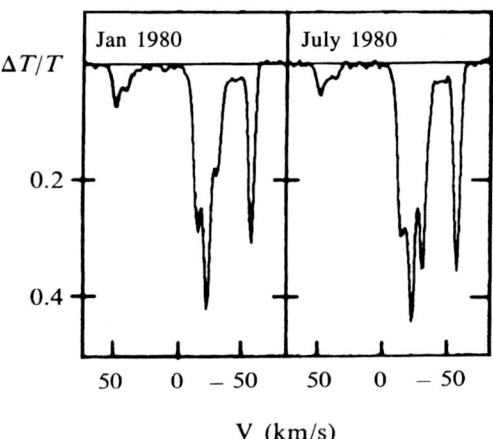

Fig. 10.4. The radio spectrum of the source 0235+164 around the frequency $\nu = 932$ MHz ($z_\text{a} = 0.524$). Five components spanning 100 km s^{-1} are present. The profile is variable in time (this is due to changes in the morphology of the continuum source). (From Wolfe et al. 1982)

of the Lyα line) we can estimate T_s to be less than 10^3 K in the few cases for which the appropriate data are available.

The velocity structure of the gas can also be investigated, since a resolution of the order of 1 km s^{-1} is easily obtained in the radio range. In general, several components separated by about 10 km s^{-1} are observed, each with an internal dispersion of the order of 1 km s^{-1}. Taken together, these results are in good agreement with the assumption of an intervening galactic disc.

Note that in the radio range, measurements of Faraday rotation are also possible; unfortunately their interpretation is not unambiguous since part of the effect may occur in the source itself (see Chap. 8). Nevertheless, a correlation is observed between the presence of optical lines and the Faraday rotation measured indicating that a significant fraction of the Faraday effect is related to intervening material. The results obtained are consistent with a size $l \approx 50$ kpc, a magnetic field $B \approx 2\,\mu$G, and an electron density $n_e \approx 10^{-3}$cm^{-3}.

The Search for Associated Molecules: H_2 and CO. The most abundant molecule expected, H_2, can in principle be detected through its UV absorption lines. In practice all these transitions occur at $\lambda_r < 1215$ Å and are therefore expected in the Lyman α forest where the large number of lines makes all identifications difficult and uncertain. Up to now, convincing results have been obtained in a few cases only; the first system discovered is that at $z_a = 2.812$ in PKS 0528−250 ($z_e = 2.77$), which has $N(\text{H\,\sc i}) \approx 1.9\times 10^{21}$ cm^{-2}. The method used to avoid difficulties related to the crowding of lines consists in calculating the cross-correlation of the observed spectrum with a theoretical absorption spectrum of H_2 at a redshift z; a well-defined peak appears at $z = z_a = 2.812$.

For the CO molecule (not detected by its UV lines in the above system) attempts have been made to observe absorption from the rotation lines $J = 1 \to 0$ (115 GHz), $J = 2 \to 1$ (230 GHz), used as tracers of molecular gas in our Galaxy. The required conditions (satisfied by very few QSOs) are the existence of a substantial continuum emission (several 0.1 Jy) at about 100 GHz and a value of z_a leading to a frequency where the atmosphere is transparent. Only upper limits have been obtained for those systems detected in the optical range. However, four molecular absorption-line systems have been discovered in the mm domain and redshift range 0.25–0.89. Two of them arise from the galaxy hosting the continuum source itself, while two others are due to intervening galaxies that act as gravitational lenses on the background source. Many transitions from several species have been detected, as illustrated in Fig. 10.5.

It is certainly significant that none of these four systems could have been studied in the optical, the background source being too faint owing to the extinction implied by the large amount of foreground material. Thus, the absence of H_2 or CO in damped Lyman α systems cannot be taken as an

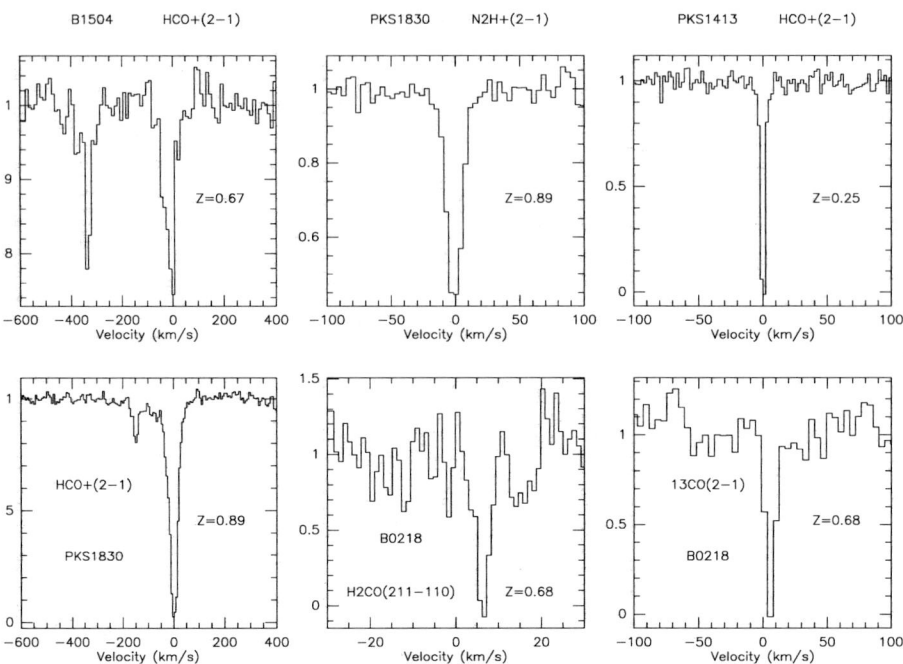

Fig. 10.5. Synoptic view of four detected molecular absorption systems, in various representative molecules. Notice that the line widths are quite different, from 1 to 100 km s^{-1}. The signal is normalized to the detected continuum (From Combes and Wiklind 1996)

indication that the ISM in distant galaxies is less rich in molecules compared to nearby galaxies.

Dust. For systems with $N(\text{H{\sc i}}) \geq 10^{21}$ cm^{-2}, significant associated extinction is expected if the intervening material is of galactic type. Indeed, when we observe at λ_{obs} a QSO with an absorption system at redshift z_a, the intervening matter induces an extinction $A[\lambda_{\text{obs}}/(1+z_a)] > A(\lambda_{\text{obs}})$, since the extinction increases from the visible to the ultraviolet. Three different signatures are expected. The clearest one is the absorption at 2200 Å. This very broad feature (from 2000 to 2500 Å) is present everywhere in our Galaxy, but its origin remains poorly understood. On the other hand, this signature is absent in the extinction curve of the Small Magellanic Cloud and it may be associated with a particular type of dust grains. The two other deviations expected involve the overall shape of the spectrum; assuming that intrinsic emission satisfies a power law $F_\nu = k\nu^{-\alpha}$, we find that selective dust extinction causes to the first order an increase in α (difficult to confirm since the intrinsic spectral index α is unknown) and to the second order a curvature of the spectrum. The existence of a reddening excess for QSOs having at least one system with $N(\text{H{\sc i}}) \geq 2 \times 10^{20}$ cm^{-2} has been established statistically.

The presence of dust associated with absorption systems has important implications for large-scale opacity in the universe. Even if the dust to gas ratio is smaller than Galactic, extinction can significantly modify the luminosity function measured for QSOs at large z (and may in part explain their apparent rarity at $z > 3.5$). As noted above, intervening dust can also bias studies of the highest column density systems.

10.2.4 Direct Searches for Absorbing Galaxies at Low z

When $z \leq 1$, it is possible to detect galaxies responsible for Mg II systems. Indeed for $H_0 = 50\,\mathrm{km\,s^{-1}\,Mpc^{-1}}$ and $q_0 = 0$, a projected linear separation l at a redshift z corresponds to an angle

$$\Delta\theta\,['']= 3.4\times 10^{-2}\frac{(1+z)^2}{[z+(z^2/2)]}l\;[\mathrm{kpc}].$$

$\Delta\theta$ is about $5''$ at $z=0.5$ for $l=40\,\mathrm{kpc}$. In good seeing conditions (better than $1.5''$) we thus expect in general to separate the image of the QSO from that of the absorber. The image of an ordinary galaxy (with an optical diameter of 20 kpc) should also appear extended. Finally the expected magnitude is about 22, which is easily detectable with CCD imaging. The search for absorbing galaxies is conducted in two steps: (a) detection by broad-band imaging (R or V) of an extended object in the neighbourhood of a QSO with a low-redshift absorption system; (b) spectroscopy at low resolution ($\approx 15\,\mathrm{\AA}$) of this object to verify that its redshift coincides with the absorption redshift. The success of this second test strongly relies on the presence of well contrasted spectral signatures (for example [O II]$\lambda 3727$ in emission or the doublet at 3934, 3963 Å from Ca II in absorption). Figure 10.6 shows the results obtained for the system at $z_\mathrm{a} = 0.4299$ for PKS 2128−12. A successful identification has now been obtained for several tens of cases; for a large fraction of them the galaxy shows strong emission lines and is thus of spiral type. Furthermore these objects appear relatively blue (F_λ is constant below the Balmer discontinuity at $\lambda_\mathrm{r} = 3626\,\mathrm{\AA}$), which suggests strong stellar activity.

These results directly show that the Mg II systems at small z are due to galaxies and thus confirm the intervening-halo hypothesis (it remains, however, to determine what fraction of galaxies are surrounded by a halo). Observations of this type are somewhat similar to studies concerning 'quasar–galaxy' pairs, chance associations on the sky of a nearby galaxy and a QSO with a much larger redshift. About fifteen cases of this type have been listed. The first pair for which a positive result has been obtained comprises the spiral galaxy (Sb) NGC 3067 ($V = cz = 1500\,\mathrm{km\,s^{-1}}$) and at $1.9'$ from its centre (which corresponds to $17(50/H_0)$ kpc in projected distance or $64(50/H_0)$ kpc in the plane of the disc inclined at $75°$) the radio-loud quasar 3C 232 ($z = 0.534$). 21 cm absorption was first found in the H I emission spectrum and then the 3934, 3969 Å doublet of Ca II was detected in the visible. A small fraction of observed pairs shows absorption of Ca II or Na I. In

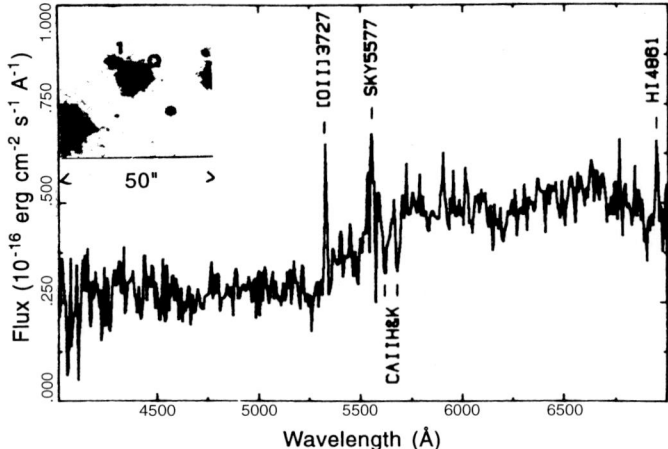

Fig. 10.6. The spectrum (resolution: 14 Å) of the galaxy (indicated by the numeral 1 at the top left) responsible for the system at $z_a = 0.4299$ of the quasar 2128−12 (indicated by Q). The projected distance 1–Q is $64(50/H_0)$ kpc ($8.6''$ on the sky). (From Bergeron 1986)

total the extent of Ca II – of the order of $25(50/H_0)$ kpc – appears to be less than that of Mg II, which is consistent with the ionization structure of these elements. UV observations of the Mg II and C IV doublets from the galaxy in the spectrum of the nearby quasar would be important for investigating the halo properties at $z \approx 0$.

10.2.5 Narrow-Line Systems at $z_a \approx z_e$

When the redshifts of the QSO and the absorption-line system become close to each other, the origin of the absorption is more difficult to establish because new possibilities arise: absorption due to the interstellar medium of the host galaxy or to gas ejected by the quasar itself. To decide whether these systems are of the same nature as those at $z_a < z_e$ it is possible to study the distribution of the parameter $\beta = V_r/c$, where V_r is the relative velocity of the gas with respect to the QSO (see Sect. 10.1.3). Absorption by ejected gas should result in an excess of line systems for some typical value of β (for example $V \approx$ several 10^3 km s^{-1}). In fact the situation is not simple because an excess at $\beta \approx 0$ does not necessarily imply an intrinsic origin: if, as is currently believed, nuclear activity is favoured by tidal interactions, the probability of finding one additional galaxy along the line of sight is increased. A statistical study has been made on two different samples and leads to opposite conclusions. The first sample comprises few radio-loud quasars with a steep spectrum ($\alpha > 0.7$) and has no excess at $z_a \approx z_e$, unlike the second, which contains many radio sources (an excess of systems appears with

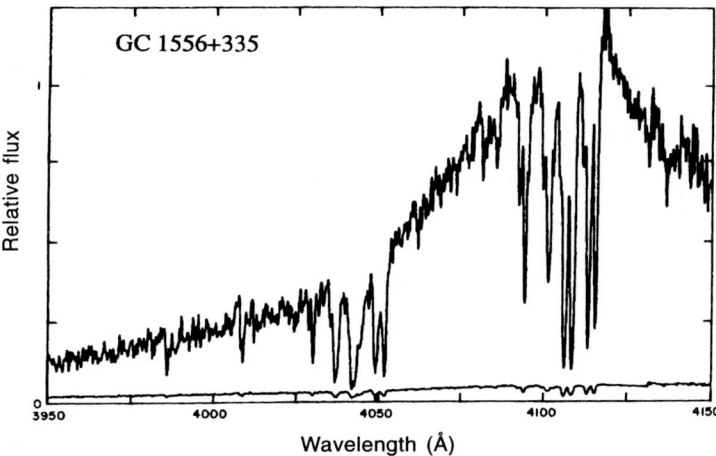

Fig. 10.7. The blue spectrum of the quasar GC 1556+335 ($R = 0.5\,\text{Å}$). On the C IV emission line two groups of C IV absorption doublets can be seen at $z_a = 1.61$ and 1.64. (From Morris et al. 1986)

$V_\text{r} < 5000\,\text{km\,s}^{-1}$). These results suggest that the presence of 'associated' absorption is directly linked to radio properties; this is not surprising because detailed observations of radio sources clearly reveal the existence of ejected matter (jets, superluminal motion, and so on). Figure 10.7 shows such an example of associated absorption in the spectrum of GC 1556+335. In general these systems are characterized by a higher degree of ionization than usual (with N V lines present).

At small redshifts several attempts have been made to identify the possible absorbing galaxy at $z_a \approx z_e$, as described above (Sect. 10.2.4). Unfortunately the positive results obtained in several cases are not completely convincing because the quasar is often surrounded by a cluster of galaxies at its redshift. This increases the chance of finding a galaxy close to the QSO but unrelated to the absorber (this galaxy can, for instance, lie behind the QSO).

A possible continuity between these systems and those with broad lines discussed below has been suggested. In many cases, such as that of GC 1556+336, an accumulation of narrow lines is observed shortward of the emission line corresponding to the same transition. An additional indication favouring an intrinsic origin for some of these systems (matter ejected by the QSO as in the case of the broad lines) comes from the remarkable coincidences observed between lines of systems having slightly different redshifts (Fig. 10.8). Thus for some doublets (λ_1, λ_2) (from C IV, Si IV, or Mg II), the second line (λ_2) of the first system (z_1) coincides with the first line (λ_1) of the second system (z_2); that is,

$$(1 + z_1)\lambda_2 = (1 + z_2)\lambda_1.$$

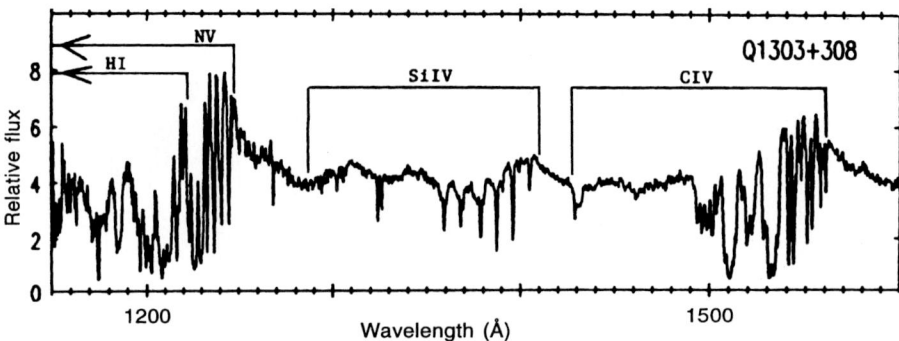

Fig. 10.8. The spectrum of the broad-absorption-line quasar 1303+308. The absorption shows much structure. Remarkable coincidences are observed between various lines of the Si IV doublets (the phenomenon of 'line locking'). (From Turnshek 1988)

With a good spectral resolution ($\approx 1\,\text{Å}$) the probability of seeing such a configuration by chance becomes small, which establishes the reality of the phenomenon. These results imply the existence of well-defined values for the relative velocities of the clouds. The origin of this quantization would reside in the fact that the dynamics of the gas is governed by the radiation of the QSO (via radiation pressure; one thus speaks of *line locking*). The few systems for which this phenomenon has been observed correspond to ejection velocities reaching $0.3c$.

10.3 Broad Absorption Line Systems

In contrast to the preceding systems, the systems we shall now consider are characterized by a very broad absorption profile (Fig. 10.2). This induces a sometimes very significant distortion in the emission lines: for example the absorption of N V modifies the emission line of the same ion at $\lambda_r \approx 1240\,\text{Å}$ and the Lyman α line. A complex profile results that varies greatly from one object to another. The occurrence of these lines at redshifts always close to z_e, as well as the large velocity dispersion of the gas, clearly indicates that this phenomenon is directly associated with the quasar itself. We shall first describe the main properties of broad absorption lines and of the QSOs in whose spectra they are detected before briefly discussing models.

10.3.1 The Characteristics of QSOs with Broad Absorption Lines

The absorption has a 'P Cygni-type' profile (from the name of a star in the constellation Cygnus having similar-looking lines) in the shape of a trough located on the blue side of the corresponding emission line; sometimes it overlaps the latter. In other cases it appears detached from 5000 to $10\,000\,\text{km}\,\text{s}^{-1}$.

The terminal velocity in general lies between 10×10^3 and 30×10^3 km s^{-1} but can reach 65×10^3 km s^{-1}. We observe mainly the lines of C IV, Si IV, N V, and O VI, and more rarely H (Lyα) or Mg II. The profiles are generally asymmetric and have an abrupt edge on the red side. The presence of narrow features varies greatly from one object to another (Figs. 10.2 and 10.8). Small time variations of the profiles have been detected on the scale of a few years.

The analysis of the column densities shows that $N(\text{H})$ is of the order of 10^{19} to 10^{21} cm^{-2}. By comparison with narrow-line systems the degree of ionization is very high (strong N V and O VI lines are systematically present). The abundance of metals seems to be 10 to 100 times the solar abundance, which is difficult to understand because the regions responsible for the emission lines have almost solar abundances.

One of the fundamental questions concerning the broad absorption lines is the following: is the underlying phenomenon present in all QSOs but observed in only some because the absorbing region covers only a fraction of all directions as seen from the quasar, or by contrast is it specific to a particular class of objects? In the first hypothesis the fraction of space covered by the gas with broad lines is directly given by the fraction of QSOs displaying broad lines. This latter quantity is estimated to be $f \approx 3$ to 10% for $z \leq 2.5$; at high redshifts it might exceed 10% (in fact one QSO with the largest emission redshift known ($z_e = 4.47$) has broad absorption lines).

Almost all QSOs with broad absorption lines are not intense radio sources. On average their optical flux (magnitudes B, V, and R, for example) is more variable and more polarized than that of the other, radio-quiet QSOs. The presence of broad absorption lines seems to be correlated with the intensity of the emission lines, QSOs with broad absorption lines having a weaker C IV emission line and a stronger N V emission line than other quasars. There is also a relation between the contrast ($I_{\text{line}}/I_{\text{cont}}$), the width, and the profile of the absorption lines: PHL 5200, one of the prototypes of this category, has clearly defined emission lines, and at the same time the absorption profile is very smooth; on the other hand for 1303+308 (Fig. 10.8) the emission lines are poorly defined and the absorption shows many narrow features.

10.3.2 Modelling the Broad-Line Systems

It is difficult to imagine a model that accounts for all the observed properties. Initially, the idea was proposed that the gas of the interstellar medium of the host galaxy is expelled by radiation from the active nucleus located at its centre. Such a scenario naturally explains the value of f observed, the QSOs with broad absorption lines corresponding to the case where the host galaxy is seen edge-on. The observation of this latter phenomenon for several objects with broad absorption lines at small z shows an almost circular image (thus a disc seen face-on), which is incompatible with this hypothesis.

For several quasars the Lyα emission line completely disappears: this shows that the absorbing gas must cover the region that produces this line.

The region where the absorption is formed might thus be located between the BLR and the NLR (at $r > 0.1\,\mathrm{pc}$ from the central nucleus). Models attempting to reproduce the observed profile suggest a global disc-type structure for the absorbing gas and an inhomogeneous and decelerated radial flow.

10.4 Lyman α Systems

On the blue side of the Lyα emission line of all quasars a large number of absorption lines appear: this is the 'Lyman α forest' (Figs. 10.1 and 10.2). Their presence in this part of the spectrum, associated with the fact that hydrogen is by far the most abundant element in the universe, suggested very early on that they were Lyman α lines. A direct confirmation of this hypothesis is given by the detection of the Ly β, Ly γ transitions associated with the strongest Lyα lines. We are clearly dealing here with systems in the usual sense of the term, even if, in the majority of cases, only the Lyα line is observed. These systems may correspond to the signature of a new type of object (intergalactic clouds?), which has motivated many observations and much theoretical work in the past decade. Before giving some details of the physical conditions in 'Lyman α clouds', we briefly describe the statistical information available about such absorptions.

10.4.1 The Redshift Distribution

We have already mentioned in the context of C IV and Mg II systems the interest in measuring $\mathrm{d}N/\mathrm{d}z$ and its variation with z (Sect. 10.2.1). Lyman α systems are very well suited to this type of study because of their large number: the observation of a limited number of QSOs allows us to easily obtain significant results. To count only Lyα systems it is necessary first to exclude (a) Lyα lines having associated metal lines and (b) the metal lines already identified (systems having other lines beyond the Lyman α forest). Spectra with a resolution of 1 to a few Å allow lines to be clearly separated and listed. For a decade several contradictory results were obtained for the exponent γ, which describes the variation of $\mathrm{d}N/\mathrm{d}z$ ($\mathrm{d}N/\mathrm{d}z \propto (1+z)^\gamma$). Recently it has been possible to show that these discrepancies originated in a systematic reduction of the number of lines when z_a becomes close to z_e. The usual relation is therefore inadequate and can be replaced by

$$\frac{\mathrm{d}N}{\mathrm{d}z} = \left(\frac{\mathrm{d}N}{\mathrm{d}z}\right)_0 (1+z_0)^\gamma \left(\frac{1+z_\mathrm{a}}{1+z_\mathrm{e}}\right)^{-\delta},$$

with $\gamma \approx 2.3$ and $\delta \approx 2.1$ (in the range $1.5 < z < 3.8$). We then have positive evolution (recall that for $q_0 = 0$ we expect, without evolution, $\mathrm{d}N/\mathrm{d}z \propto (1+z)$, that is, $\gamma = 1$) but simultaneously, along each line of sight, fewer lines at $z \approx z_\mathrm{e}$. The most probable origin of this 'inverse effect' is a larger degree of ionization of the material, owing to the additional UV radiation provided by the QSO. Indeed when the analysis is restricted to lines formed beyond the sphere of influence of the QSO, the inverse effect is no longer detectable (the same argument explains why the narrow-metal-line systems at $z_a \approx z_e$ have lines from N V; see Sect. 10.2.5). In the future it will be possible to test this interpretation directly by examining the spectrum of QSOs with large z but smaller luminosities (the inverse effect should then be less pronounced). It is noteworthy that detailed modelling of the inverse effect leads to an estimate of the UV background (due to the cumulated emission of all the QSOs and galaxies), a quantity of primary importance in studies of the intergalactic gas.

Another approach has also been used to investigate the evolution of Lyα systems; it relies on the fact that at low resolution the existence of the Lyman α forest results in a lower apparent level of the continuum. The depression is measurable by comparison with the continuum estimated redward of the Lyα emission line and in a first approximation it is proportional to dN/dz (Fig. 10.9). This method is less rigorous because it does not allow us to remove the contribution from metal-line systems; conversely it is easier to implement, and data from the *IUE* satellite have thus enabled us to extend the domain explored down to $z = 1$. Recently, *HST* observations have been able to address in a more direct way this important question of Lyman α forest evolution.

In the UV spectra of low z QSOs, there appear to be far more Lyman α forest lines than would be expected from an extrapolation of the evolution

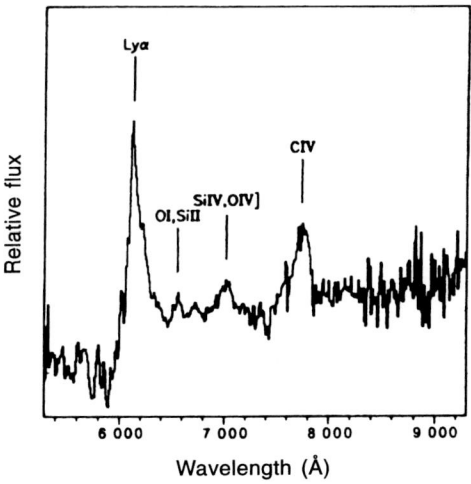

Fig. 10.9. The spectrum of a QSO with a high redshift (0046−293, $z_e = 4.01$). The presence of the Lyman α forest at $\lambda < 6000$ Å causes, at low resolution (20 Å here), a strong depression of the continuum. (From Warren 1987)

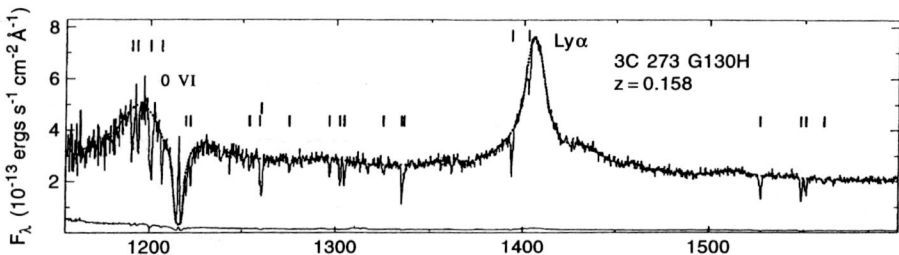

Fig. 10.10. The spectrum of 3C 273. Several weak Lyman α lines are detected in the range 1215–1400 Å. (From Bahcall et al. 1993)

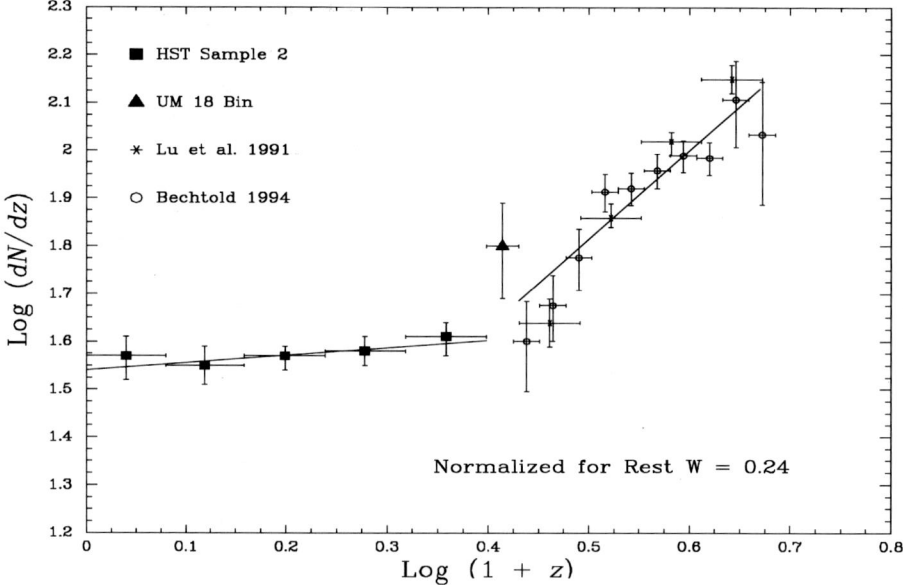

Fig. 10.11. Variation of the number density of Lyman α lines with redshift ($z < 1.5$: *HST* data; $z > 1.5$: ground-based data). (From Weymann et al. 1998)

law determined at high redshift. This is illustrated in Figs. 10.10 and 10.11; in the latter, a break clearly appears in the distribution of the number density at $z \approx 1.6$.

Initial studies of the two-point correlation function of Lyman α systems showed a small excess of pairs for separations $\Delta V \approx 200$–$300 \,\mathrm{km\,s^{-1}}$. This is a notable difference with respect to C IV systems (see Sect. 10.2.1) and an argument in favour of an intergalactic origin. However, more recent investigations have shown that the strength of the clustering increases with H I column density, indicating some continuity between the properties of the metal absorbers and those of Lyman α systems.

10.4.2 Column Densities and the Velocity Dispersion

The distribution of H I column densities is an important characteristic of the population of absorbing objects because any feature may give some information about their nature. Unfortunately for the majority of lines the determination of $N(\text{H I})$ is tricky because we are not dealing with one of the simple cases where the velocity dispersion plays no role: optically thin lines ($N \propto W$) or very saturated lines ($N \propto W^2$). The large number of lines in the forest introduces additional difficulties: the position of the continuum is poorly defined, and furthermore the partial overlapping of 2 or 3 lines is frequent. With a resolution of the order of $20 \,\text{km}\,\text{s}^{-1}$ (giving $\Delta\lambda = 0.4\,\text{Å}$ to $\lambda = 6000\,\text{Å}$) most of the lines are resolved; we can thus obtain $N(\text{H I})$ and σ_v by fitting to each observed line a Voigt profile convoluted to the instrumental profile. When Lyβ is unambiguously detected, a more accurate estimate is obtained by making a simultaneous fit. Further, with the improvements in the quality (S/N and resolution) of the spectra, the above difficulties has been largely overcome. The resulting distribution for $N(\text{H I})$ is close to a power law, $N(\text{H I})^{-\beta}$, with $\beta \approx 1.5$ (a slight deficiency with respect to this law is present at $N(\text{H I}) \approx 10^{15}$ cm^{-2}). It is remarkable that this is in good agreement with the distribution obtained for the Lyα lines of metal systems. This can be understood as evidence for a link between the two populations of objects responsible for these systems. The velocity dispersion inside clouds is on average $\sigma_v \approx 20\,\text{km}\,\text{s}^{-1}$. Moreover note that the equivalent-width distribution (more or less exponential with a mean value $\langle W \rangle = 0.30\,\text{Å}$) does not strongly depend on z, at least at high z. We can deduce that the column density $N(\text{H I})$ associated with an individual cloud should evolve as $(1+z)^6$ and then decrease rapidly with time (this is true only if the same types of absorber are present in the z range over which the evolution is determined).

10.4.3 The Heavy-Element Abundance

By definition, Lymanα systems have no associated metal lines. However, since $N(\text{H I})$ is clearly smaller ($N(\text{H I}) \approx 10^{12}$ to 10^{16} cm^{-2}) than for metal-rich absorbers, we cannot conclude a priori that the relative abundance of heavy elements is really smaller in the Lymanα clouds. This question is crucial with respect to their origin: the absence of metals would imply that the gas has never been enriched by stellar nucleosynthesis of C, N, and O and that it therefore has a primordial origin.

One of the methods originally used to place a limit on the abundance of metals takes advantage of the large number of Lymanα systems; it consists in adjusting the spectrum to the reference frame of the absorber for each Lymanα line of the spectrum (just by dividing the wavelength by $1 + z_\text{a}$). We then obtain a large number of spectra (all drawn from the same one) which can be added 'coherently': all the Lyα lines and in particular the O VI, C IV, . . . lines are thus 'stacked', which may render their detection easier.

A positive result has been obtained in this way for C IV. Individual lines allowing a strong constraint on the metallicity to be imposed have also been looked for in individual systems, for example Si IIIλ1206. This line is observed in many cases. Furthermore, sensitive observations of associated C IV lines, performed with 10 m-class telescopes have shown that most systems with $N(\text{H\,I}) > 10^{15}$ cm^{-2} have a typical metal abundance which is about 10^{-2} that of the Sun. We then see that the frontier between Lyman α and metal systems is not as clear-cut as was initially thought and indeed, that some continuity in metal abundance is present.

10.4.4 Other Properties. Possible Models

The velocity dispersion mentioned above can be related to a temperature. A typical value, $b_v = \sqrt{2}\sigma_v \approx 30\,\text{km}\,\text{s}^{-1}$, thus corresponds to $T \approx 5 \times 10^4$ K. This is in reality only an upper limit on the real temperature of the medium because part of the observed dispersion is undoubtedly due to macroscopic motion (turbulence in the clouds).

Valuable information about the structure or size of the absorbers is given by observation of gravitationally lensed quasars and pairs of quasars. In the former case, at $z \approx 2$, the two images formed by gravitational deflection appear to have in common in their spectra some Lyman α lines from which sizes of the order of 200 kpc can be derived. Pairs of quasars studied at lower redshift indicate the presence of common absorption lines for separations exceeding 1 Mpc.

Various models have been proposed to explain Lyman α absorptions. If the discrete lines observed are assigned to discrete clouds occupying a small fraction of the total volume, then a key question is to identify what mechanism is responsible for the confinement of these 'Lyman α clouds'. It was initially suggested that the pressure of a hot low-density phase could explain the observed properties; however, such a phase has not been found, and furthermore, this model cannot reproduce the broad H I column density distribution. Shocks induced in the intergalactic medium by violent episodes of star formation in galaxies could have generated shells of compressed gas, but in such a scenario the mutual correlation between Lyman α lines should be similar to that of galaxies, whereas it appears to be much fainter. In the framework developed during recent years to account for structure formation, Lyman α 'clouds' are confined by the gravity of cold-dark-matter minihaloes. The kind of moderate (dark-matter) overdensities required to generate gas concentrations with characteristics similar to those inferred from Lyman α lines are naturally predicted by hydrodynamical simulations of structure formation in the universe, and more specifically, the statistics of these haloes is fully consistent with the observed column density distribution. Moreover, the time evolution of dN/dz is also easily understood as a result of both the Hubble expansion and the gas flow in the potential wells of the dark matter.

Mass estimates indicate that at $z > 2$, the intergalactic medium thus modelled contained most of the baryons in the universe. This fact, together with the success of the above picture in explaining the Lyman α forest data, suggests that the latter contain information of great cosmological importance. Indeed, the kind of theory needed to make predictions concerning Lyman α absorptions is relatively simple, whereas the link between galaxies and the dark-matter distribution is much more complex; this is because galaxies are associated with large overdensities, well beyond the linear regime and in which modelling of the star formation activity is extremely difficult and thus highly uncertain.

Exercises

10.1 Establish the relation giving the ejection velocity of absorbing matter detected at a redshift z_a when expelled by a quasar at a redshift z_e (see Sect. 10.1.3). What form does this relation take if $z_a \approx z_e$?

10.2 An observation made at a resolution of 0.5 Å (at half maximum) has provided a measurement of the width at half maximum of an H I line equal to 0.9 Å. What constraint can be inferred on the temperature of the gas? Establish as a preliminary the relation between the velocity dispersion *along the line of sight* and the temperature. Answer the same question for a C IV line. Draw conclusions.

10.3 The number of quasars of magnitude m_B less than m is (per square degree on the sky)
$$\log N = 0.88(m - 18.3).$$
Derive, for an observer situated at a latitude ϕ of $20°$, the number of QSOs for which it is possible to obtain a spectrum with a good resolution (a few Å) and a good signal-to-noise ratio, that is, $m_B < 17.5$ (assume that no observations are made below $30°$ above the horizon). If these were distributed at random on the sky, how many apparent pairs of quasars separated by less than $1'$ would you expect? Draw some conclusions.

10.4 Owing to the expansion of the universe, the redshift of an object varies in time according to the relation
$$\frac{dz}{dt} = H_0 \left[1 + z - \sqrt{1 + 2(q_0 + 1)z + (4q_0 + 1)z^2 + 2q_0 z^3}\right].$$

Discuss the possibility of observing this effect for an absorption-line system and of deriving the deceleration parameter q_0. Estimate the required spectral resolution, the optimal class of line system for such a study, the minimum interval Δt to detect the effect, and so on. (From Davies and May 1978.)

References

Bahcall, J. N., et al. (1993) *Astrophys. J. Suppl.* **87**, 1.
Bergeron, J. (1986) *Astron. Astrophys.* **155**, L8.
Bergeron, J. and Boissé, P. (1991) *Astron. Astrophys.* **243**, 344.
Bergeron, J., Kunth, D., Rocca-Volmerange, B., and Tran Thanh Van, J. (editors) (1987) *High Redshift and Primeval Galaxies. Proceedings of the 3rd IAP Meeting* (Editions Frontiéres, Paris).
Blades, C., Turnshek, D., and Norman, C. A. (editors) (1987) *Absorption Lines: Probing the Universe* (Cambridge University Press, Cambridge).
Boksenberg, A., Carswell, R. F., Smith, M. G., and Whelan, J. A. J. (1978) *Mon. Not. Roy. Astron. Soc.* **184**, 773.
Combes F., Wiklind T. (1996) in *Cold Gas at High Redshift*, edited by M.N. Bremer, P.P. van der Werf, H.J.A. Röttgering and C.L. Carilli (Kluwer Academic Publishers). p.215
Davies, M. M., and May, L. S. (1978) *Astrophys. J.* **219**, 1.
Morris, S. L., et al. (1986) *Astrophys. J.* **310**, 40.
Petitjean, P. (1988) in *Formation and Evolution of Galaxies*, Proceedings of the Les Houches School, edited by O. Le Fevre and S. Charlot (Springer).
Pettini, M., and Blades, J. C. (1983) *Astrophys. J.* **273**, 436.
Rauch, M. (1998) *Annu. Rev. Astron. Astrophys.* **36**, 267.
Songaila A. (1998) *Astron. J.* **115**, 2184.
Turnshek, D. (1988) in *Absorption Lines: Probing the Universe*, edited by C. Blades, D. Turnshek, and C. A. Norman (Cambridge University Press, Cambridge).
Warren, S. J. (1987) *Nature* **325**, 131.
Weymann, R. J., Carswell, R. F., and Smith, M. G. (1981) *Annu. Rev. Astron. Astrophys.* **19**, 41.
Weymann, R. J., et al. (1998) *Astrophys. J.* **506**, 1.
Wiklind, T., and Combes, F. (1997) in *Highly Redshifted Radio Lines*, edited by C.L. Carilli, S.J.E. Radford, K.M. Menten and G.I. Langston) ASP Conf. Series, Vol. 156, p. 202.
Wolfe, A. M., Davis, M. M., and Briggs, F. H. (1982) *Astrophys. J.* **259**, 495.

11. The Universe on a Large Scale

To know the distribution of matter in the universe has been one of astronomers' constant quests. It is pursued by studying more and more distant galaxies. The idea of a remote universe had been held since the middle of the eighteenth century. In the nineteenth century Herschel and Dreyer catalogued the many bright nebulae and noted how they were grouped. Recognition of the extragalactic realm of nature, mainly due to Shapley and Hubble in the years 1920–1930 (by the identification of Cepheids in Andromeda), like the discovery of the universal recession of galaxies by Hubble, opened the way to the development of extragalactic astronomy and to what would become observational cosmology. On extragalactic scales the galaxies, owing to the contrast in density that they represent (the mean density of our Galaxy within 10 kpc is approximately 2×10^{-24} g cm^{-3}, while the mean density of the universe lies (probably) between 10^{-29} and 10^{-31} g cm^{-3}), are the most immediately discernible entities. Until recently they were thought to provide a reliable indicator of matter on a large scale, and their study led to the idea of a globally homogeneous and isotropic universe.

11.1 Structure and Homogeneity

In furthering this research many astronomers tackled the enormous task of compiling catalogues that give both the positions and magnitudes of galaxies brighter than some limiting magnitude and counts of galaxies as a function of their magnitude. The Shapley–Ames catalogue (1932), for example, gives the coordinates and magnitudes of 1250 galaxies brighter than the 13th magnitude over the whole sky and was extended by de Vaucouleurs and de Vaucouleurs (*Reference Catalogue of Bright Galaxies*). Zwicky and his collaborators (1961–1968) likewise made a catalogue of around 5000 galaxies in the northern hemisphere brighter than the 15th magnitude from photographic plates of the Palomar Sky Survey, which corresponds to a distance or 'depth' of about 130 Mpc, or $13\,000\,h$ km s^{-1}.[1] Finally the Lick catalogue, extended by Shane and Wirtanen (1967), contains around 10^6 galaxies up

---

[1] The factor $h = H_0/100$ km s^{-1} Mpc^{-1} expresses the uncertainty of a factor of 2 in the value of the Hubble constant (see below).

to a distance of approximately 220 Mpc. For a catalogue limited to apparent magnitude m_0 the depth D is given by

$$D^* = 10^{[0.2(m_0 - M^*) - 5]} \text{ Mpc},$$

where M^* is the absolute magnitude of the Schechter luminosity function (see Sect. 1.4): this is the largest distance at which sample galaxies having a typical absolute magnitude M^* can lie.

Shane and Wirtanen's catalogue has been one of the most extensive and most used galaxy catalogues: it covers about 70% of the sky and gives the number of galaxies in cells of $10' \times 10'$ up to a limiting magnitude of 18.8. Since the 1980s automated measuring machines such as COSMOS and APM have led to the development of modern catalogues containing more than 2×10^6 galaxies with reliable photometry down to $B = 20.5$.

The study of such a sample reveals that on large distance scales ($\approx$ 100 Mpc) the density of galaxies is independent (to within a factor of two) of the direction of observation. Likewise the study of the distribution of quasars and the most distant radio sources shows that the distribution is homogeneous and isotropic. Nevertheless the most recent work on catalogues that are arranged according to the distance of galaxies as well as their position on the sky (with a 'depth' up to $60\,000 \text{ km s}^{-1}$) shows that these properties are now observed indicating that the 'size of homogeneity' has been reached. Finally Hubble's law seems to be independent of the direction of observation.

Similar indications come from the study of the diffuse electromagnetic background (Fig. 11.1). From the point of view of cosmology the most important parts of the spectrum are the millimetre and X-ray domains since they are the least contaminated by emission from our Galaxy. The main characteristic is the observation of a diffuse millimetre background whose properties cannot be attributed to the superposition of discrete sources and are well described by those of a black body at 2.7 K.

The degree of isotropy (except for a dipole moment due to the proper motion of our Galaxy) is about 5×10^{-6}, as recently determined by the COBE satellite. This isotropy is also observed in the diffuse X-ray background, whose probable origin is the integrated emission from distant objects such as the quasars (whose contribution is estimated at between 40 and 80%) superposed on that of a hot intergalactic gas.

This hypothesis of isotropy and homogeneity, which is verified observationally, constitutes the cosmological principle, on which the standard model of the universe (Chap. 13) rests.

On a smaller scale the distribution of galaxies is far from uniform. They tend to arrange themselves in pairs, groups, clusters, superclusters, filaments, and so on. Herschel had already observed (at least in projection) in certain regions of the sky true 'clusters' of nebulae. These clusters of galaxies are not simple statistical fluctuations of a uniform background. Zwicky, for example, studied the distribution of galaxies in N cells on the sky containing a total

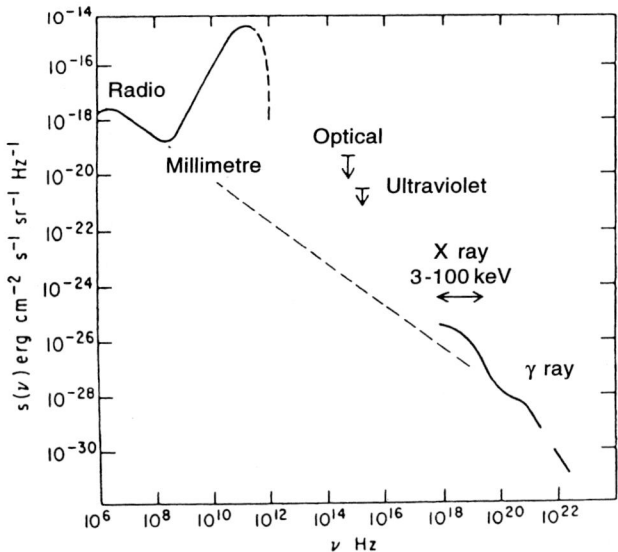

Fig. 11.1. The diffuse electromagnetic background as a function of frequency. Observations are indicated for the different domains: radio, millimetre, X ray, and γ ray. Upper limits are given for the optical and UV domains. (From Boldt 1987)

of $N_{\rm tot}$ galaxies. He calculated the observed standard deviation $\sigma_{\rm obs}$ (which measures the departure from the mean):

$$\sigma_{\rm obs} = \left(\frac{1}{N}\sum_{i=1}^{N}(n_i - \bar{n})^2\right)^{1/2},$$

where $\bar{n} = N_{\rm tot}/N$, and compared this with that corresponding to a uniformly random distribution:

$$\sigma_{\rm calc} = \bar{n}^{1/2}$$

(for large N and $N_{\rm tot}$).

This calculation, made in three fields (Coma, Pegasus, and Corona Borealis), contains respectively 36 000, 9000, and 76 000 galaxies, leading to values of the ratio $\sigma_{\rm obs}/\sigma_{\rm calc} > 15$ and demonstrating the existence of three important condensations (clusters of galaxies) in these regions.

In studying a surface of 30 206 square degrees on the photographic plates of the Palomar Sky Survey, Abell identified 2712 clusters of galaxies. He defined such a system as an association of galaxies comprising at least 50 members in an interval of 2 magnitudes (from the third brightest galaxy in the cluster) and within a circle of $1.5h^{-1}$ Mpc. From the same data Zwicky and his collaborators defined analogous structures using different criteria. The limit of a cluster is determined by the contour (isopleth) where the numerical surface density of galaxies is twice that of the 'local background'. This contour

318   11. The Universe on a Large Scale

Fig. 11.2. The distributions of groups of galaxies in the Local Supercluster. There are about fifty groups (bearing the names of constellations or catalogues) like the Local Group. The centre of the Supercluster lies in the Virgo Group, the Local Group being at the periphery. (From Luminet 1982)

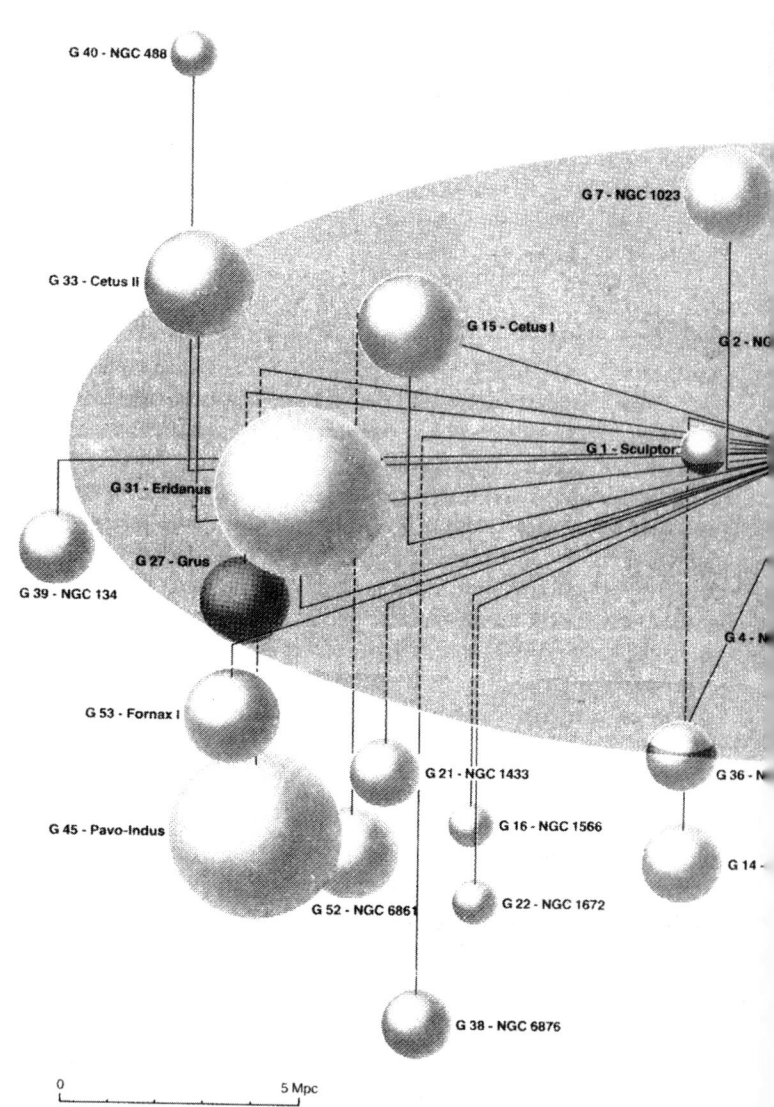

11.1 Structure and Homogeneity 319

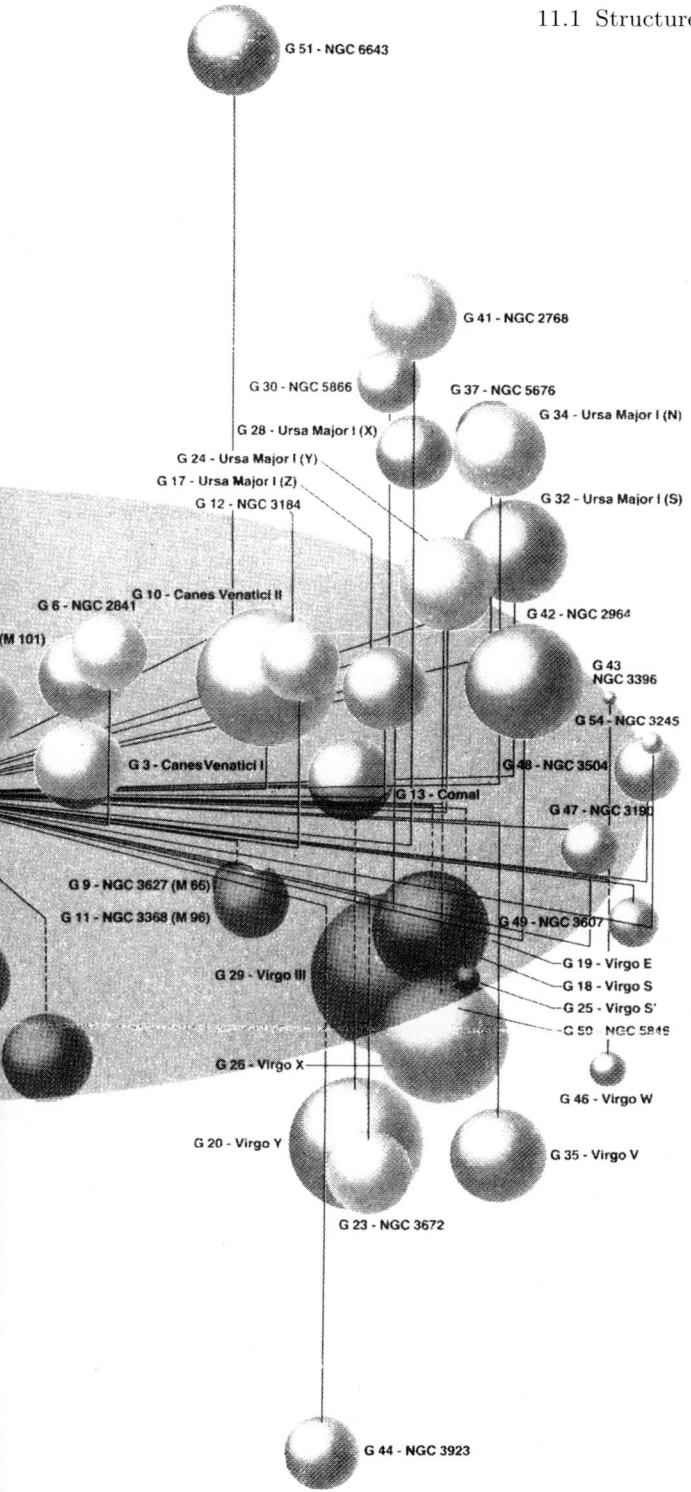

must contain at least 50 galaxies in a magnitude interval $(m_1, m_1 + 3)$, where m_1 is the magnitude of the brightest galaxy. These objects are collected in the catalogues of Abell and Zwicky.

An equally important step was the identification in the 1950s of the Local Supercluster, a grouping of about fifty groups of galaxies centred on the Virgo cluster (Fig. 11.2). Our Galaxy belongs, with Andromeda (M31), to the Local Group. This group extends to about 1 Mpc and contains more than a dozen dwarf objects situated at the periphery of the Local Supercluster, or Virgo Supercluster.

11.2 The Distance of the Galaxies: Probing the Universe

Determining the distance of objects in the universe remains a key problem in modern astronomy. This is clearly essential for cosmology in order to map the realm of the galaxies and probe the universe. In fact the extragalactic distance scale is established in a series of steps, each of which is susceptible to errors that call into question the whole edifice.

11.2.1 Parallaxes and Trigonometric Methods

Historically distances first had to be known on the Earth in order to determine the scale of the Solar System. We now use modern techniques such as radar to measure the distances of the planets. Later by using triangulation the distance of nearby stars could be determined by measuring the trigonometric parallax.

If one observes a nearby star E at six-month intervals, owing to the movement of the Earth in its orbit around the Sun, its apparent position on the background of the sky varies over an angle 2θ (θ is called the trigonometric parallax: see Fig. 11.3). By definition a star whose parallax is $1''$ is at a distance of 1 parsec (or 3×10^{16} m). The precision of trigonometric parallax is around $0.03''$; in other words this method cannot be used beyond 30 pc. The *HIPPARCOS* satellite (High-Precision Parallax-Collecting Satellite), launched in 1989, has enabled a precision of $0.001''$ obtained for several thousands of stars.

Another method consists in the study of the kinematics of a cluster of stars. All the stars in a cluster are situated at the same distance and move as a whole, at the cluster speed, to which is added the individual random motion, which has a velocity less than the group velocity. By a phenomenon analogous to the perspective effect where parallel lines appear to meet at the horizon, the proper motion of stars points in a direction known as the convergent. This direction is that of the group velocity of the cluster. By observing this at two different times we can determine the direction in which the group moves by averaging the measurement of the individual velocities of the stars.

11.2 The Distance of the Galaxies: Probing the Universe

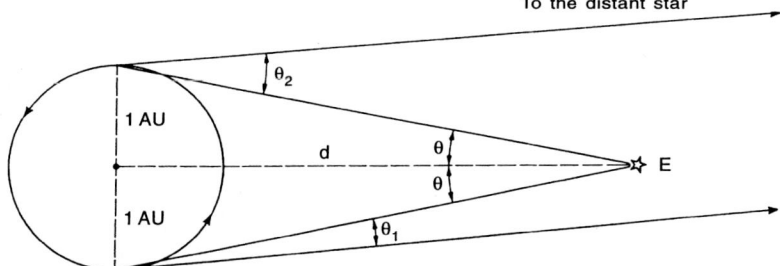

Fig. 11.3. After an interval of six months a nearby star's angular position will be seen to vary with respect to other much more distant stars. Taking these stars to be at infinity, we have: $\theta_1 + \theta_2 = 2\theta$, where θ, the parallax of E, is the angle subtended by the radius of the Earth's orbit (1 AU) at the distance of E: $\theta\,["] = 1\,\text{AU}/d\,[\text{pc}]$

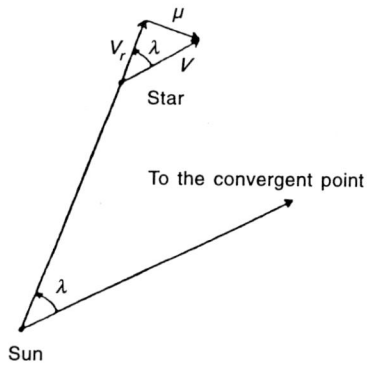

Fig. 11.4. Some definitions. λ is the angle between the direction of the star and the convergent point. μ is the proper (tangential) motion of the star, V_r its radial velocity, and V its total velocity. The direction of the convergent point is parallel to V

If λ is the angle between the direction of the cluster and that of the convergent (see Fig. 11.4 for definitions), we have

$$\tan\lambda = \frac{V_t}{V_r} = \frac{\mu d}{V_r}$$

or

$$d = V_r \frac{\tan\lambda}{4.74\mu}\,[\text{pc}],$$

where V_r is in km s^{-1} and μ is in arcseconds per year. This can be determined for each star for which one measures the radial velocity (the Doppler effect) and the proper motion by observation at two different times. The

322   11. The Universe on a Large Scale

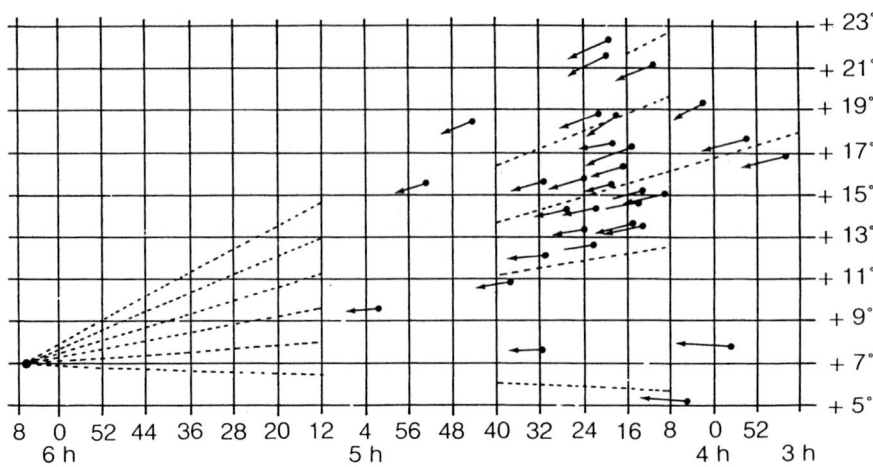

Fig. 11.5. The first determination of the convergent point for the Hyades cluster (by Boss in 1908). The arrows indicate the direction and amplitude of the proper motion of 41 individual stars. (From Gougenheim and Bottinelli 1987)

measurement is repeated for many stars of the cluster once one has determined the motion of the cluster as a whole and its distance. The Hyades cluster (Fig. 11.5), whose distance measured by this method is 46 pc, is the touchstone of the edifice that constitutes the extragalactic distance scale.

To obtain the distance of other clusters similar to the Hyades cluster we make use of a 'fitting' method for clusters of stars on the main sequence of the HR diagram. For this we construct the HR diagram for the Hyades cluster in terms of the absolute magnitude M_V or the logarithm of the luminosity as a function of the spectral type or colour. For a more distant cluster, for example the Pleiades, we construct a similar diagram with an apparent magnitude m_V. By lining-up the main sequences of the two supposedly similar clusters we determine the distance modulus ($m - M = 5 \log d_{\mathrm{pc}} - 5$) for the Pleiades and, step by step, that of other clusters (Fig. 11.6).

For the most distant stars we must abandon trigonometric methods. We then use a method whose principle is as follows: we are interested in objects whose absolute magnitude M is correlated with a measurable physical property. Measurement of the apparent magnitude m then enables us to determine the distance of the object by means of the distance modulus.

11.2.2 Cepheids and Standard Candles

Variable stars, in particular the Cepheids, play an important role here. Massive stars in an advanced stage of evolution reach a phase of instability. This leads to pulsation of the external layers and results in an important variation in their apparent brightness with periods going from 2 to 150 days. In

11.2 The Distance of the Galaxies: Probing the Universe

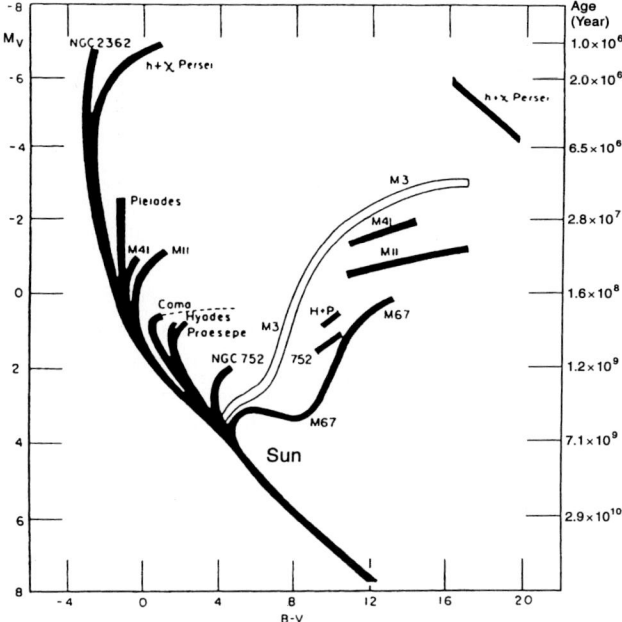

Fig. 11.6. Fitting the HR diagrams of different clusters to that of the Hyades cluster allows their distance to be determined. (From Sandage 1957)

1908 Leavitt discovered a relation between the period P of Cepheids in the Small Magellanic Cloud and their luminosity (or absolute magnitude M) of the type

$$M = a + b \log P.$$

We can determine the slope b by observing Cepheids in the Magellanic Clouds: their absolute magnitude, which is unknown, only differs from their apparent magnitude by the same constant amount. We determine the constant a by observing Cepheids at a known distance (or parallax) in our own Galaxy. Once the relation is calibrated, measurement of the period P of an object determines M, which gives the distance d through measurement of the apparent magnitude m.

This method saw a resurgence of interest with the use of infrared techniques. One could therefore expect to obtain a dispersion of only 0.02 magnitudes for the P–L relation of the Cepheids.

Other variable objects, such as RR Lyrae stars or novae, are equally useful as distance indicators. The advantage of RR Lyraes is that they have a small dispersion for the P–L relation, but on the other hand these are objects intrinsically less bright than the Cepheids and their luminosity is a function of their metallicity, which might lead to an important error in the distance. Novae show a correlation between the absolute magnitude and the rate of decrease of their luminosity. However, this relation is also subject to a degree of caution. The group of objects of this type, known as 'standard candles', are called 'primary calibrators', because we establish a relation for objects belonging to our own Galaxy. For example the calibration of the period–luminosity relation of Cepheids depends on the distance of the Hyades cluster and enables distances of around 7 Mpc to be reached.

This primary stage allows us to determine the distances of around a dozen galaxies sufficiently close for us to detect these standard candles. We shall next define the secondary and tertiary calibrators and so on, calibrated with respect to these objects. We need to observe objects that are sufficiently bright to be observable at large distances and that possess a property correlated with the luminosity or whose absolute magnitude is assumed to be universal. We thus use the diameter of H II regions, the luminosity of supernovae at their maximum, and the luminosity–width correlation of the arms (normalized to the diameter) of spiral galaxies. It is of course important that the intrinsic dispersion of these relations be as small as possible.

11.2.3 The Tully–Fischer Relation

A very useful recent method is based on the so-called Tully–Fischer relation. The study of rotation curves of spiral galaxies whose distance is known allows us to determine an empirical relation between the value ΔV of the width of the 21 cm line emitted by the hydrogen in these galaxies and the luminosity:

$$L = k\Delta V^\alpha,$$

where k is a constant of proportionality and $\alpha \approx 4$.

This method has been extended to the domain of infrared magnitudes, which allows problems of internal extinction to be neglected. Figure 11.7 shows, for example, the velocity–distance relation for eleven clusters of galaxies from the infrared Tully–Fischer relation. In a similar way one can use the Faber–Jackson relation, which links the internal velocity dispersion σ of elliptical and lenticular galaxies to their luminosity L. This relation is expressed as

$$L = k'\sigma^\beta,$$

where

$$3 \leq \beta \leq 4.$$

Thus the determination of ΔV or σ by observation leads to a measurement of L (and then the absolute magnitude M), which, compared with the apparent magnitude m, then leads to the distance of the object.

It is important to note that all systematic errors associated with a given step are passed on to successive steps. We therefore seek to get rid of calibrators that support each other, at least in part.

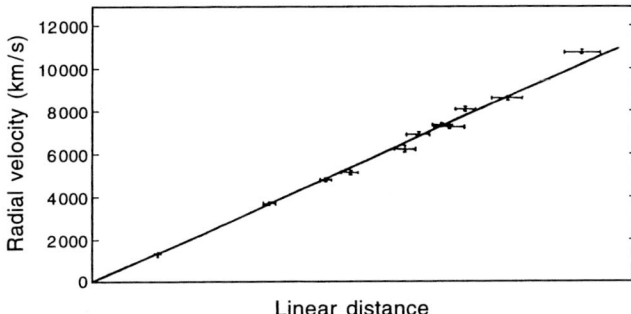

Fig. 11.7. The velocity–distance relation for eleven clusters of galaxies from the infrared Tully–Fischer relation. The value deduced for H_0 is $90 \,\mathrm{km\,s^{-1}\,Mpc^{-1}}$. (From Aaronson 1986)

An important path was cleared by Hubble in the 1930s. From observations made with the telescopes at Mount Wilson he showed that galaxies' velocities of recession v inferred from the redshift ($v = c(\lambda_{\mathrm{obs}} - \lambda_0)/\lambda_0) = cz$, z being the redshift) of their spectral lines are correlated with their distances d estimated with the help of the preceding calibrators. The expression of this relation (which is linear for small velocities) constitutes Hubble's law:

$$v = H_0 d,$$

where H_0 is the 'Hubble constant', *which is independent of the direction in the sky*. This law then indicates that the universe is expanding isotropically.

Such a relation is a very valuable tool since, in principle, one need only obtain the spectrum of an object and measure its shift with respect to that of an object at rest to obtain its distance. In fact things are not so simple because the value of the Hubble 'constant', in spite of decades of effort, has not been determined to better than within a factor of two. This uncertainty is expressed by the relation

$$H_0 = h 100 \,\mathrm{km\,s^{-1}\,Mpc^{-1}},$$

where $1/2 \leq h \leq 1$. These two values (50 and $100 \,\mathrm{km\,s^{-1}\,Mpc^{-1}}$), defended respectively by Sandage and Tammann (ST) on the one hand and by de Vaucouleurs (DV) and his collaborators on the other, are called the long ($H_0 = 50$) and the short ($H_0 = 100$) distance scales. The reasons for this uncertainty stem from the difficulties inherent in the calibration techniques outlined above. They are of several orders and have already caused astronomers to revise the extragalactic distance scale several times. For example the discovery that the period–luminosity relation for the Cepheids was not unique modified the distances by a factor of two. On the other hand, relations that link a physical property (size, maximum brightness, and so on) with the luminosity or the absolute magnitude are applied to a class of objects that are all assumed to be of the same luminosity L_0. In fact there is an *internal dispersion* around L_0, and the larger this dispersion, the more it affects distance determinations. These methods are essentially statistical; in other words, we are interested in *samples of objects* to which the procedures described above can be applied. However, the use of samples from a catalogue that only contains objects brighter than a given apparent limiting magnitude m_L leads to a bias, the so-called Malmquist bias, commonly encountered in astrophysics. For such a cutoff m_L one tends to select more intrinsically bright objects than faint ones. These bright objects are not necessarily representative of the average properties of the population to be studied. In this case the absolute magnitude of the sample will be less than the true average (Fig. 11.8). The overestimation of the luminosity will lead as a consequence to an underestimation of the distance and thus an overestimation of H_0.

Another difficulty stems from extinction; this is due to dust present within our Galaxy and also in the galaxy observed which modifies the apparent brightness of that galaxy. In the observed galaxy the extinction depends on the inclination: it is almost zero when the galaxy is viewed face-on and at a maximum when it is viewed in profile. Estimates of this correction remain very uncertain, which leads to errors in the value of H_0. The extinction in a particular direction due to our Galaxy is corrected with the aid of extinction maps established from galaxy counts. Assuming that the distribution of distant galaxies is uniform, we find that the difference in the mean number of objects per unit surface brighter than a limiting magnitude in principle

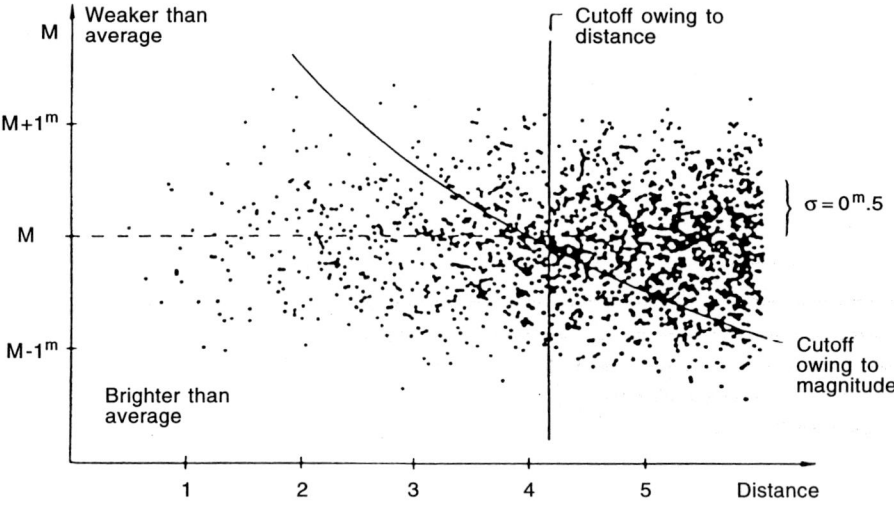

Fig. 11.8. An illustration of the Malmquist bias. The points show the distribution of objects as a function of absolute magnitude and distance. A constant space density and a Gaussian luminosity function of width $\sigma(m) = 0.5$ m are assumed. The difference in the samples obtained by limiting either the distance or the apparent magnitude is striking. (From Tammann 1979)

enables corrections for extinction to be evaluated. This is where the difference between the two estimates begins. ST consider a zero absorption at the galactic pole and DV an absorption of 20% (recent measurements from the satellite *IRAS* lead to intermediate values!). Then the use of different calibrations for the *P–L* relation for the Cepheids already lead, at the distance of Andromeda, to a *cumulative* difference of approximately 30% in the value of H_0. For the most distant objects each group uses different calibrations. Although for each distance scale the error in the distance will be within reasonable bounds, the cumulative effect is finally a factor of 2.

The recent and future use of infrared techniques will allow a better determination of the distances of nearby galaxies, the first rung in the intergalactic distance ladder. Indeed observations at $\lambda \approx 2\mu m$ in principle allow the problems of intergalactic extinction and internal absorption to be eliminated.

The application of Hubble's law itself is not exempt from problems. The correspondence between velocity and distance presupposes that the galaxies are only moving by virtue of the general expansion of the universe or at least that their own velocities are small compared with the velocity of recession. As will be seen below, this is not always true. For example the Local Group, to which our Galaxy belongs, moves with a group motion that constitutes a difference from the Hubble flow of the order of $400\,\mathrm{km\,s^{-1}}$ (compare the motion towards Virgo). Other large-scale motions have been discovered. It is clear that this can significantly affect the estimation of distance and thus the determination of the Hubble constant.

11.2.4 The Sunyaev–Zel'dovich Effect

One quite different method is based on the Sunyaev–Zel'dovich effect. X-ray observations have allowed the discovery of a hot intercluster gas ($T \approx 10^8$ K) in clusters of galaxies. The electrons of this hot plasma of density n_e may diffuse low-frequency photons by Thomson diffusion (the effective cross-section being $\sigma_T = (8\pi/3)(e^2/m_e c^2)^2$). The corresponding optical thickness is

$$\tau_T = \int \sigma_T n_e \, dl.$$

For a typical cluster of galaxies $n_e \approx 10^{-3}\,\mathrm{cm^{-3}}$, $l \approx 1$ Mpc and thus $\tau_T \approx 10^{-(2-3)}$. Therefore a fraction τ_T of the photons of a radio source situated behind an X-ray-emitting cluster will experience this diffusion in passing through the intercluster medium.

A possible source is constituted by the photons of the cosmic microwave background radiation. The hot electrons with a temperature T_e of the intercluster medium will 'heat' the photons of the 2.7 K radiation and thus alter the spectrum of the cosmological black body in the direction of the cluster. For each mean diffusion a photon of frequency ν will have its frequency altered by $\Delta \nu / \nu = 4kT_e/m_e c^2$. The change in 'brightness temperature' of the cosmological black body due to these interactions (the Sunyaev–Zel'dovich effect) is

$$\frac{\Delta T}{T} = -\int \frac{2kT_e}{m_e c^2} \, d\tau_T \quad (h\nu \ll kT_e).$$

The result is illustrated in Fig. 11.9. It can be seen that at low frequencies (the Rayleigh–Jeans part) the effect is to *reduce* the black-body temperature. Measurement of the black-body temperature in the direction of an X-ray-emitting cluster compared with that measured far from all sources allows $\Delta T/T$ to be determined.

The combination of the characteristics of X-ray emission of a cluster with the measurement of $\Delta T/T$ then leads in principle to a new estimate of the distance to this object. Consider this for a spherical region of gas (assumed to be isothermal and of uniform density) of radius R at a distance D and subtending an angle θ. The X-ray luminosity of the cluster is given by

$$L_X \propto n_e^2 T_e^{1/2} R^3$$

and

$$\frac{\Delta T}{T} = -\frac{4kT_e}{m_e c^2} \sigma_T n_e R,$$

while an observer sees a flux

$$F_X \propto \frac{L_X}{D^2} \propto n_e^2 T_e^{1/2} \theta^3 D.$$

From this we get, on substituting for n_e,

$$F_X \propto \left(\frac{\Delta T}{T}\right)^2 \frac{\theta}{T_e^{3/2} D}.$$

Measurements of F_X, $\Delta T/T$, and T_e thus give the distance D, leading to an *absolute* determination of the Hubble constant H_0 by comparison with the velocity of recession of the cluster.

Future measurements to be made by new instrumentation will give, in principle, a determination of H_0 with a precision of 10%.

Since the discovery by Hubble of the phenomenon of the recession of galaxies, the determination of H_0 and distances is one of the constant goals of observational cosmology. However, as time has gone on, it has become apparent that that way is strewn with pitfalls. In fact studying the universe as a whole is indissoluble from the study of its constituents (stars, galaxies, and so on) and their properties. One cannot understand one without the other.

11.2.5 The Surface Brightness Fluctuation Method

An interesting approach to measure distances is the use of surface brightness fluctuations (SBF) able to yield accurate distances of elliptical galaxies and

11.2 The Distance of the Galaxies: Probing the Universe

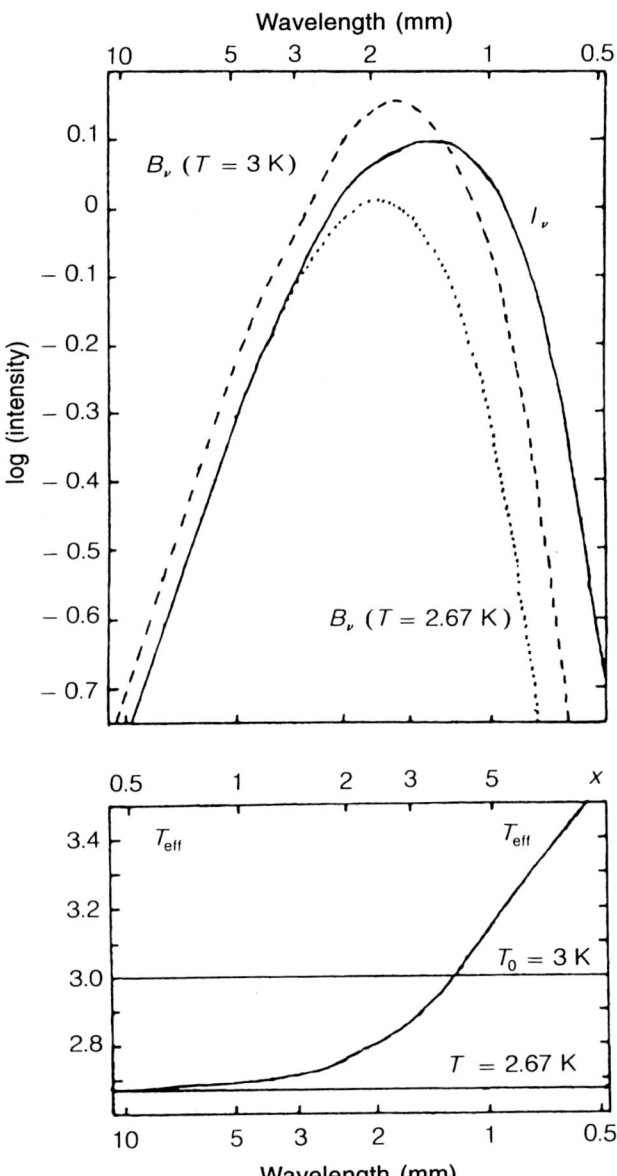

Fig. 11.9. The Sunyaev–Zel'dovich effect on the 2.7 K radiation. The broken line represents the initial black body, the solid line the result after Compton diffusion with the electrons. The dotted line shows the spectrum of a black body simulating the spectrum produced by the Compton effect at long wavelengths. In the lower part of the diagram the effective temperature of the observed spectrum as a function of wavelength is given. The effective temperature is defined as that of a black body having the same intensity at a given wavelength. (From Sunyaev and Zel'dovich 1980)

bulges of spirals out to about 20 Mpc. The method considers that the light emitted by galaxies in fact comes from discrete but unresolved stars, and although one observes a smooth surface brightness profile, Poissonian fluctuations are present. If surface brightness (SB) is independent of distance, the amplitude of the surface brightness fluctuations (SBF) is not. Indeed, the number of stars (N) in a given resolution element (pixel) of the detector increases as D^2 when the distance D of a galaxy increases while the flux f of each star decreases as D^{-2} (so the surface Brightness Nf does not depend on D). However, the statistical fluctuation (r.m.s.) of the flux from one pixel to another is of the order of $N^{1/2}f$ (which decreases as D^{-1} when D increases). The variance (Nf^2) of the SB normalized by the mean SB (Nf) decreases then with distance; as a consequence, distant galaxies will have a smoother appearance than nearby ones.

Consider, for instance, a galaxy made up of only one star family, such that the mean number of stars per pixel on the detector is 100. Fluctuations from pixel to pixel will be of the order of 10% of the mean signal, so one will see r.m.s. fluctuations that vary as the square root of the local brightness. The number of stars responsible for the signal is related to the ratio between the r.m.s. fluctuation and the square root of the local brightness, which provides the galaxy distance if the intrinsic luminosity of the stars is known. Suppose now that the galaxy is two times further away. On the same pixel as above there are now 400 stars, and the flux is four times less (so the surface brightness is constant) but the fluctuations are now at a level of 5%, scaling as the inverse of the distance.

Actually there are many other sources of fluctuations in a CCD image of a galaxy (instrumental as well as foreground and background stars or galaxies ...) so the analysis has to take them into account. However, if one knows the luminosity function of the stellar population, it is easy in principle to derive the 'fluctuation luminosity' (in fact the luminosity-weighted luminosity of the population $\bar{L} = \sum n_i L_i^2 / \langle L \rangle$ with n_i the number of stars of luminosity L_i). The corresponding 'fluctuation magnitude' $\bar{M}$ (in a given band) follows, allowing a direct distance determination that is independent of the global dynamics or the environment of the galaxy.

Extensive surveys have been done at optical, and recently at IR, wavelengths in particular in the Local Group and the Virgo Cluster, to test and calibrate this method, since measuring SBFs in galaxies with known distances (for example those in the Virgo cluster with Cepheid distances measured by the *HST*) provides an empirical calibration of the SBF distance scale. Comparisons between SBF distances and a variety of other estimators, including Cepheid variable stars, the planetary nebula luminosity function (PNLF), Tully–Fischer (TF), Dn-σ, SN II, and SN Ia, demonstrate the robustness of the SBF calibration. Figure 11.10 shows, for instance, the comparison between the values of the SBF parameters derived for several groups of galaxies and the distances to the groups according to the above six methods.

11.2 The Distance of the Galaxies: Probing the Universe

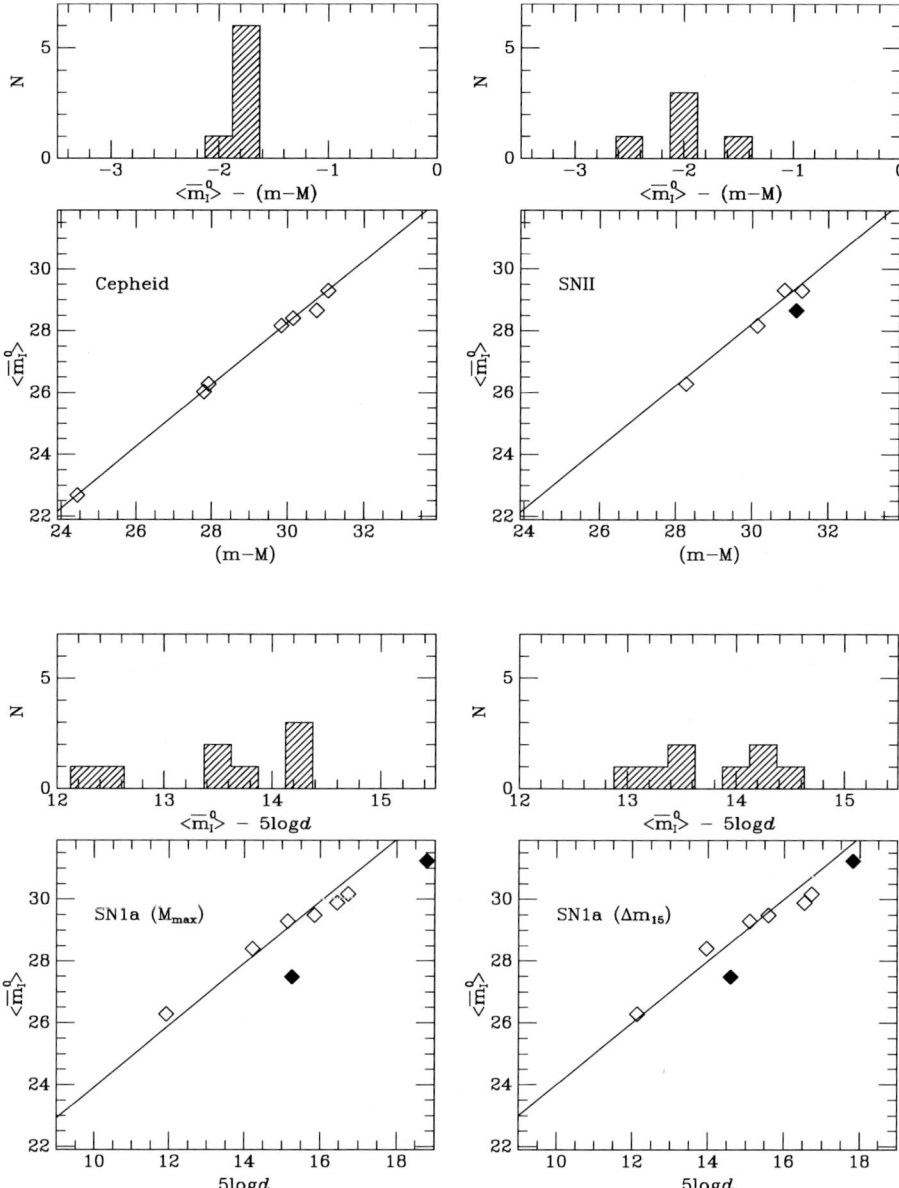

Fig. 11.10. The mean apparent fluctuation magnitude in the I–band $\langle \overline{m}_I^0 \rangle$ derived for some groups of galaxies compared to other distance estimators: Cepheids, SN II, and SN Ia. The other estimators' distances d are displayed either in terms of Mpc, expressed as a distance modulus $(m-M)$, or in terms of $\mathrm{km\,s^{-1}}$, plotted as five times the logarithm. The lines are drawn according to a weighted fit of unity slope between each set of distances. Above each distance comparison is a histogram of the differences $\langle \overline{m}_I^0 \rangle - (m-M)$ or $\langle \overline{m}_I^0 \rangle - 5\log d$

As a result of this comparison, an estimate of the Hubble constant could be determined, leading to the value $H_0 = 81 \pm 6$ km s^{-1} Mpc^{-1}.

11.3 The Third Dimension

In spite of the uncertainties in the value of the Hubble constant, the measurement of the redshift in the spectrum of galaxies remains the best method for obtaining their distance and thus enriches the various catalogues of galaxies by adding a third spatial dimension, based on photographic negatives.

Recent technological progress with regard to telescopes, detectors, and reduction methods also means that it is now possible to acquire relatively quickly the radial velocities of many, possibly faint, objects. The measurement of recession velocities began as long ago as 1912 by Slipher, and the number of velocities measured grew without end ($N < 1000$ in 1975, $N \geq 20\,000$ in 1990).

During the 1980s an effective survey by the Cambridge Center for Astrophysics (CFA) measured the radial velocities of about 2400 galaxies over 2.7 steradians up to a magnitude of about $m_B = 14.5$, which corresponds to a distance of 105 Mpc, thus giving access to a vast fraction of the real universe. Figures 11.11, 11.12 and 11.13 show, for different velocity intervals (and thus distance intervals), the distribution of galaxies as a function of the declination (δ) and right ascension (α). Figure 11.14 gives views *along the line of sight* (thus perpendicular to the view of Figs. 11.11–13) in declination slices. In this way the existence of structures known already as the Virgo and Hercules (A2151) clusters and the Coma supercluster is confirmed.

In cone diagrams (which show velocity plotted against right ascension) clusters of galaxies (such as Virgo) are indicated by a narrow and elongated 'finger-shaped' structure. This is due to the fact that structures are linked gravitationally. In such a system the galaxies are localized within a restricted volume in real space having an 'agitation' velocity (due to their kinetic energy) of the order of 10^3 km s^{-1}. Their radial velocity is then not indicative of distance: it is the mean velocity (of the centre of mass) that gives the distance to the cluster. In this type of diagram each galaxy is represented by a point corresponding to the cluster distance to which is added (or subtracted) the projection of its own velocity in a small solid angle.

Examining these data shows the existence of many superclusters and their structure, sometimes extending over $60h^{-1}$ Mpc and appearing in the form of ribbons or filaments. Another striking point is the presence of voids reaching sizes the order of those of the superclusters. Figure 11.15 shows an example of the large void in Boötes. Assuming spherical symmetry, it would appear that there is a sphere about $60h^{-1}$ Mpc in diameter that is empty of bright galaxies, whereas about twenty or so would be expected in this same volume for a uniform distribution.

11.3 The Third Dimension   333

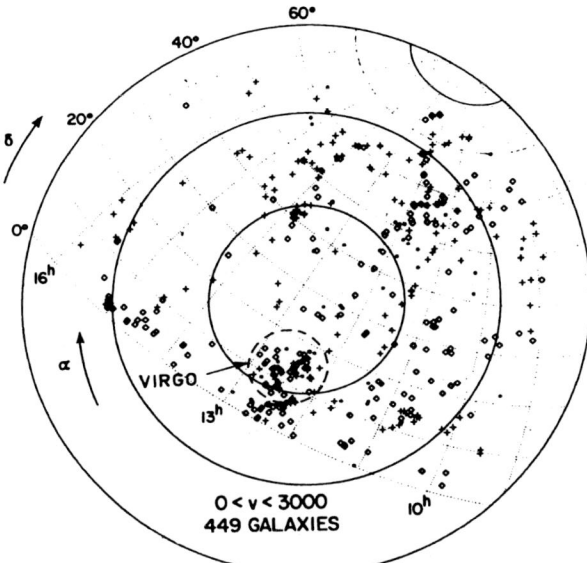

Fig. 11.11. A projection on the sky of galaxies brighter than $m_B = 14.5$. The view corresponding to depth or distance into space is obtained from the radial-velocity intervals. The contours indicate structures that are already known (groups, clusters, and so on). The symbols have the following meanings: $\circ = 0\text{--}1000\,\text{km}\,\text{s}^{-1}$; $\diamond = 1000\text{--}2000\,\text{km}\,\text{s}^{-1}$; $+ = 2000\text{--}3000\,\text{km}\,\text{s}^{-1}$. (From the CFA survey)

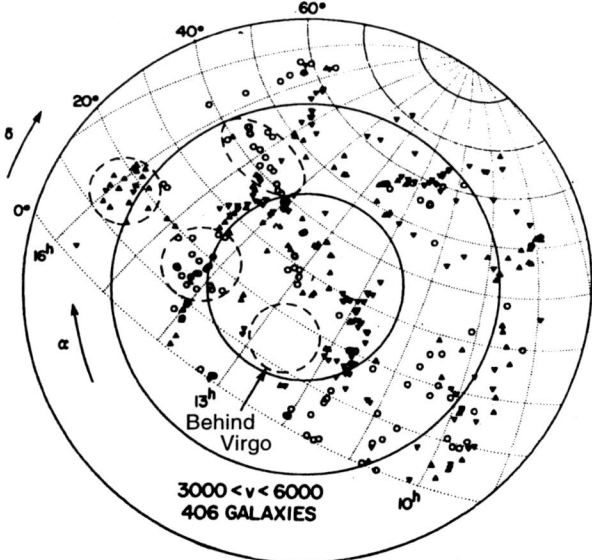

Fig. 11.12. The same as Fig. 11.11, with the following meanings for the symbols: $\triangledown = 3000\text{--}4000\,\text{km}\,\text{s}^{-1}$; $\triangle = 4000\text{--}5000\,\text{km}\,\text{s}^{-1}$; $\circ = 5000\text{--}6000\,\text{km}\,\text{s}^{-1}$

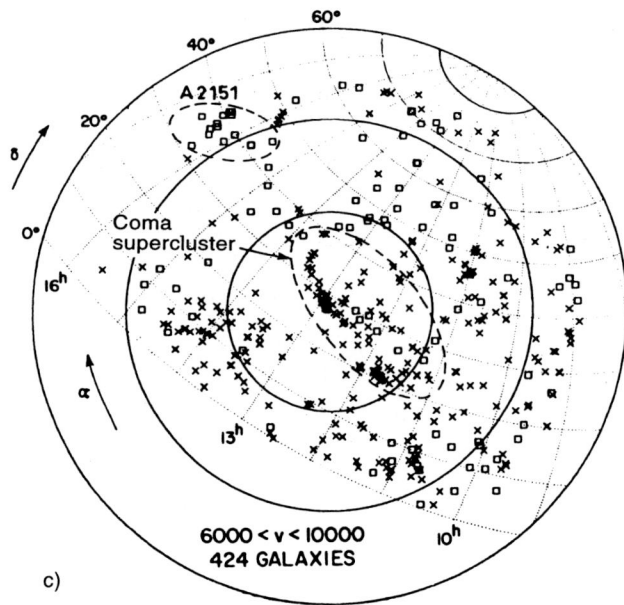

Fig. 11.13. The same as Fig. 11.11, with the following meanings for the symbols: $\times$ = 6000–8000 km s^{-1}; $\square$ = 8000–10 000 km s^{-1}

A supplementary step has been achieved by the CFA group, who measured the velocities of galaxies up to a magnitude $m = 15.5$ in a declination slice 6° thick and over 9 hours of right ascension ($\approx 100°$) across. The interest of such a sample is that at least two dimensions (breadth and depth) are of the order of magnitude of the largest structures observed. Figure 11.16 shows the results obtained. Except for the central structure corresponding to the Coma cluster, the galaxies are preferentially localized at the edge of more or less circular regions void of bright galaxies and with a typical diameter of $25h^{-1}$ Mpc. The radial velocities available in the adjacent declination slices show that similar structures persist. Furthermore for a given void the diameter grows, reaches a maximum, and then decreases when one passes continuously from one slice to another. This suggests in fact a 'soap-bubble' topology, the structures of Figs. 11.11–13 being a cross-section of these. In this view the long individual filaments observed in the catalogues mentioned above are only cross-sections perpendicular to the surface of these bubbles. Such a 'cellular structure' has been confirmed by recent surveys on wide fields: the *ESO Slice Project* (ESP), and the *Las Campanas Redshift Survey* (LCRS) up to 60 000 km s^{-1} and in narrower pencil beams (for example the *ESO Sculptor Survey* up to $z=0.7$). In particular, a periodic structure appears which could correspond to the intersection of the beam with bubbles or cells regularly spaced by 128 Mpc.

To recap: from galaxies to clusters, superclusters, filaments, bubbles, and voids, the universe is structured on scales going from 20 kpc to 100 Mpc. The

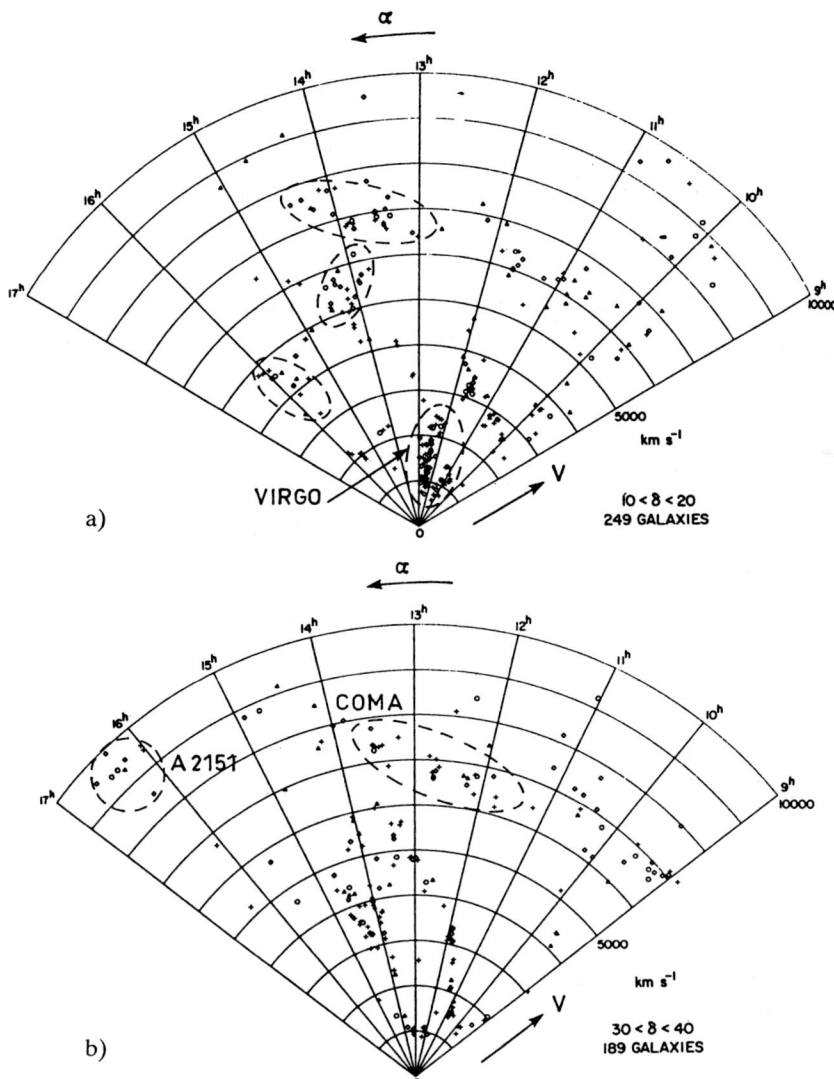

Fig. 11.14. The distribution of galaxies as a function of the right ascension (α) and the velocity for two declination slices (δ). The views are perpendicular to those in Figs. 11.11–13, and the structures seen there can be identified here

question of the exact spatial grouping of matter on large scales at higher redshifts is still a challenge to observers, it is also a challenge to theoreticians who have to construct models to take account of the observations.

To anticipate a little: there were a long debate between two scenarios concerning the formation of galaxies: the 'pancake', or top–down, model, and the

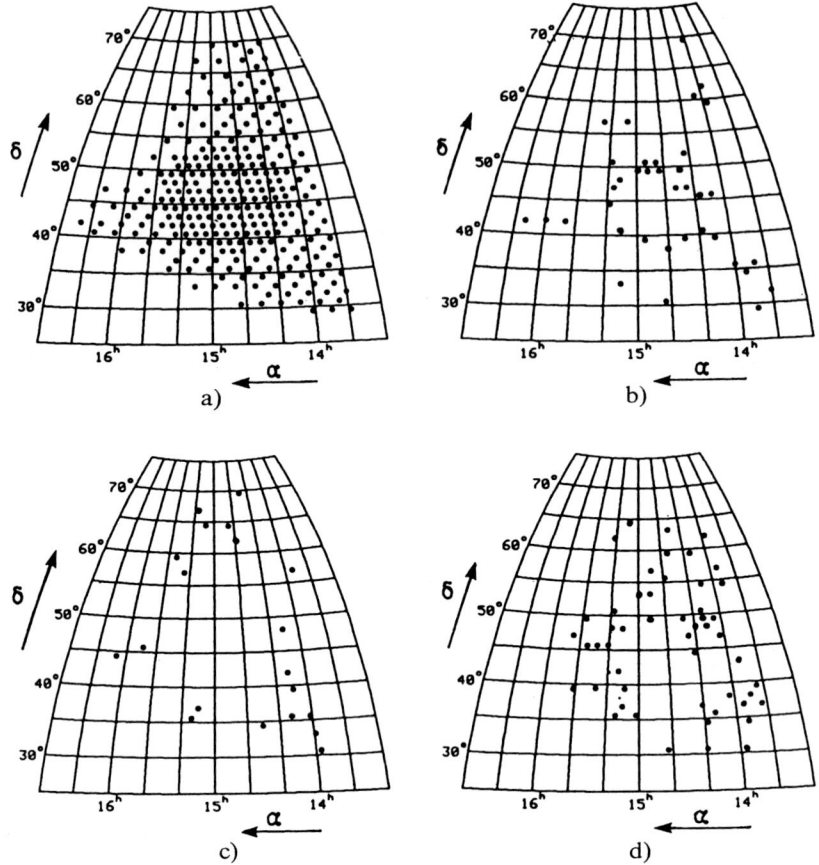

Fig. 11.15. (a) Positions on the sky of 28 fields where radial velocities have been measured (in the constellation Boötes). (b–d) Positions of galaxies whose apparent velocities belong to three domains: 6000–12 000 km s^{-1} (b), 12 000–18 000 km s^{-1} (c), and 18 000–24 000 km s^{-1} (d)

hierarchical, or bottom–up, model (see Chap. 12 on the formation of galaxies). In the first scenario the first structures that form have characteristic scales corresponding to those of superclusters and elongated forms (filaments) or flattened filaments (pancakes). The galaxies form *subsequently* by fragmentation of these superclusters. On the other hand in the second model the first structures that appear correspond to subgalactic scales. The other sustems form *after* the galaxies by gravitational attraction. As the process does not have a privileged scale, this leads to a *hierarchical* structure, the galaxies regrouping into clusters, and the clusters into superclusters. The density profile of these systems thus has the same appearance whatever the scale considered.

11.3 The Third Dimension   337

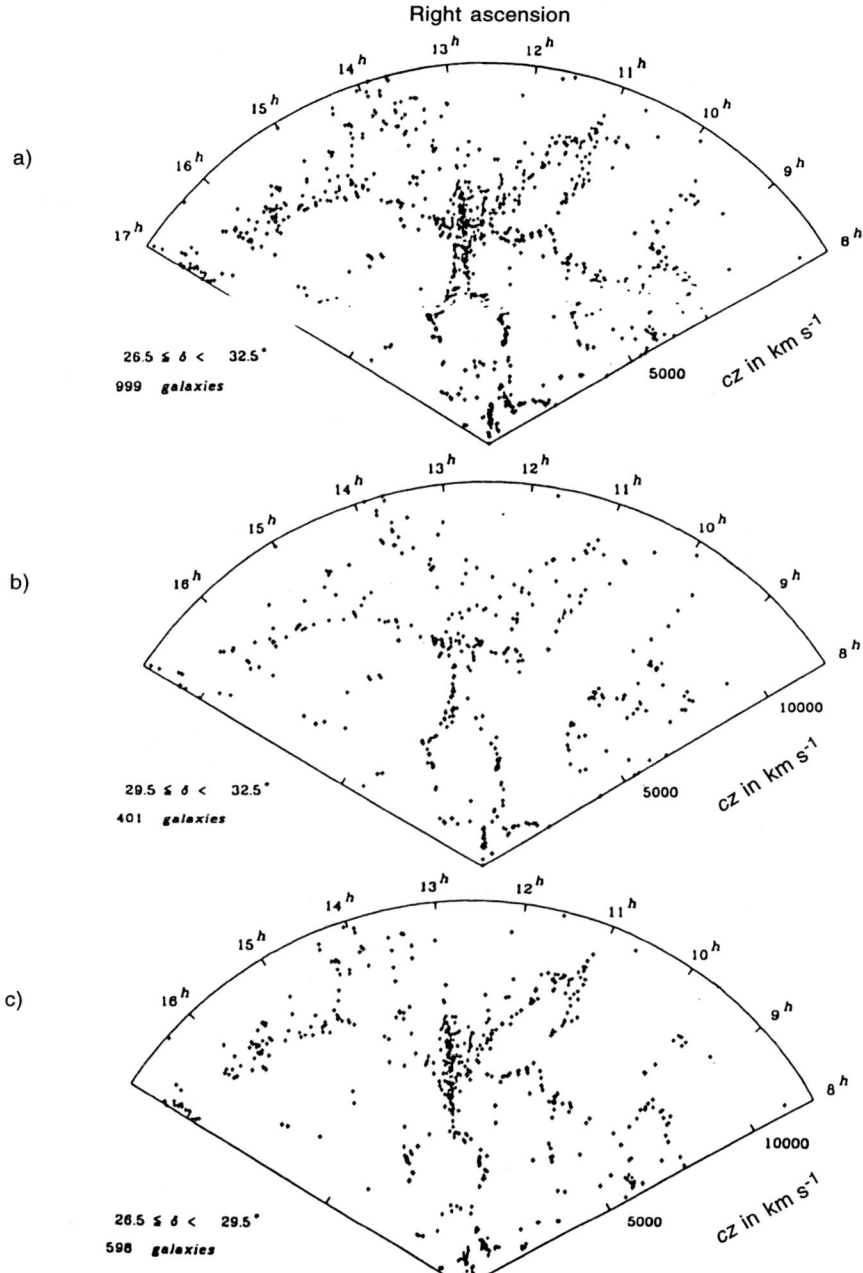

Fig. 11.16. (a) A map of the recession velocities measured by the CFA for which $m_B \leq 15.5$; (b, c) 3° declination slices whose superposition gives (a). The important structure at $\alpha = 13\,\mathrm{h}$ corresponds mainly to the Coma cluster. (From de Lapparent 1986)

It is immediately apparent that the presence of filaments and empty regions of galaxies separating the filaments finds a natural explanation in the top–down schema, unlike the bottom–up model, where an inhibitory process has to be invoked for the formation of galaxies so as to explain a void such as that in Boötes or others. This is why formation models call for a universe dominated by nonbaryonic cold matter where only density perturbations beyond a certain threshold (biased formation) show characteristics similar to those of Fig. 11.16 (see Chap. 12).

11.4 Statistical Methods: Correlation Functions and Percolation

11.4.1 Correlation Functions

In the previous section we gave an essentially qualitative description of structures in the universe. Simple and concrete arguments were developed in relation to currently proposed theoretical models. This approach is subjective, however, and more quantitative methods have to be developed to invalidate one or other of these theories.

The application of correlation methods to the study of extragalactic data has been developed largely by Peebles and his collaborators. This statistical tool measures the correlations between the objects studied and will possibly enable structures such as filaments to be revealed in a given distribution. It evidently constitutes a fundamental test for all theoretical models. The most widely used, because it is the easiest to extract from the data, is the two-point correlation function $\xi(r)$. Assuming that the distribution of galaxies in the universe is the realization of a stationary random process (in space) of mean density $\bar{n}$, the *spatial* correlation function $\xi(r)$ can be defined in three equivalent ways. Firstly it can be defined by means of the joint probability of having two objects in volume elements dV_1 and dV_2 separated by r_{12}:

$$\mathrm{d}P(r) = \left(\frac{\bar{n}}{N}\right)^2 \left(1 + \xi(r_{12})\right) \mathrm{d}V_1 \mathrm{d}V_2,$$

where N is the total number of galaxies in the chosen sample, and $r_{12} = |\boldsymbol{r}_1 - \boldsymbol{r}_2|$.

Secondly it can be defined as the correlation function of the density $n(r)$:

$$\bar{n}^2 \left(1 + \xi(r)\right) \stackrel{\mathrm{def}}{=} \langle n(\boldsymbol{R}+\boldsymbol{r}) n(\boldsymbol{R}) \rangle_{\boldsymbol{R}},$$

where

$$\bar{n}^2 \xi(r) \stackrel{\mathrm{def}}{=} \langle [n(\boldsymbol{R}+\boldsymbol{r}) - n](n(\boldsymbol{R}) - n) \rangle_{\boldsymbol{R}},$$

where $\langle\ \rangle_{\boldsymbol{R}}$ indicates an average over the volume.

11.4 Statistical Methods: Correlation Functions and Percolation

Thirdly it can be defined in a more practical way as

$$1 + \xi(r) \stackrel{\text{def}}{=} \frac{N_P(r)}{N_P^{\text{Poisson}}},$$

where $N_P(r)$ is the number of pairs of galaxies whose separations r lie in the interval $(r - \Delta r, r + \Delta r)$, and $N_P^{\text{Poisson}}(r)$ is the number of pairs corresponding to a Poisson distribution for the same volume considered. The function ξ thus measures the tendency of the objects under consideration to group together. It is zero for a uniform random distribution and positive (negative) for a more (less) concentrated distribution. In practice, from the above formula, when a distribution of N points of density $n(r)$ in a spherical volume is analysed, the number of pairs (for small distances) separated by $r + dr$ will be $\frac{1}{2} N 4\pi r^2 n(r) \, dr$. To obtain ξ, therefore, one divides by the number of pairs of the Poisson distribution having the same number of points in the same volume.

By way of example, we shall estimate the correlation functions in two models: one schematizing a model of a pancake or filament, the other a hierarchical model.

Suppose we start with a homogeneous distribution of points in a sphere of radius R and density $\bar{\rho}$; hence, by definition, $\xi(r) = 0$. Imagine now that all the particles are 'sampled' to form an infinite, flat disc (a pancake) of the same radius and density $\sigma = \frac{4}{3} R \bar{\rho}$. By definition

$$1 + \xi(r) = \frac{N_P(r)}{N_P^{\text{Poisson}}(r)} = \frac{\sigma \pi R^2 \sigma 2\pi r \, dr/2}{\bar{\rho}(4\pi/3) R^3 \bar{\rho} 4\pi r^2 \, dr/2} = \frac{\sigma}{2\bar{\rho}r}.$$

An analogous calculation in the case where the particles are 'sampled' according to a diameter (a filament) of density $\theta(2R\theta = \bar{\rho} V)$ gives

$$1 + \xi(r) = \frac{\theta}{4\pi \bar{\rho} r^2}.$$

The correlation function can thus acquire the form of a power law by a simple geometric effect.

Let us now look at a hierarchical model. To this end, consider a sphere of radius R and place in this N spheres of radius R/λ. In each of these place N spheres of radius R/λ^2, and so on to L levels. At the final level L there are thus N^L points in N^{L-1} spheres each of radius $r_0 = R/\lambda^{L-1}$. Figure 11.17 illustrates this process (a fractal process) for $N = 3$, $\lambda = 2$, and $L = 3$. A hierarchy is then constructed of groups (clusters) with a characteristic scale r.

Whereas each 'cluster' is constructed as an individual entity, the ensemble forms a continuous structure in space, without precise boundaries at each scale of clustering. Increasing scales correspond to more and more complex structures that can be identified with galaxies, clusters, clusters of clusters, and so on.

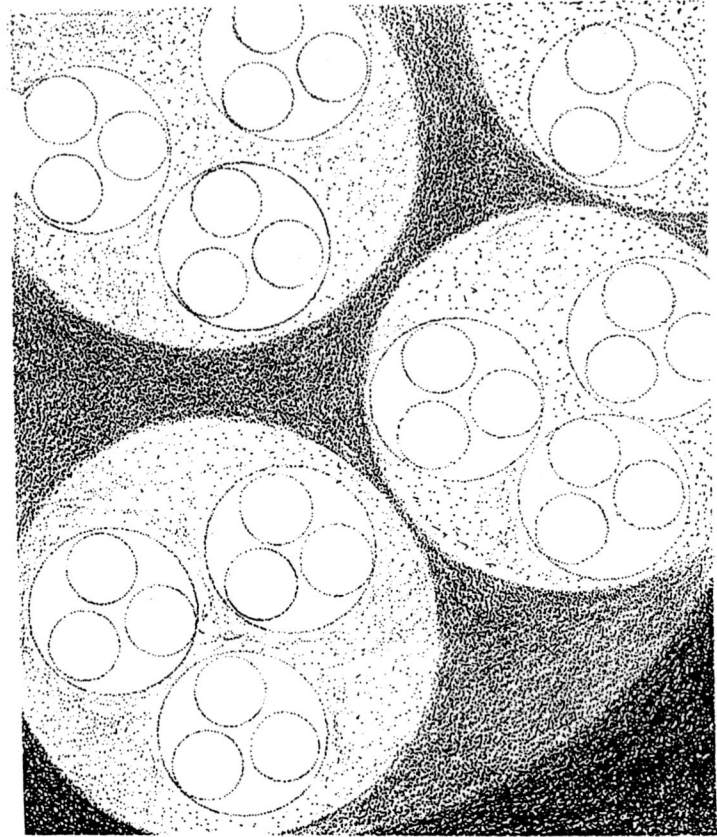

Fig. 11.17. An example of the hierarchical structure constructed with spheres. The parameters are $N = 3$, $\lambda = 2$, and $L = 3$

The scale of clusters at the level K (counted from the scale r_0) will be

$$r \approx r_0 \lambda^K$$

and contain N^K particles. The mean density in each of these clusters is thus

$$n \approx N^K r^{-3} \approx N^{\log(r/r_0)/\log \lambda} r^{-3}$$

or

$$n \approx r^{-\gamma} \quad (r_0 \leq r \leq R),$$

setting $\lambda = N^{1/(3-\gamma)}$.

Once N is chosen, λ is fixed for all values of γ wanted. In such a model the density will then vary as a power law with the cluster size (in other words, the resolution at which the system is examined). We now estimate the correlation function at a given scale r.

11.4 Statistical Methods: Correlation Functions and Percolation

At this scale the hierarchy can be regarded as being constituted of independent clusters of dimension r and density profile $n(r)$. We calculate ξ by using the expression for the density autocorrelation function. It will soon be seen that in the expression $\bar{n}^2(1+\xi)$ the term $\bar{n}^2$ corresponds to pairs realized by taking points belonging to two different clusters (with a uniform distribution). As a consequence ξ is thus given by the number of pairs coming from the same cluster multiplied by the number of clusters. This will be realized for the distances between pairs s less than the size r of a cluster occupying the volume $N_{\text{cluster}} \times \frac{4}{3}\pi(r-s)^3$ of the total volume $\frac{4}{3}\pi R^3$. Hence

$$\bar{n}^2 \xi = n^2(r)\left(\frac{r-s}{R}\right)^3 N_{\text{cluster}} \approx n^2(r)\frac{r^3}{R^3}N_{\text{cluster}} \quad \text{(for small } s\text{)}$$

or

$$\bar{n}^2 \xi = n(r)\frac{n(r)r^3 N_{\text{cluster}}}{R^3} = n(r)\bar{n}.$$

From this we get

$$\xi \propto n(r) \propto r^{-\gamma};$$

that is, the correlation function of such a hierarchy is proportional to the density profile. Here again power-law behaviour is obtained, it being possible to adjust the power by the choice of parameters defining the hierarchy.

What then is the correlation function resulting from the data?

Before the existence of large surveys involving the measurement of radial velocities, $\xi(r)$ was deduced from the angular correlation function $w(\theta)$ which measures in a similar way the correlations of galaxies in the plane of the sky and can be calculated from two-dimensional catalogues. The function $w(\theta)$ expresses the joint probability of finding pairs separated by an angle θ in the solid angles $\mathrm{d}\theta_1$ and $\mathrm{d}\theta_2$:

$$\mathrm{d}P = \left(\frac{\bar{\sigma}}{N}\right)^2 (1+w(\theta))\,\mathrm{d}\Omega_1 \mathrm{d}\Omega_2,$$

where σ ($=\bar{n}D^{*3}$ if D^* is the depth of the catalogue) is the mean surface density of galaxies on the sky.

With the assumption of homogeneity and isotropy, and for small angles, it is possible to deduce $\xi(r)$ from $w(\theta)$ by 'deprojection'.

To simplify matters, assume that the galaxies have the same absolute magnitude M^*. The depth from a catalogue limited to the apparent magnitude m_0 is given by

$$m_0 - M^* = 5 \log D^*_{\text{pc}} - 5.$$

Hence $w(\theta)$ is obtained by simple integration along the two lines of sight up to the depth D^*:

$$\mathrm{d}P = \left(\frac{\bar{\sigma}}{N}\right)^2 (1+w(\theta))\,\mathrm{d}\Omega_1 \mathrm{d}\Omega_2,$$

$$dP = \left(\frac{\bar{n}}{N}\right)^2 \int_0^{D^*} r_1^2 \, dr_1 \int_0^{D^*} r_2^2 \, dr_2 \bigl(1 + \xi(r_{12})\bigr) \, d\Omega_1 d\Omega_2,$$

which gives

$$w(\theta) = 9 D^{*-6} \int_0^{D^*} dr_1 \int_0^{D^*} dr_2 \, r_1^2 r_2^2 \xi\left(\sqrt{r_1^2 + r_2^2 - 2 r_1 r_2 \cos\theta}\right).$$

Moreover notice that only galaxies with a small angular distance are correlated and that ξ will differ significantly from zero only for small distances $r_{12} \ll r_1 + r_2$. This allows us to approximate the cosine by $\cos\theta = 1 - \frac{1}{2}\theta^2$, write

$$4 r_1 r_2 = (r_1 + r_2)^2 - (r_1 - r_2)^2 \approx (r_1 + r_2)^2,$$

and extend the domain of integration up to infinity. Hence setting

$$u = \frac{r_1 + r_2}{2 D^*} \quad \text{and} \quad v = \frac{r_1 - r_2}{D^* \theta},$$

we obtain

$$w(\theta) = D^{*-1} F(D^* \theta),$$

with

$$F(x) = 9 \int_0^\infty u^4 \, du \int_0^\infty \xi\left(\sqrt{v^2 + u^2}\, x\right) dv.$$

The above expression for $w(\theta)$ in fact expresses a scaling law, since for another catalogue of depth $D^{*\prime} = x D^*$ we get

$$w'(\theta) = D^{*\prime -1} F(D^{*\prime} \theta) = x^{-1} D^{*\prime -1} F(D^* x \theta).$$

It is thus possible to superimpose the curves $w(\theta)$ and $w'(\theta)$ if the angular scale of the second catalogue is multiplied by x and the scale of correlation is divided by x. This is an important point since the verification in practice of such a property is a measure of the hypotheses underlying homogeneity and isotropy.

A first verification can be made with the aid of Abell's catalogue of clusters.

The distances of clusters and the number of galaxies they contain (corrected for the background) allow us to classify these clusters into different groups of 'distance' and 'richness'. These classes are defined in Table 11.1.

Figure 11.18 shows $w(\theta)$ for clusters of galaxies belonging to the Abell catalogue for different distance classes. The first thing to note is that the grouping of galaxies does not stop at the scale of clusters and confirms that these are not distributed uniformly. In other respects the scale law is clearly satisfied, which confirms the initial hypothesis.

Figure 11.19 shows the correlation functions $w(\theta)$ for different catalogues. Figure 11.20 shows the same data reduced to the depth of Zwicky's catalogue. On a large scale, global homogeneity is satisfied, since $w(\theta)$ clearly tends

11.4 Statistical Methods: Correlation Functions and Percolation

Table 11.1. Richness Classes and Distance Classes of Abell Clusters

Richness Class	0	1	2	3	4	5
Number	30–49	50–79	80–129	130–199	200–299	> 300
Distance class	1	2	3	4	5	6
Distance [Mpc]	80	115	200	270	420	540
Magnitude of the tenth brightest galaxy	13.1–14.0	14.1–14.8	14.9–15.6	15.7–16.4	16.5–17.2	17.3–?

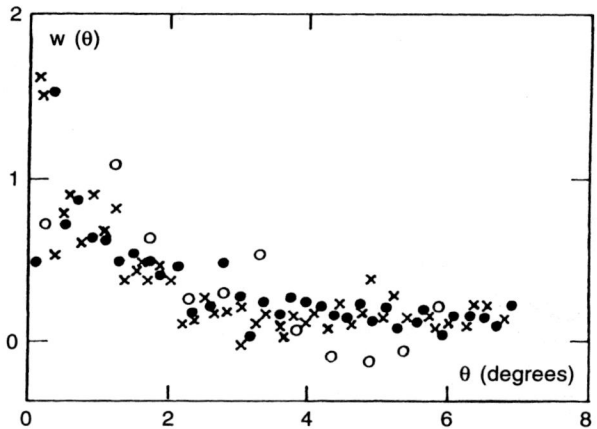

Fig. 11.18. A test of the scale law of the angular correlation function $w(\theta)$ for Abell clusters arranged in different distance classes. The symbols have the following meanings: • = classes 1–6; × = classes 1–5; ○ = classes 1–4. The different classes have been reduced to the same scale

towards zero for large θ. The scale law is equally satisfied for different depths. Finally the data indicate the absence of a characteristic length and can be fitted by a power law:

$$w(\theta) \propto \theta^{-0.8},$$

which by deprojection gives

$$\xi(r) \approx (r/r_0)^\gamma,$$

with $\gamma \approx 1.8$ and $r_0 \approx 5h^{-1}$ Mpc. Notice also a cutoff at about $10h^{-1}$ Mpc. A recent application to the APM catalogue shows that the scale law is also clearly satisfied for these data.

Other authors have also studied the correlation function at four points from the analysis of the Lick and Zwicky catalogues in order to test for

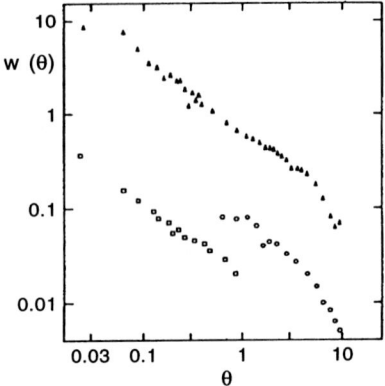

Fig. 11.19. Angular correlation functions $w(\theta)$ for different catalogues of galaxies: □ = the Jagellonian catalogue; ○ = Shane and Wirtanen's catalogue; △= Zwicky's catalogue

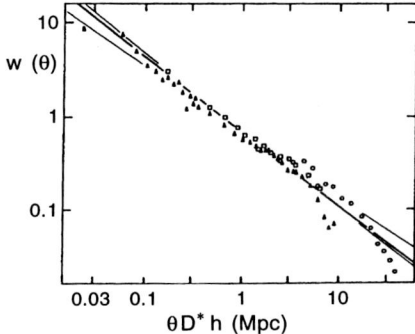

Fig. 11.20. A test of the scale law for galaxies. The data of Fig. 11.15 have been reduced to the same scale (that of Zwicky)

the presence of filaments. To this end, the quadruplets define an elongation parameter ϵ:

$$\epsilon \approx 2S/D_{\max}^2,$$

where S is the surface included between the lines joining the four galaxies, and $D_{\max}$ is the distance between the most elongated galaxies; thus for a square $\epsilon \approx 1$ and for a line $\epsilon = 0$. Figure 11.21 shows the quantity δ as a function of ϵ, where

$$\delta \approx (V_{\text{obs}} - V_{\text{mod}})/V_{\text{obs}}$$

is the relative departure of the observed four-point correlation function from that of a model having no physical filaments (the hierarchical model). The presence of filaments could be detected by the fact that $\delta(\epsilon)$ will increase at small ϵ, which is not the case (Fig. 11.21). Thus either the possible filaments

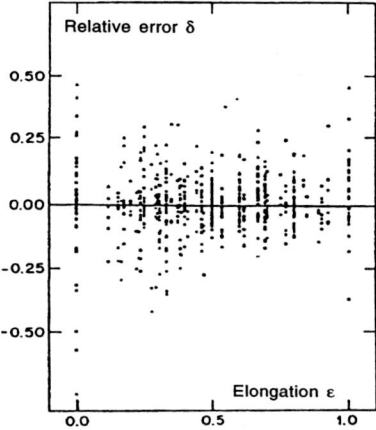

Fig. 11.21. A test to discover filaments. These are detected if δ increases when $\epsilon \to 0$ (see the text)

do not exist or this type of information is difficult to extract from projected data.

In fact obtaining radial-velocity surveys allows a more direct evaluation of $\xi(r)$ by using the velocities as distance indicators. The function $\xi(r)$ obtained from radial-velocity surveys such as the CFA one closely follows the power law given above. Other data taken from recent wide or deep surveys indicate similar results with slight differences: in particular an anticorrelation ($\xi < 0$), that is, the presence of *voids* in the domain 10–$20h^{-1}$ Mpc, while $\xi \to 0$ at larger scales. Moreover, data from the *Canada–France Redshift Survey (CFRS)* indicate an evolution with redshift: the correlation function maintaining its power-law shape but with a change of its amplitude for z between 0 and 1.

What bearing do the different scenarios have on all this? In the case of the hierarchical model the slope predicted for the nonlinear regime ($\delta\rho/\rho \gg 1$) is (see Chap. 12)

$$\gamma_{\rm nl} = (9 + 3n)/(5 + n),$$

and $\gamma \approx 1.8$ is obtained with $n = 0$, where n is the index of the power spectrum of fluctuations, whereas the value favoured by theoreticians is $n = 1$.

We saw earlier on that the influence on $\xi(r)$ of such structures as pancakes or filaments can be understood easily with a simple model. For plane or linear structures of size R the expression $1 + \xi(r)$ is given by

$$1 + \xi(r) = \begin{cases} R/2r & \text{for a pancake,} \\ (R/2r)^2 & \text{for a filament.} \end{cases}$$

Numerical simulations show that the behaviour of $\xi(r)$ can be fitted by top–down scenarios if one inserts into the spectrum a break at small scales so as to take account of dissipative phenomena to which adiabatic fluctuations below a length L_{min} are subjected. These simulations show that the slope of $\xi(r)$ varies over time and that a good fit to the data is only obtained for $z \approx 1$. Hence we must either completely eliminate the underlying model, since quasars and galaxies are observed at much larger z, or maintain that the galaxies and the superclusters form independently at different times.

11.4.2 Percolation

Another method in principle allowing the grouping of galaxies and their geometric structures to be revealed has been developed by the Russian school: percolation. The principle is simple: for a given sample of galaxies, with coordinates known within a 'box' of dimension D, a sphere of radius R can be traced around each galaxy. If R is too small, each galaxy remains isolated, whereas if R is too large, the corresponding sphere surrounds the entire system. In practice there exists a critical radius R_c that allows galaxies to be linked together in systems or chains of length D: there is thus percolation. Figure 11.22 shows an example of two different distributions having the same mean density corresponding schematically to filamentary and hierarchical structures. The percolation is obtained more easily in the filamentary case than in the hierarchical case, where R_c must be of the order of the intercluster distance. Figure 11.23 shows how, in making R vary, subsystems of different multiplicities are detected.

Several parameters can be defined by this method.

- The length L_{max} of regions connected as a function of R. The percolation is said to be critical when $L \approx D$.
- $L(R_c) = D$ defines the critical percolation radius. For example for a model of particles distributed uniformly on a grid $L(\tilde{R}) = 0$ for $\tilde{R} < \tilde{R}_c$, where we have introduced the dimensionless radius $\tilde{R}$ such that

$$\tilde{R} = R/\bar{R},$$

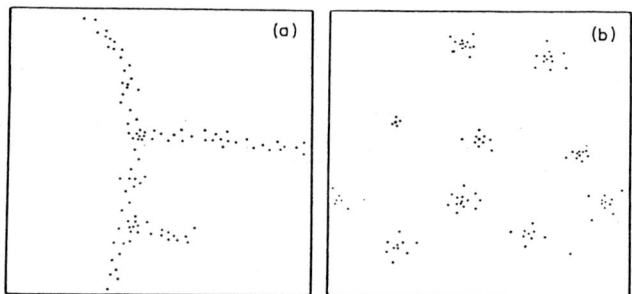

Fig. 11.22. Two-dimensional examples of **(a)** the filamentary distribution and **(b)** the hierarchical distribution

11.4 Statistical Methods: Correlation Functions and Percolation

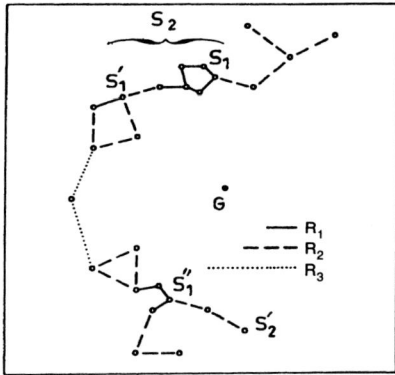

Fig. 11.23. Systems of galaxies formed by varying the distance R to them. For $R = R_1$ systems S_1, S_1', and S_1'' are obtained with 6, 4, and 2 components respectively. For $R = R_2$, S_1 and S_1' fuse into S_2. For $R = R_3$, S_2 and S_2' fuse, while G remains separate

where
$$\tfrac{4}{3}\pi \bar{R}^3 \bar{n} = 1.$$

For $\tilde{R}_c = 1$ (that is, $R_c = \bar{R}$) the percolation is *abrupt*. For a Poisson distribution $\tilde{R}_c = 0.86$. Thus R_c is a revealing parameter of the distribution studied.

This method has been applied in the region of the Local Supercluster (in a cube of $40h^{-1}$ Mpc) from CFA data. As an illustration, we shall now compare the information extracted from the data (O in Fig. 11.24) with three theoretical models. The first (H) is a schematic version of the scenario where the galaxies are distributed according to a hierarchy without a privileged scale, that is, where the density is given by

$$n(r) \propto r^{-\gamma},$$

whatever the scale considered. The second (A) is a top–down model, and the third (P) is a Poisson distribution. Figure 11.24 shows the evolution of L with R. Observe that the curves O and A start the same way and that the model H seems to be unacceptable. The distribution of multiplicities of connected systems expressed as a percentage of the total number of galaxies contained in the sample for $R \approx 5$ Mpc is more striking. Figure 11.25 shows a peak for O and A corresponding to a structure containing close to half of the considered objects, unlike models H and P. However, the other galaxies of A seem to be distributed randomly (see the comparison of A and P), unlike the data where intermediate systems containing 2, 4, 6, ... galaxies are present. Finally the values determined for R_c, or rather

$$B_c \stackrel{\text{def}}{=} \frac{4\pi}{3}\left(\frac{R_c}{D}\right)^3,$$

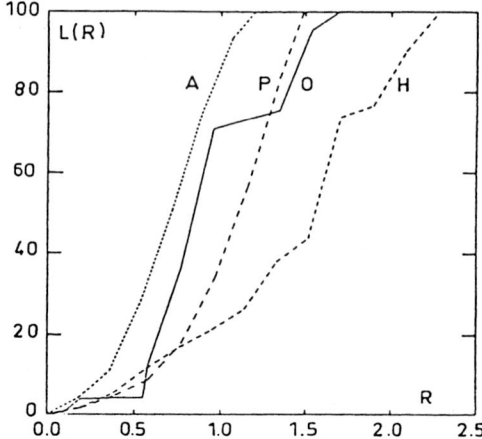

Fig. 11.24. The size L of the largest structures found as a function of the distance R for observational (O), adiabatic (A), hierarchical (H), and Poisson (P) catalogues

are
$$B_c \approx \begin{cases} 2.7 & \text{for P,} \\ 1.0 & \text{for O,} \\ 1.0 & \text{for A,} \\ 5.0\text{--}11 & \text{for H.} \end{cases}$$

It appears as a consequence that this type of study (with $L(R)$, B_c, and multiplicities) seems to favour a 'filamentary' vision of the universe and thus the top–down models. In fact 'filament'-type, 'pancake'-type, or hierarchical models are archetypical. In reality there can be a combination of these modes of distribution and the structures are perhaps not as simple as those in Fig. 11.22. A plausible hierarchical model does not embody well-defined islands of galaxies. In fact there are galaxies between clusters and they bias the percolation as the results of numerical simulations indicate.

11.5 Estimating the Mass of Groups and Clusters of Galaxies

To estimate the global mass (visible and invisible) of groups and clusters of galaxies, it is necesary to begin with their kinematics when framing hypotheses to do with their dynamic state. The clusters of galaxies represent such contrasts of density with respect to the intercluster medium that they cannot just be the fortuitous presence of objects but probably gravitationally bound structures. Assuming that they are in a stationary state of equilibrium, it is then possible to apply the virial theorem, linking the kinetic energy E_c to the potential energy U:

11.5 Estimating the Mass of Groups and Clusters of Galaxies

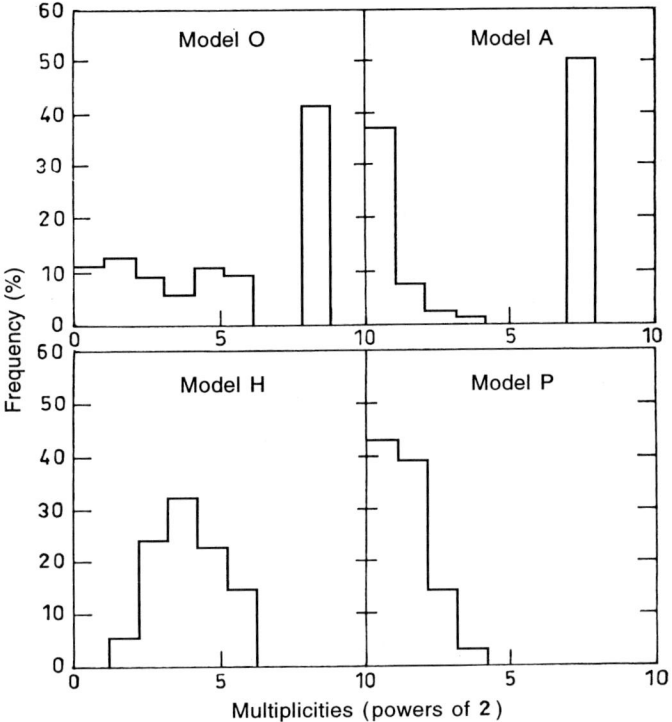

Fig. 11.25. Histograms (by percentages) of the number of components (in powers of 2) for a given distance R between neighbours

$$\langle 2E_c + U \rangle_t = 0,$$

where $\langle \ \rangle_t$ indicates an average over time. This leads to a relation between the total mass of galaxies, the velocity dispersion (that is, the measurement of velocities of galaxies with respect to that of the centre of mass), and a characterstic dimension of the system. The mass of the system and the mass–luminosity ratio, M/L, can then be deduced.

For N galaxies assumed to be point masses m_i with positions r_i and velocities v_i

$$\sum_i^N m_i \langle v_i^2 \rangle_t = \sum_{i=l}^N \sum_{j<l} G m_i m_j \left\langle \frac{1}{r_{ij}} \right\rangle_t$$

(assuming no correlations between m, v, and r). In practice the mean over time is replaced either by averaging over several systems or by calculating an estimate for a system comprised of many particles.

In fact what are measured are the velocities along the line of sight v_{proj} and the distances projected on the plane of the sky. Assuming that the velocity distribution is isotropic (that is, $v_x^2 = v_y^2 = v_z^2$) and there is spherical symmetry, as well as a ratio M/L identical for each galaxy, we get for the mass–luminosity ratio

$$\frac{M}{L} = \frac{3\pi}{2G} \left(\sum_{i=1}^{N} L_i v_{i\,\text{proj}}^2 \Big/ \sum_{i=1}^{N} \sum_{j=1}^{N} L_i L_j r_{ij\,\text{proj}}^{-1} \right).$$

The main difficulty with this method comes from the definition of the groups themselves, the inclusion of galaxies not belonging to the group, or vice versa, which seriously biases the estimate of the velocity dispersion.

Estimating M/L is very sensitive to the inclusion or exclusion of members with high velocities and at large interparticle distances, which can be galaxies seen in projection or real members. We are forced to define groups by the use of automatic algorithms that select, for example, the density excess beyond a certain threshold. The application of this technique to the CFA catalogue led to an estimate of $M/L \approx 260h$ (in solar units), indicating the presence of mass trapped in these systems (for an individual galaxy the visible mass gives $M/L \approx 5\text{--}10h$; see Chaps. 1 and 3).

Clusters of galaxies have the advantage of containing several tens of members and are, in principle, less sensitive to the statistical uncertainties that characterize the groups, but the problem of contamination by background galaxies remains. However, in practice the measurement of many radial velocities is telescope-time consuming, and complete samples are rarely achieved. Finally the presence of subgroups in various clusters, such as Coma, also has the effect of complicating the estimate of the velocity dispersion.

The amount of data available in general enables a more subtle study than the application of the virial theorem. It is possible in particular to construct a projected numerical density profile of the galaxies $\sigma(s)$. If $n(r)$ is the spatial density of the galaxies at a distance r from the centre, n_0 the central density (at $r = 0$), and s the projected radius, then

$$\sigma(s) = \int_{-\infty}^{+\infty} \frac{n(r) r \, dr}{\sqrt{r^2 - s^2}}.$$

We can then fit this observed profile to different theoretical models. The simplest is the 'isothermal' model. The spatial density $n(r)$ is thus linked to the gravitational potential $\Phi(r)$ (such that $\Phi(0) = 0$) and to the velocity dispersion $\langle v^2 \rangle$ by

$$n(r) = n_0 e^{-\Phi(r)/\langle v^2 \rangle}.$$

If the mass responsible for the potential is distributed like the galaxies, the Poisson equation for the potential is thus written as

$$\frac{1}{r^2} \frac{d}{dr} \left(r^2 \frac{d\Phi}{dr} \right) = 4\pi G n_0 m \, e^{-\Phi/\langle v^2 \rangle},$$

where m is the mass (assumed identical) of the galaxies. By changing the variables $\psi \stackrel{\text{def}}{=} \Phi/\langle v^2 \rangle$, $\xi \stackrel{\text{def}}{=} r/\beta$, and $\beta \stackrel{\text{def}}{=} \langle v^2 \rangle/(4\pi G n_0 m)^{1/2}$, the above

11.5 Estimating the Mass of Groups and Clusters of Galaxies

equation becomes the classical Emden equation for an isothermal sphere of gas in hydrostatic equilibrium:

$$\frac{1}{\xi^2}\frac{d}{d\xi}\left(\xi^2\frac{d\psi}{d\xi}\right) = e^{-\psi},$$

with boundary conditions $\psi = 0$ and $d\psi/d\xi = 0$ at $\xi = 0$.

However, at large radii ($\xi \gg 1$) $n(\xi) \propto 2n_0/\xi^2$, and as a consequence the total number and the total mass diverge as r; also the isothermal model is known not to be a good representation of the cluster as a whole. To overcome this problem King developed cut-off models by introducing a distribution function in phase space:

$$f(\boldsymbol{r},\boldsymbol{v})\,d^3r\,d^3v \propto e^{(\Phi(0)-\Phi(r))/\langle v^2\rangle_\infty}\left(e^{-v^2/2\langle v^2\rangle_\infty} - e^{\Phi(r)/\langle v^2\rangle_\infty}\right)d^3r\,d^3v,$$

where $\langle v^2\rangle_\infty$ is the velocity dispersion of a non-cut-off cluster. The velocity dispersion is thus cut off at the escape velocity $v_{esc} = (-2\Phi(r))^{1/2}$ giving $f(\boldsymbol{r},|\boldsymbol{v}| \geq v_{esc}) = 0$. The potential tends towards zero at infinity and the density $n(r)$ becomes zero at a finite radius.

Unfortunately none of the above models has an exact analytic representation. However, the following expressions:

$$n(r) = n_0\left[1 + (r/r_c)^2\right]^{-3/2},$$

$$\sigma(r) = \sigma_0\left[1 + (s/r_c)^2\right]^{-1},$$

where r_c is the 'core radius' and $\sigma_0 = 2n_0 r_c$, give a satisfactory approximation of the central parts of the isothermal model and are the ones currently used.

The use of models similar to the above with variations (galaxies having different masses, the velocity dispersion being dependent on the radius, the distribution of dark matter being different from that of the galaxies, and so on) leads to conflicting results. However, recent works has led to a 'mean' value of $(M/L)_B \approx 300h$ for galaxy clusters implying $\Omega \sim 0.3$ if representative of total matter distribution.

Like individual galaxies, many clusters of galaxies are X-ray sources. This radiation is emitted by a hot intercluster gas (10^8 K), the probable residue of the cloud of primordial gas giving birth to the system. The gas is then enriched by that dragged away from galaxies in the course of their movement in the cluster.

The gas that is in hydrostatic equilibrium (the crossing time of the cluster for a sound wave is very much less than the age of the universe) is a good 'tracer' of the gravitational potential and thus of the mass distribution. This technique is similar to that used for elliptical galaxies.

If the X-ray-emitting gas is in hydrostatic equilibrium in the global potential of the cluster, we get the relation

$$\frac{dp}{dr} = -G\frac{M(r)}{r^2}\rho,$$

where ρ and p are the density and the pressure and $M(r)$ is the mass inside the radius r. By using the perfect-gas relation we get for the mass

$$M(r) = -\frac{kT}{\mu G m_\mathrm{p}} r \left(\frac{d\ln\rho}{d\ln r} + \frac{d\ln T}{d\ln r}\right),$$

where μ is the mean molecular weight, m_p is the mass of the proton, and T is the temperature of the gas. Fitting a brightness profile of observed X rays (the brightness being proportional to the square of the density) and the X-ray spectrum allow us to determine the density gradient and the mean temperature. For a long time, the absence of spatial resolution for the X-ray data impeded the determination of a temperature gradient. As a consequence the determinations of masses and thus the ratios M/L depended on the different hypotheses made on the thermodynamic behaviour of the gas. Now, data from the various X-ray satellites ASCA (Japan), XMM-Newton (Europe) and CHANDRA (USA) have begun to give insights on the temperature gradient. In spite of those uncertainties, it is worth noting that the use of X-ray data led to a 'low' value of Ω via the 'baryon catastrophe' argument.

The Baryon Catastrophe. A limit on the value of the density parameter Ω_0 comes from the estimation of the baryonic contribution to the mass of clusters of galaxies. Several analyses (for example that of the well-studied Coma cluster) lead to a baryonic mass fraction which, if not sufficient to give $\Omega_0 = 1$, corresponds to an estimate of Ω_b (on the scale of clusters of galaxies) much greater than the baryon quantity predicted by the models of primordial nucleosynthesis in a $\Omega_0 = 1$ universe. This is the so-called *baryon catastrophe*.

Indeed, the baryonic mass in clusters is dominated by that of the X-ray emitting gas, which is superior to that of galaxies by a factor of 4 or 5. The fraction of baryons in clusters is therefore the fraction of the X-ray gas mass to the total virial mass and is of the order of

$$f_\mathrm{b} = (6 \pm 0.5)\% h^{-3/2}$$

If one assumes that the fraction of baryons in clusters is representative of that in the entire universe, which is supported by N-body simulations, then $\Omega_\mathrm{b} = f_\mathrm{b} \times \Omega_0$. Combining this with the limits on $\Omega_\mathrm{b} \times h^2$ imposed by the primordial nucleosynthesis (see Chap. 13), one gets:

$$0.007 < 0.06 h^{1/2} \Omega_0 < 0.024$$

or

$$0.11 h^{-1/2} < \Omega_0 < 0.4 h^{1/2}$$

From these values it follows that *if* $\Omega_0 = 1$ it is necessary to have a Hubble constant H_0 as low as 16 km s^{-1} Mpc^{-1}, very far from current estimates! On the other hand, using an H_0 value around 70 km s^{-1} Mpc^{-1} leads to an Ω_0 value smaller than 1 and so incompatible with the usual inflation paradigm.

However this discussion needs very precise estimates of the emitting gas fraction, and space missions such as CHANDRA or XMM will probably lead to new considerations. We should note also that the existence of a non zero cosmological constant allows Ω_0 to be much lower than 1 and solves the puzzle.

Finally, although the various estimates described above still entail uncertainties, the presence of matter trapped in groups and clusters is likely. It is probable that there is a common constituent of the haloes of spiral galaxies, groups, and clusters. However, its nature and precise grouping (similar or otherwise to that of the galaxies) remain open questions (see Chap. 13).

11.6 Large-Scale Motions. The Virgo Infall

11.6.1 The Great Attractor

If the distribution of galaxies in the universe is globally isotropic, we have seen, nevertheless, that on a small scale they will form into condensations such as groups and clusters. It is clear that the presence of a concentration of mass like that of a cluster of galaxies will create around it a more or less radial velocity field and play the role of an attractor. One such idea can be applied naturally to our Galaxy, since the Virgo cluster is situated in our immediate neighbourhood and has a recession velocity $U_{\text{recession}}$ of $\approx 1000 \, \text{km s}^{-1}$.

As a consequence the Galaxy undergoes an acceleration due to this local overdensity which gives to it a 'peculiar' velocity. This velocity field constitutes the difference with the Hubble flow, or the 'Virgo infall'. In other words, our recession velocity with respect to Virgo is smaller than $H_0 d_{\text{Virgo}}$, where d_{Virgo} is the real distance to Virgo and H_0 is the unperturbed Hubble constant, that is, calculated away from the perturbation caused by Virgo. Measurement of this difference with the Hubble flow enables us to measure the total mass contained in d_{Virgo}, the mass–luminosity ratio, and thus the local density Ω_0.

In Chap. 12 (on the formation of galaxies) we shall study, using a linear approximation, the evolution of density perturbations that have led to clusters like Virgo.

With these density perturbations is associated the velocity field $\boldsymbol{u}(\boldsymbol{x},t)$ such that
$$\frac{\mathrm{d}}{\mathrm{d}t}\big(\boldsymbol{u}(\boldsymbol{x},t)\big) + \frac{1}{R}\frac{\mathrm{d}R}{\mathrm{d}t}\boldsymbol{u} = \boldsymbol{g}\big(\boldsymbol{x}(t),t\big),$$
where R is the scale factor, $\boldsymbol{x}$ the comoving coordinate (that is, in a reference frame that accompanies the Hubble flow) of a galaxy, and $\boldsymbol{g}(\boldsymbol{x},t)$ the peculiar acceleration measured by an observer at the coordinate $\boldsymbol{x}$. The source of this

acceleration is the potential due to the density excess $\rho(\boldsymbol{x},t)$ with respect to the mean density $\langle\rho\rangle$ with contrast δ defined by

$$\delta(t) = \frac{\rho(\boldsymbol{x},t) - \langle\rho(t)\rangle}{\langle\rho(t)\rangle}.$$

This acceleration is given by

$$\boldsymbol{g}(\boldsymbol{x},t) = GR(t) \int d^3x' \, \delta(t') \langle\rho(t')\rangle \frac{\boldsymbol{x}' - \boldsymbol{x}}{|\boldsymbol{x}' - \boldsymbol{x}|^3}.$$

In the case where the velocities are small, such that the corresponding displacement is small with respect to the scale where g varies, that is, $|\boldsymbol{u}(\boldsymbol{x},t)|t \ll xR(t)$, $\boldsymbol{g}(\boldsymbol{x}(t),t) \approx \boldsymbol{g}(\boldsymbol{x},t)$ and the equation for the velocity is

$$\boldsymbol{u}(\boldsymbol{x},t) = \frac{1}{R(t)} \int_0^t R(t') \boldsymbol{g}(\boldsymbol{x},t') \, dt'.$$

Taking account of the definition of δ and the conservation relation

$$\langle\rho\rangle R^3 = \text{constant},$$

we thus find that $\boldsymbol{g}(\boldsymbol{x},t)$ is linked to its present value by

$$\boldsymbol{g}(\boldsymbol{x},t) = \left(\frac{R_0}{R(t)}\right)^2 \frac{\delta(t)}{\delta_0} \boldsymbol{g}(\boldsymbol{x},t_0),$$

which leads to

$$\boldsymbol{u}(\boldsymbol{x},t_0) = \boldsymbol{g}(\boldsymbol{x},t_0) \frac{R_0}{\delta_0} \int_0^{t_0} \frac{\delta(t')}{R(t')} \, dt'.$$

Knowledge of the temporal variation of the density perturbations thus determines the velocity field $\boldsymbol{u}(\boldsymbol{x},t)$. The density contrast $\delta(t)$ is given by the linearized equation (see Chap. 12)

$$\frac{d^2\delta}{dt^2} + \frac{2}{R}\frac{dR}{dt}\frac{d\delta}{dt} = 4\pi G \langle\rho\rangle \delta,$$

or

$$\frac{d}{dt}\left(R^2 \frac{d\delta}{dt}\right) = 4\pi G \langle\rho\rangle R^2 \delta$$

or again

$$\int_0^{t_0} \frac{\delta}{R} \, dt = \frac{1}{4\pi G \rho_0 R_0} \frac{d\delta_0}{dt}.$$

We then have

$$u = \frac{g_0}{4\pi G \rho_0} \frac{1}{\delta_0} \frac{d\delta_0}{dt} = \frac{g_0}{4\pi G \rho_0} R_0 H_0 f(\Omega),$$

where $f(\Omega) = (R/\delta)(d\delta/dt)$ depends on the solution δ and is approached by

$$f(\Omega) = \Omega^{0.6}.$$

We therefore end up with

$$u = \frac{2}{3} g_0 \frac{f}{H_0 \Omega_0} = \frac{2}{3} \frac{g_0}{H_0 \Omega_0^{0.4}}.$$

Assuming the potential is created by a spherical overdensity δ, this gives for the radial velocity

$$u = \frac{1}{3} \delta d_{\text{Virgo}} H_0 \Omega_0^{0.6}.$$

In this equation $H_0 d_{\text{Virgo}} = U_{\text{recession}} + u$ is the 'true' recession velocity of Virgo, that is, the observed recession velocity corrected for the peculiar velocity.

The problem is thus to estimate the local density contrast δ. *Assuming that the mass responsible is distributed like the galaxies* one estimate is given by the excess in the galaxy counts $\delta N/N = \delta \rho/\rho = \delta$ inside d_{Virgo} (with respect to a homogeneous universe). Under these conditions knowledge of the various parameters v, $\delta N/N$, and H_0 fixes the density parameter Ω_0.

Estimates of $\delta N/N$ are made by studying catalogues of galaxies, but differences exist between the authors. A mean estimate is $\delta N/N \approx 3$. It remains to determine the velocity of the Virgo infall. We have seen in Sect. 11.2 that there exists a relation between the intrinsic luminosity L of the galaxies and the internal velocity dispersion σ:

$$L \propto \sigma^4$$

(the Faber–Jackson relation).

If the radial velocity of a galaxy in Virgo were used (incorrectly) as a distance indicator to convert apparent magnitudes into absolute magnitudes, the luminosity would be systematically affected, it being underestimated with respect to the real value. Therefore the comparison of galaxies at the heart of Virgo with a sample of galaxies far away following the unperturbed Hubble flow allows us to determine the velocity of the motion (Fig. 11.26) and gives $u \approx 500 \text{ km s}^{-1}$.

11.6.2 Motion with Respect to the Cosmic Microwave Background Radiation

The cosmic black-body radiation at 2.7 K and the diffuse X-ray background together provide a reference frame with respect to which we can measure the differences from a uniform expansion. For an observer moving with respect to a system of comoving coordinates, the energy received per unit frequency, unit surface, unit time, and unit solid angle from the cosmic background differs from that received by a stationary observer. Firstly the number of photons is increased by

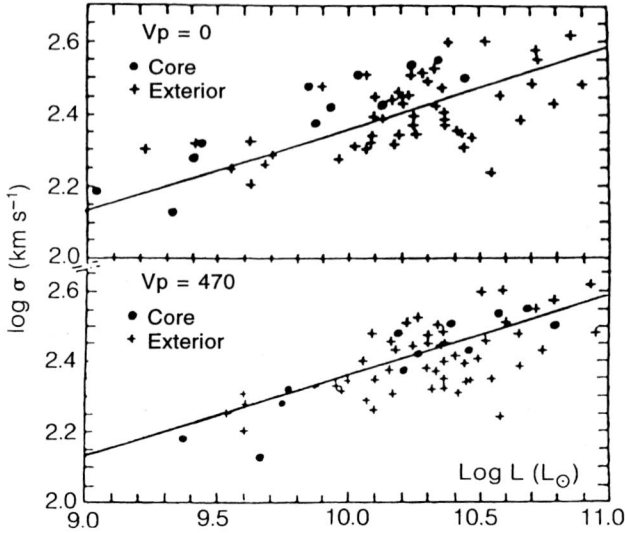

Fig. 11.26. The luminosity of elliptical galaxies as a function of their velocity dispersion in the heart of Virgo (•) and beyond (+). (**a**) Assuming a Virgo-infall velocity of zero; the galaxies of Virgo do not satisfy the correlation $L \propto \sigma^4$. (**b**) Assuming a Virgo-infall velocity for the Local Group of $\approx 450\,\mathrm{km\,s^{-1}}$

$$\frac{c\,dt + u\cos\theta\,dt}{c\,dt} = 1 + \beta\cos\theta,$$

where θ is the angle between the direction of reception and that of the motion, and $\beta = u/c$. Secondly the energy of each photon is increased (by the Doppler effect) by $1 + \beta\cos\theta$, while the frequency band is enlarged by the same amount. The two effects cancel out in the final analysis. Finally the solid angle at the receiver is smaller by $(1 + \beta\cos\theta)^{-2}$.

As a consequence the intensity in the direction θ for a moving observer $i'_{\nu'}$ is linked to the isotropic intensity i_ν by the relation

$$i'_{\nu'}(\theta) = (1 + \beta\cos\theta)^3 i_\nu.$$

For the intensity of a black body we have

$$i'_{\nu'}(\theta) = \frac{2h}{c^2} \frac{\nu'^3}{e^{h\nu'/k(1+\beta\cos\theta)T} - 1},$$

which is that of a black body at a temperature

$$T' = T(1 + \beta\cos\theta).$$

For a moving observer the radiation will appear hotter in the direction of motion and colder in the opposite direction with a 'dipolar' variation of the form

$$\frac{\delta T}{T} = \frac{u}{c}\cos\theta.$$

This variation in temperature is actually detected ($\Delta T/T \approx 10^{-3}$) and it can thus be used to determine the velocity of the observer. The velocity measured by this method is about $600\,\text{km}\,\text{s}^{-1}$ and the direction of motion is at about 45° from the centre of Virgo. The addition of these two motions implies a supplementary component of attraction to a large complex: the Hydra–Centaurus supercluster. Recent work suggests that this gigantic complex itself has a motion towards the 'Great Attractor' (Fig. 11.27), a galaxy concentration 20 times more massive than Virgo and located at 45 Mpc. We have presented here a schematic analysis. More sophisticated methods (*POTENT*, wavelet reconstruction ...) led to the determination of a combination of Ω and the bias parameter b and then to a degeneracy in the parameter space!

Studies of large-scale motions (Virgo, the Great Attractor, ...) represent the largest scale on which a dynamical measurement of Ω_0 has been possible. According to whether the hypothesis of similar distributions of galaxies and

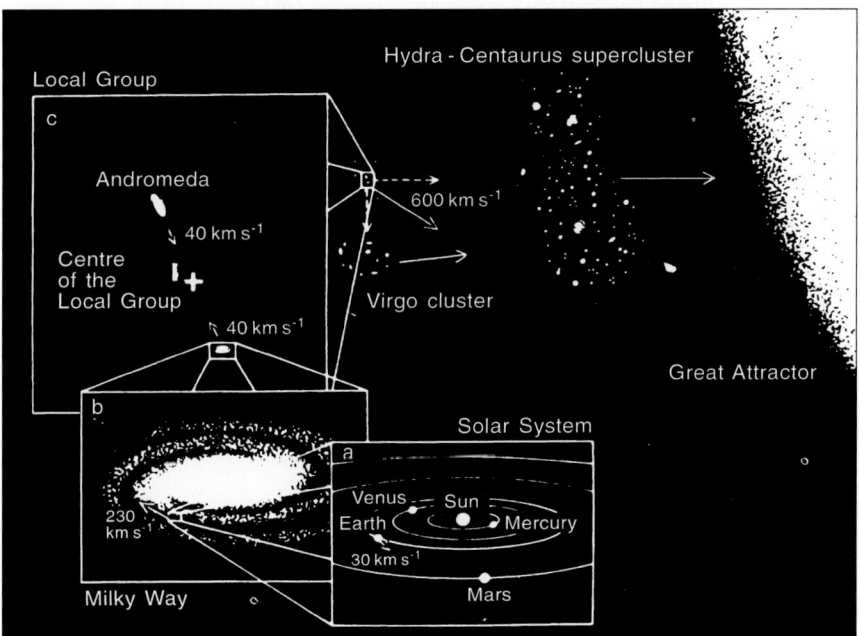

Fig. 11.27. The absolute motion of the Local Group. The attraction due to the Virgo cluster constitutes a component of the motion; however, another component is directed towards the Hydra–Centaurus supercluster, itself moving. Recent results suggest that the Local Group, Virgo, and Hydra–Centaurus are falling towards a vast concentration of galaxies at about two times the distance of the supercluster: the Great Attractor. (From Dressler 1987)

Exercises

11.1 Knowing that Sirius has an apparent magnitude $m_V = -1.45$ and an absolute magnitude $M_V = 1.41$, calculate its distance in parsecs.

11.2 Show that the mass–luminosity ratio of a group of clusters determined from the virial theorem is proportional to H_0.

11.3 The analysis of catalogues of galaxies assumed to constitute representative samples of the distribution of matter in the universe gives for an estimate of the total luminosity density (in the V band)

$$L_V \approx 2 \times 10^8 h \, L_\odot \, \text{Mpc}^{-3}.$$

Deduce the value of the critical mass–luminosity ratio $(M/L)_\text{crit}$ corresponding to the critical density ρ_crit (that is, $\Omega_0 = 1$).

11.4 Calculate for a typical cluster of galaxies ($kT_e \approx 5\,\text{keV}$) the order of magnitude of the Sunyaev–Zel'dovich effect.

References

Aaronson, M. (1986) in *Proceedings of the NATO Advanced Workshop on Galaxy Distances and Deviations from Universal Expansion*, edited by B. F. Madore and P. B. Tully (NATO–ASI) p. 57.
Binney, J., and Tremaine, S. (1986) *Galactic Dynamics* (Princeton University Press, Princeton NJ).
Boldt, E. (1987) *Phys. Rep.* **145**, 215.
Bottinnelli, L., Gougenheim, L., and Paturel, G. (1984) *Ann. Phys. France* **9**, 503.
De Lapparent, V. I. (1986) Thesis, University of Paris.
Dressler, A. (1987) *Scientific American* July 1987.
Gougenheim, L., and Bottinelli, L. (1987) in *Aux Confins de l'Univers*, edited by J. Schneider (Fayard, Paris).
Luminet, J. P. (1982) in *Grand Atlas d'Astronomie*, edited by J. Audouze and G. Israël (Encyclopedia Universalis, Paris).
Peacock, J.P. (1999) *Cosmological Physics* (Cambridge University Press, Cambridge, MA)
Peebles, P.J.E.,1993 *Principles of Physical Cosmology* (Princeton University Press, Princeton NJ)
Sandage, A. (1957) *Astrophys. J.* **125**, 435.
Sarazin, G. L. (1986) *Rev. Mod. Phys.* **58**, No. 1.
Sunyaev, R. A., and Zel'dovich, Ya. B. (1980) *Ann. Rev. Astron. Astrophys.* **18**, 537.

Tammann, G. A. (1979) in *Physical Cosmology*, edited by R. Balian, J. Audouze, and D. N. Schramm (Les Houches Summer School, NATO, North Holland, Amsterdam), p. 66.

Tonry, J. L., Blakeslee, J. P., Ajhar, E. A., Dressler, A. (1997) *Astrophys. J.* **475**, 399.

Trimble, V. (1987) *Annu. Rev. Astron. Astrophys.* **25**, 425.

12. The Formation of Galaxies and Large Structures in the Universe

The formation of galaxies and large structures in the universe remains an unresolved problem. The starting point for current scenarios based on the idea of 'gravitational instability' is, however, simple. In a medium of uniform density a local density excess (overdensity) of matter will attract nearby matter by the effect of its own gravitation and this effect will accelerate. This type of model, which takes up the ideas of Jeans on the evolution of inhomogeneities of a static gravitational medium, predicts a very rapid (exponential) increase of such irregularities. In fact the application of it to a homogeneous, expanding universe of density $\bar{\rho}$ with some primordial irregularities $\rho_{\text{pert}} = \bar{\rho} + \delta\rho$ (where $\delta\rho/\rho \ll 1$) shows that the increase is not as rapid. There is in effect competition between the growth of the perturbations, characterized by a gravitational collapse time $(t_{\text{eff}} \propto (G\rho_{\text{pert}})^{-1/2})$, and the expansion $(t_{\text{exp}} \propto (G\bar{\rho})^{-1/2})$, which tends to dilute all local overdensities. The result is that the growth of fluctuations in an expanding universe is slow. For baryons this increase only begins at recombination ($z_{\text{rec}} \approx 1000$) and in the linear phase the density contrast only grows by a factor $(1 + z_{\text{rec}}) \approx 10^3$. Now to obtain an excess $\delta = \delta\rho/\bar{\rho} = (\rho - \bar{\rho})/\bar{\rho} \geq 1$ corresponding to existing objects (galaxies, clusters, and so on), an initial inhomogeneity δ_i of the order of 10^{-3} must be assumed. However, there is at least one observational constraint on the initial amplitude δ_i. It is given by observation of the 2.7 K cosmic background radiation. This radiation is homogeneous to better than 10^{-5} ($\delta T/T \leq 10^{-5}$), and as density fluctuations lead to temperature fluctuations ($\delta T/T \approx \delta\rho/\rho$), it follows that at the epoch of recombination the matter must have the same degree of inhomogeneity. This is the main difficulty that confronts models of galaxy formation. The appearance of grand unification theories (GUTs) and inflationary models (Chap. 13) partly resolves these problems by predicting in particular the existence of nonbaryonic particles ('-inos') of various masses. On the one hand there is the hidden contribution to the density parameter Ω_0 (predicted equal to 1 by inflationary theories), and on the other there is the source of fluctuations leading to large structures in the real universe. The growth of these fluctuations can begin before recombination; it is thus possible to postulate a δ_i compatible with observations of the cosmic microwave background. However, even in this setting, many difficulties still persist, but we shall nonetheless try to give a review of this rapidly evolving field.

The formulation of the problem of the evolution of primordial fluctuations, while simple in principle (linearize Einstein's equations for small δ), is complex in practice (owing to the formalism of general relativity, the interactions of various fluids: baryons, photons, nonbaryonic particles, and so on). However, if the characteristic scale of the perturbations is much smaller than the horizon and if the gravitational effects of the pressure can be neglected, we can limit ourselves to a Newtonian formalism. (In what follows we shall put the emphasis on physical ideas rather than the mathematical formalism.)

12.1 The Jeans Mass and the Growth of Perturbations

12.1.1 The One-Component Model

The mathematical problem of the instability of a homogeneous distribution of matter was first worked on by Jeans. This theory takes account of two factors: gravity, which forces the matter to condense, and pressure, which tends to reduce the inhomogeneities. In a collisional fluid the behaviour of waves (characterized by the velocity of sound c_s) is subject to the competition between the pressure, which causes the propagation of sound waves of constant amplitude, and the gravitation, which seeks to increase this amplitude. The greater the size of the system considered, the more gravity tends to overcome the pressure and will lead to the system collapsing in on itself. The size of the system that sets a limit on these two components is known as the Jeans length, L_J. In a collisionless fluid the same description holds, gravity thus overcoming the velocity dispersion of the particles.

More precisely if we linearize the hydrodynamic equations:

$$\frac{\partial \boldsymbol{u}}{\partial t} + (\boldsymbol{u} \cdot \nabla)\boldsymbol{u} = -\nabla \Phi - \frac{1}{\rho}\nabla p,$$
$$\frac{\partial \rho}{\partial t} + \nabla \cdot \boldsymbol{u} = 0,$$
$$\nabla^2 \Phi = 4\pi G \rho,$$

where p, ρ, u, and Φ are respectively the unperturbed pressure, density, velocity, and gravitational potential, we obtain for $\delta = \delta\rho/\rho$ (the unperturbed solution corresponding to a static and uniform fluid $\rho = \mathrm{constant}$, $p = \mathrm{constant}$, $u = 0$)

$$\ddot{\delta} = 4\pi G \rho \delta + c_s^2 \nabla^2 \delta,$$

where c_s is the velocity of sound.

The qualitative behaviour described above is thus easily obtained. It can be seen that in the absence of a pressure term δ will grow exponentially with a time scale $t_{\mathrm{eff}} \approx (G\rho)^{-1/2}$. With the pressure term alone we have sound waves with a period $t_s \approx c_s^{-1} L$. The above limit from which gravitational instability will appear is thus given by $t_{\text{eff}} \approx t_s$, which gives

$$L_J \approx c_s (G\rho)^{-1/2},$$

to which the Jeans mass is connected by

$$M_J \approx L_J^3 \rho \approx c_s^3 G^{-3/2} \rho^{-1/2}.$$

The Jeans mass is then the minimum mass of a system for which the pressure cannot counteract the increase in the density contrast. As a consequence a matter distribution having several inhomogeneities will condense into discrete objects with a size greater than or equal to M_J. It is important to note that for masses of about M_J the pressure is not negligible and the systems formed will have a more or less spherical shape.

The above description applies to a static medium, whereas we are interested in the formation of structures in an expanding universe whose characteristic time is (Chap. 13)

$$t_{\text{exp}} \approx (G\rho)^{-1/2} \approx (H)^{-1}.$$

This time is comparable to the characteristic time of exponential growth when gravity dominates and thus leads to a modification of the above result. We now consider the situation within an expanding universe and introduce in place of the variables (r, v) the coordinates (x, u) defined by

$$r = R(t)x,$$

$$v = \dot{R}(t)x + R(t)u,$$

where $R(t)$ is the scale factor (Chap. 13).

In this coordinate system the above equations can be written as

$$\frac{\partial u}{\partial t} + \frac{1}{R}(u \cdot \nabla)u + \frac{\dot{R}}{R}u = -\frac{1}{\rho R}\nabla p - \frac{1}{R}\nabla \Phi,$$

$$\frac{\partial \delta}{\partial t} + \frac{1}{R}\nabla \cdot (1+\delta)u = 0,$$

$$\nabla^2 \Phi' = 4\pi G R^2 (\rho - \bar{\rho}) = 4\pi G R^2 \bar{\rho} \delta.$$

In the case of small perturbations ($\delta \ll 1$), linearizing these equations gives

$$\frac{\partial^2 \delta}{\partial t^2} + \frac{2\dot{R}}{R}\frac{\partial \delta}{\partial t} = \left(\frac{c_s}{R}\right)^2 \nabla^2 \delta + 4\pi G \bar{\rho} \delta.$$

By writing the solution as a sum of plane waves,

$$\delta = \delta(t) e^{i\mathbf{k} \cdot \mathbf{x}},$$

$$\lambda = 2\pi R(t)/k,$$

we obtain

$$\frac{d^2\delta}{dt^2} + \frac{2\dot R}{R}\frac{d\delta}{dt} = \left[4\pi G\bar\rho - (c_s k/R)^2\right]\delta.$$

The term on the right shows the effect of the competition between gravity and the pressure gradient mentioned before.

The pressure and gravitational terms will be equivalent when the wavelength is equal to the Jeans length:

$$L_J = c_s(\pi/G\bar\rho)^{1/2}.$$

In the case $\lambda \ll L_J$ the pressure term dominates, which leads to a sound wave whose amplitude slowly decreases owing to the expansion term

$$\delta = A(t)e^{-i\Phi(t)} \propto \frac{1}{\sqrt{c_s(t)R(t)}} e^{-i\int(c_s k/R)\,dt}.$$

In the case $\lambda \gg L_J$ the gravitational term dominates and we get

$$\frac{d^2\delta}{dt^2} + \frac{2\dot R}{R}\frac{d\delta}{dt} = 4\pi G\bar\rho\delta.$$

In this case δ also increases but the increase is no longer exponential owing to the expansion term, which plays the role of a friction term.

The type of growth will depend on the type of universe chosen by means of the temporal dependence of $R(t)$. In the case $\Omega_0 = 1$ (the Einstein–de Sitter universe) $R(t) \propto t^{2/3}$ and it can be easily shown that δ expands as

$$\delta = A(\boldsymbol{x})t^{2/3} \propto R(t) \propto (1+z)^{-1}.$$

In the case of a universe of low density (such that $\Omega_0 \ll 1$) the scale factor behaves as follows:

$$R(t) \propto \begin{cases} t^{2/3} & \text{for} \quad 1+z \gg 1/\Omega_0, \\ t & \text{for} \quad 1+z \ll 1/\Omega_0. \end{cases}$$

We thus have

$$\delta \propto \begin{cases} t^{2/3} & \text{for} \quad 1+z \gg 1/\Omega_0, \\ \text{constant} & \text{for} \quad 1+z \ll 1/\Omega_0. \end{cases}$$

The result is that in a universe of low density the perturbations stop growing for $z_F \propto 1/\Omega_0$. Indeed beyond this the universe has too rapid a growth rate for inhomogeneities to develop. It would then be difficult a priori for galaxies to form in an open universe.

In the case where the fluid is composed of noninteracting particles (a collisionless fluid) the description is similar to replacing the competition between gravity and pressure by that between gravity and kinetic energy. The Jeans mass can thus be calculated by using the velocity dispersion $\langle v^2\rangle^{1/2}$. When the energy density is dominated by radiation (and neutrinos), and in fact in the case where the wavelength of the perturbation is larger than the horizon distance (ct), it can be shown that the increase is proportional to $(1+z)^{-2}$ (see the exercises at the end of the chapter).

12.1.2 The Two-Component Model

Up to now we have presented the case of a medium with one component. In fact a realistic evolutionary model of the perturbations of a baryonic or 'dark' component must take account of the presence of a sea of photons.

In the case of a universe composed of photons of density ρ_γ ($\propto R^{-4}$) and (nonrelativistic) baryonic or dark matter of density ρ_x ($\propto R^{-3}$) there is no increase in density contrast for the nonrelativistic components as long as the latter constitutes a negligible fraction of the total density (the Zel'dovich–Mezsaros effect). If the pressure is negligible, the linearized equation can always be written as

$$\frac{\partial^2 \delta}{\partial t^2} + \frac{2\dot{R}}{R}\frac{\partial \delta}{\partial t} = 4\pi G \rho_x \delta,$$

where δ is the (fractional) density contrast of the nonrelativistic material. If we set $y = \rho_x/\rho_\gamma = R/R(t_{eq})$, where $R(t_{eq})$ is the value of the scale factor at the epoch of equilibrium t_{eq} (where $\rho_x = \rho_\gamma$; see Chap. 13), we have

$$\frac{d^2\delta}{dy^2} + \frac{2+3y}{2y(1+y)}\frac{d\delta}{dy} - \frac{3\delta}{2y(1+y)} = 0.$$

We thus obtain a growing mode such that

$$\delta_x \propto 1 + 3y/2.$$

For $y < 1$ the perturbation remains constant. The physical reason is that in this case ($\rho_x < \rho_\gamma$) the time of expansion ($\propto (G\rho_\gamma)^{-1/2}$) is much smaller than the time of growth of the component x ($\propto (G\rho_x)^{-1/2}$) up to t_{eq}.

Table 12.1 summarizes the different regimes.

Table 12.1. Radiation and Matter Regimes in the Universe

	Radiation-dominated era $z > z_{eq}$ $t < t_{eq}$	Matter-dominated era $z < z_{eq}$ ($z \gg 1/\Omega_0$) $t > t_{eq}$
$\lambda > ct$	$\delta \propto (1+z)^{-2}$	$\delta \propto (1+z)^{-1}$
$\lambda < ct$	$\delta \approx$ constant	$\delta \propto (1+z)^{-1}$

Note first that in the case of a purely baryonic universe the increase of perturbations will be a maximum for $\Omega_0 = 1$, but when $t_{eq} \approx t_{rec}$, this increase will only be (in the linear phase) by a factor of approximately 10^3 ($z_{rec} \approx 1000$), which implies initial perturbations of approximately 10^{-3} to reach the nonlinear stage today. As we shall see later on in detail this does not agree with the observation of fluctuations in the 2.7 K radiation. This leads us to consider the possible evolution of dark-matter fluctuations. We then assume in addition to the baryons and photons the existence of a (nonbaryonic) component contributing to Ω and thus to the dynamics of the universe (one possibility is the existence of massive neutrinos, photinos, and so on or completely different particles).

12.1.3 Dark Matter

The candidates for dark matter are in general divided into two large classes: hot dark matter (HDM) and cold dark matter (CDM). This distinction derives from their behaviour (due to their mass) at the time of the decoupling phase at a temperature T_D. Each component remains in equilibrium so long as the characteristic time of the reactions that couple it to the rest of the matter is less than the expansion time. If this is not the case there is decoupling and the number of particles per unit comoving volume is then 'frozen'. If this occurs while the particles are still relativistic, the species is said to be 'hot'; when they are nonrelativistic, the species is said to be 'cold'.

For a relativistic component 'x' at the time of decoupling ($kT_D \gg m_x c^2$) the abundance n_x is of the order of magnitude of that of the photons. If these particles dominate the mass of the universe and give $\Omega \approx 1$, we then have

$$\sum m_x \approx 100\,\text{eV}.$$

The velocity dispersion of this 'hot matter' is thus (in the absence of subsequent gravitational interaction)

$$\langle v^2 \rangle_x^{1/2} = \frac{kT_D}{m_x c} \frac{1}{1+z}.$$

For a massive neutrino of mass $m_\nu \approx 30\,\text{eV}$ this velocity is of the order of several km s^{-1}. When $kT_D \ll m_x c^2$, the particles are no longer relativistic at the time of decoupling and their abundance relative to that of the photons is thus exponentially small:

$$\frac{n_x}{n_\gamma} \propto \left(\frac{m_x}{T}\right)^{3/2} e^{-m_x c^2 / kT}$$

and will depend on the details of the interaction. We can show that the resulting mass density depends on the mass as follows:

$$\rho_x \approx 10^{-28}\,\text{g cm}^{-3}\,(m[\text{GeV}])^{-1.85},$$

whence, if these particles are to contribute significantly to ensuring that $\Omega_0 = 1$,

$$\rho_x \leq \rho_c \approx 2 \times 10^{-29}\,\text{g cm}^{-3}$$

and

$$m_x \geq 2\,\text{GeV}.$$

The velocity dispersion of such particles,

$$\langle v^2 \rangle^{1/2} \approx \left(\frac{kT_D}{m_x} \right)^{1/2} \frac{1}{1+z},$$

is thus very small ($T_D < 1\,\text{MeV}$). A possible candidate for this cold matter is the photino.

12.2 The Origin, Spectrum, and Nature of the Fluctuations

12.2.1 The Origin of the Fluctuations

An immediate question concerns the origin of the fluctuations as well as the value of their amplitude for a given scale and the dependence as a function of this scale (in other words, what is the *initial spectrum of the fluctuations?*). The homogeneity of the universe, which is the basis of cosmological models, is observed on a large scale ($D > 100\,\text{Mpc}$). We thus assume the presence, in its primordial phase, of small fluctuations that will subsequently develop. The nature of the physical process that is the source of these fluctuations is still unknown. We can imagine a statistical origin or the existence of a primordial turbulence of cosmic fluid, or maybe something else. In the standard model these fluctuations are present ab initio and this model does not propose a single process peculiar to the creation of these perturbations.

In fact our present knowledge of physics does not allow us to go back to the time before the Planck era or the era of quantum gravity ($t \approx 10^{-43}$ s). In all probability before this time as yet unknown mechanisms caused the sudden appearance from quantum chaos of a homogeneous universe peppered with fluctuations with a gravitoquantum origin. As regards the spectrum the principle of 'maximum ignorance', and also the fact that gravity does not have a characteristic scale, leads us to postulate a power-law spectrum:

$$\frac{\delta\rho}{\rho} = \frac{\delta M}{M} = A M^{-\alpha}.$$

The necessity of not having too significant perturbations on a small scale (leading to an abundant population of primordial black holes) or on a large scale (corresponding to an inhomogeneous universe) leads us to imagine values of α close to 0 or 1. In this setting the initial amplitude is simply adjusted to give the density contrasts currently observed. One particular and very important value of the exponent is $\alpha = 2/3$; this is the 'Harrison–Zel'dovich' spectrum. It is such that the primordial fluctuations enter into the horizon (see Sect. 13.1) with an amplitude independent of the scale. One such spectrum is predicted by inflationary models with an amplitude of approximately 10^{-4}–10^{-5} compatible with the constraints imposed by galaxy-formation models.

12.2.2 Isothermal and Adiabatic Fluctuations

A distinction must finally be made between fluctuations according to whether they perturb the entropy per baryon or not. On the one hand we can consider the isothermal or entropic fluctuations associated with variations of the baryonic density ($\delta n_B \neq 0$) in a uniform sea of photons ($\delta n_\gamma = 0$, $T = $ constant); on the other hand adiabatic fluctuations where the entropy remains constant. There are in this case perturbations of both the photons and the baryons (or other particles that contribute to the entropy).

In the adiabatic case we have $s = $ constant, and thus

$$\frac{\delta s}{s} = \frac{3\delta T}{T} - \frac{\delta n_B}{n_B} = 0,$$

which signifies that at each density fluctuation there will correspond a fluctuation of the 2.7 K black-body temperature

$$\frac{\delta T}{T} = \frac{1}{3}\frac{\delta \rho}{\rho},$$

unlike the isothermal case for which

$$T = \text{constant},$$
$$\frac{\delta s}{s} = \frac{\delta n_B}{n_B},$$
$$\frac{\delta T}{T} = 0.$$

Recent developments in particle physics (GUTs) give priority particularly to adiabatic fluctuations. Indeed these theories predict a constant value for the ratio $n_\gamma/n_B \approx 10^{11}$. As a consequence this forbids the presence of isothermal fluctuations (where the entropy, and thus the ratio n_γ/n_B, varies). In this theory only adiabatic fluctuations can lead to the formation of galaxies.

From the point of view of models, the repartition of fluctuations at a given instant is regarded as a random homogeneous process in space. The function $\delta(\boldsymbol{x},t) = \delta = (\rho - \bar\rho)/\bar\rho$ is a random function of $\boldsymbol{x}$ with a mean value of zero ($\langle \delta \rangle = 0$) and whose correlation function is given by

$$\langle \delta(\boldsymbol{x})\delta(\boldsymbol{x}+\boldsymbol{r}) \rangle = \int \delta(\boldsymbol{x})\delta(\boldsymbol{x}+\boldsymbol{r})\,\mathrm{d}^3\boldsymbol{x}$$

($\langle \delta^2(\boldsymbol{r}) \rangle$ is thus a measure of the fluctuations at a point $\boldsymbol{r}$).

Normally one introduces the Fourier transform of δ:

$$\delta_{\boldsymbol{k}} = \int \delta(\boldsymbol{x},t) e^{i\boldsymbol{k}\cdot\boldsymbol{x}}\,\mathrm{d}^3\boldsymbol{x},$$

where $|\boldsymbol{k}| = 2\pi R(t)/L$ is the comoving wave number associated with the scale L.

Like δ, $\delta_{\boldsymbol{k}}$ is a random function of $\boldsymbol{k}$ defined by its modulus $|\delta_{\boldsymbol{k}}|$ and the phase $\Phi_{\boldsymbol{k}}$ such that
$$\delta_{\boldsymbol{k}} = |\delta_{\boldsymbol{k}}| e^{i\Phi_{\boldsymbol{k}}}.$$
In general we hypothesize that δ is a Gaussian process whose phases are initially random.

From the preceding definitions the correlation function of the random processes δ turns out to be
$$\xi(r) = \langle \delta(\boldsymbol{x}+\boldsymbol{r})\delta(\boldsymbol{x})\rangle \propto \int d^3\boldsymbol{k}\, e^{-i\boldsymbol{k}\cdot\boldsymbol{r}} \langle |\delta_{\boldsymbol{k}}|^2 \rangle,$$
that is, the Fourier transform of the 'power spectrum' $\langle |\delta_{\boldsymbol{k}}|^2\rangle$ (that is, the variance of $|\delta_{\boldsymbol{k}}|$). This expresses a link between a theoretical value (the spectrum of fluctuations) and an observable (the correlation function).

Other quantities (connected with the spectrum) are also currently used. Thus the measurement of differences from the mean density (or the mean mass) at a given scale $L \approx 1/k$ is
$$\delta M = \left(\left\langle \frac{\delta M}{M}\right\rangle^2\right)^{1/2} = \delta = \left\langle \left(\frac{\delta\rho}{\rho}\right)^2\right\rangle^{1/2}$$
and is connected to $|\delta_{\boldsymbol{k}}|^2 = k^n$ (a power-law spectrum) by
$$\delta M \propto \left(\int |\delta_{\boldsymbol{k}}|^2 k^2\, dk\right)^{1/2} \propto \left(k^3 |\delta_{\boldsymbol{k}}|^2\right)^{1/2}$$
(with $k \propto 1/L$), and as a result various dimensional dependences follow:
$$\delta = \delta M \propto k^{(n+3)/2} \propto L^{-(n+3)/2} \propto M^{-\alpha},$$
with $\alpha = (1/2) + (n/6)$, a relation which expresses the connection between the exponents α and n. If we now look at the velocities induced by the growth of density fluctuations, we obtain from the continuity equation (written in Fourier space) $v_k \propto \delta_k/k$, from which
$$v = \langle v^2\rangle^{1/2} \propto \left(\int k^2\, dk |v_k|^2\right)^{1/2} \propto \left(k^3 |v_k|^2\right)^{1/2} \propto \left(k|\delta_{\boldsymbol{k}}|^2\right)^{1/2}.$$

These are quantities calculated in models that have been compared with observations.

12.3 The Linear Evolution of Perturbations

We can now examine the behaviour of the perturbations and their spectrum over the thermal history of the universe. As we shall see, many mass scales will be introduced naturally into the spectrum (we shall limit ourselves here to calculations of orders of magnitude). In a first step we work on the assumption of a purely baryonic universe. The Jeans mass fixes the lower limit of perturbations that will grow in the course of time: above M_J there is a growth of perturbations; below it, acoustic-type behaviour. From its beginnings the universe was dominated by radiation and the velocity of sound was thus of the order of that of light, $c_s \approx c/\sqrt{3}$, while the density varied as T^4. Hence

$$L_J \propto c t_{\exp} \propto L_{\text{horizon}},$$
$$M_J \propto c^3 T_\gamma^{-2},$$

which *increases* with the expansion (approximately as the mass within the horizon, M_{horizon}) since T_γ decreases. An abrupt change appears at recombination. At this time the universe is dominated by matter; the density varies as

$$\rho_m \propto R^{-3} \propto T_\gamma^{-3}$$

and the velocity of sound as

$$c_s \propto R^{-1},$$

from which we obtain

$$M_J \propto R^{-3/2}.$$

After recombination the Jeans mass decreases. Calculations of the values of M_J just *before* and *after* recombination show that they are of the order of

$$M_J(\text{before}) \approx 1.4 \times 10^{18}\, M_\odot \left(\Omega h^2\right)^{-1/2} \left[1 + 30 \left(\Omega h^2\right)^{3/2}\right],$$
$$M_J(\text{after}) \approx 7.9 \times 10^{4}\, M_\odot \left(\Omega h^2\right)^{-1/2};$$

that is to say, M_J suddenly falls by about 12 orders of magnitude at this epoch. We now distinguish between isothermal and adiabatic modes. In principle we have seen the existence of an acoustic mode below the Jeans mass. However, before recombination, baryons and photons are strongly coupled and this leads to dissipative phenomena. On the one hand in the isothermal case the electrons (and the protons) at the time of the growth phase must move (with a velocity v) with respect to a uniform background of photons. This results in a frictional force (per unit mass of the matter) equal to

$$f = -\gamma_f v,$$

where $\gamma_f = 4\sigma_T \rho_\gamma c / 3 m_p$ (σ_T is the effective Thomson cross-section), which is introduced into the equation of evolution of δ. It can be easily verified that the time associated with this phenomenon (γ_f^{-1}) is much shorter than that associated with the growth of perturbations and so prevents them. On the other hand up to $z = z_{\text{eq}} \approx z_{\text{rec}}$ no growth occurs at all, the matter not being dominant (the Mezsaros effect). As a consequence the initial matter distribution (the spectrum of fluctuations) is preserved up to recombination in the isothermal case. At the start of recombination, masses greater than the Jeans mass ($\approx 10^6 \, M_\odot$) can thus increase.

In the adiabatic case the interaction between photons and electrons (viscosity and thermal conduction) will suppress the oscillatory behaviour below a minimum scale. Indeed whereas the material particles tend to agglomerate under the effect of their mutual gravitation, the photons that diffuse within the overdensity will have the effect of dispersing them owing to the radiation pressure. This diffusion of photons in the interior of the fluctuation can be conceived of as a random walk. The time necessary for a photon to diffuse over a distance L is thus $t \approx L^2/cl_\gamma$ (l_γ designating the mean free distance, $(\sigma_T n_e)^{-1}$). As a consequence, for a perturbation on a given scale L to survive the dissipation it is necessary that the associated time t is *greater* than the time of expansion. This fixes a minimum scale $L \approx (l_\gamma c t_{\exp})^{1/2}$ and a mass $M_S \approx 10^{12}(\Omega h^2)^{-5/4} \, M_\odot$ (also known as the Silk mass) below which the perturbations are 'rubbed out'. As a result, the spectrum of adiabatic perturbations at the end of recombination will be characterized by a *cutoff at small scales*. After recombination the masses capable of growing will be of the order of $10^{13} \, M_\odot$, much greater than the Jeans mass ($\approx 10^6 \, M_\odot$). In this case the large structures (clusters of galaxies) form first, the galaxies forming afterwards by fragmentation. This is the so-called pancake scenario.

We now examine the evolution of perturbations for (noncollisional) dark matter formed out of neutrinos. In an analogous manner to the case of baryons small fluctuations of neutrinos will develop by 'gravitational instability' for masses greater than the Jeans mass M_{J_ν}. As long as the neutrinos are relativistic the Jeans mass increases over time (as approximately M_{horizon}). Below the Jeans mass the behaviour is, however, different from that of the baryons. The neutrinos are in effect almost without interaction, which leads to the absence of acoustic behaviour, by means of a damping effect. Indeed as long as they are relativistic the neutrinos have very high velocities and cannot remain trapped in the overdensities. They easily escape these regions and, owing to the random direction of their velocities, tend to make the density distribution uniform by 'rubbing out' the fluctuations. On the other hand once they become nonrelativistic (NR),

$$kT_{\text{NR}} \approx m_\nu c^2 \approx 6 \times 10^3 m_\nu \, [\text{eV}] \approx 1.8 \times 10^5 m_\nu \, [\text{eV}]/30,$$

their velocity decreases, thus allowing the growth of perturbations. The distance that they can traverse freely, L_{FS} (the free streaming length), up to t_{NR} is therefore the minimum size of fluctuations that can survive the phenomenon of damping. We then have

$$L_{\text{FS}} \approx \begin{aligned} & ct_{\text{NR}} \\ & \approx c(T_{\text{Planck}}/T_{\text{NR}})^2 t_{\text{Planck}} \\ & \hbar M_{\text{Planck}}/c^2 m_\nu^2 \\ & \approx 40(m_\nu/30 \, [\text{eV}])^{-1} \, \text{Mpc} \\ & \approx 13(\Omega_0 h^2)^{-1} \, \text{Mpc}. \end{aligned}$$

Only those perturbations of neutrinos having scales greater than L_{FS}, to which there corresponds a mass

$$M_{\text{FS}} = M_{\nu\,\text{max}} \approx 4 \times 10^{15} (m_\nu/30\,[\text{eV}])^{-2}\,M_\odot$$

characteristic of that of superclusters, can grow.

After the decoupling of the photons the increase of baryonic fluctuations (which up until then have been in the acoustic phase) starts and is accelerated by the presence of large inhomogeneities in the neutrino background. It can be shown that δ_B grows rapidly (even for fluctuations that are initially absent) in such a way that

$$\delta_B - \delta_\nu \xrightarrow[t\,\text{increasing}]{} 0.$$

In fact this model is indistinguishable from a purely baryonic model beginning with recombination with an amplitude $10^3(1+z_{\text{rec}})(1+z_{\text{NR}})^{-1}$.

Figure 12.1 shows such evolution. It is therefore now possible to obtain $\delta_B \approx 1$ from $\delta_{B\,\text{initial}} \approx 10^{-5}$, which does not agree with the case of a uniquely baryonic universe.

In a 'cold' (or CDM – cold dark matter) scenario the particles are not relativistic and the damping effect will no longer operate, unlike in the 'hot' (or HDM) scenario. The spectrum of fluctuations will, however, also entail a characteristic length. As we have seen, as long as one component does not dominate the dynamics of the universe ($z > z_{\text{eq}}$), perturbations of the scale L within the horizon ($L < L_{\text{horizon}}$) cannot grow, while those beyond the horizon grow as $(1+z)^{-2}$.

As a consequence all the scales δ_k still beyond the horizon grow by a factor

$$(1+z)^{-2} \propto k^{-2}$$

with respect to those entering the horizon. As a result the initial spectrum will for these scales be modified by a factor k^{-2}. We then have

$$\delta M = \delta M_{\text{initial}} k^{-2} \propto k^{(n-1)/2} \propto M^{3\alpha-2}$$

and similarly

$$(|\delta_k|^2) \propto k^{n-4}$$

up to a scale $L_{\text{eq}}(M_{\text{eq}})$, which is the last to enter the horizon when the period of equilibrium is reached.

For $z < z_{\text{eq}}$ the fluctuations then increase by the same factor $((1+z)^{-1})$ at all scales. This spectrum thus shows a break at $L_{\text{eq}}(M_{\text{eq}})$ such that

$$M_{\text{eq}} = \frac{4\pi}{3}\left(\frac{ct_{\text{eq}}}{1+z_{\text{eq}}}\right)^3 \rho_{\text{CDM}} \approx 2 \times 10^{15}\left(\Omega h^2\right)^{-2} M_\odot.$$

The spectrum will then have the following features:

12.3 The Linear Evolution of Perturbations   373

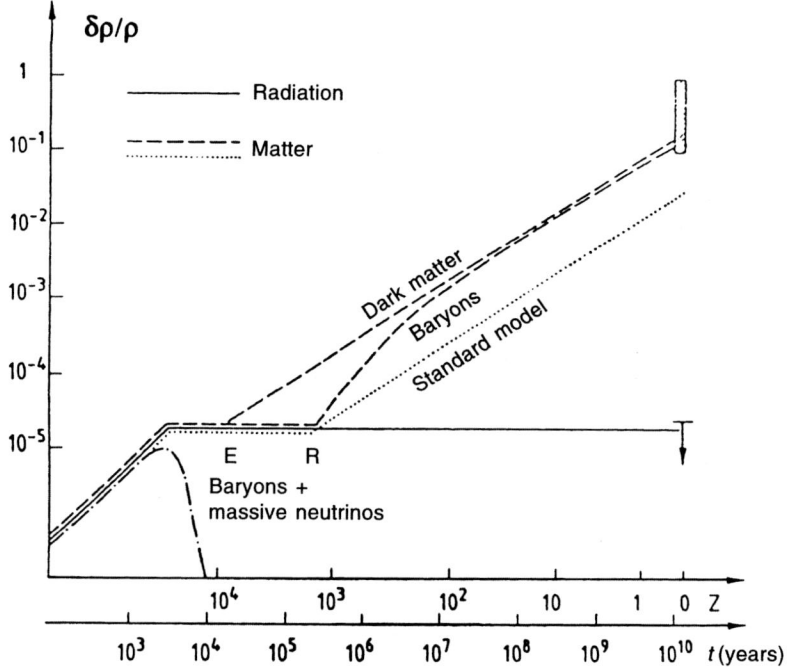

Fig. 12.1. The growth of adiabatic fluctuations at a scale of $10^{14}\,M_\odot$ (8 Mpc). The fluctuations of radiation initially grow up to the point when the mass contained within the horizon (a sphere of radius ct) is equal to their own mass. They remain constant thereafter and their amplitude can be fitted so as to be compatible with current observations (arrow), which give an upper bound for this amplitude. The fluctuations of matter in the standard model (dots) follow those of the radiation as long as the temperature is sufficient to ionize the matter. From the moment when the electrons recombine (R) the fluctuations of matter grow, but too slowly to reach the amplitude observed (hatched rectangle). In models where dark matter is not composed of baryons but of different neutral particles, only the gravitational interaction imposes on the matter fluctuations following those of the radiation, and they can grow from the time (E) when the matter supplants the radiation as the source of the gravity. The amplitude of the dark-matter fluctuations is therefore larger, and the baryons that, in this model, at most represent 10% of the total mass finish by having the same fluctuations as the dark matter. In the massive-neutrino scenario the fluctuations of these particles are destroyed by dissipative effects, at the scale considered. (From Turner 1984)

$$\delta M \propto \begin{cases} k^{(n-1)/2} \propto M^{-(n-1)/6} & \text{for } M < M_{\text{eq}}, \\ k^{(n+3)/2} \propto M^{-(n+3)/6} & \text{for } M > M_{\text{eq}}, \end{cases}$$

regardless of the nature of the particles constituting the dark matter.

Detailed calculations indicate an analogous but less constrasted component. Figure 12.2 summarizes the spectra obtained according to the different types of fluctuation and matter (baryonic or not) regarded as subsequent (nonlinear) evolution.

374   12. The Formation of Galaxies and Large Structures in the Universe

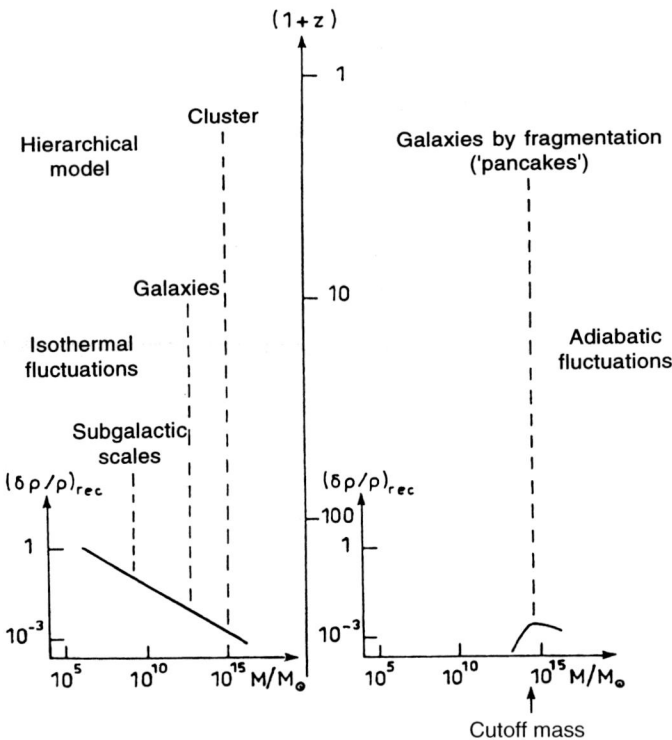

Fig. 12.2. Two possible forms of the spectrum of fluctuations at recombination. For isothermal fluctuations the objects in the domain 10^6–10^8 $M_\odot$ form a little after t_{eq}. The galaxies and the clusters then form and constitute a hierarchical structure. For adiabatic fluctuations the first scales to condense are intermediate between those of galaxies and clusters. For massive neutrinos the scenario is similar to that on the right, the characteristic scale thus corresponding to that of superclusters. (From Rees 1984)

The advantage of using 'cold' matter resides in the fact that the inhomogeneities can increase from z_{eq}. As a consequence, as the fluctuations of the baryons afterwards 'follow' the dark matter, 'initial' fluctuations smaller than in the purely baryonic case are sufficient, in principle, to lead to the present structures and thus satisfy the constraints imposed by the 2.7 K radiation.

Looking at the spectra reveals a fundamental difference. In the case of isothermal baryons or cold matter there are small scales that first become nonlinear, that is, that will form condensations (with $\delta > 1$) decoupling from the general expansion, at around the Jeans mass after recombination ($\approx 10^5$–10^6 $M_\odot$). The first objects formed would be subgalactic (globular clusters, dwarf galaxies, and so on), the largest structures forming successively by gravitational grouping: this is the hierarchical, or 'bottom–up', scenario. In the case of adiabatic baryons or 'hot' matter the scales available

(10^{13}–$10^{15}\ M_\odot$) are characteristic of those of clusters or superclusters. The galaxies form afterwards by fragmentation of these large structures; this is the pancake, or 'top–down', scenario.

12.4 Nonlinear Evolution

12.4.1 The Pancake Model

We have seen how in the case of adiabatic baryons or 'hot' matter the baryon–photon interaction or the decrease in perturbations of the neutrinos will imprint a characteristic scale on the initial spectrum of the fluctuations. According to the values of Ω these scales broadly correspond to those of clusters of galaxies or superclusters. The important point is that these scales are very much greater than those of the Jeans mass after recombination, and the pressure thus plays a totally negligible role. As these regions of excess density have a small chance of being spherical, if the damping is more rapid in one direction, the inequality of the velocities will be maintained, and we end up with one- or two-dimensional objects (filaments, pancakes).

In this scenario the galaxies form by fragmentation from the large gaseous initial structure (and therefore *after* the supercluster or cluster), following on from a shock formed by the collapse of the system during which the dissipative phenomena (radiation, and so on) will be important.

12.4.2 The Hierarchical Scenario: The Spherical Model

In the case of isothermal baryonic fluctuations or of cold matter the spectrum at recombination does not show a cutoff at small scales. The scale for which an initial fluctuation δM_i now becomes nonlinear is such that (for $\Omega_0 = 1$)

$$\frac{\delta M}{M} = \left(\frac{\delta M}{M}\right)_i t^{2/3} = \left(\frac{M_{\rm rec}}{M}\right)^\alpha (1+z)^{-1} \approx 1.$$

The spectrum is normalized by noticing that the correlation function $\xi(r)$ takes the value 1 for $r = r_0 \approx 5h^{-1}$ Mpc. To this scale there corresponds a recombination mass $M_{\rm rec}$ of the order of 10^5–$10^8\ M_\odot$ (according to the values of n), which is smaller than that of a galaxy but comparable to the Jeans mass after recombination. For a spectrum declining with scale the evolution is thus nonlinear at small scales. We therefore expect the formation of systems of about 10^6 solar masses (of the same order as that of stellar clusters), these clouds fragmenting to form stars. But the process of gravitational instability continues to larger scales; these systems will group together to form galaxies which will themselves form clusters and then superclusters. In this schema we can thus imagine a hierarchical structure from galaxies to superclusters. In this setting the structures (galaxies, clusters, and so on) form from bodies

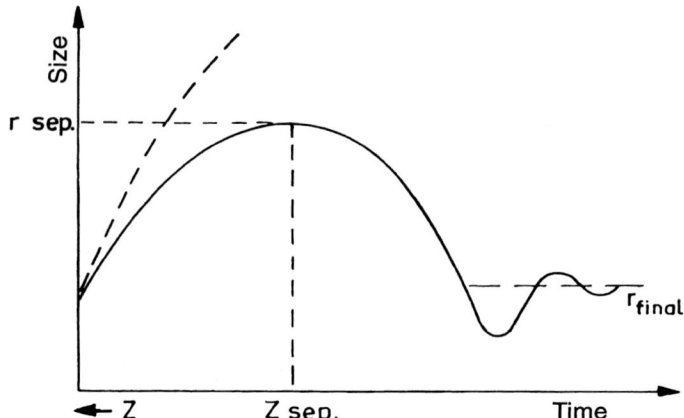

Fig. 12.3. The schematic evolution of a spherical inhomogeneity. The short-dashed line represents the evolution over time of the scale factor $R(t)$. The long-dashed line indicates the value of the radius of final equilibrium reached after several oscillations of the system in the phase of 'virialization'. The point corresponding to the separation of the inhomogeneity with respect to the rest of the expanding universe is also indicated ($z_{\text{sep}}, r_{\text{sep}}$)

already constituted by purely gravitational interaction and not by dissipative fragmentation processes of a gas. The characteristics of the objects formed can be evaluated in a simple case: the spherical model (Fig. 12.3). We consider an overdensity of density ρ' immersed in a universe described by a Friedmann model of density $\bar{\rho}$ ($\rho' > \bar{\rho}$). The gravitational effect of layers external to the sphere being zero, we can describe this perturbation by a Friedmann model of density ρ'. The history of this miniuniverse is therefore described by the usual equations of cosmology after recombination. Hence with an overdensity of radius R we have

$$\left(\frac{\dot{R}}{R}\right)^2 = \frac{8\pi G\rho}{3} - H_{\text{rec}}^2 (\Omega' - 1) \left(\frac{R_{\text{rec}}}{R}\right)^2$$

$$\left(\frac{\dot{R}}{R}\right)^2 = \Omega' H_{\text{rec}}^2 \left(\frac{R_{\text{rec}}}{R}\right)^3 - H_{\text{rec}}^2 (\Omega' - 1) \left(\frac{R_{\text{rec}}}{R}\right)^2.$$

The maximum radius of expansion R_{max} that the fluctuation will reach will be (writing $\dot{R} = 0$)

$$R_{\text{max}} = R_{\text{rec}} \delta^{-1},$$

to which corresponds the density

$$\rho_{\text{max}} = \delta^3 \rho_{\text{rec}}.$$

The maximum radius is attained at a redshift z_{max} such that

$$(1+z_{\max}) = \begin{cases} (1+z_{\rm rec})\dfrac{R_{\max}}{R_{\rm rec}} \\ (1+z_{\rm rec})\delta^{-1}, \end{cases}$$

and the corresponding density contrast is therefore

$$\delta_{\max} = \frac{9\pi^2}{16}.$$

At this time of maximum expansion of the miniuniverse the total energy is equal to the potential energy,

$$E_{\rm tot} \approx -\frac{GM^2}{R_{\max}},$$

and once equilibrium is reached

$$2E_{\rm eq} + V_{\rm eq} = 0,$$

whence

$$V_{\rm eq} = 2E_{\rm tot} \approx -\frac{2GM^2}{R_{\max}}.$$

It follows that the final structure will be such that

$$R_{\rm final} \approx \tfrac{1}{2} R_{\max},$$

$$\rho_{\rm final} \approx 8\rho_{\max}.$$

12.4.3 Numerical Simulations

The basic idea behind the formation of structures is that the universe has evolved from a simple homogeneous state to the complex, well-structured present state.

Such a process can be followed using semi-analytic modelling or numerical simulations. The first efficient numerical codes appear in the 1970s. It was then only possible to follow the collisionless gravitational collapse of a single fluid using an N-body code. But with the advent of very powerful computers, it is now possible to follow the coupled history of multiple fluids (dark matter, baryonic gas, stars, galaxies, ...). Progress in this domain can be measured, for instance, by the increase in the number of resolution elements (particles or cells), which grew from $\sim 10^2$ in the 1970s to $\sim 10^8$ at the beginning of 2001.

The aim of N-body simulations is to follow realistically the (non linear) evolution of the density distribution. The most direct way to achieve this aim is to fully account for the particle–particle (PP) gravitational interaction by summation of all particle–pair interactions. The main advantage of this direct approach is that hypotheses are kept to a minimum. However it is very time consuming, since it scales as N^2, where N is the number of particles involved. It has thus been restricted for a long time to a limited number of particles (a few thousand) until now when dedicated machines such as GRAPE (GRAvity PipE) computers have appeared which hardwire the calculation of the forces in the direct N-body code on special–purpose chips that are set on a printed circuit board coupled to the computer where the code is executing.

More efficient algorithms have also been devised in order that the computation of the interparticle force scales with $N\log N$ rather than N^2. Indeed, when collective (mean field action) and not binary effects are considered, the direct computation of each pair of interactions can be avoided by defining a mean field due to all the particles of the considered system. It is therefore possible to describe large ensembles (e.g. 10^6) of particles. To determine this field, the Particle-Mesh (PM) method divides the configuration space into cells defining a mesh for which it is possible to calculate the potential from the mean density and thus the resulting field, which can be interpolated at the position of each particle. When both types of interaction have to be addressed, another approach is the Particle-Particle, Particle-Mesh method (P^3M). In this hybrid method, short–range (the range being of the size of the cell) interactions are calculated from particle to particle (using the PP method) while the PM method is used for the long–range ones. Finally another performant method is the Tree–code method, where one uses the multipolar expansion of the gravitational interactions (see Chap. 6). Such simulations need ingredients to be run. One is the size of the box in which particles are put. This size is related to the problem being considered: formation and evolution of a galaxy, formation and evolution of a cluster of galaxies, or evolution of a part of the universe. The second ingredient concerns initial conditions. Again this depends on what one is interested on. One assumes, for example, a random density field for cosmological simulations or a given density profile for the evolution of galaxies or clusters. Finally one has to ask when simulations have to be stopped. For cosmological simulations (the simulation of a part of the universe), a 'clock' is defined by the behaviour of the two-point correlation function of particles which is identified with that of galaxies. When the correlation function in the simulation is identical to the present observed one, the time elapsed in the simulation is then identified with the age of the universe. For bound systems, the evolution is followed until equilibrium is reached and one stops the calculation when, for example, the virial ratio attains its asymptotic value.

Figure 12.4 shows the distribution of galaxies at $z=0$ for four different CDM models using a P^3M code run on Cray T3D parallel supercomputers with 256^3 particles.

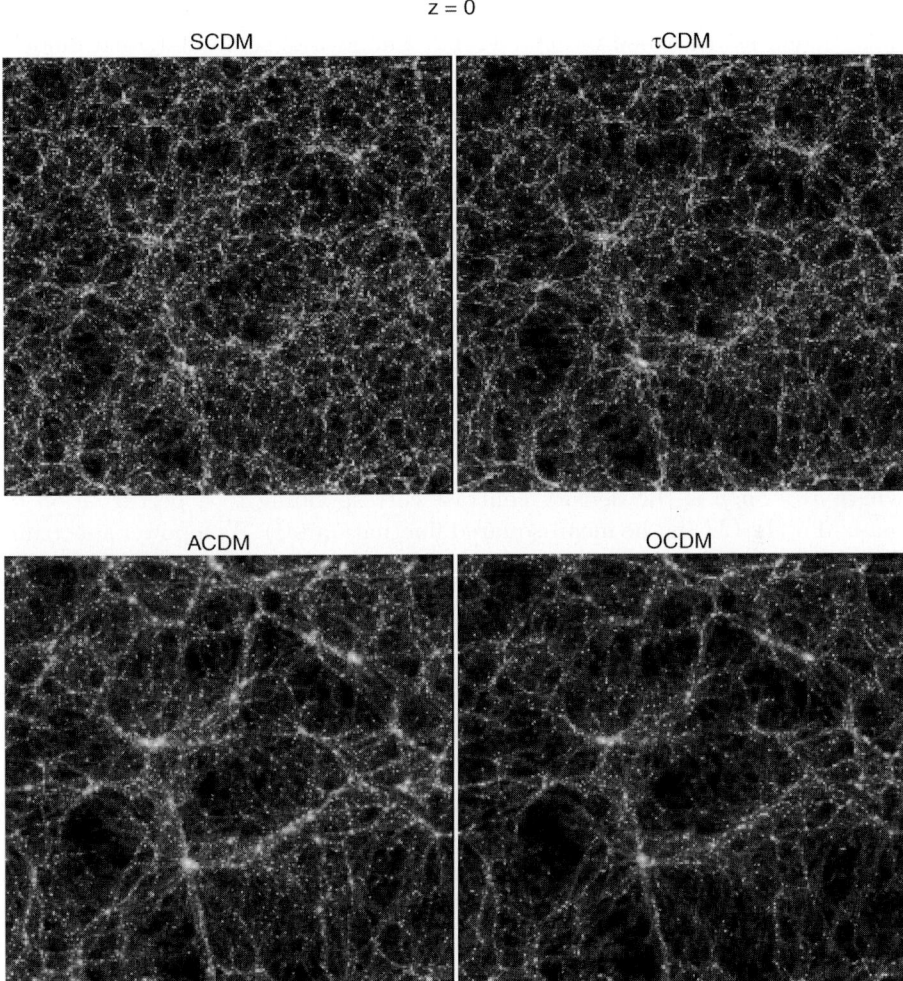

The VIRGO Collaboration 1996

Fig. 12.4. Comparison of four CDM models at redshift 0, showing projected mass distributions in slices of the Virgo simulation boxes

12.4.4 Semi-analytical Approaches

Numerical simulation is a very powerful method because in some ways it is a means of experimenting with the universe itself, since one is able to build several realizations of this unique object. However, it needs sophisticated computers, and codes are generally very time consuming. Another way is to use analytical or semi-analytical approaches. In the CDM scenario, dark matter dominates and small objects form first, followed by subsequent hierarchical building up. An initial perturbation (dominated by non baryonic dark matter) grows and collapses. This collapse can be described by the spherical (top hat) model, as described in Sect. 12.4.2. The idea of semi-analytical models is thus to start by assuming a power spectrum of linear fluctuations $P(k)$, choosing a given cosmology (that is cosmological parameters H_0, Ω_0, Λ_0) and some dark matter recipe (CDM, HDM, ...) and then using the Press and Schechter prescription. Indeed Press and Schechter (1974) derived a formalism which enables one to derive the number of virialized objects of a given mass M formed at a given epoch and their evolution with cosmic time. In this formalism, perturbations are assumed to grow until they reach amplitudes greater than a critical value δ_c after which they lead to bound objects of mass M. Assuming that primordial density perturbations are Gaussian fluctuations, the distribution of the amplitudes of the perturbations of a given mass M is

$$p(\delta) = \frac{1}{\sqrt{(2\pi)}\sigma(M)} \exp[-\delta^2/2\sigma^2(M)]$$

where $\delta = \delta\rho/\rho$ is the density contrast corresponding to a perturbation of mass M and $\sigma(M)$ is the mean-squared fluctuation $\langle\delta^2\rangle$. As a power spectrum P(k) is chosen and the way that perturbations grow is known (for example $\delta \sim R \sim t^{2/3}$ in an Einstein–de Sitter model) then the fraction of the mass in the universe that is in bound structures with amplitudes greater than a given δ_c can be calculated and is given by

$$F(M) = \frac{1}{\sqrt{(2\pi)}\sigma(M)} \int_{\delta_c}^{\infty} \exp[-\delta^2/2\sigma^2(M)]d\delta$$

From this expression it is then possible using scaling relations to derive the space density of objects of mass M

$$N(M) = \frac{1}{2\sqrt{\pi}}(1+n/3)(\bar\rho/M^2)(M/M^*)^{(3+n)/6} \exp[-(M/M^*)^{(3+n)/6}]$$

where n is the index of the power spectrum, and M^* is a reference mass related to δ_c which depends on time as

$$M^* = M_0^*\left(\frac{t}{t_0}\right)^{\frac{4}{(3+n)}}$$

It is necessary to add a factor 2 to this expression in order to account for the total mass being in bound objects. This formalism is a very powerful one for studying the evolution of structures (such as clusters of galaxies) in the context of hierarchical scenarios, since it turns out that the results obtained using it are in a very good agreement with various CDM N-body simulations. The above expression thus gives, for example, the number of dark-matter haloes of a given mass M which collapse at a specific redshift z. To go further one has to assume that the collapse can be described by the spherical model (see Sect. 12.4.2), with the baryonic phase cooling in and forming stars which then enrich the interstellar medium with metals and dust or blow it away. It is then possible to derive from such prescriptions some statistical properties of galaxies such as the luminosity function $\Phi(L)$ in both the optical and infrared domains. Figure 12.5 shows the contribution of galaxies to the extragalactic background light using such semi–analytical models from the optical to submillimetre bands. This calculation is important for validating the approach used since it is a global confrontation with observations. The two lines are the prediction of the contribution of galaxies to the diffuse extragalactic background light at $z = 0$ from a semi–analytical model compared with observations.

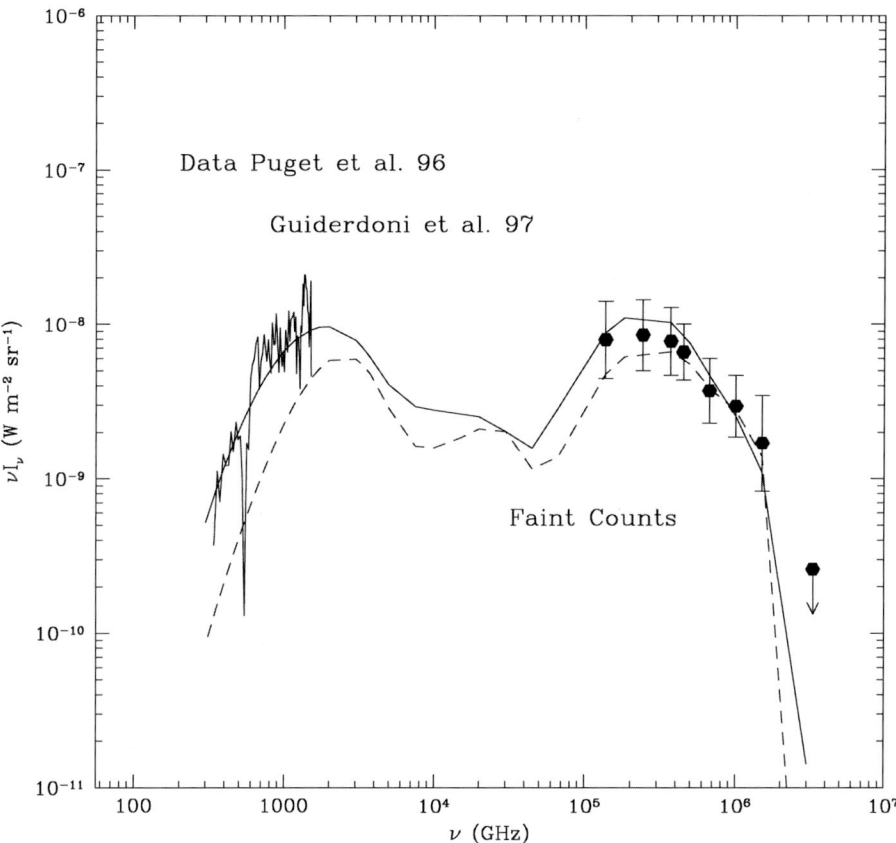

Fig. 12.5. The contribution of galaxies to the diffuse extragalactic background light at $z \approx 0$. The data on the left are from Puget et al. (1996), and on the right from Guiderdoni et al. (1997). The solid line is the predicted contribution from the semi-analytic models with a 'burst–like' star formation rate at intermediate and high–redshift. The dashed line is the background flux predicted by the same models with a 'quiescent' star formation rate (From Devriendt et al. 1999)

12.5 The Models in the Light of Observations

There are essentially three observational tests that enable constraints to be placed on the different theoretical models. One concerns the grouping of large structures (that is, superclusters, clusters, and galaxies) in the universe, characterized, for example, by the correlation function. The second concerns the peculiar velocities (otherwise referred to as departures from the Hubble flow) following on from the growth of density perturbations. The third is given by the fluctuations of the 2.7 K radiation induced by the density fluctuations. To these three quantitative constraints we can add the upper limit of the redshift for the formation of galaxies, which must be at least equal to that of the most distant quasars or galaxies currently known ($z \approx 6 - 7$).

We know that in the nonlinear regime ($\xi > 1$) the slope of the correlation function is approximately 1.8. In the pancake scenario we obtain flat or linear structures. It can be easily shown (see Chap. 11) that, for ideal objects of one or two dimensions and size R, the expression $1 + \xi(r)$ is given by

$$1 + \xi(r) = \frac{N_{\mathrm{p}}(r)}{N_{\mathrm{p}}^{\mathrm{Poisson}}(r)} = \begin{cases} R/2r & \text{flat,} \\ R^2/2r^2 & \text{'cigar';} \end{cases}$$

that is, $\xi(r)$ takes the form of a power law as a direct consequence of the transition from a three-dimensional system to a smaller dimension. Numerical simulations show that the behaviour of $\xi(r)$ can be fitted by 'top–down' models. However, in these models, the slope of $\xi(r)$ varies over time and a good fit is only obtained for $z_{\mathrm{formation}} \approx 1$, which contradicts observations.

In the hierarchical scenario an idea of the behaviour of ξ can be had from the spherical model. We have seen that if we begin with a (spherical) fluctuation δ_{rec} of size R and mass M, the final density ρ is such that

$$\rho \propto \rho_{\max} \propto \delta_{\mathrm{rec}}^3.$$

With a spectrum of the form

$$\delta_{\mathrm{rec}}^2 \propto R^{-(3+n)} \propto M^{-(n+3)/3}$$

we obtain

$$\rho \propto M^{-(n+3)/2},$$

which with

$$\rho \propto r^{-3}$$

gives

$$\rho \propto r^{-(9+3n)/(5+n)}.$$

In the case of a continuous hierarchy the correlation function is such that (see Chap. 11)

$$\xi(r) \propto \rho(r),$$

whence
$$\xi(r) \propto r^{-(9+3n)/(5+n)}.$$

We see that with $n = 0$ or 1 we obtain behaviour close to that observed.

The second constraint concerns the proper velocities. From the (Fourier-transformed) continuity equation we get
$$\langle v^2 \rangle \approx k^3 |v_k|^2 = k|\delta_k|^2.$$

We thus obtain values predicted by the linear theory for the peculiar velocities obtained in the case of different possible spectra (once the spectrum has been *normalized* with the aid of the correlation function).

With 'cold' matter we expect that no coherent motion will develop on a large scale, unlike in the case of neutrinos (owing to the respective cutoffs in the spectrum).

Figure 12.6 shows the results obtained in the two cases for a spectrum characterized by $n = 1$, compared with observed values of the velocity (compared with the Hubble flow), for the Virgo supercluster ($d \approx 15h^{-1}$ Mpc) and for a sphere of galaxies at approximately $25h^{-1}$ Mpc. We see that hot dark matter leads to too high velocities, unlike cold matter. Nevertheless motions of the order of $1000\,\mathrm{km\,s^{-1}}$ could have been discovered on scales of 100 Mpc, which are equally difficult to incorporate in the 'cold' scenario; but these results are still very controversial.

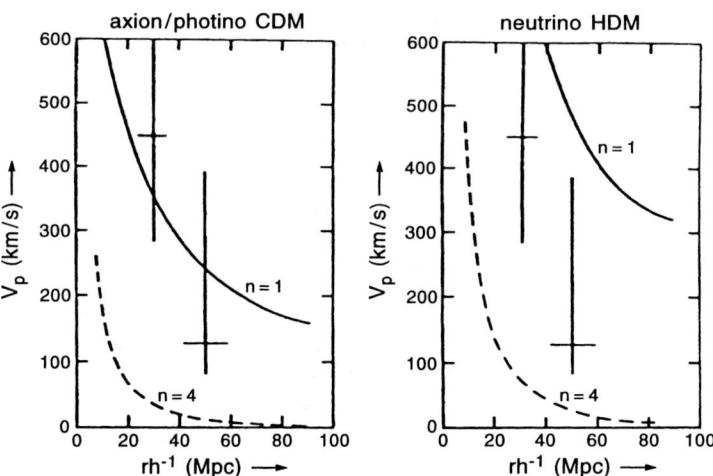

Fig. 12.6. The peculiar velocity (with respect to the Hubble flow) of a local condensation of 30 and $50h^{-1}$ Mpc measured with respect to the reference frame of the 2.7 K radiation compared with theoretical models of cold dark matter (axions or photinos) and hot dark matter (neutrinos) for different values of n (see the text). (From Silk 1984)

The last constraint is due to the 2.7 K radiation. The cosmic microwave background radiation gives us direct information about the structure of the universe and its isotropy on a very large scale and equally at very large redshifts. The first indication of anisotropy has come with the confirmation of the existence of a dipolar moment due to the motion of the Sun and the Galaxy, brought about by the presence of large-scale mass fluctuations. This detection is consistent with the assumed cosmological origin of the 2.7 K radiation. In fact in the framework of gravitational-instability models the existence of small inhomogeneities in the primordial universe whose growth will lead to the present large structures observed implies the presence of fluctuations of temperature in the radiation. A detailed calculation of these fluctuations requires the solution of equations coupling the evolution of the matter and the radiation. We have restricted ourselves here to estimating the different possible contributions. This coupling between matter and radiation is effected in various ways. A first contribution appears in the case of adiabatic fluctuations, since we have

$$\left(\frac{\delta T}{T}\right)_{\text{rec}} = \frac{1}{3}\left(\frac{\delta\rho}{\rho}\right)_{\text{rec}},$$

which will be zero for isothermal fluctuations. In fact this equation assumes that the decoupling of matter and radiation at recombination is instantaneous, which is not the case. The small degree of residual ionization is nonetheless sufficient to obliterate the fluctuations at small scales, and only those having a scale greater than $10^{15}\, M_\odot$ approach the above limit.

However, other contributions appear. One comes from the gravitational spectral shift: the photons that are received from regions with a density greater or less than average (the Sachs–Wolf effect). An estimate of this effect for a scale L is given by

$$\left(\frac{\Delta T}{T}\right)_g \approx \frac{\Delta\nu}{\nu} \approx \frac{\delta\Phi}{c^2}$$
$$\approx \frac{G\,\delta M}{Lc^2}$$
$$\approx \left(\frac{\delta\rho}{\rho}\right)_{\text{rec}}\left(\frac{L}{ct}\right)^2_{\text{rec}}.$$

It is particularly important at large scales ($L > (ct)_{\text{rec}}$).

Another contribution is due to motions induced by the velocity perturbations, which create a Doppler effect:

$$\left(\frac{\Delta T}{T}\right)_v \approx \frac{v}{c},$$

where v is the velocity of the fluctuation relative to the expansion; it obeys

12.5 The Models in the Light of Observations

the relation
$$v \approx \left(\frac{\delta\rho}{\rho}\right)_{\text{rec}} \left(\frac{L}{ct_{\text{rec}}}\right).$$

To estimate the angular dependence of the different contributions to $\Delta T/T$ it is necessary to fix the initial spectrum of the fluctuations:
$$\frac{\delta\rho}{\rho} = \frac{\delta M}{M} = M^{-(1/2)-(n/6)}.$$

Bearing in mind that a sphere of mass M subtends (at recombination) an angle θ,
$$\theta \approx \left(\frac{M}{10^{12} M_\odot}\right)^{1/3} (\Omega h^2)^{2/3} h^{-1} \text{ arcmin},$$

we have
$$\left(\frac{\Delta T}{T}\right)_{\Delta S=0} \propto \theta^{-(3/2)-(n/2)},$$
$$\left(\frac{\Delta T}{T}\right)_{v} \propto \theta^{-(1/2)-(n/2)},$$
$$\left(\frac{\Delta T}{T}\right)_{g} \propto \theta^{(1/2)-(n/2)}.$$

The order of magnitude of the fluctuations can be easily estimated. In the purely baryonic case the growth of the fluctuations begins at recombination, and to get $\delta\rho/\rho \approx 1$ now we have
$$\left(\frac{\delta\rho}{\rho}\right)_{\text{rec}} \approx \frac{10^{-3}}{\Omega_0},$$

to which corresponds
$$\frac{\Delta T}{T} \approx \frac{3 \times 10^{-4}}{\Omega_0}.$$

The corresponding mass scale is of the order of $10^{15} M_\odot$, to which an angular scale of the order of $\theta \approx 10$ arcmin corresponds.

In fact, the variation of the amplitudes of the fluctuations with respect to their size on the sky corresponds to the *power spectrum* of these fluctuations. These quantities are decomposed on the basis of spherical harmonics and then expressed in terms of multipoles characterized by a number l, each l corresponding to an angular scale. Figure 13.8 gathers the most recent results.

These observations exclude a baryonic model, and the presence of the main peak is consistent with a flat universe. Moreover, details of the structures in the power spectrum give strong constraints on the underlying physics. It is in principle possible to discriminate between a 'cosmic strings' scenario and an 'inflation' one. The precise adjustment of the observed peaks will lead to a very good determination of the various cosmological parameters as well as the nature of the dark matter. This will be the task of the future space missions *MAP* and *PLANCK*.

12.6 The Quest for Primordial Galaxies

When did the galaxies and the large–scale structures in the universe form? This is still a fundamental question for cosmologists who are looking for the very first generation of objects that then led to systems similar to our Galaxy.

During the 1980s a population of very faint galaxies (B $\sim 27-28$: the 'Tyson population') was discovered. Large in number, (more than 100 per arcmin^2), they seemed very distant and appeared as objects with bursts of stellar formation present. At almost the same epoch, gravitational lensing phenomena (arcs in clusters of galaxies) were also discovered. In fact, clusters of galaxies appear to act as gravitational telescopes, allowing the observation of faint galaxies (amplified by lensing). The modelling of several arcs and arclets (see Chap. 9) in the vicinity of clusters reveals that the above population was effectively a high redshift one ($z \sim 1$). Multicolour photometry (for example in the B, V, R, I, J and K bands, in some way equivalent to very low resolution spectroscopy) was later performed on some objects lensed by clusters and led to an estimate of the galaxy energy distribution over a very large wavelength range.

Otherwise, using models for the evolution of galaxies (which accounts for the morphological type of the galaxies, the fraction of stellar type, the time evolution of the stellar formation rate, metallicity and so on) it is possible to predict the energy distribution of galaxies at different redshifts.

Fitting this theoretical or synthetic energy distribution (SED) to the observations enables one to derive the redshift (and in fact the type) of a given galaxy (for example using a χ^2 technique). This technique is known as the 'photometric redshift technique' and is illustrated in Figs. 12.7 and 12.8.

Later on, in 1995, the HST provided the deepest view of the universe ever obtained by observing the same field (the Hubble Deep Field: HDF) during 10 days (150 orbits), accumulating exposures in four filters (UV, B, R, IR). These deep images reached magnitudes around 29-30 (Fig. 12.9) revealing a very distant (and therefore very young) population of galaxies. To estimate their redshift, the technique described above has been applied to these extraordinary data confirming that a large number of these objects are very distant ones ($z = 2$ to perhaps 6) and leading also to the discovery that clustering is already present at these epochs. However, the field of the HDF is rather small (~ 9 arcmin^2), and another method has been recently developed which makes it possible to undertake large surveys of galaxies at high and very high redshifts. This is called the 'Lyman break' technique and has been followed by similar ones. It is based on the fact that whatever the history of a galaxy is, a 'break' in its spectrum is expected at 912 Å in the rest frame due to photo-electric absorption both in the galaxy and in the intergalactic medium by neutral hydrogen. Thus, the flux at wavelengths below $912(1+z)$ Å is almost zero. This 'break' is sufficiently strong that it can be observed using broad–band filters. Thus, for a given set of filters, there is a redshift value for which the spectrum of the galaxy is so shifted that only the (almost zero

12.6 The Quest for Primordial Galaxies 387

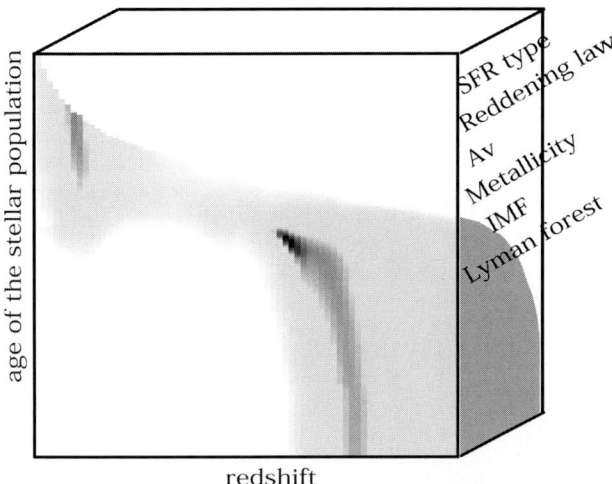

Fig. 12.7. The figure displays an artist's view of the SED fitting procedure. The figure presents a likelihood map for a $z = 4$ object. The shaded area corresponds to the highest confidence level region according to the χ^2 probability. (From Bolzonella et al. 2000)

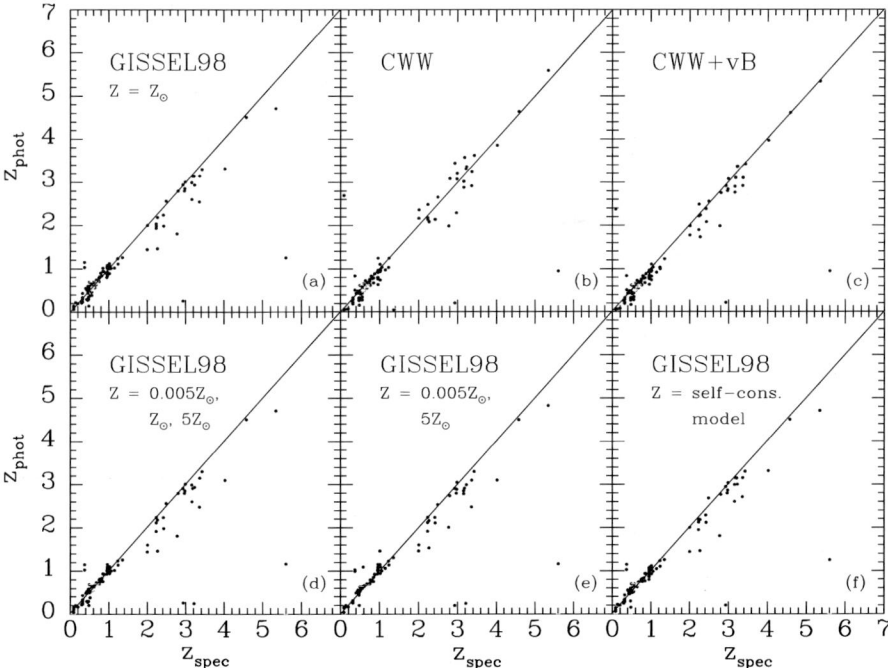

Fig. 12.8. A comparison of photometric versus spectroscopic redshifts within the HDF. Different theoretical models are used with, for example, various metallicities. (From Bolzonella et al. 2000)

388   12. The Formation of Galaxies and Large Structures in the Universe

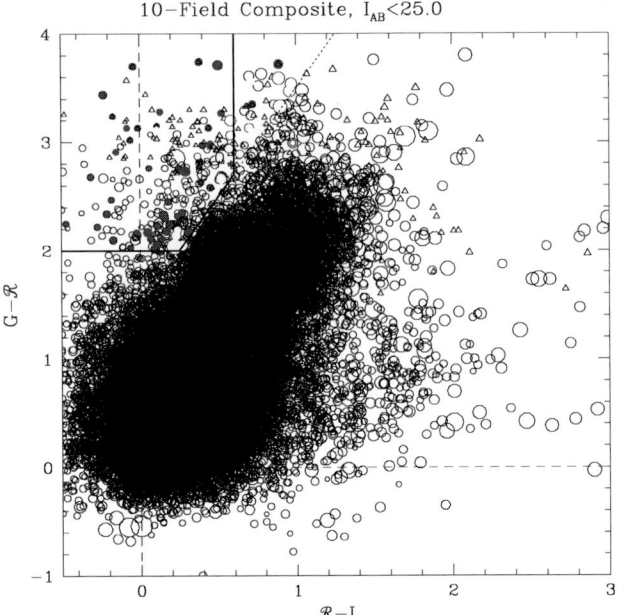

Fig. 12.9. A colour–colour diagram ($G - R$ versus $R - I$), where the spectroscopic G-band-break object–selection region is shown enclosed with a dotted line, and the region used for the statistical samples is indicated with a heavy line. There are approximately 29 000 objects represented here, of which 207 satisfy the primary colour selection criteria. Symbol size scales inversely with apparent magnitude, and triangles represent objects with limits only on the $G-\mathcal{R}$ colour. Filled symbols show objects with spectroscopic redshifts, where lighter shading indicates 'interloper' galaxies, and darker shading indicates objects with spectroscopic redshifts in the range $3.7 < z < 4.8$. (From Steidel et al. 1999)

flux) below 912 Å (in the rest frame) enters in the bluer filter while there is non zero flux in the redder ones (these are the so-called UV and B or G dropouts). So, using selected filters (e.g. U, B, R) and colour-colour diagrams ($U - B, B - R$), it is possible to identify high–redshift candidates, since they will be located in a very particular region of such diagrams.

Figure 12.9 shows such a diagram for 29 000 objects (I< 25) observed in the G (4870 Å), R (6930 Å) and I (6930 Å) bands. Here with the filters used, the expected redshift is around 4. High–z candidates ($3.7 < z < 4.8$) are those with low R-I and high G-R and spectroscopy at Keck telescope on some of the 207 candidates confirms the estimated high redshift.

Even if the energy distribution of these galaxies is *observed* in the red part of the electromagnetic spectrum, it corresponds in fact to the *intrinsic* UV or blue part in the rest–frame of these objects. As this energy domain is characteristic of star formation, gathering large samples of galaxies at various redshifts and measuring their UV rest frame energy distribution allows one

12.6 The Quest for Primordial Galaxies

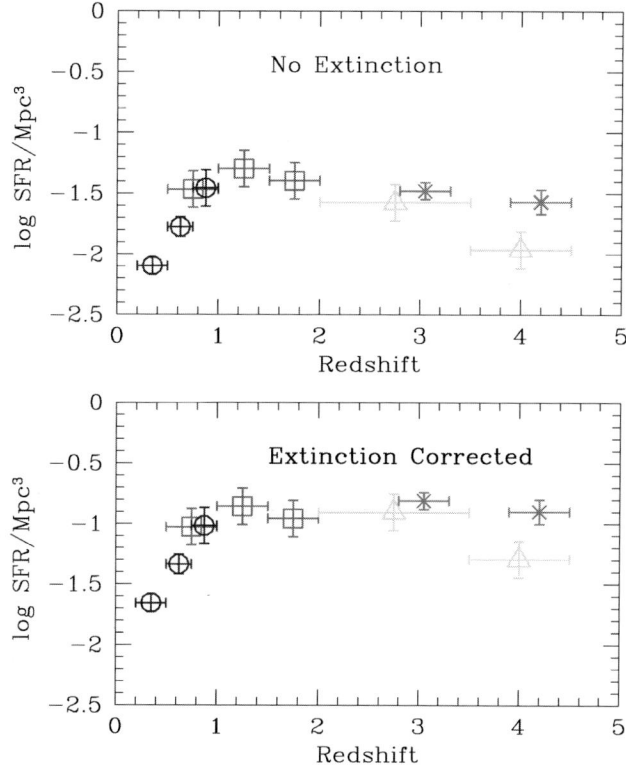

Fig. 12.10. The UV luminosity density as a function of redshift, following Madau et al. (1996). The two diagrams show the effect of extinction. (From Steidel et al. 1999)

to follow with time the star formation history in the universe. To do that, one estimates the star formation density in given redshift bins by calculating the UV luminosity (integrating the luminosity function), which can be easily related to the star formation rate (SFR), and the comobile volume at the given redshifts.

Figure 12.10 shows the evolution with time (redshift) of the SFR between $z = 0$ and $z = 5$. There is still (in 2001) a debate about the exact contribution of the extinction processes, which strongly affect the UV radiation emitted by these distant galaxies, and how they modify the conclusions drawn from from such a diagram concerning a possible peak of star formation rate at $z = 2$.

12.7 Conclusions

The usual cosmological models, and in particular their consequences concerning the formation of galaxies, have been profoundly revised by the contribution of new ideas from particle physics. These predict:

- a primordial phase of inflation taking account of the large-scale structure (isotropy) of the universe;
- $\Omega_0 = 1$;
- the existence of many possible candidates for dark matter.[1]

On the other hand these ideas lead to predictions also concerning the form and amplitude of the spectrum of the initial fluctuations (the Harrison–Zel'dovich spectrum).

There remains, of course, confrontation with observations. These contribute constraints on both the baryon density and the amplitude of the primordial fluctuations by means of the degree of organization of matter on a large scale and the departures created with respect to the Hubble flow. Finally observation of the cosmic background radiation leads to estimates of the initial amplitude of fluctuations on many scales.

The purely baryonic model has been (for now) abandoned owing to its inability to predict the formation of galaxies without the creation of too significant fluctuations of the 2.7 K radiation (and with a zero cosmological constant Λ). While the model with neutrinos seems to take good account of the structures observed on large scales, this nevertheless implies a formation time for the galaxies that is difficult to reconcile with that observed for the quasars. The model with cold matter is unfortunately not without its faults. There does not seem to exist a single natural mechanism allowing baryons to be separated from cold matter on a large scale. As a result, measurements of Ω_0 must be close to 1, which is still controversial.

The last trick up the sleeves of theoreticians to get out of this impasse is therefore to imagine that the formation of galaxies is biased. We can, for example, conjecture that (by a still unknown mechanism) among the initial (Gaussian) fluctuations only those that are above a certain level led to the formation of the galaxies that we see. In this case it is clear that the distributions of galaxies and matter are not identical, which can be reconciled with $\Omega_{obs} < \Omega_0 = 1$.

However, things are not clear cut, and future progress in this fast-evolving field will be made firstly through astronomical observations, which will impose

---

[1] Note, however, that two new possibilities have recently appeared concerning the nature of the dark matter. Firstly two groups have detected a microlensing effect in the direction of the Magellanic clouds. This effect could be the tracer of brown dwarfs or MACHOs (massive compact halo objects), which could account for at least part of the missing mass. Secondly some astronomers suggest that molecular hydrogen organized like a fractal would form the bulk of the dark matter in spiral galaxies characterized by a flat rotation curve.

stricter constraints on models, secondly through the refinement of galaxy-formation models, which will lead to a more detailed description of the properties of galaxies (type, size, kinetic moment, and so on), and finally through the direct detection of dark matter.

Exercises

12.1 From the equation of evolution for a perturbation δ find the behaviour

$$\delta \propto R(t) \propto t^{2/3}$$

for a universe dominated by matter.

12.2 For a radiation-dominated universe the equation of evolution for a perturbation δ is written as

$$\ddot{\delta} + (2\dot{R}/R)\dot{\delta} = 8\pi G\rho\delta.$$

Deduce that the solution increases as

$$\delta \propto t.$$

Express this result as a function of $R(t)$ and then of z.

References

Audouze, J., and Tran Thanh Van, J. (editors) (1984) *Fundamental Interactions and Cosmology* (Editions Frontières, Paris).
Bolzonella, M., Miralles, J.M., Pello, R. (2000) *Astron. Astrophys.* **363**, 476
Börner, G. (1988) *The Early Universe, Facts and Fiction* (Springer, Berlin, Heidelberg).
Davies, R. L., Efstathiou, G., Falls, S. M., Illingworth, G., and Schechter, P. L. (1983) *Astrophys. J.* **266**, 41.
Davies, R. D. et al. (1987) *Nature* **326**, 462.
Devriendt J., Guiderdoni B., Sethi S.K. (1999) in "Dwarf Galaxies and Cosmology", proceedings of the XVIIIth Rencontres de Moriond, eds T.X.Thuan, C. Balkowsky, V. Cayatte, J. Tran Thanh Van, Editions Frontieres, Paris.
Fabian, A. C., Geller, M., and Szalay, A. (1987) *Large Scale Structures in the Universe* (Saas-Fee Advanced Course 17, Geneva Observatory, Geneva).
Guiderdoni B., Bouchet F.R., Puget J.L., Lagache G., Hivon E., (1997) *Nature* **390**, 257.
Madau, P., et al. (1996) *Mon. Not. Roy. Ast. Soc.* **283**, 1388.
Partridge, R. B. (1988) *Rep. Prog. Phys.* **51**, 647.
Peebles, P. J. E. (1980) *The Large Scale Structure of the Universe* (Princeton University Press, Princeton NJ).
Press W.H., Schechter P. (1974) *Astrophys. J.* **187**, 425
Puget J.L., Abergel A., Boulanger F., et al. (1996), *Astron. & Astrophys.* **308**, L5.

Rees, M. J. (1984) in *The Very Early Universe*, edited by G. W. Gibbons, S. W. Hawking, and S. T. G. Silklos (Cambridge University Press, Cambridge).

Silk, J. (1984) in *Fundamental Interactions and Cosmology*, edited by J. Audouze and J. Tran Thanh Van (Editions Frontières, Gif-sur-Yvette), p. 413.

Smoot, G. F. et al. (1992) *Astrophys. J.* **396**, L1.

Steidel, C. C., Adelberger, K. L., Giavalisco, M., Dickinson, M., Pettini, M. (1999) *Astrophys. J.* **519**, 1.

Turner, M. S. (1984) in *Fundamental Interactions and Cosmology*, edited by J. Audouze and J. Tran Thanh Van (Editions Frontières, Gif-sur-Yvette), p. 300.

Uson, J. M., and Wilkinson, D. J. (1984) *Astrophys. J.* **277**, L1.

Weinberg, S. (1973) *Gravitation and Cosmology* (Wiley, New York).

Zel'dovich, Ya. B., and Novikov, I. D. (1983) *Relativistic Astrophysics II: The Structure and Evolution of the Universe* (University of Chicago Press, Chicago IL).

13. Cosmology

Cosmology rests on two hypotheses: that the universe is homogeneous and that it is isotropic, on large scales. The verification of these hypotheses is therefore essential. Homogeneity is guaranteed if it is confirmed that the universe is isotropic at all points. Observations allow us to draw conclusions about the isotropy only from our Galaxy, and this is not sufficient to prove global homogeneity. The universe may be strongly inhomogeneous but nevertheless have a spherical symmetry. An observer situated at the centre of this distribution would see an isotropic universe and conclude, falsely, that it is homogeneous. But for others, situated elsewhere, it would appear anisotropic. It is therefore useful to introduce the so-called Copernican principle, which states that we are not privileged observers in the universe. The verification of isotropy is therefore sufficient to guarantee the homogeneity of the universe.

Locally the distribution of galaxies is clearly not homogeneous. On large scales, however, the situation seems more favourable. Before we continue this discussion it is therefore important to be more specific about the spatial scales concerned beyond which the properties of homogeneity and isotropy must be satisfied. So as to keep the scope of our discussion as flexible as possible we shall restrict ourselves, in the first instance, to purely dimensional arguments. Newtonian theory introduced a characteristic gravitation time (instability time):

$$t_c \approx \frac{1}{\sqrt{G\rho}} \approx 10^{10} \text{ years},$$

where G is the gravitational constant and ρ the mean density.

Special relativity introduces a limiting speed: that of light, c. We may then construct a natural distance scale:

$$L \approx \frac{c}{\sqrt{G\rho}} \approx 3\,000 \text{ Mpc}.$$

Before we introduce a more complete theory of gravitation – general relativity – note that the homogeneity and isotropy of the universe are confirmed on scales of the order of 3000 Mpc. At such distances, counts of galaxies, radio sources, and quasars indicate that large variations ($\delta\rho/\rho \gg 1$) are ruled out. Unfortunately these observations alone cannot confirm with sufficient precision ($\delta\rho/\rho \ll 1$) that homogeneity and isotropy are guaranteed.

Much more detailed information is given by the 3 K cosmic microwave background radiation. The 'source' of the 3 K radiation lies at a distance of the order of at least 3000 Mpc. On the other hand if we acknowledge that a photon undergoes a change in frequency through the Einstein effect ($\delta\nu/\nu \approx GM/c^2 l$) when it escapes from a local density excess of mass M and characteristic size l, it is possible to obtain much stricter bounds. One can imagine two types of inhomogeneity: those including the background source at 3 K and others situated between the background source and us.

The isotropy of the background radiation shows, for inhomogeneities of the first type, that

$$\frac{\Delta T}{T} \approx \frac{G\,\delta M}{c^2 l} \leq 10^{-5},$$

that is,

$$\frac{\delta\rho}{\rho}\left(\frac{l}{c/H}\right)^2 \leq 10^{-5}.$$

For those situated between the background and us the effect is differential: the photons cross the structure in a characteristic time

$$t = \frac{l}{c}.$$

During this time the structure (owing to expansion) changes in size by

$$\delta l = \dot{l}t = Hlt = \frac{Hl^2}{c}$$

and the differential effect becomes

$$\frac{G\,\delta M}{c^2 l}\frac{\delta l}{l} = \frac{\delta\rho}{\rho}\left(\frac{l}{c/H}\right)^3 \leq 10^{-5}.$$

These constraints turn out to be very strong and all the more interesting since it is at the largest scales that they are the most severe; as such, they are complementary to those obtained by galaxy counts.

The fundamental limitation to this line of argument is clear: based on a simple dimensional analysis it can turn out to be quite inexact. (In fact, in the case of general relativity, these limits point to a homogeneous and isotropic universe at a very distant epoch, leaving the possibility of inhomogeneities with an amplitude $\delta\rho/\rho \approx 1$ for the present universe.) One can also question the validity of Newtonian theory on large scales: if the gravitational force has a finite scale-length (smaller than 3000 Mpc), the foregoing analysis is incorrect. In spite of these restrictions, the isotropy of the 3 K background, without prejudice to its origin, remains a very important argument in favour of homogeneity and isotropy.

These arguments also show that the homogeneity and isotropy of the universe are not stable characteristics. Moreover the instability time that we have found is comparable to the typical age of old stars. A cosmological theory might then be capable of simultaneously explaining homogeneity at large scales, inhomogeneity at small scales, and the apparent equality of the age of the oldest stars and the instability time of the universe.

The linear relation between the distance of the nearby galaxies and their redshift established by Hubble was a major discovery for cosmology. This redshift was interpreted as a Doppler shift, indicating that the universe is expanding. The subject of cosmology was then firmly established.

A homogeneous and isotropic universe may be described geometrically using the Robertson–Walker metric. In a system of coordinates (t, x, y, z) this is written as

$$ds^2 = c^2 dt^2 - R^2(t) \frac{dx^2 + dy^2 + dz^2}{(1 + kr^2/4)^2},$$

where $r^2 = x^2 + y^2 + z^2$ and where $R(t)$ is the scale factor.

It is always possible to choose the system of coordinates in such a way that the constant k takes one of the three values -1, 0, $+1$. The determination of $R(t)$ requires a complete theory of gravitation and an equation of state. General relativity, which is compatible with observations, is still by far the most used framework. The Hubble constant H_0 allows us to introduce a new characteristic time fundamental to cosmology, the Hubble time:

$$t_H \approx \frac{1}{H_0}.$$

The order of magnitude of the age of the universe in the models with expansion is our starting point. Now observations show that

$$t_H \approx \frac{1}{\sqrt{G\rho}} \approx 10^{10} \text{ years}.$$

This is altogether natural if the expansion is responsible for the redshift but may be a coincidence in another type of interpretation. The approximate equality of the age of the oldest stars, the instability time of the universe, and the Hubble time constitutes a persuasive argument in favour of Friedmann–Lemaître models, that is, models based on general relativity.

13.1 The Geometrical Description of the Universe

We have just seen in the previous section one of the forms of the Robertson–Walker metric. In spherical coordinates it takes the form

$$ds^2 = c^2 dt^2 - R^2(t) \left(\frac{dr^2}{1 - kr^2} + r^2 d\Omega^2 \right), \tag{13.1}$$

where
$$d\Omega^2 = d\theta^2 + \sin^2\theta \, d\phi^2. \tag{13.2}$$

Note that this metric constitutes only a local description of the universe. The value of k, for example, indicates the local structure of space: spherical ($k = +1$), flat ($k = 0$), or hyperbolic ($k = -1$). However, the global form of the universe, its topology, is not absolutely determined by the metric nor by the theory of gravitation. Thus the total volume of space is not known a priori even if the metric is determined. A spherical space ($k = +1$) is necessarily a finite volume; by contrast a flat or hyperbolic space may be finite or infinite.

In a metric theory of gravitation that obeys special relativity the equation for the trajectory of photons is written
$$ds^2 = 0. \tag{13.3}$$

This equality allows the interpretation of observations in a homogeneous and isotropic universe before we even have available a complete theory of gravitation and an equation of state for the matter, which are necessary for obtaining the dependence of R as a function of time.

Matter is described by the energy–momentum tensor $T^{\mu\nu}$; this may be expressed as a function of the momentum four-vector p_i, the energy E_i, and the position $r_i(t)$ of each particle i:

$$T^{\mu\nu}(r,t) = \sum_i \frac{p_i^\mu p_i^\nu}{|E_i|} \delta(r - r_i(t))$$

($\mu = 0$ for the temporal coordinate and $\mu = i = 1, 2, 3$ for the spatial coordinates.)

On the other hand in the case of a homogeneous and isotropic space, it is possible to show that $T^{\mu\nu}$ may always be put into the diagonal form

$$T^{00} = \rho(t)c^2$$

and
$$T^{ij} = -\delta_{ij} P(t),$$

where $P(t)$ is the pressure, and written as

$$T^{\mu\nu} = (\rho c^2 + P) U^\mu U^\nu + P g^{\mu\nu}, \tag{13.3a}$$

where U^μ is the velocity vector, $g^{\mu\nu}$ the metric tensor, and $U^0 = 1$, $U^i = 0$.

The current vector is $J^\mu = nU^\mu$, where n is the particle number density. The above tensor $T^{\mu\nu}$ corresponds to that of a perfect fluid. On the other hand the spatial components of the velocity vector are zero. The matter contained in this universe is then at rest with respect to a system of spatial coordinates (r, θ, ϕ) known as comoving coordinates.

13.1 The Geometrical Description of the Universe

The conservation equations of $T^{\mu\nu}$ allow us to write

$$\frac{\mathrm{d}}{\mathrm{d}t} R^3(t)\left(\rho(t) + \frac{P(t)}{c^2}\right) = +R^3(t)\frac{\mathrm{d}}{\mathrm{d}t}\frac{P(t)}{c^2}. \qquad (13.3\mathrm{b})$$

In the absence of creation or annihilation processes, the conservation of the current four-vector implies that

$$nR^3(t) = \text{constant}. \qquad (13.3\mathrm{c})$$

This shows that the number of particles per comoving volume is conserved. For example in the actual universe we have $P \approx 0$, which enables us to write that the number of galaxies per comoving volume is constant:

$$n_\mathrm{g} R^3(t) = \text{constant}$$

(this is on condition that the possible fusion or formation of galaxies is ignored).

13.1.1 The Redshift

We have seen that galaxies show a redshift in frequency approximately proportional to their distance. This property is a general one in geometrical models based on the Robertson–Walker metric (see Exercise 13.1). In such models the universe is expanding or contracting, and the redshift at small distances may be interpreted as a simple Doppler effect. At large distances this interpretation is no longer correct. The change in observed frequency then has to be calculated according to the position of the source. To do this we assume that an observer is situated at the origin of the spatial coordinates ($r = 0$) at a time t_0 and that he or she sees the light of a source with coordinates $r_1, \theta_1, \phi_1, t_1$. A received photon has, in the course of its trajectory, the coordinates $r(t), \theta_1, \phi_1, t$. This trajectory is a null geodesic:

$$c^2 \mathrm{d}t^2 - R^2(t)\frac{\mathrm{d}r^2}{1 - kr^2} = 0.$$

One thus obtains the coordinate r_1 of the source:

$$\int_{t_1}^{t_0} \frac{c\,\mathrm{d}t}{R(t)} = \int_0^{r_1} \frac{\mathrm{d}r}{\sqrt{1 - kr^2}} = S_k^{-1}(r_1), \qquad (13.4)$$

where

$$S_k(r_1) = \begin{cases} \sin r_1 & \text{if } k = +1, \\ r_1 & \text{if } k = 0, \\ \sinh r_1 & \text{if } k = -1. \end{cases}$$

Hence we know the coordinate r_1 (which does not change over time if the source is effectively comoving). Suppose now that we are interested in two events separated by δt_1 at the nucleus of the source. These are observed at times separated by δt and we have

$$\int_{t_1}^{t_0} \frac{c\,dt}{R(t_1)} = \int_{t_1+\delta t}^{t_0+\delta t} \frac{c\,dt}{R(t)}.$$

From this we get

$$\frac{\delta t_0}{R(t_0)} = \frac{\delta t_1}{R(t_1)}.$$

Then, if the source emits at a frequency $\nu_1 = 1/\delta t_1$,

$$\frac{\nu}{\nu_1} = \frac{R(t_1)}{R(t)}. \tag{13.5}$$

The redshift is then

$$z = \frac{\lambda - \lambda_1}{\lambda_1} = \frac{\nu_1 - \nu}{\nu} = \frac{R(t_0)}{R(t_1)} - 1 \tag{13.6}$$

(a priori this quantity might just as well be positive as negative, but in our universe it is positive: hence the terminology).

For nearby galaxies the value of z is very small and the corresponding velocity is tiny with respect to that of light. The shift may then reasonably be interpreted as a classical Doppler effect. By contrast, for the most distant galaxies the value of z may exceed 1, which necessitates a complete theory of gravitation.

13.1.2 The Concept of Distance

There are many ways of defining the distance between two points, depending on the different physical methods deployed. The measurement may be made with the aid of a series of rulers, by the method of parallaxes, by the measurement of the angular diameter of an object whose size is known, by the measurement of the apparent luminosity of a standard source, by radar echo, and so on. In everyday life all of these methods give the same result. This is no longer true in general in a metric theory of gravitation. The differences are mainly felt in the case $z \geq 1$. In general a distance may always be written as

$$d = \frac{z}{H_0} + O(z^2).$$

We shall now explain some of these distances and see how they depend on each other. First of all we shall calculate the precise distance from the angular diameter (see Fig. 13.1a). This is defined by

$$d_\theta = D/\theta,$$

where D is the real size of a given object (a standard ruler, for example) and θ is its apparent diameter in the sky (assuming that there is no correction due to the projection on the sky).

One can choose the origin of the system of coordinates in such a way that the coordinates of the ends of the ruler are $(r_1, 0, 0)$ and $(r_1, \theta, 0)$.

The distance between the ends of the ruler is given by $D^2 = -ds^2 = +R^2(t_s)r_1^2\theta^2$. We then have $D = R(t_s)r_1\theta$, yielding

$$d_\theta = R(t_s)r_1, \qquad (13.7)$$

r_1 being given by (13.4).

In the same way, we can estimate the power originating from a source received by a telescope. We place the source at the origin of the coordinates.

During the time interval dt_s, the source emits energy $L\,dt_s$; if its luminosity is L, then only the fraction $L\,dt_s\,(\pi\theta^2/4)(1/4\pi)$ will reach the telescope. As before, $\theta = D/R(t_R)r_1$, where r_1 is the coordinate of the observer when the origin is chosen at the nucleus of the source. The corresponding photons reach the telescope during an interval $dt_R = (1+z)\,dt_s$. Furthermore, as the energy of each photon is multiplied by a factor $1/(1+z)$, the energy received by the telescope is

$$\frac{L}{4\pi}\frac{\pi D^2/4}{R^2(t_R)r_1^2}\frac{dt_R}{(1+z)^2}.$$

Finally the apparent luminosity l is

$$l = \frac{L}{4\pi R^2(t_R)r_1^2(1+z)^2} = \frac{L}{4\pi R^2(t_s)r_1^2(1+z)^4} = \frac{L}{4\pi d_L^2}.$$

The 'luminosity distance' (see Fig. 13.1b) is then defined by

$$d_L = R(t_s)r_1(1+z)^2.$$

Moreover the coordinate r_s of the source seen by the observer is the same as the coordinate r_1 of the observer seen from the source. The luminosity distance is then

$$d_L = (1+z)^2 R(t_s)r_s = (1+z)^2\,d_\theta,$$

where $R(t_s)$ is the scale parameter at the epoch of emission and r_s is the coordinate of the source (given by (13.4)).

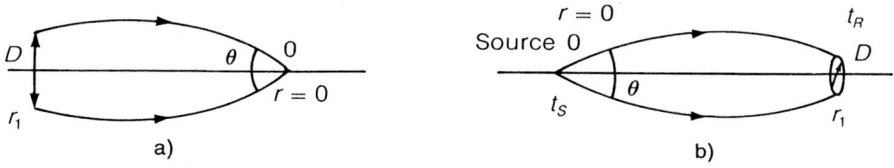

Fig. 13.1. (a) The distance defined with respect to the angular diameter. (b) The definition of 'luminosity distance'

13. Cosmology

Notice that L represents here the *total* luminosity of the source (the bolometric luminosity). A receiver being generally sensitive only in a frequency interval $\Delta\nu$, account must be taken of the fact that the energy received in a band $\Delta\nu$ around a frequency ν has been emitted in a band $\Delta\nu' = \Delta\nu(1+z)$ at a frequency $\nu' = (1+z)\nu$ (the K correction; see Exercise 13.5).

The 'proper' distance may be introduced in a like way. It is obtained with the help of small rulers set up beforehand. For two points with coordinates $(0, 0, 0, t)$ and $(r_1, 0, 0, t)$ the proper distance is given by the sum of the lengths of each ruler. Since the integral is taken over a surface $t = $ constant, we have $\mathrm{d}s^2 = \mathrm{d}l^2$; that is,

$$d_\rho = \int \mathrm{d}l = R(t)\int_0^{r_1} \frac{\mathrm{d}r}{\sqrt{1-kr^2}} = S_k^{-1}(r_1)R(t).$$

This proper distance is not in practice accessible to observation; it is therefore of much less interest than the two distances mentioned above.

In general it is clear that each way of measuring a distance gives a different result. This is due to the expansion (R varies with t); on the plus side, one can easily verify that in general the various distances coincide in a static universe.

13.1.3 The Concept of Horizon

The horizon demarcates events that are observable at a certain instant in the lifetime of the universe; these events belong to the 'past light cone'. If the observer has coordinates $(0, t)$, the coordinate r_1 of a point emitting a light signal at a time t_1 is given by the relation

$$\int_0^{r_1} \frac{\mathrm{d}r}{\sqrt{1-kr^2}} = \int_{t_1}^{t} \frac{\mathrm{d}t}{R(t)}.$$

If the integral on the right diverges when t tends towards 0, one can in principle receive a signal from all points in space (because the system of coordinates covers the entire group of points). If, on the other hand, this integral converges, there exists a maximum finite value r_h for the coordinate r_1. The observer cannot receive information from points situated at $r > r_h$, r_h being defined by

$$\int_0^{r_h} \frac{\mathrm{d}r}{\sqrt{1-kr^2}} = \int_0^{t} \frac{c\,\mathrm{d}t}{R(t)}.$$

13.1.4 The Evolution of the Cosmic Microwave Background Radiation

In 1964, while in the process of adjusting a radio horn antenna, Penzias and Wilson could not eliminate a persistent background noise in the instrument, in spite of many attempts. The background did not vary according to the direction of observation nor the time of day or year. This indicated an extragalactic origin. Following contact with Dicke's team at the University of Princeton, the existence of the background radiation became the subject of a publication, accompanied by another article by the team at Princeton, proposing that the background was of cosmological origin. This origin implied that it should be a black body, after measurements of the temperature at around 3.5 K. As a result of the discovery of the background radiation, observers wanted to measure its spectrum. The first measurements were made in the Rayleigh–Jeans part of the spectrum. Measurements in the Wien part were much more sensitive. At the beginning of the 1970s these measurements, taken together, revealed a spectrum compatible with a black body at 2.7 K. During the 1980s more precise measurements seeming to indicate an important distortion in the Wien part provoked many studies of the possible sources of the distorted black body (the inverse Compton effect, dust, and so on). It is only recently that a very high-quality measurement of the spectrum at 50 different wavelengths was made by the *COBE* satellite (with results announced in 1990) (see Fig. 13.2).

These results showed that the cosmic background is a very good approximation of a black body at $T = 2.735$ K. This measurement, made 25 years after the discovery of the cosmic background radiation, is a brilliant confirmation of the 'big bang' cosmological model (the processes capable of ther-

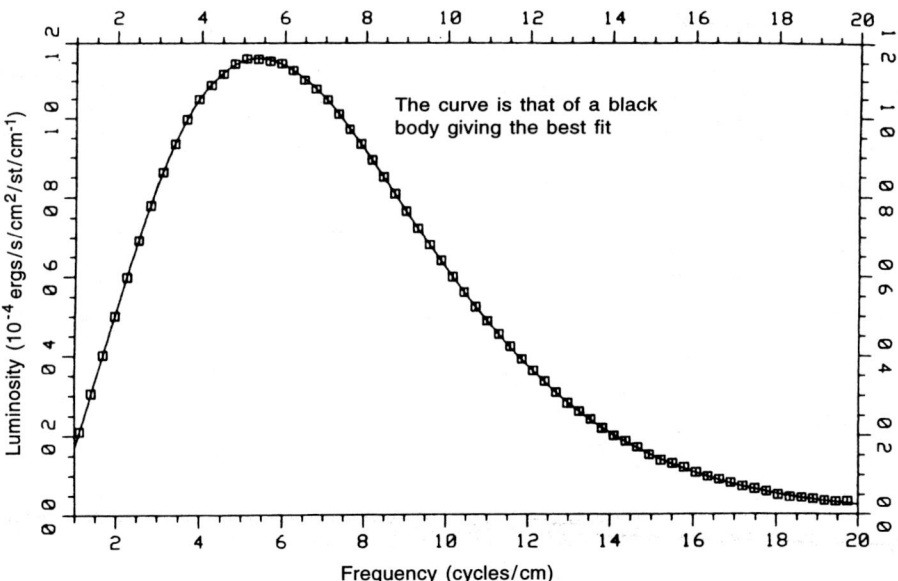

Fig. 13.2. The spectrum of the cosmic background radiation obtained by the *COBE* satellite in 1990. (From Mather et al. (1990); courtesy of the *COBE* Science Working Group)

malizing the cosmic background radiation could have taken place only in the first year of the universe).

The cosmic radiation no longer interacts with matter. We shall see that its distribution remains that of a black body for the duration of the expansion.

In the absence of creation or annihilation processes the quantity $nR^3(t)$ is conserved.

Suppose that the universe were filled with black-body radiation at a temperature T at a time t. If we are interested only in photons whose frequency lies in a small interval $d\nu$, the above relation becomes

$$\mathrm{d}n(\nu)R^3(t) = \text{constant}.$$

On the other hand as

$$\mathrm{d}n(\nu) = \frac{8\pi}{c^3} \frac{\mathrm{d}\nu^3}{e^{h\nu/kT} - 1}$$

at a time t', the frequency of these photons will be $\nu' = R(t)\nu/R(t')$. Their distribution becomes

$$\mathrm{d}n'(\nu') = \left(\frac{R(t)}{R(t')}\right)^3 \frac{8\pi}{c^3} \frac{\mathrm{d}\nu^3}{e^{h\nu/kT} - 1},$$

if we set $T' = TR(t)/R(t')$; we then end up with the density

$$\mathrm{d}n'(\nu') = \frac{8\pi}{c^3} \frac{\mathrm{d}\nu'^3}{e^{h\nu'/kT'} - 1},$$

which is exactly that of a black body at a temperature T'.

Hence we see that the type of distribution (that of a black body) is preserved as the universe evolves, although these photons take part in no interactions. The conservation of the black-body radiation then corresponds to a kinetic 'freezing-out'.

To obtain the above result we have made use of the conservation of the number of photons. In a situation of thermodynamic equilibrium one gets the same result owing to the conservation of entropy, S, the expansion being adiabatic:

$$SR^3 = \text{constant};$$

that is, $T^3 R^3 = \text{constant}$, or $T \propto 1/R$.

13.2 Friedmann–Lemaître Models

The foregoing description is limited by the fact that $R(t)$ remains an arbitrary function as long as one does not have a particular theory of gravitation and an equation of state. In what follows we use just one theory: that of general relativity. Although other, rival, theories exist, general relativity remains by far the most satisfactory. To this day there is not a single observation that contradicts it.

Einstein's equations allow us to establish the evolution equations of the parameter $R(t)$ for a Robertson–Walker universe:

$$\frac{2\ddot{R}}{R} + \left(\frac{\dot{R}}{R}\right)^2 = -\frac{kc^2}{R^2} + \Lambda - \frac{8\pi G}{c^2} P,$$

$$\left(\frac{\dot{R}}{R}\right)^2 = -\frac{kc^2}{R^2} + \frac{\Lambda}{3} + \frac{8\pi G}{3}\rho. \quad (13.8)$$

The constant Λ is the 'cosmological constant'. It is the only free parameter possible in Einstein's theory. The principle of economy of hypotheses would want one a priori to choose $\Lambda = 0$. However, as we shall soon see, the observational constraints on Λ are too weak for it to be neglected, and it may even be indispensable if the value of $H_0 = 100\,\mathrm{km\,s^{-1}\,Mpc^{-1}}$ were confirmed. On the other hand, recent theories of the primordial universe ('inflationary' theories) require the introduction of a huge cosmological constant (representing the energy of empty space). The possibility of a redundant term is then imaginable. In what follows, the following notation is used:

$$H = \frac{\dot{R}}{R} \qquad \text{Hubble's constant,}$$

$$\Omega = \frac{8\pi G \rho}{3H^2} \qquad \text{the density parameter,}$$

$$\lambda = \frac{\Lambda}{3H^2} \qquad \text{the reduced cosmological constant,}$$

$$q = -\frac{\ddot{R} R}{\dot{R}^2} \qquad \text{the deceleration parameter,}$$

$$\alpha = \frac{kc^2}{H^2 R^2} = \Omega + \lambda - 1 \qquad \text{the reduced curvature.}$$

All these parameters are functions of time t. The quantities capable of being observables are evidently the present values of these parameters. They will be distinguished with an index '0' when the present values are meant (the age of the universe today is t_0, the present density parameter is Ω_0, and so on).

The observational redshift z is therefore

$$1 + z = \frac{R_0}{R} = \xi.$$

13.2.1 The Matter-Dominated Universe

The amount of radiation and relativistic particles known at present do not seem to be sufficient to contribute in a significant way to the energy density of the universe. In the equation of state of the universe we can therefore take the pressure $P = 0$. The conservation equations thus become

$$\rho(t)R^3(t) = \rho_0 R_0^3,$$

from which we get $H^2(t)$:

$$H^2(t) = H_0^2 \left(-\alpha_0 \xi^2 + \lambda_0 + \Omega_0 \xi^3\right).$$

Similarly we obtain $\dot{R}(t)$ for $k \neq 0$ by noting that $R_0^2 = c^2/H_0^2|\alpha_0|$:

$$\dot{R}^2(t) = \frac{c^2}{|\alpha_0|} \left(\Omega_0 \xi + \frac{\lambda_0}{\xi^2} - \alpha_0\right),$$

which, with (13.4), allows us to obtain the coordinate r_1 of a source as a function of its observed redshift z_1 (the generalized Mattig relation):

$$r_1 = S_k \int_1^{1+z_1} \frac{\alpha_0^{1/2} \, d\xi}{(\Omega_0 \xi^3 - \alpha_0 \xi^2 + \lambda_0)^{1/2}}.$$

In the case where $k = 0$, R_0 is indeterminate. In practice this does not pose a problem, since one needs only the expression for $R_0 \times r_1$ in the calculation of the observed values. One can then write directly

$$R_0 r_1 = \frac{c}{H_0} \int_1^{1+z_1} \frac{d\xi}{(\Omega_0 \xi^3 + \lambda_0)^{1/2}}.$$

Similarly the time past since the epoch when the light was emitted is

$$\Delta t = \frac{1}{H_0} \int_1^{1+z_1} \frac{d\xi}{\xi \left(\Omega_0 \xi^3 - \alpha_0 \xi^2 + \lambda_0\right)^{1/2}}.$$

If this integral does not diverge at infinity (that is to say, when $\Omega_0 \xi^3 - \alpha_0 \xi^2 + \lambda_0$ does not cancel for some $\xi > 1$), we obtain the age of the universe:

$$t_0 = \frac{1}{H_0} \int_1^{+\infty} \frac{d\xi}{\xi \left(\Omega_0 \xi^3 - \alpha_0 \xi^2 + \lambda_0\right)^{1/2}}.$$

This expression indicates that the age t_0 is of the order of $1/H_0$, except, possibly, if the parameters Ω_0 and λ_0 are close to the values for which the integral diverges.

13.2.2 Models with a Zero Cosmological Constant

The expression for H^2 becomes

$$H^2 = H_0^2 \left(\Omega_0 \xi^3 - \alpha_0 \xi^2\right)$$
$$= H_0^2 \xi^2 (\Omega_0 \xi + 1 - \Omega_0),$$

and consequently

$$\left(\frac{\dot{R}}{R_0}\right)^2 = H_0^2(\Omega_0 \xi + 1 - \Omega_0) = H_0^2(1 + \Omega_0 z). \tag{13.9}$$

Case 1: $\Omega_0 = 1$. In other words $\rho = 3H_0^2/8\pi G = \rho_c$, where $\rho_c = 1.9 \times 10^{-29} h^2$ g cm^{-3}, and $h = H_0/100$ km s^{-1} Mpc^{-1}. The above equation may be easily integrated (knowing that $\xi = R_0/R$) to give

$$\frac{R(t)}{R_0} = \left(\frac{3}{2} H_0 t\right)^{2/3}.$$

The universe then has a finite age with the value

$$t_0 = \tfrac{2}{3} H_0^{-1}$$

This universe undergoes an infinite expansion, and its spatial curvature is zero ($k = 0$). This is the Einstein–de Sitter universe.

Case 2: $\Omega_0 < 1$. Relations (13.8) and (13.9) show that neither $\dot{R}(t)$ nor $\ddot{R}(t)$ may be zero; the behaviour of $R(t)$ is thus monotonic and regular. In the case we are interested in $(\dot{R}/R_0)_0 > 0$, and the expansion is indefinite. For such a universe the spatial curvature is negative ($k = -1$), and one speaks of the universe being open. The expression for $R(t)$ can only be easily obtained in the limit of a young universe, that is, for small R; we then simply have

$$\dot{R} \propto \xi \quad \text{and} \quad R(t) \propto t^{2/3}$$

so long as $\xi \gg (1 - \Omega_0)/\Omega_0 \approx 1/\Omega_0$.

The initial behaviour of $R(t)$ will then be the same as in the Einstein–de Sitter universe. By contrast for large t and R

$$\dot{R} \propto \xi \quad \text{and} \quad R(t) \propto t.$$

There are thus two regimes for the behaviour of $R(t)$, the transition occurring at the time when $\xi \approx 1/\Omega_0$.

It is possible to obtain a parametric representation of $R(t)$:

$$H_0 t = \frac{\Omega_0}{2(1-\Omega_0)^{3/2}}(\sinh\psi - \psi),$$

$$\frac{R(t)}{R_0} = \frac{\Omega_0}{2(1-\Omega_0)}(\cosh\psi - 1).$$

This representation allows us to obtain the following explicit expression for the age of the universe:

$$t_0 = \frac{1}{H_0}\left[\frac{1}{(1-\Omega_0)} - \frac{\Omega_0}{2(1-\Omega_0)^{3/2}}\text{arccosh}\left(\frac{2-\Omega_0}{\Omega_0}\right)\right].$$

Case 3: $\Omega_0 > 1$. Here the spatial curvature is positive ($k = +1$), and the universe is said to be closed. Its volume is necessarily finite. The expression for $\dot{R}^2$ allows us easily to see that this parameter goes to zero,

$$\dot{R}^2 = 0,$$

for a maximum value of the radius

$$R_{\max} = \frac{\Omega_0}{\Omega_0 - 1}R_0.$$

On the other hand when R is small, it behaves in the same way as in the case $\Omega_0 = 1$; that is,

$$R(t) \propto t^{2/3} \quad \text{for} \quad \xi \gg \frac{\Omega_0 - 1}{\Omega_0}.$$

The expression for $\ddot{R}(t)$ is always negative, which shows that $R(t)$ effectively goes to zero.

The universe therefore 'begins' at $t = 0$ with a singularity, and then undergoes an expansion that stops at $R_{\max}$. A contraction occurs next in a symmetric way and the universe 'ends' with a singularity.

As in the case where $\Omega_0 < 1$ we can obtain a parametric representation of $R(t)$:

$$H_0 t = \frac{\Omega_0}{2(\Omega_0 - 1)^{3/2}}(\phi - \sin\phi),$$

$$\frac{R(t)}{R_0} = \frac{\Omega_0}{2(\Omega_0 - 1)}(1 - \cos\phi).$$

These three models of the universe are called Friedmann–Lemaître models. They have the common characteristic of beginning with a singular state for which the curvature, like most of the physical quantities, diverges (we have neglected here the pressure term, although this approximation is no longer valid when $R(t)$ tends towards zero; we shall soon see that the divergence

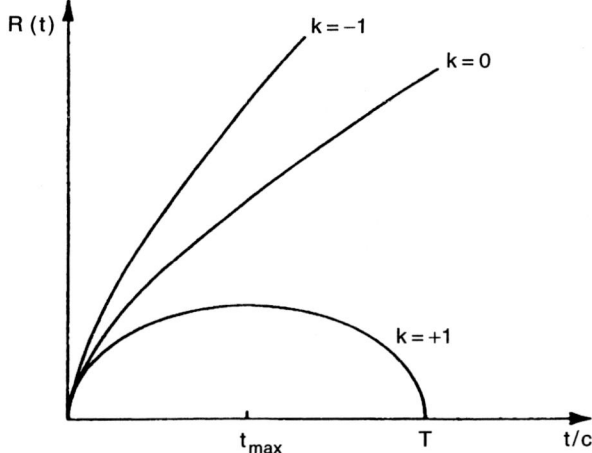

Fig. 13.3. The appearance of the parameter $R(t)$ in Friedmann–Lemaître models ($\Lambda = 0$). Models for which the density is greater than the critical density ($\rho > 3H_0^2/8\pi G = \rho_c = 1.9 \times 10^{-29} h^{-2}\,\mathrm{g\,cm^{-3}}$) have a closed spatial geometry (known as a hypersphere). On the other hand models with a lower density are open ones with continuous expansion. For $\rho = \rho_c$ the expansion is infinite ($R(t) \propto t^{2/3}$) and the geometry of the space is Euclidean. The equation for the parameters $R(t)$ may be obtained by a simple Newtonian calculation. These curves also give the form of evolution for a perturbation after recombination: for a perturbation that has evolved in a bound system (a galaxy, cluster, etc.), its density would exceed the density that today corresponds to the critical density. Calculation of the evolution for perturbations shows that they grow as $1/(1+z)$ up to the redshift corresponding to $1/\Omega_0$

remains when one takes account of the radiation). A plot of the parameter $R(t)$ for the three values $k = -1$, $k = 0$, and $k = +1$ is given in Fig. 13.3.

For these three models one can explicitly obtain the quantity $R_0 r_1$ given by (13.4):

$$R_0 r_1 = \frac{2c}{H_0} \frac{\Omega_0 \xi + 2 - 2\Omega_0 + (\Omega_0 - 2)\sqrt{\Omega_0 \xi + 1 - \Omega_0}}{\Omega_0^2 \xi}.$$

This equality is the Mattig relation; it allows the different distances given in Sect. 13.1.2 to be calculated.

13.2.3 Models with a Nonzero Cosmological Constant

We give here only a partial account; however, all the models compatible with present observations are included.

First, the expression for $\dot{R}^2$,

$$\dot{R}^2 \propto \Omega_0 \xi - \alpha_0 + \lambda_0/\xi^2,$$

shows that models with $\lambda_0 < 0$ have to recontract, since ξ cannot tend towards zero, and that, furthermore, the second derivative,

$$\ddot{R} \propto \Omega_0 - 2\lambda_0/\xi^3,$$

remains greater than a positive value.

On the other hand if $\lambda_0 > 0$, note that the second derivative goes to zero for a particular redshift z_p given by

$$1 + z_p = \left(\frac{2\lambda_0}{\Omega_0}\right)^{1/3}.$$

Examination of the function $\dot{R}^2(\xi)$ shows that there exists a critical value of λ_0 corresponding to the equation $\dot{R} = \ddot{R} = 0$ and satisfying

$$\lambda_c = \tfrac{4}{27}(\Omega_0 + \lambda_c - 1)^3 \Omega_0^2.$$

Hence one can qualitatively describe the components of the different possible models of the universe according to the values of λ_0 and Ω_0 by plotting the variation of $\dot{R}^2$ with ξ.

Figure 13.4 represents the evolution of the parameter $R(t)$ for different regions of the Ω_0–λ_0 plane. The variety of possible models of the universe is much greater than in the case $\lambda_0 = 0$; in particular notice that if $\lambda_0 > \lambda_c$ (for a given value of Ω_0), the universe did not experience an initial singularity. Thus there exists a minimum value of R and hence a maximum value of the redshift, which is less than $(2\lambda_0/\Omega_0)^{1/3}$. This characteristic allows us to exclude this type of model, since the existence of quasars with large redshifts ($z \approx 4$) requires that $\lambda_0 \approx 100$ (since $0.01 \leq \Omega_0 \leq 1$). This remarkably strong and simple constraint implies that

$$\Lambda < 10^{-54} \, \text{cm}^{-2}, \tag{13.10}$$

whereas the stricter (noncosmological) constraints obtained from studies of Solar System dynamics give

$$|\Lambda| < 10^{-42} \, \text{cm}^{-2}.$$

In fact constraint (13.10) has been slightly improved by studying the lenses responsible for multiple QSO images.

In Friedmann–Lemaître models, that is, those with $\lambda_0 = 0$, the age of the universe is necessarily less than the Hubble time:

$$t_0 \leq \frac{1}{H_0} \int_1^{+\infty} \frac{d\xi}{\xi^2} = \frac{1}{H_0}$$

(this limit also applies to universes with a negative cosmological constant). We have already seen that the Hubble constant may be equal to

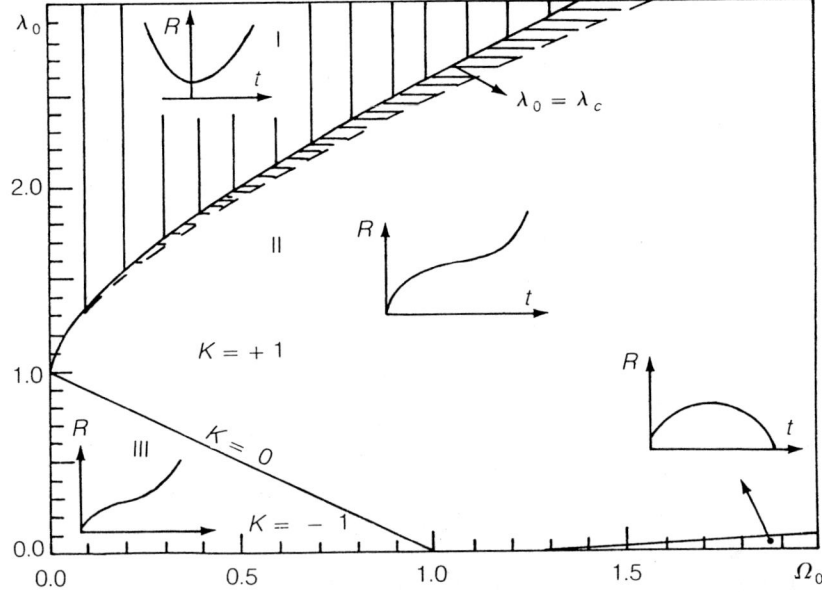

Fig. 13.4. The evolution diagram for the scale parameter $R(t)$ for models with a positive cosmological constant according to the different regions of the Ω_0–λ_0 plane. Region I: These models do not have an initial singularity; the existence of quasars with large redshifts totally excludes this type of model. Region II: For these models the geometry is closed but the expansion is infinite. These models, for a given value of H_0, can be older than those where $\Lambda = 0$. In this region the observational limits only exclude a very small region (horizontal hatching). The models of region III are very close to those with the same density but with $\Lambda = 0$. The models with $\Lambda < 0$ are not excluded by observations, but they do not help in the slightest with the problems of cosmology

$100 \, \mathrm{km^{-1} \, s^{-1} \, Mpc}$. An upper limit for the Hubble age can therefore be obtained:

$$t_0 \leq 10^{10} \text{ years.}$$

Now the age of the universe estimated by other methods (stellar populations, for example) is seen to be rather greater than this value, being

$$t_0 \approx 14\text{–}20 \times 10^9 \text{ years.}$$

It should not be concluded, however, that these observed values definitely require the introduction of a cosmological constant, the reason being the uncertainty associated with the above values. But it seems reasonable to imagine the possibility of a nonzero positive cosmological constant, since in that case the age of the universe may be greater than the Hubble time. On the other hand although the observations cannot exclude a negative cosmological constant, the introduction of such a constant is not suggested by any argument.

The plausible values are then

$$0 \leq \lambda_0 \leq \lambda_c$$

(for a given value of Ω_0).

We shall now give a final argument in favour of a positive cosmological constant: inflationary models of the primordial universe imply that the curvature is zero; that is,

$$\Omega_0 + \lambda_0 = 1.$$

The value $\Omega_0 = 1$ is not excluded by observations (see Sect. 13.2.4); however, the values deduced from dynamical measurements indicate rather that

$$\Omega_0 \approx 0.1\text{--}0.3.$$

The prediction of inflationary scenarios would then be

$$\lambda_0 = 1 - \Omega_0 \approx 0.7\text{--}1.0.$$

13.2.4 The Radiation-Dominated Universe

The universes that we have studied above are those for which the equation of state is $P \approx 0$. As soon as we go back in time, that is, as z increases, this approximation is no longer valid, in particular when the particles become relativistic. In fact we shall see that the pressure is important well before, because of the contribution of photons to the energy density of the universe. These are dominated by the photons of the cosmic background radiation. Today their density is

$$\rho_{\gamma_0} = \frac{4\sigma}{c^3} T^4 \approx \frac{4\sigma}{c^3} \times (2.7\,\text{K})^4 \approx 4.4 \times 10^{-34}\,\text{g cm}^{-3},$$

where σ is the Stefan constant. The density parameter for radiation is then

$$\Omega_{\gamma_0} = \frac{4.4 \times 10^{-34}}{h^2 \times 1.9 \times 10^{-29}} \approx \frac{4 \times 10^{-5}}{h^2}.$$

Furthermore the matter density is

$$\rho_{M_0} = \Omega_0 h^2 \times 1.9 \times 10^{-29}\,\text{g cm}^{-3}.$$

In the course of time the contribution to the density of each type of matter evolves in a different manner. On the one hand

$$\rho_M(z) = \rho_0(1+z)^3;$$

on the other

$$\rho_\gamma(z) = \rho_{\gamma_0}(1+z)^4.$$

There is therefore a value of $1+z$ for which the contribution of radiation to the density of the universe is of the same order of magnitude as that of matter:

$$1 + z_{\text{eq}} \approx \Omega_0 h^2 \times 2.3 \times 10^4.$$

At $z > z_{\text{eq}}$ the equation of state $P = 0$ can no longer be used. From the expressions for the density and the pressure we obtain

$$\rho(T) = \sum_i \int \frac{1}{c^2} E_i(p)\, n_i(p, T)\, \mathrm{d}p^3,$$

$$P(T) = \sum_i \int c^2 \left(\frac{p^2}{3 E_i(p)} \right) n_i(p, T)\, \mathrm{d}p^3,$$

these integrals being evaluated in momentum (p) space and account being taken of all the particles present. In the regime that concerns us it is the photons, and also possibly other relativistic particles such as neutrinos, that interest us. For these $E = pc$, which gives us the equation of state:

$$P = \tfrac{1}{3}\rho c^2.$$

The conservation equation (13.3b) then allows us to write

$$R\dot{\rho} + 4\dot{R}\rho = 0;$$

that is,

$$\rho R^4 = \text{constant}$$

at times for which $z \geq z_{\text{eq}}$. It is clear to see that for reasonable models of the universe ($\lambda_0 \approx 1$, $\Omega_0 \approx 1$) equations (13.8), which give the scale parameter, then reduce to

$$2\frac{\ddot{R}}{R} + \left(\frac{\dot{R}}{R}\right)^2 \approx -\frac{8\pi G}{c^2} P,$$

$$\left(\frac{\dot{R}}{R}\right)^2 \approx \frac{8\pi G}{3}\rho.$$

Taking account of the conservation equation, we get

$$\frac{R}{R_1} = \sqrt[4]{\frac{32\pi G \rho_1}{3}}\, t^{1/2},$$

where ρ_1 and R_1 are the values of the parameters at a certain arbitrary time t_1 ($t_1 \leq t_{\text{eq}}$). This enables us to understand why, in the theory of general relativity, the equation of state $P = 0$ or $P = \tfrac{1}{3}\rho c^2$ leads to a singularity.

Indeed in both cases we have

$$R \to 0 \quad \text{when} \quad t \to 0.$$

In fact this remains true so long as $3P/c^2 > \rho$: equations (13.8) allow us to write

$$\frac{\ddot{R}}{R} = -\frac{4\pi G}{3}\left(\rho + \frac{3P}{c^2}\right).$$

$\ddot{R}$ therefore remains negative and R becomes equal to zero (except if the equation of state allows P to become strongly negative). This singularity of the radius of curvature leads to the divergence of the physical quantities

$$T \propto \frac{1}{R} \quad \text{and} \quad \rho \propto \frac{1}{R^4}.$$

Note that the reality of this singularity is not so well established, because it is no longer legitimate to use classical physics (like general relativity) 'before' the Planck time

$$t_{\text{Pl}} \approx \left(\frac{G\hbar}{c^5}\right)^{1/2} \approx 5 \times 10^{-44}\, \text{s},$$

where quantum effects become important (the Planck time is when the horizon of the universe becomes equal to its minimum size, prescribed by Heisenberg's uncertainty principle in quantum mechanics).

In conclusion we see that, for the primordial universe, the behaviour of the scale factor remains very simple, as long as the density is dominated by relativistic particles ($z \geq z_{\text{eq}}$).

13.3 The Hot Phase of the Universe

We are now going to study the properties of the universe in its 'young' phase. The mass density then dominates owing to the contribution of the relativistic particles, and the behaviour of the scale factor $R(t)$ as a function of time may therefore be considered independent of the cosmological model used. As we have seen, $R(t)$ varies as $t^{1/2}$:

$$R(t) = R_1 \left(\frac{t}{t_1}\right)^{1/2}.$$

The temperature of the universe increases indefinitely as soon as one approaches a singularity. It is sometimes useful to associate an energy or a temperature with an epoch rather than a redshift. The correspondence between the two is the following:

$$E = 2.32 \times 10^{-4}(1+z)\, \text{eV}.$$

13.3.1 The Thermal History of a Particle

We shall now analyse the evolution of a particle, highlighting the principal stages. The characteristic time associated with interactions for a relativistic particle is written as

$$\tau \propto \frac{1}{n\sigma c},$$

where n is the density of the interacting particles and σ is the corresponding effective cross-section. In the absence of annihilation or creation reactions, the comoving density is conserved, and therefore

$$n \propto \frac{1}{R^3}.$$

We can compare τ with the age of the universe:

$$\frac{\tau}{t} \propto \frac{R}{\sigma}.$$

It is only if the ratio τ/t is smaller than 1 that collisions play a role. Thus before applying results that are valid in a situation of equilibrium, we should examine if the reactions now are effective in equilibrium. Following on from the above equations, the ratio τ/t will always be less than 1 at epochs where the scale parameter is sufficiently small (except if the effective cross-section decreases rapidly at high energies). On the other hand during expansion the ratio τ/t becomes much greater than 1. The evolution of a given type of particle then necessarily leads to a situation far from equilibrium. The history of the universe may then seem astonishing, in that a situation far from equilibrium follows an initial equilibrium. This is due to expansion, which tends to reduce the density of the types of particle present. By contrast when one goes back in time, every type of particle is in equilibrium if the radius of the universe is sufficiently small. It is then useful to suppose that 'initially' all the particle types are in equilibrium. This might, however, be inexact if an infinity of types of particle were to appear (in a situation far from equilibrium) as soon as the energy were raised. This eventuality is not anticipated in modern theories of particle physics, which, on the contrary, predict a finite total number of particles. In what follows we suppose then that the 'initial' state corresponds to a situation where all the types of particle present are in equilibrium.

13.3.2 A Description of the Initial State of Equilibrium

The distribution law of a particle in equilibrium allows us to determine the corresponding number density:

$$n_i = \int \frac{g_i}{2\pi^2 \hbar^3 c^3} \frac{E^2 \, \mathrm{d}E}{\mathrm{e}^{E/kT} \pm 1},$$

g_i being the statistical weight of the species and E designating the total energy:
$$E^2 = m^2 c^4 + p^2 c^2.$$
The chemical potential μ of the species has been assumed to be zero, which is always the case when the distributions of particles and antiparticles are identical (that is, in the absence of asymmetry). For a fermion gas (which has half-integer spin) the sign is $+$; for a boson gas the sign is $-$. For massless particles, for example in the very important case of photons, $g = 2$, and their density is given by
$$n_\gamma = 16\pi \zeta(3) \left(\frac{kT}{\hbar c}\right)^3,$$
where $\zeta(3) \approx 1.202\ldots$ Similarly we may calculate the energy density of the photons:
$$u_\gamma = aT^4 \quad \text{with} \quad a = \frac{2\pi^2 k^4}{15 c^3 \hbar^3} = \frac{4\sigma}{c}$$
and thus their contribution to the pressure:
$$P_\gamma = \tfrac{1}{3} a T^4.$$

For other particles it is interesting to distinguish two regimes. The first (the ultrarelativistic regime) corresponds to energies much greater than the rest energy of the particle: the relation $E = pc$ is a good approximation. We can then neglect the fact that the particle would have a nonzero mass. The particle density n, the energy u, and the pressure P are then given, for bosons, by
$$n_b = \tfrac{1}{2} g_b n_\gamma,$$
$$u_b = \tfrac{1}{2} g_b a T^4 = \tfrac{1}{2} g_b u_\gamma,$$
$$P_b = \tfrac{1}{3} \tfrac{1}{2} g_b a T^4 = \tfrac{1}{2} g_b u_\gamma$$
and, for fermions, by
$$n_f = \tfrac{3}{8} g_f n_\gamma,$$
$$u_f = \tfrac{7}{16} g_f a T^4 = \tfrac{7}{16} g_f u_\gamma,$$
$$P_f = \tfrac{7}{16} \tfrac{c^2}{3} g_f a T^4 = \tfrac{7}{16} g_f u_\gamma.$$

The second regime is that where the particles are nonrelativistic but always in equilibrium (situations far from equilibrium are studied below). The bosons, like the fermions, then obey the Maxwell distribution function:
$$n_i = \frac{g}{2\pi} \left(\frac{kT}{\hbar c}\right)^3 \left(\frac{mc^2}{kT}\right)^{3/2} e^{-mc^2/kT}$$
$$= \frac{g}{2\pi^2} n_\gamma \left(\frac{mc^2}{kT}\right)^{3/2} e^{-mc^2/kT},$$
$$\rho_i = n_i m_i,$$
$$P_i = n_i kT.$$

13.3.3 The Chemical Decoupling of a Particle

The situation described above assumes that the characteristic reaction times bringing a given particle into play are much shorter than the age of the universe. In particular there must be one reaction at least that allows the equilibrium of a particle, A, and its antiparticle, $\bar{\text{A}}$, that is, the reaction

$$\text{A} + \bar{\text{A}} \rightleftharpoons \gamma + \bar{\gamma}$$

or

$$\text{A} + \bar{\text{A}} \rightleftharpoons \text{p} + \bar{\text{p}},$$

where p is some particle. Nevertheless all these processes end up sooner or later by no longer being effective. The comoving density is then frozen from this epoch onwards, if the particle is stable. The density evolves simply as follows:

$$n = n_{\text{D}} \frac{R_{\text{D}}^3}{R^3(t)}.$$

The epoch t_{D} at which the chemical decoupling is produced is very important. Indeed if this decoupling is produced in the relativistic regime, the density would be approximately that of the photons:

$$n_{\text{D}} \approx n_\gamma,$$

neglecting differences in the statistical factors. On the other hand if the decoupling is produced in the nonrelativistic regime, the density will become extremely low because of the term $e^{-mc^2/kT}$.

Subsequently particle A may no longer take part in reactions with any of the other particle types present; in this case it interacts with the rest of the universe through gravitation alone. This phase is called the thermal freeze. The particle is then entirely decoupled from the rest of the universe.

To illustrate the above we shall consider the decoupling of an electron neutrino (or more generally all particles that interact through the weak force). As long as there is equilibrium, the density is simply

$$n_\nu = \tfrac{3}{8} n_\gamma.$$

Chemical decoupling occurs when the temperature is of the order of 10^{10} K. The corresponding energy is very much greater than the mass energy of the neutrino (which is less than 10 eV). At the moment of decoupling, $T_\nu = T_\gamma$. During expansion the form of the distribution is, as for the photons, conserved, and the 'temperature' (this is no longer a thermal temperature but a kinetic one) evolves as $1/R$. However, the actual temperature of the neutrinos is not exactly that of the photons. Indeed after the decoupling of the neutrinos the annihilation $e^+ - e^-$ will produce a 'reheating' of the photons. Before annihilation the entropy density is

$$S = \frac{1}{T}(u+P),$$

which for relativistic particles corresponds to

$$S = \frac{4}{3}\frac{1}{T}u.$$

As the neutrinos are decoupled, they can be ignored in the calculation for entropy:

$$S = \frac{4}{3}\frac{1}{T}(u_\gamma + u_{e^+} + u_{e^-}) = \frac{11}{3}aT^3.$$

After the annihilation of the electrons, the entropy is uniquely determined by the photons:

$$S = \frac{4}{3}aT_\gamma^3.$$

The conservation of entropy then implies a 'reheating' of the photons whereas the 'temperature' of the neutrinos remains unchanged:

$$\left(\frac{T_\gamma}{T_\nu}\right) = \left(\frac{11}{4}\right)^{1/3};$$

that is,

$$T_\nu = 0.7 T_\gamma.$$

In the same way, we can calculate the decoupling temperature of the protons a priori. This decoupling occurs later than for a neutral or charged lepton because of the strong interaction. As a result the ratio of the actual density of the protons or the antiprotons to that of the photons will become of the order of 10^{-19}. Now the observed abundance is much greater, of the order of 10^{-9}. This number also agrees with the predictions of nucleosynthesis. Furthermore it seems that observations exclude the existence of a significant amount of antimatter. We then have to conclude that the 'initial' state was asymmetric and composed of a slight excess of baryons with respect to antibaryons. The origin of this asymmetry has received a satisfactory explanation within the framework of recent models from particle physics.

13.3.4 Primordial Nucleosynthesis

We are now going to briefly present some aspects of primordial nucleosynthesis. Let us first consider the evolution of the relative density of neutrons and protons. The reactions that maintain proton–neutron equilibrium are the following:

$$n + \nu \rightleftharpoons p + e^-,$$
$$p + \bar{\nu} \rightleftharpoons n + e^+.$$

At very high energies the rates of these reactions are equal and the densities of the two species are then the same. The neutrinos, although decoupled, continue to contribute to the equilibrium when the temperature becomes less than 10^{10} K because of the low density of baryons and neutrons. As soon as the temperature becomes lower, the equilibrium will alter according to the value of the effective cross-sections. As the neutron is slightly heavier than the proton, the equilibrium will move in favour of the protons. Thus as long as the reactions take place, the ratio f of the densities evolves as a function of the temperature:

$$f = \frac{n_\mathrm{N}}{n_\mathrm{p}} = e^{-\Delta m c^2/kT},$$

Δm being the difference in mass between the proton and neutron. If this equilibrium is maintained for a long time, neutrons vanish to the benefit of protons. Now the freezing of the reactions occurs precisely when the neutron density begins to diminish, that is, towards 1.3×10^9 K. Then because the reactions stop at a temperature that corresponds to the difference in mass between the proton and neutron, the neutron–proton ratio is fixed at a value that is neither 0 nor 1 but

$$f = \frac{n_\mathrm{N}}{n_\mathrm{p}} \approx 0.15.$$

At this energy neutrons are free and begin to decay. In fact they will rapidly fuse with protons to form the light elements, mainly helium 4 (the primordial abundance of the other elements being very low). We can thus estimate the abundance of helium 4:

$$Y = \frac{m_{^4\mathrm{He}} n_{^4\mathrm{He}}}{m_\mathrm{H} n_\mathrm{H} + m_{^4\mathrm{He}} n_{^4\mathrm{He}}} = \frac{2f}{1+2f} \approx 0.25.$$

A more precise calculation of the abundance of light elements requires all the reactions to be taken into account; the most rapid are those which lead to the production of helium 4:

$$\mathrm{p} + \mathrm{n} \rightleftharpoons \mathrm{d} + \gamma,$$
$$\mathrm{d} + \mathrm{d} \rightleftharpoons {}^3\mathrm{He} + \mathrm{n},$$
$$^3\mathrm{He} + \mathrm{n} \rightleftharpoons {}^3\mathrm{He} + \mathrm{p},$$
$$^3\mathrm{He} + \mathrm{d} \rightleftharpoons {}^4\mathrm{He} + \mathrm{n}.$$

The other reactions, in particular those which produce lithium and beryllium, are much slower. In return, deuterium and helium 3 are destroyed as quickly as they are produced in the above reactions. Eventually only helium 4 is produced in significant quantities. The other elements cannot be formed in appreciable quantities during primordial nucleosynthesis. Certain poorly known values such as the lifetime of the neutron, certain effective cross-sections, or the number of species of relativistic neutrinos (determining the expansion rate) enter into the calculation of the abundances, but their influence on the result is small. The main parameter is the ratio η of the number of protons to the number of photons, or, more usefully,

$$\eta_{10} = 10^{10} \frac{n_B}{n_\gamma}.$$

The results concerning nucleosynthesis constitute a brilliant confirmation of the cosmological model known as the 'hot big bang'. Indeed the observed abundances are in remarkable agreement with the observations, taking account of the uncertainty affecting certain values and the fact that the subsequent chemical evolution has been able to modify partially the primordial abundances. Furthermore, within the framework of this model, nucleosynthesis will allow us to determine the value of the parameter η and, as a consequence, to estimate the contribution Ω_B of the baryons to the density of the universe, thanks to the relation

$$\Omega_B = 0.0035 \eta_{10} h^{-2} \left(\frac{T}{2.7}\right)^3.$$

Figure 13.5 shows the abundance of the primordial elements predicted by the models. The abundance of helium 4 is expressed as a function of that of D and ^{3}He:

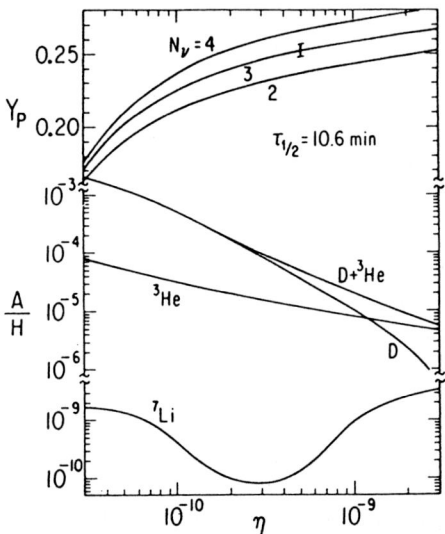

Fig. 13.5. The primordial abundance of light elements as a function of the parameter η (see the text), as predicted in the 'standard' models. Y_p is the fraction of the mass in the form of helium. The abundance of the other elements is the numerical abundance with respect to hydrogen. The curves relative to helium are given according to the number of types of stable and relativistic neutrinos at the epoch of nucleosynthesis. (From Yang et al. 1984)

13.3 The Hot Phase of the Universe   419

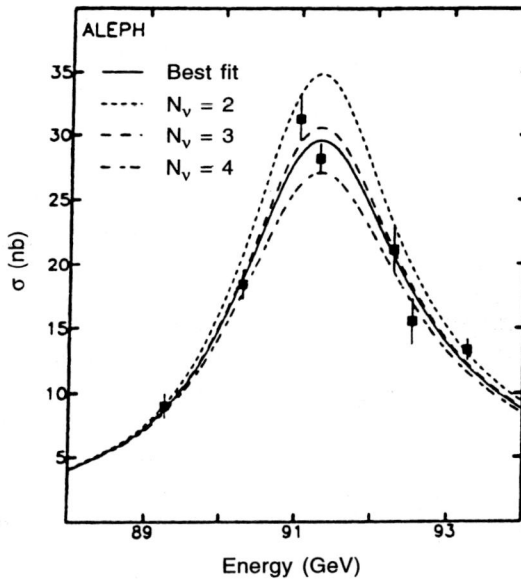

Fig. 13.6. In 1989 the starting-up of LEP enabled physicists to measure the number of types of neutrino. The results of three different experiments, OPAL, ALEPH, and L3, showed conclusively that there are 3 neutrinos. The results are in good agreement with primordial nucleosynthesis. This figure represents the effective crosssection for the reaction $e^+e^- \to$ hadrons, as a function of the energy in the centre of mass, and the models for different numbers of neutrinos. (From Decamp et al. 1990)

$$Y = 0.261 - 0.018 \log \left(10^5 \frac{n_\mathrm{D} + n_{^3\mathrm{He}}}{n_\mathrm{H}} \right) + 0.014(N_\nu - 3) + 0.014(\tau_{1/2} - 10.6),$$

where $\tau_{1/2}$ is the lifetime of the neutron in minutes and N_ν is the number of types of neutrino. This last quantity remained uncertain until 1990: the number of types observed in the laboratory was 2, but theoreticians suspected the existence of a third neutrino (the τ neutrino). Measurements of Z_0 made at LEP confirmed the result: there are three types of neutrino (Fig. 13.6).

It is mainly the dependence of the primordial abundances of ^{3}He, ^{4}He, D, and ^{7}Li with respect to the ratio η that is studied, since the other quantities cannot be determined with sufficient precision. In principle these observations allow us to verify that the abundances of each of the elements correspond to a unique value of this parameter. In practice we are far from this ideal situation, since these abundances, in particular for ^{3}He and D, are modified by an order of magnitude by the later chemical evolution in galaxies. The example of deuterium is significant: its abundance is very sensitive to the ratio η and would allow a precise measurement. Unfortunately galactic evolution tends to destroy this fragile element. If the models of chemical evolution described were well established, it would be possible to determine the primordial abundance.

In fact it is rather the observed abundance that sets a constraint on the models of chemical evolution. The situation is similar for helium 3; moreover the determination of its abundance remains tricky, and the dispersion of the results obtained is very large (an order of magnitude). Similarly estimation of the abundance of helium 4 remains uncertain. A value currently used is

$$Y \approx 0.245 \pm 0.003,$$

but the divergence in the measurements by different authors is of the order of 1%, which illustrates the difficulty.

Lithium plays a special role: the predicted abundance has a minimum, which is not in fact a priori a good indicator of the ratio η. Furthermore it is a fragile element, and it may simply be destroyed or created within stars. This is clearly illustrated by the great divergence in the measured abundances, which are sometimes very much less than the very lowest allowed by nucleosynthesis. Nevertheless the constraint that derives from the abundance of lithium in metal-poor stars has been shown to be extremely fruitful. Since these stars have no convective envelope, we can hope that evolution has not modified the primordial value too much. Now the results are remarkably consistent from one star to another. If the lithium undergoes significant evolution in these stars, it is probable that it manifests itself as a dispersion of the measured abundances, as is the case for metal-rich stars. The abundance of lithium then allows a constraint on η more certain, in fact, than that which we can deduce for other elements. The recent detection of ^{6}Li argues in favour of the absence of significant destruction of ^{7}Li. This is because ^{6}Li is a much weaker element than ^{7}Li. In fact the main sources of error in η derive from the effective cross-sections of the creation and annihilation reactions of lithium (where there is a factor of about 2 in uncertainty). Taking account of these uncertainties and of those affecting abundance measurements, we end up with the following inequality:

$$1.2 \leq \eta_{10} \leq 6.4.$$

The uncertainties in the value of the Hubble constant, and thus those in the temperature of the cosmic background radiation, are reflected in the value of Ω_B. The temperature of the cosmic background radiation is now well known; we get

$$0.0040 \leq \Omega_B h^2 \leq 0.025.$$

We obtain a narrower range in looking for the domain of η_{10} compatible with the abundance of all the light elements:

$$\eta_{10} = 4.1,$$

which leads to a very narrow range for $\Omega_B h^2$:

$$\Omega_B h^2 \approx 0.015.$$

A value of Ω_B of the order of 0.1 is then just compatible with primordial nucleosynthesis. Now dynamical measurements of the density of the universe indicate rather higher values. This constitutes the only solid observational argument that justifies the introduction of a nonbaryonic component. However, the uncertainties concerning nucleosynthesis and its dynamical measurements are too large for this conclusion to be inevitable. It is important to note that if the Hubble constant is of the order of $100\,\text{km}\,\text{s}^{-1}\,\text{Mpc}^{-1}$, the above conflict is once again more troublesome. This will be discussed in detail in the next section.

13.4 The Very Early Universe

The study of homogeneous cosmology, that is (in a very broad sense), everything that is not directly concerned with the formation of galaxies, shows that it is perfectly possible to give an account of observations. We recall the three great tests of 'classical' cosmology: the redshift of galaxies, which varies linearly with distance for $z \ll 1$; the abundances of helium and other elements; and finally the existence of the black-body cosmic microwave background radiation. It is clear, moreover, that these three facts could be predicted on theoretical grounds. In fact no 'rival' theory that is as simple can compete with this model.

13.4.1 Problems of the Classical Big-Bang Theory and Grand Unified Theories

When one considers the inhomogeneous aspect of the universe, that is, when one seeks to explain the fact that the galaxies exist and that their distribution is not homogeneous but hierarchical, it might seem that the successes of the model are diminished. Yet, for one thing, the characteristic time of gravitational instability is precisely of the order of magnitude of the inverse of the Hubble constant. Newtonian gravitation is then sufficient to confirm that the distribution of matter, at least for the densest regions, has to evolve in a time of the order of the Hubble time t_H (this point being a simple coincidence in a stationary theory). Just as interesting is the fact that the ages of the oldest stars are of the order of t_H, while older stars may be observed a priori.

On the other hand certain consequences of the model appear to be problematic. Thus since observations show that the ratio η of the number of protons to the number of photons is of the order of 4×10^{-10}, the constraints stemming from nucleosynthesis enable very precise bounds to be placed on this figure. Nevertheless the origin of this number is in no way settled by the model. We have equally seen that we can predict a background of neutrinos and antineutrinos such that

$$n_\nu = n_{\bar\nu} \approx n_\gamma.$$

It is possible to predict in the same way what must be the abundance of protons and antiprotons. What is determined in this calculation is the epoch in which protons and antiprotons decouple. Now the decoupling is very 'slow', unlike for neutrinos, and one would expect that ($\bar{\mathrm{p}}$ designates the antiproton)

$$n_{\mathrm{p}} = n_{\bar{\mathrm{p}}} \approx 10^{-19} n_\gamma,$$

which is wrong according to observations by ten orders of magnitude! Furthermore a universe symmetric with respect to matter–antimatter appears unrealistic on many grounds. Of course, it is clear that the Solar System contains almost no antimatter. Likewise cosmic rays that come from the distant stars contain almost no antiprotons. The small fraction observed is very probably a secondary product of collisions. Our Galaxy is then uniquely made of matter. On large scales the indications are less certain. Nevertheless if certain galaxies in a cluster were made of antimatter, they would emit gamma radiation strongly because of the exchange of gas between the interstellar medium and the galaxies. In the same way if there existed regions of the universe made of matter and others made of antimatter, an annihilation front would exist in the zone of contact. This front would constitute an intense source of anisotropic gamma radiation, unless the regions of matter and antimatter were well separated from one another. The only possibility would then be of having regions whose size is greater than that of the large structures actually observed. Now the epoch of proton decoupling is 10^{-3} s. The mass bounded by the horizon is then $10^{-5} M_\odot$. There could not then be, at that epoch, a causal mechanism responsible for the separation of matter and antimatter on the scale of large structures. As the epoch at which the horizon reached this scale is recent (it corresponds to a redshift of around 10), it seems that such a mechanism can be ruled out. Before this problem, cosmologists had for a long time given up trying to understand the matter–antimatter asymmetry. This situation has been profoundly modified by the so-called grand unified theories (GUTs) of particle physics, which allow the unification of fundamental interactions. The electroweak theory unifies quantum electrodynamics (the U(1) theory) and the SU(2) theory of weak interactions. One of the more spectacular predictions of this model is the existence of new particles playing the role of vectors in the unified interaction. Thus in the theory of quantum electrodynamics the boson of interaction is the photon. In the electroweak theory there are three bosons of interaction: the W^+, its partner the W^-, and a neutral particle, the Z^0. These particles have actually been observed in the accelerators at CERN. The unification of the interactions takes place for energies greater than the mass energy of these particles.

In a grand unified theory one seeks to fully encompass the strong and electroweak interactions. This implies the existence of bosons for which the mass varies from one theory to the other. In the SU(5) model the unification occurs for energies of the order of 10^{15} GeV. This domain is completely inaccessible to accelerators, and the theory can then only be tested in a very indirect manner. For example the disintegration of a proton in 10^{30} years is one of the tests proposed.

On the other hand the primordial universe constitutes an extraordinary laboratory, since at sufficiently remote epochs the energy of the particles present was of this order of magnitude. More precisely this energy would have occurred when

$$t \leq 10^{-37} \text{ s},$$

that is, a period relatively close to the Planck time. It is at this moment that 'baryogenesis' takes place. We shall briefly examine the mechanism initially proposed by Sakharov to explain the matter–antimatter asymmetry. The first condition required is nonconservation of baryon number B. Indeed if all possible reactions conserve this number, it cannot be modified in the course of evolution. Now in GUTs we expect a violation of the conservation of B. Moreover experiments concerning the disintegration of the proton are destined to provide evidence of this violation. Another necessary ingredient is noninvariance with respect to C (conjugation of charge) and CP (conjugation of charge and parity) transformations, violations that have been observed in accelerators (although the violations of CP are poorly understood from the theoretical point of view). The reason is that the baryonic charge changes sign through these transformations, and then if the system were not modified, we would end up with $n_B = 0$. Finally it is necessary that reactions that do not conserve baryon number occur far from equilibrium. Indeed a reaction at equilibrium will tend to remove the asymmetries rather than create them. Thus the reaction (in the course of which B is not conserved)

$$\gamma + \gamma \rightleftharpoons B + B$$

implies that the chemical potential of this particle is zero, B designating a baryon. The above reaction does not necessarily play an important role, but it illustrates the characteristics of a reaction that violates the conservation of baryon number.

A simple model describing baryogenesis can now be elaborated. In the framework of the SU(5) unified theory (the simplest, for this description), bosons X have a mass energy of the order of, or greater than, 10^{14} GeV. Their disintegration rate depends on their mass M:

$$\Gamma_D \approx \frac{\alpha M c^2}{\hbar},$$

where α is the coupling constant at this energy and M is the mass of the particle. The expansion rate of the universe is given by

$$H \approx (G\rho)^{1/2} \approx \left(Gg_* \frac{E}{c^2 \lambda^3}\right)^{1/2},$$

where g_* is the sum of the statistical weights of the interacting particles and λ is the wavelength associated with the energy E:

$$\lambda = \frac{\hbar c}{E}.$$

The equality $\Gamma_D \approx H$ implies that

$$E = \frac{\alpha c}{g_*^{1/2}}\left(\frac{\hbar c^3}{G}\right)^{1/2} = \frac{\alpha}{g_*^{1/2}} m_{\text{Pl}} c^2,$$

where $m_{\text{Pl}} = (\hbar c/G)^{1/2}$ is the Planck mass.

When the temperature becomes lower than the temperature associated with the mass of particle X, we have seen that the numerical density of the particles decreases very rapidly. It is then preferable that the particle is not at equilibrium before this time, that is, that the decoupling takes place when the particle is once again relativistic. This imposes a condition on the mass:

$$M \geq \left(\frac{\alpha}{g_*^{1/2}}\right) m_{\text{Pl}}.$$

One then has

$$n_X \approx n_\gamma.$$

We shall now examine the disintegration modes for the particle X. If there exists only one such mode, no asymmetry will be observed. Indeed every quantity generated by the disintegration of X would occur in a perfectly symmetrical way by the disintegration of $\bar{X}$ (in fact a partial asymmetry exists as long as the particles X and $\bar{X}$ are not entirely disintegrated; but this asymmetry subsequently disappears). Two disintegration modes must then exist that do not produce the same number of baryons:

$$X \to B_1,$$
$$X \to B_2.$$

Let r and $1-r$ be the relative rates of each reaction. Let $\bar{r}$ and $1-\bar{r}$ be those of the symmetric reactions of $\bar{X}$. The comoving number of baryons produced is

$$n_B = n_X(r - \bar{r})(B_1 - B_2).$$

Then $r \neq \bar{r}$, which requires the violation of C and CP. The entropy at this epoch is gn_γ. In the absence of other reactions that do not conserve B, the ratio n_B/n_γ will today be

$$\frac{n_B}{n_\gamma} = \frac{(r-\bar{r})(B_1-B_2)}{g_*} = \frac{\epsilon}{g_*}.$$

A slight violation of CP is then enough to obtain the observed ratio.

The numerical integration of the Boltzmann equation allows a more precise analysis and shows that the evolution occurs as we have described above. However, there is a notable difference. An asymmetry may develop even in the regime where the decoupling is slow (that is, in the nonrelativistic regime). If the decoupling occurs even more slowly, the ratio n_B/n_γ actually becomes extremely small. This schema clearly shows that grand unification theories allow the existence of an asymmetry and a ratio n_B/n_γ compatible with observations; it is unfortunately not possible to make predictions as precise as these relative to nucleosynthesis, mainly because of the uncertainties concerning CP violation. However, this remains the most effective mechanism to explain the matter–antimatter asymmetry observed.

13.4.2 The Theory of Inflation

In grand unified theories the state of complete symmetry of the theory exists only at very high energies. Throughout the history of the universe, the temperature did not stop falling, and these symmetries were broken. This is why particles are not all governed by the same interactions today. The breaking of symmetry is accompanied by a phase transition. Now under certain conditions, which we shall return to in more detail, this transition is capable of considerably modifying the evolution of the universe as a whole. Before approaching this question, we shall describe certain consequences of the cosmological model that might pose a problem. It is without doubt more justifiable to speak of the question of 'initial' conditions. It turns out that these conditions in which we would find the universe at very primitive times are indeed remarkably simple. This initial simplicity is absolutely necessary for the universe to be what we see today, because if the corresponding parameters were modified very slightly, the resulting differences would be enormous.

The Horizon Problem. The horizon of an observer is the region within which it is possible for him or her to receive information. The horizon distance L is an increasing function of time. In the radiation-dominated regime, it is simply given by

$$L = R(t) \int \frac{c\,dt}{R(t)} = 2ct.$$

If we consider an epoch t, our present horizon is composed of many horizons of this epoch. These regions can never interact; they are causally independent. Nevertheless we can see today that they are identical. The proof is given by the cosmic background radiation. Thus when we examine two regions of the sky separated by $180°$, we see totally independent regions of the universe. Now the temperature of the cosmic background between such regions differs by less than 10^{-3}. In the same way at the time of nucleosynthesis the ratio η had become almost identical in regions with no causal connection (the mass contained within a horizon at this epoch was less than one solar mass).

The Flatness Problem. The density parameter of the universe Ω is a quantity that evolves over time. Its value tends towards 1 when we examine more and more distant epochs. We therefore examine the quantity β, where

$$\beta = \frac{1 - \Omega(t)}{\Omega(t)} = -\frac{k}{R^2 8\pi G \rho / 3}$$

(see Sect. 13.2). Now

$$\beta \propto R \propto t^{3/2} \quad \text{for the matter-dominated era,}$$
$$\beta \propto R^2 \propto t \quad \text{for the radiation-dominated era.}$$

As the value of Ω is already close to unity today ($\geq 10^{-2}$), the quantity β will have been extremely close to 0 in distant epochs. For example in the era of nucleosynthesis β was of the order of 10^{-14}; at the Planck time β would be around 10^{-60}. To say the least it is not very elegant to have a parameter of the model that might be small without at the same time being able to explain its origin. Another way of formulating the same problem is the following: in an open universe a transition occurs in the temporal dependence of the expansion parameter:

$$R(t) \propto t^{2/3} \quad \text{when} \quad z > \frac{1}{\Omega_0},$$
$$R(t) \propto t \quad \text{when} \quad z < \frac{1}{\Omega_0}.$$

This allows us to associate with the universe a characteristic time scale. Similarly if the universe is closed, the epoch at which the expansion reverses allows us to define a characteristic time. Now for all the reasonable values of Ω, the characteristic time is of the order of the actual age of the universe (except if Ω is extremely close to 1). The value of the characteristic time might then have been fixed very early on, 'initially'. Here again, it might be more agreeable to have at our disposal a theory that can predict the value of the characteristic time.

Primordial Fluctuations. The existence of galaxies and large structures in the real universe clearly shows that it is inhomogeneous on small scales. In order for it to be this way it must be the case that, in the past, the universe could not have been absolutely homogeneous, at every past epoch. It is then necessary that initially density fluctuations existed, even though they must have been extremely small, so as not to produce too significant fluctuations in the cosmic background radiation, whose homogeneity is remarkable ($\delta T/T \approx 10^{-5}$). All cosmological models require the existence of a density excess with an amplitude of 10^{-4} relative to the epoch where the fluctuations meet the horizon, so that their development would be nonlinear today. The origin of these fluctuations at present appears inexplicable. In particular, statistical fluctuations alone are completely inadequate.

13.4 The Very Early Universe

The Problem of Magnetic Monopoles. It is not correct here to speak of a problem with cosmological models, but rather of a problem concerning particle physics. Indeed in all gauge theories, of which grand unified theories are an example, the phase transition towards states of broken symmetry is accompanied by the creation of topological defects: we shall now explain in a very simplified manner how these are produced. The state that represents empty space is that of minimum energy and depends on the form of the potential that governs the theory. At the time of a phase transition the final state of empty space does not correspond to that of maximum symmetry. It then becomes degenerate, since many solutions would exist for the minimum in the final state. Now there is no reason for the vacuum state's being the same from one point to another in the universe, since there is no 'privileged direction'. If there were an interaction tending to place neighbouring points in space in the same state (which corresponds in general to a configuration of minimum energy), this would not be able to occur on small scales at the horizon distance at the time of the transition; on larger scales the directions would become independent. If such regions of space have independent orientations, there necessarily exist connecting regions where the direction is singular. The topology of these regions depends on the geometry of gauge groups. There may be points, known as monopoles, one-dimensional curves, known as cosmic strings, or surfaces, known as domain walls. If λ is the size of coherence of the local direction of the vacuum, the density produced is of the order of λ^{-3}. Now every theory that contains the group U(1), that of electromagnetism, gives as a solution monopoles that carry one magnetic charge. In SU(5)-type theories the mass of the predicted monopoles is of the order of 10^{15} GeV. The amount produced is huge. Annihilation does not allow these monopoles to be done away with in sufficient numbers, and their present density would be of the order of 10^{14} times the critical density! This is then a very serious difficulty for the theory.

A Particular Phase Transition: Inflation. The study of phase transitions in the primordial universe was begun well before a theory of inflation was formulated. Now a phase transition does not occur in a rigorously homogeneous way. The basic hypothesis is that the phase transition occurs only at several points where the temperature becomes less than the critical temperature T_c and that the rest of the universe is in a (metastable) state of excess fusion. In this situation empty space has a latent heat (representing an energy associated with the vacuum). The density of the universe is given by

$$\rho \approx aT^4 + \rho_\nu,$$

where ρ_ν stands for the energy associated with a metastable state of the vacuum. When the temperature is sufficiently low, it is this contribution that dominates the density of the universe. The above equation governs the scale parameter and then that of a cosmological model for which the cosmological constant is nonzero and the density negligible (the de Sitter model). The solution is of the form

$$R(t) = R_i e^{t/\tau} \quad \text{with} \quad \tau^2 = \frac{8\pi G \rho_\nu}{3}.$$

When the phase transition occurs at t_f, the expansion factor increases considerably:

$$\frac{R_f}{R_i} \approx e^{t_f/\tau}.$$

The horizon distance then increases in the same ratio, and if this ratio were sufficiently large, the horizon distance would already exceed our present horizon distance in the absence of inflation. Of course, our actual horizon is a little larger than that defined just after inflation, but all the matter observed today, that is, that existing after inflation, is inside a volume much smaller than the total horizon; indeed, our effective horizon remains that calculated classically. The energy liberated in the phase transition would produce a medium that one can suppose is in thermal equilibrium at a temperature of the order of T_c. Evolution subsequently takes place in a classical way. In the same way we can see that the curvature term disappears. At the end of inflation we would then have

$$1 - \Omega \approx (1 - \Omega_i) e^{-2t/\tau},$$

which leads to the first prediction of the model: $\Omega_0 = 1$. In fact the value is not exactly equal to 1, but it is very difficult, and above all contrary to the spirit of the theory, to obtain a different value. Fluctuations are produced during the phase transition. Their amplitude can be estimated, but the uncertainties are such that we cannot draw quantitative conclusions; on the other hand the spectrum of these fluctuations (the so-called Harrison–Zel'dovich spectrum) is well defined:

$$P(k) \propto k,$$

and the phases of the different modes are independent. This leads to Gaussian fluctuations. The other consequences of inflation do not directly constitute tests of the theory. The horizon problem is resolved; in fact the nature of this problem, like the flatness problem, remains controversial: it is not clear that an explanation is required. The theory of inflation greatly clarifies the difficulties associated with the initial conditions; it stands as an indispensable complement to every cosmological model. Finally, as we have seen, gauge theories necessarily lead to the production of magnetic monopoles. In a grand unified gauge theory we expect in reality a cascade of phase transitions; in the course of one of these the monopoles are created, and if inflation subsequently occurs, they will be eliminated by dilution. On the other hand baryogenesis has to occur after the period of inflation. If supplementary transitions take place after inflation, other types of topological defect may appear (in smaller amounts than the monopoles, however). Cosmic strings, for example, have been the subject of very detailed studies. These objects are linear (one-dimensional) and can form loops that disintegrate rapidly in gravitational waves. The background radiation thus created may have observational consequences because it perturbs the propagation of electromagnetic signals in an achromatic way. On this point, important constraints can be obtained by the observation of pulsars. Cosmic strings are also capable of producing anisotropies in the cosmic background radiation, the nature of which (because of their non-Gaussian character) is very different from the fluctuations expected in a more classical model. Indeed the geometry of space is modified by a cosmic string, which produces gravitational-lens effects. Finally cosmic strings move with a speed close to c; the matter traversed by such an object may be drastically altered: the passage of the string may project two particles towards one another, which, for baryonic matter, manifests itself as a violent compression. This might explain the 'leaflike' structure suggested by certain observations. By contrast, closed strings or 'loops' could have an invariant distribution during the expansion of the universe (with a related scale factor).

A complete theory of the primordial universe generally leads to a universe that has a critical density. The origin of this fact recalls the flatness problem: in the primordial universe if the density were not exactly the critical density, this would imply the existence of an extraordinarily small arbitrary parameter, the value of which, moreover, has to be adjusted very precisely to be compatible with observations; it is then easy to understand why no theory exists that accounts for values of Ω_0 of the order of 0.1, for example.

Inflation got its first support from observations with the positive detection of fluctuations in the CMB (see Fig. 13.7). In the gravitational-instability picture the fluctuations are due to inhomogeneities at $z \approx 1000$. They are consistent with a power-law spectrum with an index $n = 1$ (although the uncertainty is quite large). This corresponds to the prediction of inflation.

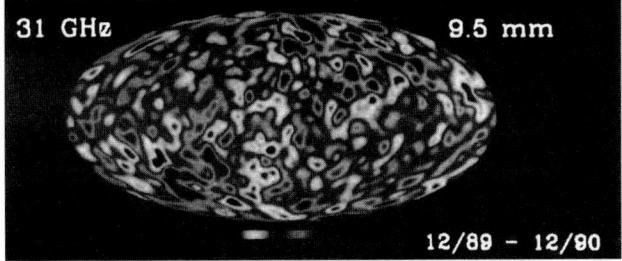

Fig. 13.7. A representation of the sky observed by *COBE* at wavelengths that are dominated by the CMB. Individual structures seen in this picture are not significant, but there is a significant global excess, whose power differs from the instrumental noise. The typical radiation seen here has an r.m.s. amplitude of the order of 10^{-5} irrespective of the scale. (From Smoot et al. 1992)

13.4.3 Nonbaryonic Dark Matter

From the observational point of view, the appeal to a nonbaryonic form of matter is not absolutely necessary; dynamical measurements made of objects such as clusters, spiral galaxies (rotation curves), or dwarf galaxies such as DDO 154, having (see Chap. 3) a wide variety of mass–luminosity ratios, comprise between 2 and 600 $M_\odot$. Nevertheless when a significant quantity of dark matter is detected, which is systematically the case when one observes regions clearly more extended than those visible optically (dwarf galaxies, normal galaxies, and clusters), the ratio of dynamical mass to the submass formed of baryons (H I gas, stars, and the X-ray-emitting gas of clusters) is of the order of 10. Thus the ratio of dynamical mass to 'visible' baryon mass (in a broad sense) seems almost constant. On the other hand the total mass observed is only a fraction of the critical density. We take as an example the Coma cluster: its M/L_B ratio is of the order of 200–500h $M_\odot/L_\odot$. If this value is characteristic of all the mass distributed throughout the universe, we obtain the value of Ω_0:

$$0.1 \leq \Omega_0 \leq 0.5,$$

account being taken of the uncertainties in the luminosity functions of galaxies (here taken to be 50%). The above framework is then marginally compatible with nucleosynthesis. Only if $H_0 = 100\,\mathrm{km\,s^{-1}}$ is recourse to nonbaryonic dark matter necessary. In this case the fraction of observed baryons is extremely low:

$$\Omega_B \approx 0.01,$$

which is also marginally consistent with the lower limit imposed by nucleosynthesis.

One can consider these observations from two different perspectives: it is clear that the dark mass will always be close to ten times the mass of baryons if the dark mass is nonbaryonic with $\Omega_0 = 1$ and $\Omega_B \approx 0.1$. On the other hand nucleosynthesis predicts for $h \approx 0.5$ an abundance of baryons rather greater than that measured (what is more, for the formation of galaxies, a model with $\Omega_B \leq 0.1$ encounters serious difficulties: we therefore generally prefer to take $\Omega_B \geq 0.1$). There are then certainly baryons that have escaped detection up to the present. A major challenge for cosmology is to understand what the dark matter is. If it is baryonic, it must then be visible to some extent. Stellar candidates pose problems because numerous constraints are imposed by the observations: they must not be visible in our Galactic halo nor in galaxies with a large redshift nor in the form of a diffuse background. They must not perturb the dynamics of our Galaxy's disc nor be perturbed by this disc; finally an excessive enrichment of the interstellar medium in light elements must be avoided. This type of constraint renders very unlikely the hypothesis of a massive halo made of stellar remnants with a mass greater than 0.1 $M_\odot$. On the other hand objects of lower mass, made up of primordial hydrogen and helium, are perfect candidates because nuclear reactions could not have been initiated there. Such stars are known as brown dwarfs. No theory exists that can predict the number of such objects capable of being formed at the same time as 'normal' stars (the initial mass function is very poorly known for small masses). The possibility of direct detection is very unlikely, the brown dwarfs being extremely faint. One possibility for their detection is that these bodies will from time to time amplify the light of distant stars through gravitational lensing.

If the dark matter is nonbaryonic, the density of the universe is without doubt equal to the critical density. Indeed nucleosynthesis does not allow Ω_B to be much larger than 0.1; as the dark matter in clusters can represent up to ten times the visible baryonic mass, the global value of Ω_0 could thus be close to 1. There again, this argument is not definitive and actually favours models with nonbaryonic matter but with $\Omega_0 < 1$. The candidates for nonbaryonic hidden mass mainly come from particle physics (a cosmological constant allows a universe of zero curvature just as required by inflation, but it cannot explain dark matter). These are particles that have practically no interaction other than gravitational. The first candidate that was proposed is the massive neutrino. Certain experiments seemed at the time to provide evidence that the electron neutrino has a nonzero mass. It now seems that this is not the case, or that in every case, if the electron neutrino is massive, its contribution to the density of the universe is minor. On the other hand one of the other types of neutrino might be massive and contribute to the density in such a way that Ω_0 would equal 1:

$$\Omega_\nu \approx \frac{m_\nu}{100\,\text{eV}} h^{-2}.$$

As they do not take part in any interaction, if these particles were concentrated in a particular region, they would diffuse, and the primordial fluctuations at small scales would be damped by this process. The spectrum would then show a cutoff at small spatial wavelengths. The galaxy-formation model that results is then a model in which the large structures are formed first. However, this scenario encounters serious difficulties; moreover the present limits on the mass of the electron neutrino have greatly reduced its interest.

There are more exotic candidates but their existence similarly remains uncertain. When their mass is close to that which would be possessed by a massive neutrino, we speak of a 'hot' model; the properties are identical to those described above and the corresponding galaxy-formation model suffers the same difficulties. Particles a little more massive than the neutrino (the mass can go up to 1 keV) might also contribute to the density of the universe. In this case we speak of a 'warm' model. The wavelength at which the cutoff occurs is then lower and the model encounters fewer difficulties in explaining the formation of galaxies. Another possibility considers particles whose mass may be of the order of GeV. They might be, for example, supersymmetric partners of a classical particle (the photino, gravitino, and so on). In this case we speak of cold matter. The spectrum then has no cutoff. This cold-dark-

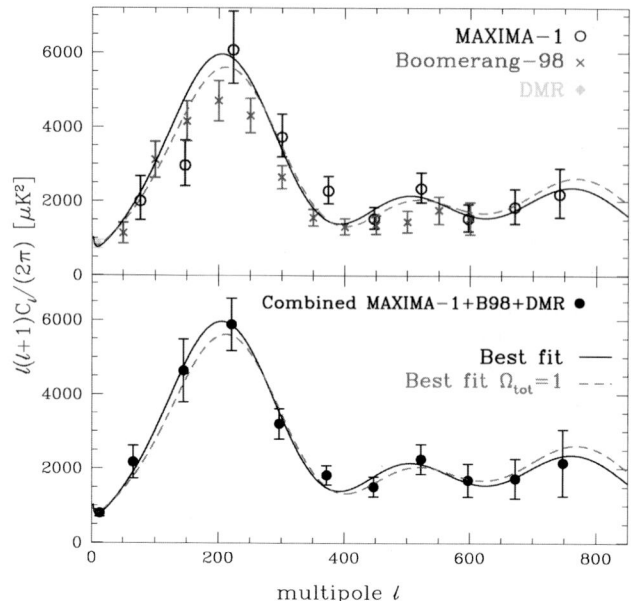

Fig. 13.8. CMB power spectra, $\mathcal{C}_\ell = \ell(\ell+1)C_\ell/2\pi$. Top: *MAXIMA*, *BOOMERanG* and *COBE*-DMR data. Bottom: maximum-likelihood fit to the power in bands for the three spectra. In both panels the curves show the best–fit model in the joint parameter estimation with weak priors and the best fit with $\Omega_{\rm tot} = 1$. These models have $\{\Omega_{\rm tot}, \Omega_\Lambda, \Omega_B h^2, \Omega_c h^2, n_s, \tau_{\rm C}\} = \{1.2, 0.5, 0.03, 0.12, 0.95, 0\}$, $\{1, 0.7, 0.03, 0.17, 0.975, 0\}$. (From Jaffe et al. 2000)

matter model has met with various difficulties and is no longer supported by the majority of cosmologists.

The exact nature of the particle has less effect on the associated galaxy-formation model. The only essential characteristics of the particle are the absence of interaction (which makes detection difficult) and the mass, which determines the modification to the initial spectrum. Thus in the case of 'hot' dark matter the model would have the same properties as would be the case for a massive neutrino or a supersymmetric light particle. In a 'cold' scenario the candidates are supersymmetric partners whose mass is of the order of GeV. But the dark matter might also be primordial black holes or light particles but produced in a cold, nonthermal way (the axion belonging to this family).

13.4.4 Recent Observational Confrontations

Nucleosynthesis. In principle the best barometer is deuterium, because it is the most sensitive to the baryon-to-photon ratio. However, this element is very fragile and consequently is easily destroyed during the chemical evolution of a Galaxy. It is therefore safe to consider its interstellar abundance as providing only a (robust) lower limit to the primordial value. With the help of large telescopes it has become possible to observe the deuterium line in some damped Lymanα systems. Unfortunately different results have been obtained along different lines of sight, which reveals either how difficult it is to infer accurate abundances from observed spectra or the evidence for inhomogeneous abundances. This has led to two different preferred values for η: one around 2×10^{-10} and one around 5×10^{-10}.

Inflation and Cosmic Background Anisotropies. The predictions of inflation at the early stages of the theory were making it a very attractive theory, which could be relatively easily tested. These predictions were the Gaussian nature of the initial fluctuations, the index of primordial fluctuations ($n = 1$), and the value of the density parameter of the universe (Ω_0). Nearly twenty years after the invention of the theory, it is interesting to look at the situation, from both the theoretical side and the observational side. The need for a theory of the very early universe, and the fact that such a theory should explain a significant number of observations in cosmology, is now well recognized. Only the topological defects model seems today to be a viable alternative, although in practice its predictive power is more limited, owing to its complexity. On the other hand many models of inflation have been developed, and several predictions of the theory have been revised. It has been recognized that the real implication of the inflation of space is to lead to a flat universe, which might not have a zero cosmological constant, although then the value of the cosmological constant is not explained. It has even been suggested that inflation can eventually lead to open models. Another important issue is that inflation is expected to produce not only density fluctuations (scalar modes) but also a significant amount of gravitational waves (tensor modes), which is now considered a key signature of inflation. Of course this depends on the detailed scenario, but inflation is not likely to produce an exact $n = 1$, but rather to produce a tilt spectrum with $n = 1 - \theta$, where θ is a number that depends on the specific inflationary scenario but which is most often a positive small number. In any scenario, the amplitude of the tensor mode of fluctuations is related to the amplitude of the scalar modes by a unique relation. Therefore the CMB offers a direct way to test the inflation idea, independently of the details of the scenario.

On the observational side, the discovery of the fluctuations by *COBE* at the observed level and the inferred value of the index of the primordial spectrum $n \sim 1$ are to be considered successes for the theory, although its predictive power has weakened and the topological defects theory achieves a similar rate of success. A supplementary success comes from the detection of fluctuations on small scales, which properties favour a flat universe. The future of the field will certainly be determined by the observations of small–scale fluctuations in the cosmological background radiation. The CMB fluctuations over the sky can be decomposed in spherical harmonics:

$$\frac{\delta T}{T} = \sum a_l^m Y_l^m$$

The angular power is then

$$C_l = <|a_l^m|^2>$$

Many experiments have now detected fluctuations on small scales which are believed to be mainly due to the CMB. Results are now commonly given in terms of the estimated C_l, which is average over several l (defined by the window function) and which has been estimated under the assumption of a flat spectrum. In inflation-like scenarios, the theoretical curve C_l is well determined once a number of parameters have been specified, including the cosmological parameters and some aspects of the inflation scenario. Therefore, the C_l potentially contains an essential piece of information for cosmology. It seems now quite well established that the main peak in the C_l curve is present at $l \sim 200-250$, which is consistent with a flat universe (Lineweaver et al. 1997) and rules out an open cosmological model. The best data today have been obtained with the balloons $MAXIMA$ and $BOOMERanG$ (see Fig. 13.8). It is likely that the forthcoming experiments on the CMB, the satellites MAP and $Planck$ will lead to a major advance in this field, allowing us to investigate the physics of the very early universe. It is fascinating that fluctuations of the CMB, detected from photons at energies of $\sim 310^{-4}$ eV, will offer a unique tool for obtaining information on physics at extremely high energies inaccessible at present from accelerators!

Exercises

13.1 (a) Beginning with the Robertson–Walker metric for the case $k = 0$, show that a region of space ($t =$ constant) is a Euclidean space of dimension 3. (b) In the same way, for the case $k = +1$ show that a region of space is a sphere in a space of dimension 4, that is, a subensemble of points such that $x^2 + y^2 + z^2 =$ constant.

13.2 Show that two neighbouring points in a homogeneous and isotropic space Δl apart have a relative velocity $v = (\dot{R}/R)\Delta l$. What is the only solution possible for a static universe?

13.3 *The K correction.* Some objects (for example galaxies) whose spectra are identical have been observed photometrically in a certain band $[\nu_1, \nu_2]$, with a profile $\phi(\nu)$ ($\phi(\nu)$ represents the fraction of energy detected at the frequency ν; $\phi(\nu)$ is always less than 1 and depends on the detector, the filter, the telescope, the atmosphere, and so on). Calculate the apparent magnitude $m^*(z)$ of the object, taking account of the shift in frequency, and then the magnitude $m(z)$ of the object if there were no change in frequency. (As we wish to compare objects that are situated at different redshifts, only $m(z)$ has a meaning in the comparison.) The difference $m^*(z) - m(z)$ is then a correction $(K(z))$ that allows a direct comparison of the apparent magnitudes of two sources with the same spectra but with different redshifts.

13.4 Show that the angular diameter of an object of fixed size always ends up by increasing when $z \to +\infty$ for $\Omega_0 > 0$. Similarly verify that the angular diameter of a comoving object is constant for large redshifts.

13.5 *Inflation.* In the early universe a phase transition may occur. The energy density of empty space then remains more or less constant over time. Show that, if this density term dominates, the expansion of the universe will become exponential. At the 'end' of the inflation, what is the value of the small-scale curvature α (specify the conditions)? Inflation thus answers the question, why is the density so close to ρ_c? In what way might this be a problem? For this case calculate $\Omega(z)$ observed at the epoch corresponding to a redshift z ($0 < z < \infty$). The introduction of a cosmological constant allows us to reconcile the observation $\Omega_0 = 0.1$ and inflation. In what way is this solution 'unaesthetic'?

References

Abbott, L. F., and So-Young Pi (editors) (1986) *Inflationary Cosmology* (World Scientific, Singapore).
Audouze, J., and Tran Thanh Van, J. (1984) *Fundamental Interactions and Cosmology* (Editions Frontières, Paris).
Balian, R., Audouze, J., and Schramm, D. N. (1980) *Physical Cosmology* (Les Houches Summer School, NATO, North Holland, Amsterdam).
Burstein, J., and Feinberg, G. (editors) (1989) *Cosmological Constants* (Columbia University Press, New York).
Decamp, D., et al. (1990) *Phys. Lett. B* **231**, 527 (the ALEPH collaboration).
Dominguez-Tenreiro, R., and Quiros, M. (1988) *An Introduction to Cosmology and Particle Physics* (World Scientific, Singapore).
Fabian, A. C., Geller, M., and Szalay, A. (1987) *Large-Scale Structures in the Universe* (Saas-Fee Advanced Course 17, Geneva Observatory, Geneva).
Gunn, J. E., Longair, M. S., and Rees, M. J. (1978) *Observational Cosmology* (Saas-Fee Advanced Course 8, Geneva Observatory, Geneva).
Jaffe A. H. et al. (2000) *Phys. Rev. Lett.*, in press (preprint astro-ph/0007333)
Kolb, E. W., and Turner, M. S. (1983) *Annu. Rev. Nucl. Particle Sci.* **33**, 645.
Kolb, E. W., and Turner, M. S. (1990) *The Early Universe* (Addison-Wesley, Redwood City).
Lineweaver C. H., Barbosa, D., Blanchard, A., Bartlett, J. G. (1997) *Astron. & Astrophys.* **322**, 365.
Mather, J. C., et al. (1990) *Astrophys. J.* **354**, L37.
Peebles, P. J. E. (1980) *The Large Scale Structure of the Universe* (Princeton University Press, Princeton NJ).
Peebles, P. J. E. (1993) *Principles of Physical Cosmology* (Princeton University Press, Princeton NJ).

Primack, J. R., Seckel, D., and Sadoulet, B. (1983) *Annu. Rev. Nucl. Particle Sci.* **33**, 645.
Raines, D. J. (1981) *The Isotropic Universe* (Adam Hilger, Bristol).
Schatzmann, E. (editor) (1973) *Cargèse Lectures in Physics* (Gordon and Breach, London).
Smoot, G. F. et al. (1992) *Astrophys. J.* **396**, L1.
Trimble, V. (1987) *Annu. Rev. Astron. Astrophys.* **25**, 425.
Weinberg, S. (1972) *Gravitation and Cosmology* (Wiley, New York).
Yang, J., Turner, M. S., Steigman, G., Schramm, D. N., and Olive, K. A. (1984) *Astrophys. J.* **281**, 493.

List of Constants, Notation, and Units Used

Fundamental Constants

Gravitational constant	$G = 6.670 \times 10^{-11} \, \text{N m}^2 \, \text{kg}^{-2}$
Velocity of light	$c = 2.9979250 \times 10^8 \, \text{m s}^{-1}$
Planck constant	$h = 6.62620 \times 10^{-34} \, \text{J s}$
Boltzmann constant	$k = 1.3806 \times 10^{-23} \, \text{J K}^{-1}$
Stefan constant	$\sigma = 5.6696 \times 10^{-8} \, \text{J m}^{-2} \, \text{K}^{-4} \, \text{s}^{-1}$
Electron charge	$e = 1.6022 \times 10^{-19} \, \text{C}$
Electron mass	$m_e = 9.1096 \times 10^{-31} \, \text{kg}$
Proton mass	$m_p = 1.6727 \times 10^{-27} \, \text{kg}$

Notation and Definitions

Apparent magnitude	$m = -2.5 \log(\text{brightness or intensity})$
Absolute magnitude	$M = m - 5 \log(\text{distance in pc}) + 5$
Apparent magnitude in the ultaviolet, blue, and visible, respectively	$U, B, V = m_U, m_B, m_V$

Units

Solar mass	$1 \, M_\odot = 1.989 \times 10^{30} \, \text{kg}$
Solar luminosity	$1 \, L_\odot = 3.826 \times 10^{26} \, \text{W}$
Parsec	$1 \, \text{pc} = 3.085678 \times 10^{16} \, \text{m}$
Year	$1 \, \text{y} = 3.2 \times 10^7 \, \text{s}$

Index

21 cm absorption   299
21 cm line   43
3C 273 (quasar)   242, 259
3D-shape   95
3D-shape of galaxies   217

Abell catalogue   317
absorbing systems: narrow-line   304
absorbing systems: broad-line systems   306
absorbing systems: dust   302
absorbing systems: molecules   301
absorption-line systems   289
absorption-line systems: broad-line systems   306
absorption-line systems: classification   293
absorption-line systems: narrow-metal-line   294
abundance gradients   37
abundances   67, 311
abundances of primordial elements   416
adiabatic fluctuations   368
AGN   65, 259
AGN: forbidden lines   266
AGN: permitted lines   263
AGN: variability and polarization   266
anaemic galaxies   9
angular correlation function   341
angular distance   398
angular momentum transfer   132
angular velocity of a bar   152
angular-momentum   142
anisotropies of the cosmic background radiation   429, 433
anisotropy of elliptical galaxies   96
annihilation   413
antenna temperature   44
antimatter   422
antispiral theorem   142
aperture synthesis   49

APM: automatic plate measuring machine   316
apocenter   213
arc-shaped structures produced by gravitational lensing   278
arm-interarm contrast   138
ASCA   352
asymmetries in galaxies   46, 75
atomic gas   43
atomic gas: radial distribution   46
axions   432

background UV radiation   298
Balmer decrement, or ratio   265
bar   79, 149
bar within bar   176
bar: response of the gas   168
baryon catastrophe   352
baryon number   423
baryonic fluctuations   368
beryllium   417
big-bang   421
BL Lac objects   230
black body   44, 400
black holes: Kerr, or rotating   286
black holes: primordial   367
black holes: supermassive   262
blazars   267
BLR: broad line region   264
BL Lac objects   283
bolometric luminosity   399
Boltzmann equation   115
bosons   413
bottom–up   335
box-shaped galaxies   166
Brandt curve   82
bremsstrahlung   38
bridge between two galaxies   187
bubbles, in structure of universe   332
bulges   14

Cambridge Center for Astrophysics (CFA)   332
Cartwheel galaxy   192
cD galaxies   9, 26
CDM: cold dark matter   366, 380
Centaurus A (radio source)   225
Cepheids   322
CFRS (Canada-France redshift survey)   29
Chandra   270, 352
Chandrasekhar formula   203, 208
chemical abundances   36
chemical decoupling   415
cigar-shaped galaxies   217
classification of galaxies   5
classification: de Vaucouleurs   7
classification: Hubble   7
classification: van den Bergh   9
clusters of galaxies: mass   348
CO molecule   55
CO-H_2 conversion ratio   55
COBE satellite   316, 401, 432
cold component of interstellar medium   33
collisions between clouds   140
collisions: excitation by   43
colour–magnitude diagram   17
colour-colour diagram   21
column densities determination   44, 290
Coma supercluster   317
comoving density   413
compact radio sources   225
companion galaxies   119, 149, 183
Compton scattering   228
concentration parameter c of King's model   104
contagious star formation   124
convergent point   320
conversion ratio CO-H_2 55
Copernican principle   393
Coriolis force   152
coronal medium   140
corotation   119, 131, 132, 144
correlation functions   338
cosmic background anisotropies   429, 433
cosmic history of star formation   30
cosmic microwave background   355, 400
cosmic strings   429
cosmological constant   405
cosmological principle   316

cosmology   221, 393
COSMOS: Coordinates, Sizes, Magnitudes, Orientation and Shapes measuring machine   316
counterarm   187
CP violation   425
critical wavelength of the Jeans instability: λ_c 131
curvature of space   403
curve of growth   291
Cygnus A (radio source)   225
C IV systems   312

damped Lyman α systems   299
damping   141
damping of shock waves   141
dark matter   221, 279, 365, 366
dark matter: nonbaryonic   430
de Sitter model of universe   405
de Vaucouleurs's $r^{1/4}$ law   11, 97
deceleration parameter   403
decoupling of neutrinos   416
decoupling of protons and antiprotons   416
density waves   75, 113, 121, 187
density-wave theory   121
depth of catalogue   315
deuterium   417
differential oscillations   196
diffuse electromagnetic background   316
disc galaxies   81
dispersion relation   127, 200
dissipation   187
domain walls   427
Doppler (acoustic) peaks   434
Doppler effect   322, 384, 395
dust   58
dust lanes in galaxies   97, 137
dynamical friction   26, 203

Eddington's model   105
Einstein X-ray satellite   263
Einstein-de Sitter universe   364, 405
elliptical galaxies   2, 83, 91
ellipticity profiles   94
emission lines   37, 259
emission lines: excitation   43
emission measure E_m 34
energy–momentum tensor   396
epicycles   113, 116
epicyclic approximation   116
equipartition of energy   103
equivalent width   291

ergodic orbits   155
Euclidean space   407
evolutionary tracks in HR diagrams   22
expansion of universe   361, 395

Faber–Jackson relation   92, 324, 355
Fanaroff-Riley FR-I and FR-II   234
Faraday rotation   228
fast Fourier transforms (FFTs)   159
fermions   413
Fe X lines in AGN spectra   262
filaments   181, 190, 375
finger-shaped structure   332
flaring of galaxy planes   85
flatness problem   426
fluctuations   367
fluctuations: adiabatic   368
fluctuations: isothermal   368
fluctuations: nature   367
fluctuations: spectrum   367
forbidden lines   36, 263
free–free emission   38, 226
freezing out   417
friction term in Jeans length   364
frictional force of gravitational encounter   203
Friedmann model   376
Friedmann–Lemaître models   402
fundamental plane   107
f(E) systems   103
f(E, J) systems   105

galaxies: bar   79, 149
galaxies: bulges   14
galaxies: classification   5
galaxies: colour distribution   20
galaxies: companion   144
galaxies: discs   14
galaxies: distance   320
galaxies: elliptical   216
galaxies: epicycles   116
galaxies: evolution   27
galaxies: formation   361
galaxies: interaction   181
galaxies: interstellar medium   138
galaxies: kinematics   69
galaxies: luminosity distributions   11
galaxies: luminosity profile   11, 14
galaxies: mass   69, 348
galaxies: merger   209
galaxies: orbits   113, 116, 152
galaxies: rotation   70

galaxies: shells   211
galaxies: shock waves   137
galaxies: spiral   209
galaxies: spiral structure   113
galaxies: stability   113, 114, 160
galaxies: statistical properties   23
galaxies: stochastic spiral   124
galaxies: tides   181
galaxies: torques   172
galaxies: vertical structure   16, 166, 196
galaxies: warp   80, 196
Gaussian fluctuations   429
gravitational lenses   274
Great Attractor   353
growth rate   362
GUT: grand unified theories   421

haloes: size   298
Harrison–Zel'dovich spectrum   390
HCN, HCO+, CS molecules   55
HDM: hot dark matter   366
head-on collision   213
Heisenberg's uncertainty principle   412
helium 3, helium 4   417
Hertzsprung–Russell (HR) diagrams   22, 322
hierarchical scenario   375
HIPPARCOS satellite   320
Holmberg, E.   181
homogeneity of universe   393
horizon   400
horizon problem   425
hot big bang   418
hot component of interstellar medium   33
HPQs (highly polarized QSOs)   267
Hubble constant   325, 395
Hubble flow   327, 383, 390
Hubble sequence   6
Hénon, M.   222
H I holes   51
H II regions   33
H I component in galaxies: global statistics   47

ILR, inner Lindblad resonance   121, 152
Index Catalogue (IC)   5
inflation   425, 427
interactions Between galaxies   181
interferometer   49
intergalactic gas   233

interstellar medium   33
intracluster gas   251
inverse Compton radiation   263
ionization state of absorbing gas   297
ionized gas   33
IRAS satellite   261, 327
irregular galaxies   5
irregular orbits   155
ISO satellite   270
isothermal fluctuations   368
isotropic oblate   94
isotropic prolate   94
isotropy of universe   393
IUE satellite   297

Jacobian   152
Jansky, K.   225
Jeans instabilities   114
Jeans length   114
Jeans mass   362
Jeans's theorem   101
jets (of radio sources)   226

K–correction   399
Kardashev effect   251
Kelvin–Helmholtz   238
Kerr, or rotating, black holes   286
kinematic waves   125
kinematics   69
kinematics of galaxies   69
King's model   104

Lagrangian points   152
Landau damping   134
large-scale structures in the universe   361
LBG Lyman break galaxies   386
leading waves   128
LEP: Large Electron Positron collider   419
Lick catalogue   315
Lin C.C.   123
Lindblad resonances   119, 152, 190
line locking   305
Liouville equation   129
lithium   417
local environment of cluster galaxies   26
Local Group   317
Local Supercluster   320
long branch of dispersion relation   131
long waves   131
Lorentz factor   226
luminosity distance   398

luminosity function   26
luminous echoes (superluminal motions)   248
Lyα forest   289
Lyman α systems   299, 308

M/L ratio   84, 349
MACHOs (Massive compact halo objects)   390
Madau curve   29
magnetic energy   227
magnetic field   38, 227
magnetic monopoles   427
Malmquist bias   326
Mattig relation   404
merger   209
mergers   66
Mezsaros effect   370
Mg II systems   292
Michie's model   105
molecular bar   61
molecular gas   55
molecular gas: radial distribution   59
morphological segregation   27
morphological types of galaxies   5

N-Body problem   158
NELGs (narrow-emission-line galaxies   282
neutrinos   366, 415
NGC 7252 galaxy   210
NGC, New General Catalogue   5
nuclei of spiral galaxies   37
nucleosynthesis   432
numerical simulations: cosmology   377

oblate galaxies   97, 219
OLR, outer Lindblad resonance   119, 154
optical depth   38
orbits   116, 152
Ostriker–Peebles criterion   164
OVVs (optically violent variables)   267

P–Cygni profiles   306
pancake model   375
pancake-shaped galaxies   219
parallaxes   320
peanut-shaped galaxies   166
Pegasus cluster   317
percolation   338, 346
perigalacticon   187

period–luminosity relation 322
periodic orbits 152
permitted lines 263
phase transition 427
photinos 366
photoionization 242
pitch angle i 128
Planck era 412
Planck mass 424
Planck time 423
Poincaré 119
Poisson distribution 339
Poisson equation 81, 104, 115, 350
polar rings in galaxies 85
polarization 228
power spectrum 369
precession 55, 131, 198
precession of orbits 126
primordial black holes 367
primordial fluctuations 426
primordial galaxies 386
primordial nucleosynthesis 416
primordial turbulence 367
prolate galaxies 217
prolate galaxies 97

QSOs 291
quantum gravity 412
quasars 259
quasars: absorption-line systems 289
quasars: emission lines 263
quasars: environment 279
quasars: evolution 272
quasars: host galaxies 279
quasars: luminosity function 271
quasars: radio-loud 299
quasars: radio-quiet 299
quasars: spatial distribution 270
quasars: twin 274

$r^{1/4}$ profile 210
radio emission 38
radio galaxies 283
radio jets 236
radio sources 225
radio sources: classification 232
radio sources: compact 229
radio sources: counts 253
radio sources: extended 229, 231
radio sources: hot spots 236
radio sources: lobes 234
radio sources: spectrum and polarization 239

radio sources: variability 244
radio sources: wide double 232
radio-loud quasars 261
radio-quiet quasars 261
Rayleigh–Jeans law 328
reacceleration of electrons 235
recollimation zone, in radio jets 240
recombination 361, 407
redshift 259, 264, 376, 380, 397
Reference Catalogue of Bright Galaxies, RC2, RC3 315
reheating of photons after chemical decoupling 415
relativistic beaming 249
relativistic electrons 226
relaxation: 2-body 101
relaxation: violent 103
rings in galaxies 7, 176, 192
Robertson–Walker metric 395
rotation 70, 93, 115
rotation curves 69, 91
RR Lyrae stars 323

S-shaped perturbations of galaxies 76
SAB galaxies 149
Sagittarius A (radio source) 237
SB galaxies 149
Schechter luminosity function 25, 316
Schmidt law 37
Schwarzschild radius 274, 285
scintillation of radio sources 236
semi-analytical simulations 379
Seyfert 1 282
Seyfert 2 282
Seyfert galaxies 282
Shane and Wirtanen catalogue 315
Shapley-Ames catalogue 315
shells 99, 211
shock waves 137
short waves 131
Shu F.H. 123
Silk mass 371
singularity 406
size–redshift relation 255
slitless spectroscopy 269
softening parameter 158
spectral index of radio sources 227
spectroscopy 91
spheroidal systems 82
spider diagram 73
spiral structure 50, 62, 75, 113, 142
stability 115, 160
standard candles: Cepheids 322
star formation 51

starburst galaxies   37, 59
stellar dynamics   113
stellar populations   17
stochastic spiral structure   149
Sunyaev–Zel'dovich effect   327
superclusters, filaments   316
superluminal velocities   246
surfaces of section   119
swing amplification   134
synchrotron   39

tails of radio sources   232
test particles (simulations)   187
thick discs of spiral galaxies   14
Thomson diffusion   327
three-body problem   187
tidal interaction   181, 202
tidal radius of a galaxy   104
time delay (radio sources)   244
Toomre A. and J.   187
Toomre criterion   115
Toomre's model of discs   82
top–down   335
topological defects   427
topology of universe   334, 427
torques   172
trailing waves   128
tree code   159
triaxial structure of elliptical galaxies   96
trigonometric methods   320
Tully–Fischer   87, 324
two-body relaxation   158

UHR: ultra-harmonic resonance   176
ultrarelativistic regime   414
universe   315
universe: distance   320
universe: distances   398
universe: geometry   395
universe: homogeneity   315
universe: hot phase   412

universe: large scale structure   315
universe: matter-dominated   403
universe: radiation-dominated   410
universe: very early   421
unsharp-masking technique   99, 211

V/σ–ϵ diagram   95
V/V_{max} test   272
van den Bergh classification of galaxies   9
vertical structure of galaxies   16, 166, 196
violation of charge and parity conjugation CP   423
violent relaxation   103
Virgo Infall   353, 355
virial relation   56, 57, 83, 88
viscosity of gas   170
Vlasov equation   101, 114
VLBI: very long-baseline interferometry   225
voids, structures in space   332

warm component of interstellar medium   33
warp   54, 80, 196
wave propagation   132
wiggle of radio jets   238
Wilson's model   105
winding problem   122
WKB (Wentzel–Kramers–Brillouin)   130, 201
wrapping in phase space   216

X-ray background   263, 316
X-ray emission   84
X-ray luminosity of a cluster   328
XMM-Newton   270, 352

Zel'dovich–Mezsaros effect   365
zero-velocity curve   121
Zwicky catalogue   315, 320

You are one **click** away from a **world of physics** information!

Come and visit Springer's
Physics Online Library

Books
- Search the Springer website catalogue
- Subscribe to our free alerting service for new books
- Look through the book series profiles

You want to order?   Email to: orders@springer.de

Journals
- Get abstracts, ToC´s free of charge to everyone
- Use our powerful search engine LINK Search
- Subscribe to our free alerting service LINK *Alert*
- Read full-text articles (available only to subscribers of the paper version of a journal)

You want to subscribe?   Email to: subscriptions@springer.de

Electronic Media
- Get more information on our software and CD-ROMs

You have a question on an electronic product?   Email to: helpdesk-em@springer.de

●········ Bookmark now:

http://www.springer.de/phys/

 Springer

Springer · Customer Service
Haberstr. 7 · 69126 Heidelberg, Germany
Tel: +49 (0) 6221 - 345 - 217/8
Fax: +49 (0) 6221 - 345 - 229 · e-mail: orders@springer.de

d&p · 6437.MNT/SFb

Printing (Computer to Film): Saladruck Berlin
Binding: Stürtz AG, Würzburg